Organic Chemistry

Organic Chemistry

Theory, Reactivity and Mechanisms in Modern Synthesis

Workbook

Pierre Vogel

Kendall N. Houk

WILEY-VCH

Authors

Prof. Pierre Vogel
EPFL
SB-DO
Avenue F.-A. Forel 2
1015 Lausanne
Switzerland

Prof. Kendall N. Houk
Dept. of Chemistry and Biochemistry
University of California
Los Angeles, CA 90095–1569
United States

Cover Credits: Fang Liu, DesignOne, Nanjing, China 210095

Library of Congress Card No.:
applied for

British Library Cataloguing-in-Publication Data
A catalogue record for this book is available from the British Library.

Bibliographic information published by the Deutsche Nationalbibliothek
The Deutsche Nationalbibliothek lists this publication in the Deutsche Nationalbibliografie; detailed bibliographic data are available on the Internet at <http://dnb.d-nb.de>.

Print ISBN: 978-3-527-34531-1
ePDF ISBN: 978-3-527-81928-7
ePub ISBN: 978-3-527-81930-0

Cover Design: SCHULZ Grafik-Design, Fußgönheim, Germany
Typesetting: SPi Global, Chennai, India
Printing and Binding: Himmer GmbH Druckerei, Augsburg, Germany

Printed on acid-free paper

10 9 8 7 6 5 4 3 2

Contents

1

Equilibria and Thermochemistry

Answers to Problems

Problem 1.1 A hydrocarbon, RH, can be isomerized in reversible reactions into two isomeric compounds **P1** and **P2** with the same heat of reaction. Both have C_1 symmetry. **P1** is a rigid molecule, and **P2** is a flexible one adopting several conformations of similar stability. Which product will be preferred at equilibrium?

Answer

Products **P1** and **P2** equilibrate with RH in reversible reactions at temperature T. The preferred product will be that for which the free energy of reaction $\Delta_r G^T$(reactants ⇆ product **P1** or **P2**) = $\Delta_r H^T - T\Delta_r S^T$ is the smallest. If $\Delta_r H^T$(RH ⇆ **P1**) = $\Delta_r H^T$(RH ⇆ **P2**), the product with the largest entropy (S^T(**P1**), S^T(**P2**)), will be favored at equilibrium at temperature T. Entropy expresses the degree of freedom available to an ensemble of molecules of a given compound. It depends upon the number of different microscopic states available to these molecules, and this depends on the size of the molecules themselves (**P1** and **P2** have the same molecular weight as they are isomers of the starting material RH), and the energy separation between the microscopic states available to them. The microscopic states that contribute most to the entropy are translational and rotational states for which the energy separation between quantum levels are the smallest. A rigid molecule is a molecule that does not equilibrate with several conformers of similar stabilities. It has only one energy minimum along all the coordinates that describe its geometrical deformations, including torsion about its σ- and π-bonds and bond elongation. A flexible molecule evolves at temperature T among several energy minima. In other words, several energy minima are populated significantly at the given temperature T. This is the case for **P2**, not for **P1**. If those energy minima correspond to different geometries (no rotation symmetry axis, $\sigma = 1$) the flexible molecule has a larger entropy than the rigid molecule; thus S^T(**P2**) > S^T(**P1**), and $\Delta_r G^T$(RH ⇆ **P2**) < $\Delta_r G^T$(RH ⇆ **P1**). Product **P2** that is composed of flexible molecules will be formed to a larger extent than product **P1** composed of rigid molecules at temperature T.

Energy hypersurface for the isomerization RH ⇄ **P1** + **P2**

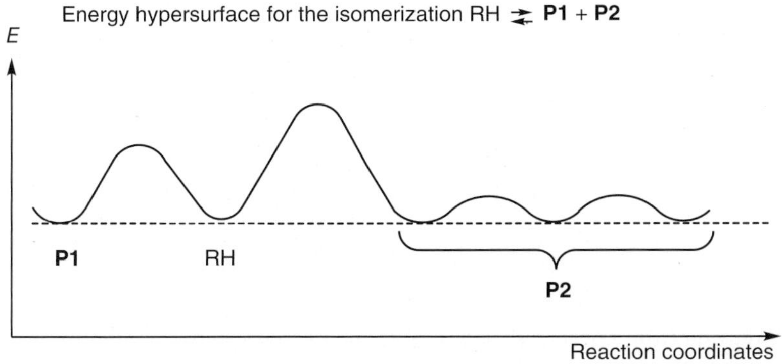

Problem 1.2 Define the symmetry numbers, σ, of ethane, propane, cyclopropane, cyclobutane, cyclohexanone, ferrocene, bicyclo[2.2.1]hepta-2,5-diene (norbornadiene), 1,4-difluorobenzene, *meso*-tartaric acid, and (*R*,*R*)-tartaric acid (see Figure 1.24 for the structure of the two latter compounds).

Organic Chemistry: Theory, Reactivity and Mechanisms in Modern Synthesis, Workbook, First Edition. Pierre Vogel and Kendall N. Houk.

Answer

Ethane is a flexible molecule that adopts a preferred staggered conformation (c. 3 kcal mol^{-1} more stable than the eclipsed conformation (Section 2.5.1), transition state of the rotation about its σ(C—C) bond). Ethane has a C_3- and three C_2-axes that are interconverted by rotation about the C_3-axis. Thus, the symmetry number $\sigma = 3 \times 2 = 6$.

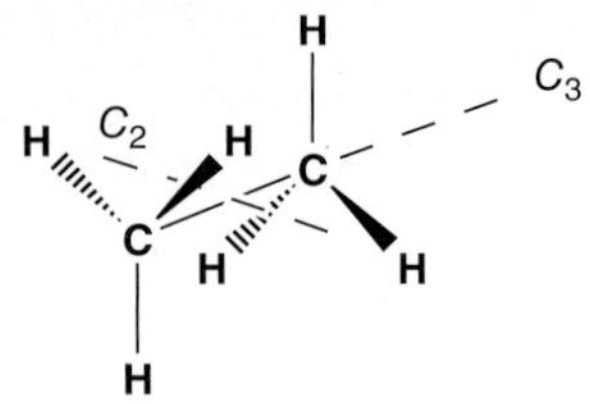

Staggered conformation

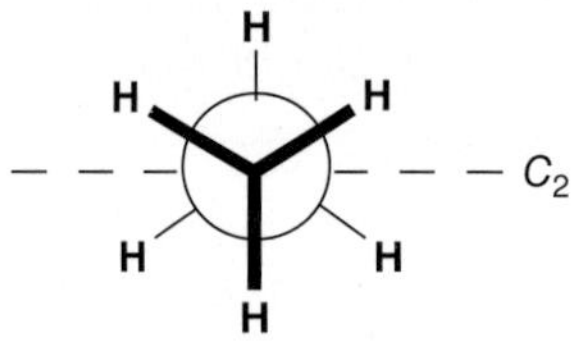

Newman projection
(perpendicular to the C_3 axis)

Propane is a flexible molecule for which the doubly staggered conformation is the most stable one (C_{2v} symmetry: one mirror plane of symmetry containing a C_2-axis of rotation) has one C_2-axis of rotation. Thus, the symmetry number $\sigma = 2$.

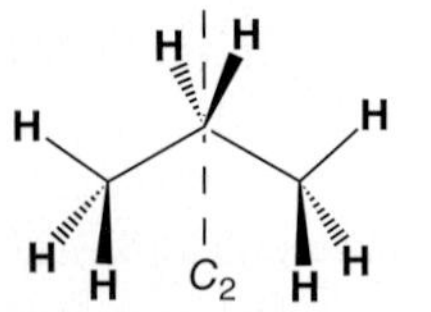

Staggered conformation

Cyclopropane is a rigid molecule (D_{3h} symmetry) with a C_3- and three C_2-axes of rotation. Rotation about the C_3 axis interconverts the C_2-axis. Thus, the symmetry number $\sigma = 3 \times 2 = 6$.

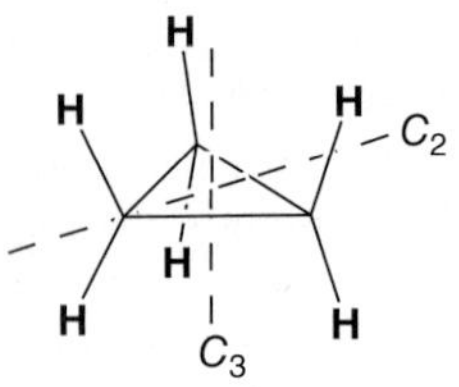

Cyclobutane adopts a puckered conformation (envelope, C_{2v} symmetry) with one C_2-axis of rotation. Thus, its symmetry number $\sigma = 2$. The planar conformation (D_{4h} symmetry) is much less stable (Section 2.6.1).

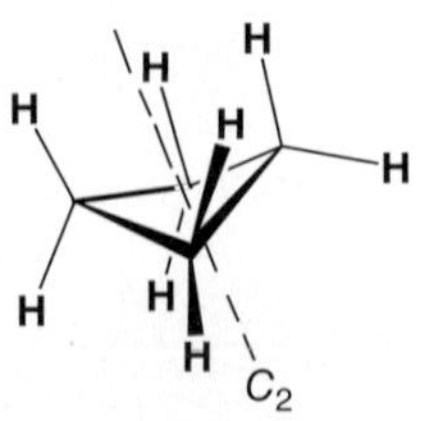

Cyclohexanone adopts a chair conformation (Section 2.6.4) of C_s symmetry (one mirror plane of symmetry), no rotation axis. Thus, the symmetry number $\sigma = 1$.

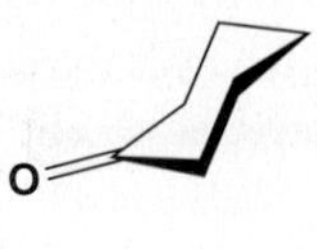

Ferrocene has a C_5- and five C_2-axes of rotation. The C_2-axes are interconverted by rotation about the C_5-axis. Thus, the symmetry number $\sigma = 5 \times 2 = 10$.

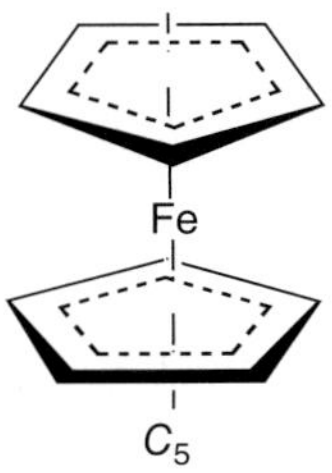

Bicyclo[2.2.1]hepta-2,5-diene (norbornadiene) is C_{2v} symmetrical (one mirror plane containing a C_2-axis of rotation). Thus, the symmetry number $\sigma = 2$.

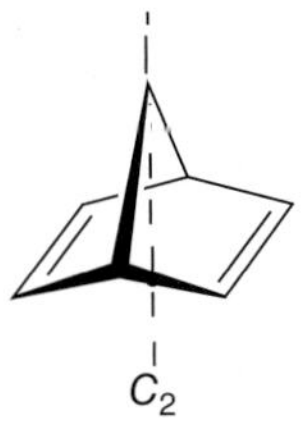

1,4-Difluorobenzene has two non-interconvertible C_2-axis of rotation. Thus, the symmetry number $\sigma = 2 \times 2 = 4$.

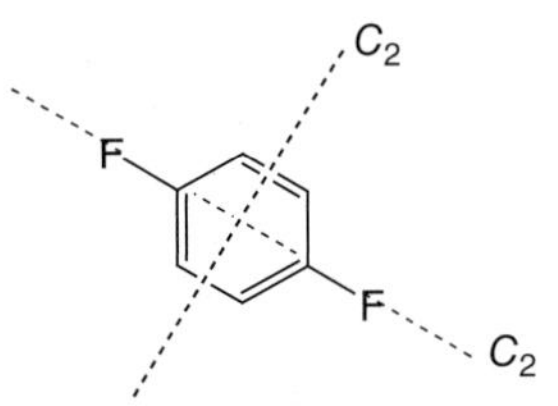

Meso-tartaric acid is (*R,S*)-tartaric acid with a mirror plane of symmetry. It does not have any axis of rotation, and thus $\sigma = 1.0$. (*R,R*)-Tartaric acid has a C_2-axis, and thus $\sigma = 2$.

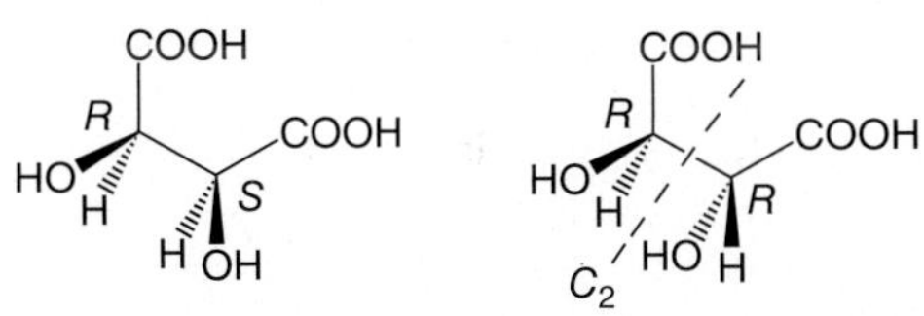

Problem 1.3 Enantiomerically pure α-amino acids such as alanine (with one stereogenic center) are isomerized into their racemic mixtures. What is the Gibbs energy of this reaction at 25 °C?

Answer

Enantiomers are stereoisomers that have the same stability and the same properties except for their optical rotation (opposite $[\alpha]$ values at any wavelength) and chiroptical properties (opposite Cotton effect in their circular dichroism spectra). Enantiomers cannot have a mirror plane of symmetry. They might have C_1-symmetry (no element of symmetry) like alanine, or have a rotation axis of symmetry like (*R,R*)-tartaric acid. The two enantiomers have the same enthalpy of formation and the same entropy. Thus,

$\Delta_r G^T$((*R*)-enantiomer ⇆ (*S*)-enantiomer) = $\Delta_r H^T$((*S*)-enantiomer) − $\Delta_r H^T$((*R*)-enantiomer) − $T\Delta_r S^T$((*S*)-enantiomer) + $T\Delta_r S^T$((*R*)-enantiomer) = 0. Thus, $RT \ln K = 0$. At equilibrium, K = [(*S*)-enantiomer]/[(*R*)-enantiomer] = 1.

Problem 1.4 What products do you expect to be formed combining HO˙ with organic compounds? What happens to NO in the air, to SO_2 in the air?

Answer

a) Water is the most stable oxide of hydrogen. Hydrogen peroxide (H_2O_2) and the radicals hydroxyl (HO˙) and hydroperoxyl (HOO˙) are expected to be reactive compounds that tend to form water on combining them with organic and inorganic compounds containing hydrogen atoms.
b) The most stable oxidized form of nitrogen is NO_2, and this is the brown gas of air pollution. Nitric oxide (NO) is an important diatomic molecule that regulates many biological functions, such as blood pressure and erectility. It is produced from arginine by NO synthase in mammals. The drug Viagra inhibits an enzyme that normally reverses the action produced by NO. Within seconds of its formation in air, colorless NO undergoes oxidation to brown NO_2, and then to nitrite and nitrate ($2NO_2 + H_2O \leftrightarrows HNO_2 + HNO_3$).
c) Sulfur trioxide (SO_3) is the most stable oxidized form of sulfur. Burning sulfur does not produce SO_3 but SO_2, a major air pollutant. The oxidation of SO_2 by air into SO_3 requires special catalysts used in industrial sulfuric acid production. Sulfate from hydrolysis of SO_3 ($SO_3 + H_2O \leftrightarrows H_2SO_4$) is involved in air pollution and acid rain, as sulfuric acid.

Problem 1.5 Propose a reaction for diimide (HN=NH) + cyclohexene and calculate its heat of reaction.

Answer
(*Z*)- and (*E*)-Diimide (HN=NH) are metastable combinations of nitrogen and hydrogen. They are sources of dihydrogen (Garbisch alkene hydrogenation, dyotropic transfer of dihydrogen, Section 5.6.2) liberating N_2. $\Delta_r H^\circ = \Delta_f H^\circ(N_2) + \Delta_f H^\circ(\text{cyclohexane}) - \Delta_f H^\circ((Z)\text{-diimide}) - \Delta_f H^\circ(\text{cyclohexene}) = 0 - 29.5 - 50.9 + 1.03 = -79.4\ \text{kcal mol}^{-1}$.

Problem 1.6 Among the amino acids serine and cysteine, which of these gives stable adducts with cyclohex-2-enone at 37 °C when they are part of a protein?

Answer
Serine ($HOCH_2CH(NH_2)COOH$) incorporated in a protein has a primary alcohol function. Its addition to an alkene is not exothermic enough to lead to a stable adduct with cyclohexenone at 37 °C as indicated with the model equilibrium:

$\Delta_r H^\circ(\text{EtOH} + CH_2{=}CH_2 \leftrightarrows \text{EtOEt}) = -51.7 - (-56.0) - 12.5 = -8.2\ \text{kcal mol}^{-1}$. The heat of reaction cannot compensate for the entropy of condensation $-T\Delta_r S^{310K} > 10\ \text{kcal mol}^{-1}$.

Cysteine is like serine but with a thiol function instead of a primiral alcohol function for which a much higher heat of reaction is estimated for its addition to an alkene as shown with $\Delta_r H^\circ(\text{EtSH} + CH_2{=}CH_2 \leftrightarrows \text{EtSEt}) = -19.9 - (-11.0) - 12.5 = -21.4\ \text{kcal mol}^{-1}$. Thus cysteine, not serine, incorporated in a protein will equilibrate with an adduct resulting from the conjugate addition of its thiol unit to cyclohex-2-enone.

Problem 1.7 Explain the difference in C=O bond stretching frequencies ($\nu_{C=O}$) between ethyl (*Z*)-3-fluorocinnamate (1736 cm^{-1}) and ethyl cinnamate ((*E*)-PhCH=CHCOOEt: 1715 cm^{-1}) [1].

Answer
The stretching frequency $\nu_{C=O}$ is proportional to the C=O bond strength. The latter depends on conjugation. Compared with saturated esters ($\nu_{C=O}$: 1750–1735 cm^{-1}), α,β-unsaturated esters have smaller $\nu_{C=O}$ (1730–1715 cm^{-1}) due to C=O bond elongation and weakening resulting from the π-conjugation.

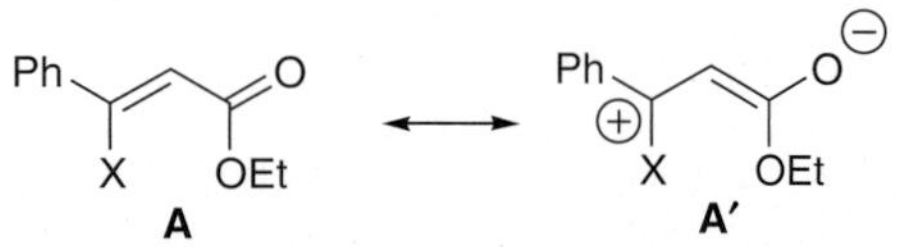

On substituting a α,β-unsaturated ester by a β-fluoro group, one destabilizes the zwitterionic limiting structure **A′** because of the strong dipole realized by the C—F bond. Retrodonation n(F:)/π^*(C=C) is inexistent since the amount of positive charge appearing at the β-center remains relatively weak (see electrostatic model of substituent effects on the stability of charged species). Thus the C=O bond in ethyl (*Z*)-3-fluorocinnamate (**A**, X = F, $\nu_{C=O}$: 1726 cm^{-1}) is stronger than the C=O bond in (*E*)-cinnamate (**A**, X = H, $\nu_{C=O}$: 1715 cm^{-1}) [1].

Problem 1.8 Estimate the standard heat of the esterification of ethanol with acetic acid. Estimate the variation of entropy of this reaction at 298.15 K and calculate the equilibrium constant at the same temperature and under one atmosphere, in tetrahydrofuran (THF) solution. Is the equilibrium constant the same under the same conditions for the esterification of anthracene-2-carboxylic acid with 2-hydroxyanthracene?

Answer

$\Delta_r H°$(EtOH+MeCOOH ⇆ MeCOOEt + H_2O) = −106.5(±0.2) − 57.8(±0.01) − (−56.0)(±0.5) − (−103.5)(±0.6) = −4.8 ± 1.3 kcal mol^{-1} (Table 1.A.4).

$\Delta_r S°$(EtOH + MeCOOH ⇆ MeCOOEt + H_2O) = 86.7 + 45.1 − 67.6 − 66.7 = −2.5 eu. One thus obtains $\Delta_r G°$(EtOH + MeCOOH ⇆ MeCOOEt + H_2O) = −4.8 − 298(0.0025) ≈ −4.0 kcal mol^{-1}, giving K = exp.($-\Delta_r G°/RT$) = exp.(4000/1.987 × 298) = 858 at 25 °C for an ideal solution.

If one considers only the entropy of translation and Eq. (1.31) for the calculation of the entropy of monoatomic molecules,

$$S_{trans} = 6.86 \log M_r(g) + 11.44 \log T - 2.31 \text{ eu}$$

$\Delta_r S°_{trans}$(EtOH + MeCOOH ⇆ MeCOOEt + H_2O) = 6.86 log[M_r(MeCOOEt)·M_r(H_2O)/M_r(EtOH)·M_r (AcOH)] = 6.86 log[88.1 × 18/46.07 × 60.05] = 6.86 log(0.5 732 145) = −1.65 eu is not nil because two molecules of similar size are equilibrated with a larger molecule and a smaller one (H_2O). Water molecule has a C_2-axis of symmetry, its symmetry number $\sigma = 2$. The other molecules intervening in this equilibrium have $\sigma = 1$. The same molecular groups and freely rotating bonds are present in reactants and products. If one consider the loss of rotational entropy by forming H_2O, a correction of $-R\ln 2 = -1.987(0.693) = -1.37$ eu has to be considered. On this basis, the calculated $\Delta_r S°_{trans+rot}$(EtOH + MeCOOH ⇆ MeCOOEt + H_2O) = −1.65 − 1.37 ≈ −3 eu, a value very close to the value of −2.5 eu estimated from the experimental entropies given for these compounds.

Experimental standard heats of formation and entropies of anthracen-2-yl anthracene-2-carboxylate (M_r = 398.463), of anthracene-2 carboxylic acid (M_r = 222.245), and of 2-hydroxyanthracene (M_r = 194.234) are not available. If one considers the model reaction PhCOOH + MeOH ⇆ PhCOOMe + H_2O for which $\Delta_r H°$(PhCOOH + MeOH ⇆ PhCOOMe + H_2O) = −64.4 − 57.8 − (−70.3) − (−49) = −2.9 kcal mol^{-1} is estimated (Table 1.A.4), one can predict that the exothermicity of the esterification of the anthracene derivatives should be less than of the formation of ethyl acetate because of steric effect. For the contribution of the entropy of translation to the equilibrium that converts anthracene-2 carboxylic acid and 2-hydroxyanthracene into the corresponding ester + water one estimates it to be more negative than in the case of the formation of ethyl acetate as given by

$\Delta_r S°_{trans}$(2-anthracenol + 2-anthracenecarboxylic acid ⇆ anthracen-2-yl anthracene-2-carboxylate + H_2O) = 6.86 log[M_r($C_{14}H_9COOC_{14}H_9$)·M_r(H_2O)/M_r($C_{14}H_9OH$)·M_r($C_{14}H_9COOH$)] = 6.86 log[398.5 × 18.02/194.2 × 222.2] = 6.8 6log(0.166 414) = −5.34 eu. If one now considers the loss of rotational entropy due to the formation of water, one estimates

$\Delta_r S°_{trans+rot}$(2-anthracenol + 2-anthracenecarboxylic acid ⇆ anthracen-2-yl anthracene-2-carboxylate + H_2O) = −5.34 − 1.37 = −6.7 eu.

The free enthalpy for this equilibrium at 25 °C is estimated to be $\Delta_r G°$ (2-anthracenol + 2-anthracenecarboxylic acid ⇆ anthracen-2-yl anthracene-2-carboxylate + H_2O) = −2.9 − 298(−0.0067) = −2.9 + 1.99 ≈ −0.9 kcal mol^{-1}

From what precedes we can predict that K(2-anthracenol + 2-anthracenecarboxylic acid ⇆ anthracen-2-yl anthracene-2-carboxylate + H_2O) < K(EtOH + AcOH ⇆ AcOEt + H_2O).

At 298 K and under one atmosphere for ideal solution in THF K(2-anthracenol + 2-anthracenecarboxylic acid ⇆ anthracen-2-yl anthracene-2-carboxylate + H_2O) = exp.(900/1.987 × 298) = 4.6.

Problem 1.9 The Newman–Kwart rearrangement is a valuable synthetic technique for converting phenols to thiophenols via their *O*- and *S*-thiocarbamates [2–4].
Explain why the *S*-thiocarbamates are more stable than their isomeric O-thionocarbamates.

Answer

In the equilibrium Ar—O—(C=S—)X ⇆ Ar—S—(C=O—)X, one exchanges a thionoester moiety into a thioester function. Since carbon and sulfur have similar electronegativities (Table 1.A.5) the C=S double bond is not as polar as the C=O double bond. According to the Pauling theory, the C=S bond is weaker than the C=O bond. The C=S bond is longer than the C=O bond, which contributes also to make it weaker. For that reason Ar—S—(C=O)—X must be more stable than Ar—O—(C=S)—X. Furthermore, part of the stability of an ester arises from the n(O)/π(C=O) conjugation (ArO—C=O ↔ ArO$^{(+)}$=C—O$^{(-)}$) and n(C=O)/σ(O) hyperconjugation (Ar—O—C=O ↔ Ar—O$^{(-)}$ + C$^{(+)}$=O ↔ Ar—O$^{(-)}$ + C≡O$^{(+)}$). In thioester the n(S)/π(C=O) conjugation and n(C=O)/σ(S) hyperconjugation intervene also but to a smaller extent than in ester as shown by the heat of the following reactions:

$$\mathrm{AcOH} + \mathrm{ROH} \underset{}{\overset{\text{Gas}}{\rightleftharpoons}} \mathrm{AcOR} + \mathrm{H_2O} \qquad (1.53)$$

R: Me, *i*-Pr

$\Delta_r H°$: −5.0, −5.3 kcal mol^{-1}

$$\mathrm{AcOH} + \mathrm{EtSH} \overset{\text{Gas}}{\rightleftharpoons} \mathrm{AcSEt} + \mathrm{H_2O} \qquad (1.55)$$

$\Delta_r H°$: +2.1 kcal mol^{-1}

Problem 1.10 A mixture of 1 mmol cyclohex-2-enone, 1 mmol of thiophenol, and 5 mg of Et_3N is kept at 25 °C in 1 ml of CH_2Cl_2. After 30 minutes at 25 °C, the ^{1}H-NMR spectra of the reaction mixture shows that the corresponding 1,4-adduct is formed almost completely. Attempted purification of the adduct by column chromatography on silica gel gives, however, only a low yield of adduct (10–20%) and recovered cyclohex-2-enone (80%) and thiophenol (80%). Why?

Answer

Thiol conjugate additions to conjugated enones are reversible in the presence of a base catalyst such as Et_3N at 25 °C. For a reaction using molar concentrations of thiophenol and cyclohex-2-enone the corresponding adduct (2-phenylthiocyclohexanone) forms to a large extent since the reaction has an exothermicity (c. −20 kcal mol^{-1}, see Problem 1.6) able to compensate for the cost of the entropy (c. 10 kcal mol^{-1}) of condensation at 25 °C. During the column chromatography on silica gel (SiO_2), the thiol, enone, 1,4-adduct, and catalyst interact with the solid support, more or less strongly. The most polar compound makes the strongest bonding with silicagel and thus is retained the longest on the column. The least polar compound is the least retained on the column and elutes the fastest. If the rate of elimination from the adduct giving thiophenol + cyclohex-2-enone is as fast as or faster than the rate of elution, the enone, thiol, and adduct will elute separately. Silica gel is acidic and catalyzes also the conjugate addition and elimination. The basic catalyst (Et_3N) remains on the column.

In order to avoid this type of problem due to the reversibility of the conjugate addition, one should extract the basic catalyst (Et_3N) from the final reaction mixture, for instance, by adding to it and at low temperature (<0 °C) a minimum amount of acidic alumina or silica gel to absorb it and then run a quick filtration at low temperature. Purification by chromatography should then be carried out on a cooled (circulating cooled [<−40 °C] ethanol in a double-walled column) column of silica gel. The extraction of Et_3N could also be done by extraction with a cooled (0 °C) diluted aqueous solution of HCl, and then washing with an aqueous solution of $NaHCO_3$ (to remove traces of HCl from the CH_2Cl_2 solution). HCl would catalyze the elimination from the adduct once the mixture is poured onto the chromatography column.

Problem 1.11 On heating 1,1-dideuteriohexa-1,5-diene equilibrates with 3,3,-dideuteriohexa-1,5-diene (reversible Cope rearrangement: Section 5.5.9.1). Similarly 1,1-difluorohexa-1,5-diene equilibrates with 3,3-difluorohexa-1,5-diene. Which of these two last isomeric compounds is most stable? [5]

Answer

At equilibrium, there is a slight preference for D at the sp3 center $K = [\mathbf{B}]/[\mathbf{A}] \approx 1.1$:

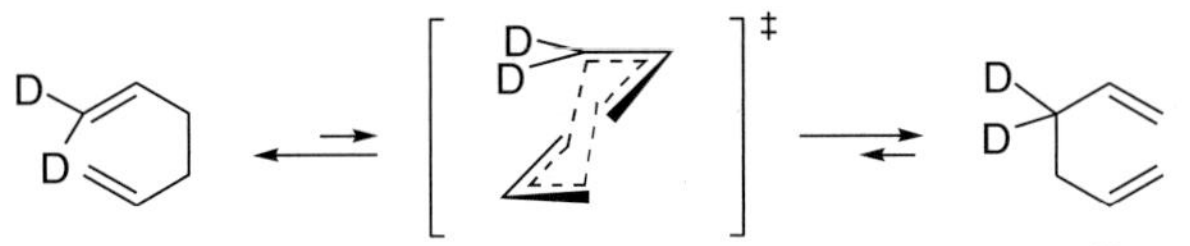

A **B**

At equilibrium $K = [\mathbf{D}]/[\mathbf{C}] \gg 1/1$:

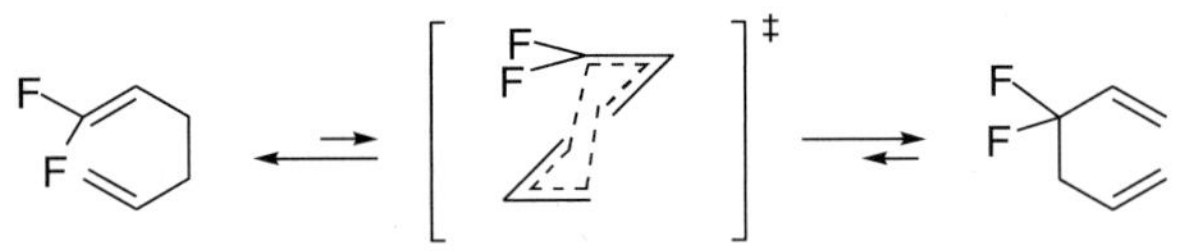

C **D**

Compound **D** is more stable than **C** since geminal difluoro substitution prefers sp^3-hybridized carbon centers rather than sp^2-carbon centers.

Problem 1.12 Is the hydrogenation of carbon monoxide into formaldehyde a feasible reaction provided that a suitable catalyst is available to catalyze this reaction?

Answer

	H_2	+ CO	$\underset{}{\overset{\text{cat.}}{\rightleftharpoons}}$	H_2CO
$\Delta_f H°$(gas):	0	−26.4		−27.7 kcal mol^{-1}
$S°$(gas):	31.2	47.2		52.3 eu (NIST Chemistry WebBook)

This reaction is endergonic ($\Delta_r G° = \Delta_r H° - T\Delta_r S°$) since it has an exothermicity ($\Delta_r H°$) of only −1.3 kcal mol^{-1}, which is not enough to compensate for the cost of entropy of condensation ($-T\Delta_r S° = -298(-26.1) = 6.5$ kcal mol^{-1}) at room temperature ($T = 298$ K). Nevertheless, formaldehyde (methanal) may form in small amounts at equilibrium and then be trimerized into 1,3,5-trioxane (metaformaldehyde), a reaction that brings an exothermicity of −28.2 kcal mol^{-1} at 25 °C and under one atmosphere. This is not enough to pay for the entropy cost of the hydrogenation of carbon monoxide (three times 6.5 kcal mol^{-1}) and that of the trimerization of formaldehyde (298(88.8) = 26.4 kcal mol^{-1}). If one considers the overall process $\Delta_r G°(3CO + 3H_2 \leftrightarrows \text{1,3,5-trioxane}) = \Delta_r H°(3CO + 3H_2 \leftrightarrows \text{1,3,5-trioxane}) - T\Delta_r S°(3CO + 3H_2 \leftrightarrows \text{1,3,5-trioxane})$ $= -111.3 - 3(-26.4) - 298(52.3 - 3(31.2) - 3(47.2) = -32.1 - 298(-157.6)) = -32.1 + 46.9 = 14.8$ kcal mol^{-1}, a highly endergonic process at 25 °C and under one atmosphere. On lowering the temperature of the reaction, it becomes less endergonic. Increasing pressure will help shift the equilibrium in favor of 1,3,5-trioxane. In theory, and for an ideal solution, the equilibrium constant $K(3CO + 3H_2 \leftrightarrows \text{1,3,5-trioxane})$ approaches unity ($\Delta_r G° = 0$) when $T\Delta_r S° = \Delta_r H°$. In this case, one estimates $T = -32\,100/(-157.6) = 203.7\ K \approx -94$ °C. As 1,3,5-trioxane is a solid, it could be eliminated from the reaction mixture by precipitation or crystallization (shifting of the equilibrium, Section 1.14.1). For such a process to be possible, one would have to find a suitable catalyst allowing one to run the condensation at low temperature and high pressure.

$3H_2C{=}O \rightleftharpoons$ 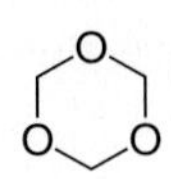 $\Delta_r H°$(gas): −28.2 kcal mol^{-1}

$\Delta_f H°$(gas):	3(−27.7)	−111.3 kcal mol^{-1}
$S°$ (gas):	3(52.3)	68.1 eu

Formaldehyde is produced in the industry by catalyzed oxidation of methanol ($2MeOH + O_2 \leftrightarrows 2H_2CO + 2H_2O$). In the "former process" the reaction occurs between 250 and 400 °C and uses a solid catalyst made of iron oxide combined with molybdenum or/and vanadium oxides.

Hydrogenation of carbon monoxide is used to produce methanol in the industry. Under one atmosphere the overall reaction $CO + 2H_2 \leftrightarrows MeOH$ has an exothermicity of -22.6 kcal mol^{-1}. Several catalysts have been developed for this reaction, which is run under high pressure and temperatures above 200 °C [6]. Under one atmosphere and at 200 °C the entropy cost $(-T\Delta_r S°(CO + 2H_2 \leftrightarrows MeOH) = -473(54 - 47.2 - 2(31.2)) =$ 26.3 kcal mol^{-1}; it is not compensated by the exothermicity of the reaction. Nevertheless, pressure insures the formation of methanol. For instance, under 100 bars the reaction occurs between 275 and 325 °C in the presence of $Pt/Co/SiO_2$ catalysts [7].

Problem 1.13 Are the hydrocarbations of carbon monoxide by alkanes, alkenes, or alkynes (or carbonylations of alkanes, alkenes, and alkynes) possible reactions provided suitable catalysts are available?

Answer

$$CH_3{-}CH_3 + CO \underset{}{\overset{\text{cat.}}{\rightleftarrows}} CH_3CH_2CHO$$
$$-20.1 \qquad -26.4 \qquad\qquad -44.1 \qquad \Delta_r H°\text{: 2.4 kcal mol}^{-1}$$

$$H_2C{=}CH_2 + CO \overset{\text{cat.}}{\rightleftarrows} H_2C{=}CHCHO$$
$$12.5 \qquad -26.4 \qquad\qquad -18\ \text{(est.)} \qquad \Delta_r H°\text{: }-4.1\text{ kcal mol}^{-1}$$

$$HC{\equiv}CH + CO \overset{\text{cat.}}{\rightleftarrows} HC{\equiv}CHCHO$$
$$54.5 \qquad -26.5 \qquad\qquad 27 \qquad \Delta_r H°\text{: }-1.1\text{ kcal mol}^{-1}$$

None of these reactions would be exergonic at room temperature (condensations with an entropy cost $-T\Delta_r S$ of 7–10 kcal mol^{-1}). However, in the case of the formation of unsaturated compounds their polymerizations are exothermic enough (<-20 kcal mol^{-1}; see Problem 1.14) to pay for the entropy cost and thus displace the reaction in the sense of polycondensations.

Problem 1.14 Are the hydrocarbations of carbon dioxide possible for alkanes, alkenes, and alkynes provided suitable catalysts are available to catalyze them?

Answer

$$CH_3{-}CH_3 + CO_2 \underset{\text{(gas)}}{\overset{\text{cat.}}{\rightleftarrows}} CH_3CH_2COOH$$
$$\Delta_f H°\text{(gas): }-20.1 \qquad -94.0 \qquad\qquad -107 \qquad \Delta_r H°\text{: 7 kcal mol}^{-1}$$

$$H_2C{=}CH_2 + CO_2 \underset{\text{(gas)}}{\overset{\text{cat.}}{\rightleftarrows}} H_2C{=}CHCOOH$$
$$\Delta_f H°\text{(gas): }12.5 \qquad -94.0 \qquad\qquad -77 \qquad \Delta_r H°\text{: 4.5 kcal mol}^{-1}$$

$$HC{\equiv}CH + CO_2 \underset{\text{(gas)}}{\overset{\text{cat.}}{\rightleftarrows}} HC{\equiv}CCOOH$$
$$\Delta_f H°\text{(gas): }+54.5 \qquad -94.0 \qquad\qquad -28 \qquad \Delta_r H°\text{: 11.5 kcal mol}^{-1}$$

The three reactions are endothermic and they all have negative entropies, being condensations. Based on their heat of reaction they are more difficult than the corresponding carbonylation (Problem 1.13). They are impossible at room temperature and under 1 atm. Under low temperature and high pressure the second reaction (formation of acrylic acid) might become feasible if a suitable catalyst is available. If the unsaturated carboxylic acids that form in the second and third reaction would undergo polymerization the polycondensation might occur since each new C—C bond generated is accompanied by a substantial exothermicity (better than 20 kcal mol^{-1}) that can compensate for the endothermicity of the hydrocarbations of CO_2 and the entropy cost ($-T\Delta_rS$) of the latter reactions and of the polymerization reactions.

$$2\,H_2C{=}CH_2 \rightleftharpoons CH_2{=}CH{-}CH_2{-}CH_3$$

$\Delta_fH°$(gas): 2(12.5) 0.0 $\Delta_rH°$: −25.5 kcal mol^{-1}

$$2\,HC{\equiv}CH \rightleftharpoons CH{=}CH{-}CH{=}CH_2$$

$\Delta_fH°$(gas): 2(54.5) 70.4 $\Delta_rH°$: −38.6 kcal mol^{-1}

One way to make the C–H carboxylation working is to quench the initial carboxylic product (RCOOH) in an exothermic reaction such as neutralization with a base to form the corresponding carboxylate salt (RCOOH + B: $\leftrightarrows$ RCOO$^-$BH$^+$). An example is given below with the CuI/Ph_3P-catalyzed carboxylation of phenylacetylene at 80 °C under one atmosphere (solvent: ethylene carbonate) and which converts the carboxylate salt into its *n*-butyl ester (overall yield: 99%). The base used is Cs_2CO_3 [8].

$$2\mathrm{Ph}-\mathrm{C}\equiv\mathrm{CH}+2\mathrm{CO}_2+\mathrm{Cs}_2\mathrm{CO}_3+2n-\mathrm{BuI}\rightarrow 2\mathrm{RC}\equiv\mathrm{CCOO}-n-\mathrm{Bu}+2\mathrm{CsI}+\mathrm{H}_2\mathrm{O}+\mathrm{CO}_2$$

Another way to have the carboxylation of alkenes and acetylenes working is to couple them with the hydrogenation of the α,β-unsaturated carboxylic that forms initially [9].

Problem 1.15 Can one fix CO_2 with epoxides? What products are expected (catalysts: mixed Mg/Al oxides, dimethylformamide [DMF] as solvent) to be formed at 100 °C?

Answer

The thermochemical data shows that the condensations of CO_2 with methyl oxirane and oxirane can occur at 100 °C. Assuming an entropy of condensation of −40 eu, the entropy cost of these reactions at 100 °C amounts to $-T\Delta^{\ddagger}S = 373(-40\,\text{eu}) \approx 15\,\text{kcal mol}^{-1}$. Depending on the reaction conditions the vinylidene carbonates obtained can be polymerized into polycarbonates [10].

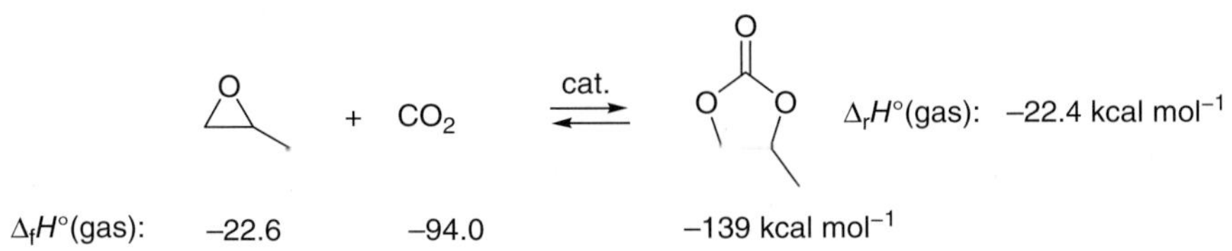

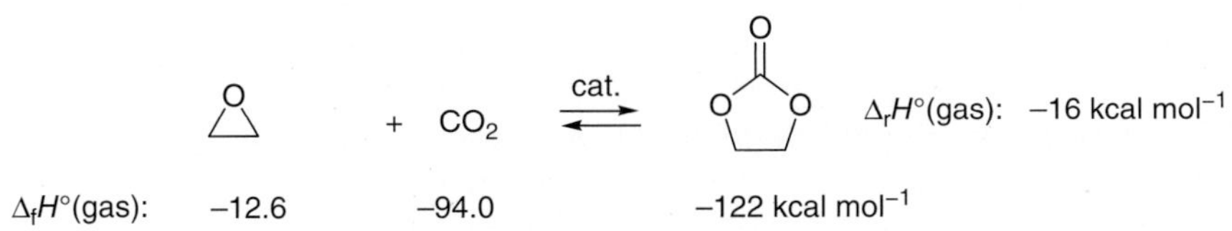

An example of catalyst for the carboxylation of methyloxirane is the Cr(III)salen complex represented below associated with 4-dimethylaminopyridine [11].

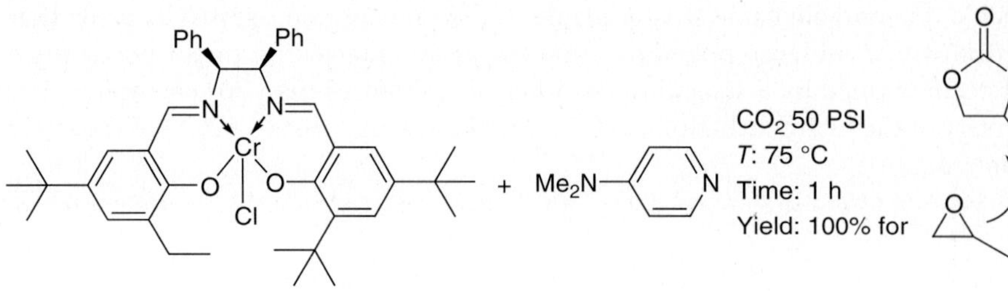

Problem 1.16 Which of the two products **B** and **C** is the favored product of cyclization of **A** catalyzed by $Bu_4N^+F^-$ (base catalyst) in DMF.

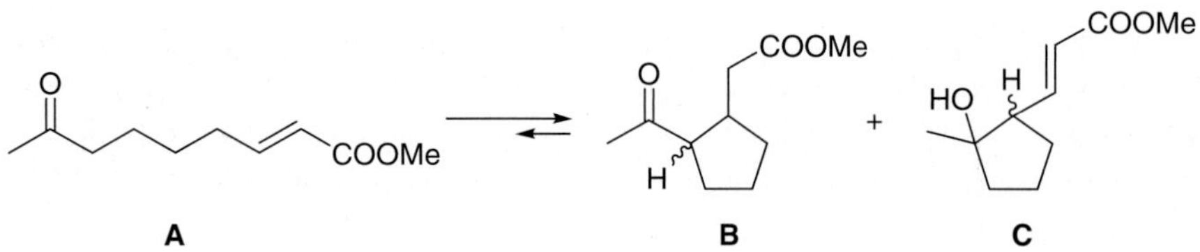

Answer

Under the action of the base (fluoride anion) ketone **A** equilibrates with two enolates (see Table 1.A.24 for pK_a values in dimethylsulfoxide [DMSO]). The ketone enolate **E** undergoes an intramolecular Michael addition onto the acrylic moiety of **E** generating the ester enolate **F**, which equilibrates with **B**, the main product of the reaction.

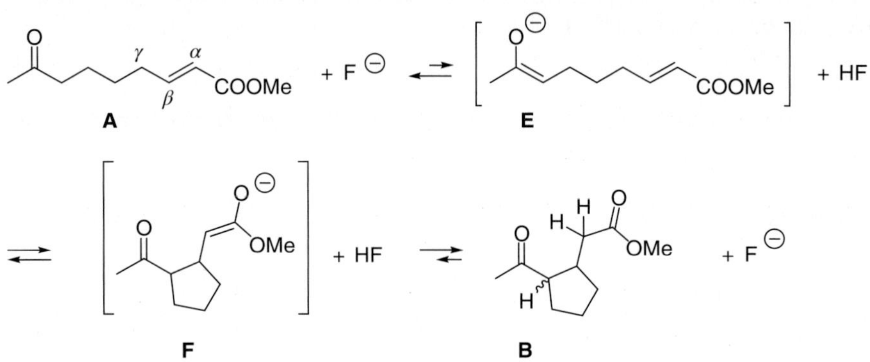

Alternatively, the acrylic ester **A** could also be deprotonated and be equilibrated with the dienolate intermediate **D**. An intramolecular 1,2-addition of **D** onto its methyl ketone moiety would generate alcoholate **G** that could equilibrate with the tertiary alcohol **C**. Since the hydrocarbation of a ketone is much less exothermic than the hydrocarbation of an alkene, this latter reaction might not be seen at room temperature. Product **C** could form under kinetic control (would form faster than **B**). On heating in the presence of the base it would equilibrate into the more stable product **B**. Both the intramolecular hydrocarbations are reversible reactions at high temperature because of their entropies of cyclization (Section 2.10) that make their exergonicity to diminish on raising temperature.

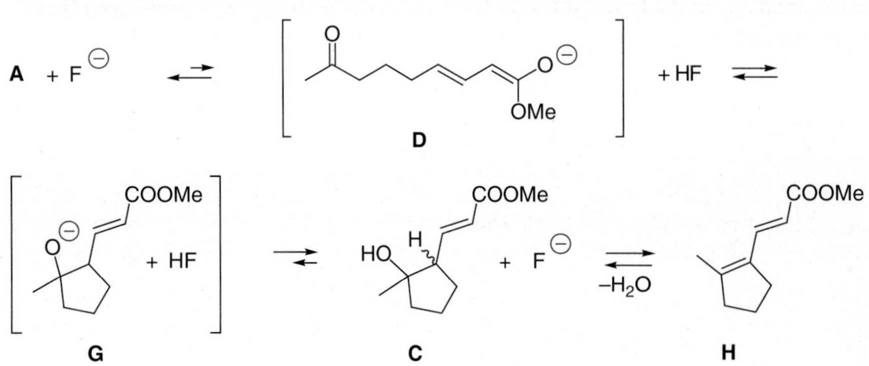

Problem 1.17 Are the Diels–Alder reactions ((4+2)-cycloadditions) of benzene with ethylene, acetylene, and allene thermodynamically possible at room temperature? Are the intramolecular versions possible? [12–15]

Answer

Using the standard heats of formation given in Tables 1.A.2 and 1.A.3 one can estimate the following standard heats of reaction:

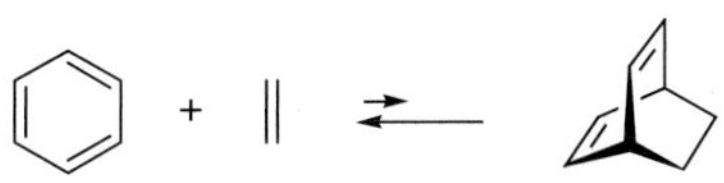

$\Delta_r H°$(benzene+ethylene⇆bicyclo[2.2.2]octa-2,5-diene)=$\Delta_f H°$(bicyclo[2.2.2]octa-2,5-diene) – $\Delta_f H°$(benzene) – $\Delta_f H°$(ethylene) = 34 – 19.8 – 12.5 = 1.7 kcal mol^{-1}. This condensation has a negative entropy of reaction similar to the Diels–Alder reaction of butadiene + ethylene giving cyclohexene (–44.8 eu, Section 1.4.4). Thus $\Delta_r G°$(benzene + ethylene ⇆ bicyclo[2.2.2]octa-2,5-diene) = 1.7 – 298(–0.045) = 1.7 + 13.4 = 15.1 kcal mol^{-1}, a very endergonic reaction. This intermelocular cycloaddition will not occur at room temperature under standard conditions. If benzene and ethylene would be attached through a tether as in **A**, and assuming the negative entropy of cyclization (blocking rotation about σ bonds of the tether) would be small, the intramolecular process will remain endergonic as it is highly endothermic.

A:

For the intermolecular cycloaddition of benzene with acetylene that would equilibrate with bicyclo[2.2.2] octa-2,5,7-triene one estimates

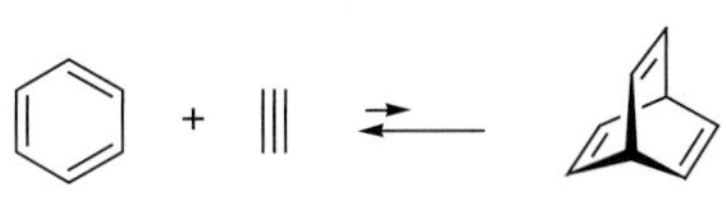

$\Delta_r H°$(benzene + acetylene ⇆ bicyclo[2.2.2]octa-2,5,7-triene) = $\Delta_f H°$(barrelene) – $\Delta_f H°$(benzene) – $\Delta_f H°$ (acetylene) = 73 – 19.8 – 54.2 = –1 kcal mol^{-1}. The situation is slightly better than for the cycloaddition of benzene to ethylene, but this exothermicity is not enough to compensate for the entropy cost (10–13 kcal mol^{-1}) of condensation. As for the hydrocarbations and the dihydrogenation, the cycloadditions of alkynes are at least 10–15 kcal mol^{-1} more exothermic than the cycloadditions of alkenes. For instance, for butadiene + acetylene → cyclohexa-1,4-diene one estimates the exothermicity –55.8 kcal mol^{-1} to be compared with –39.9 kcal mol^{-1} for the cycloaddition of butadiene + ethylene → cyclohexene. It is because of the electronic strain present in barrelene (the barrelene effect, Sections 2.9 and 4.7.6) that the intermolecular Diels–Alder reaction of benzene with acetylene (or an alkyne) is not possible. For an intramolecular version, the formation of tricyclic cycloadducts (strain increase) and the cost of entropy for blocking rotation about the σ bonds of the tether make the equilibrium favor the reactants (cycloaddents), not the product (cycloadduct).

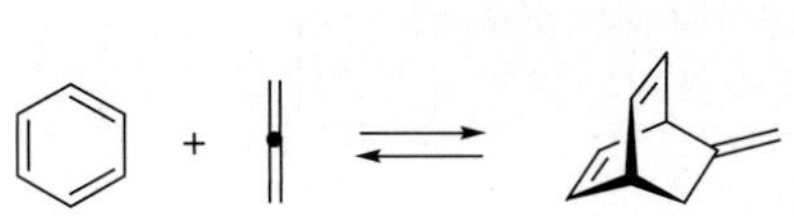

For the cycloaddition of benzene to allene one estimates

$\Delta_r H°$(benzene + allene ⇆ 7-methylidenebicyclo[2.2.2]octa-2,5-diene) = $\Delta_f H°$(7-methylidenebicyclo[2.2.2] octa-2,5-diene) – $\Delta_f H°$(benzene) – $\Delta_f H°$(allene) = 55 – 19.8 – 45.5 = –10.3 kcal mol^{-1}, just enough exothermicity to compensate the entropy cost of condensation at 25 °C under one atmosphere or an ideal solution under one atm. Alternatively, if the benzene and allene moiety are tethered as in **B**, the intramolecular version of this Diels–Alder addition is feasible at room temperature and above. It is called the Himbert cycloaddition [16, 17].

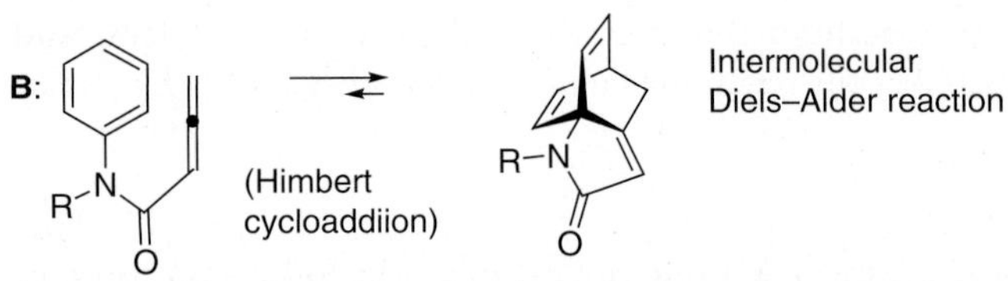

As the standard heat of formation of 7-methylidenebicyclo[2.2.2]octa-2,5-diene is not available, we can estimate it in the following way: $\Delta_f H°$(7-methylidenebicyclo[2.2.2]octa-2,5-diene) = $\Delta_f H°$(bicyclo[2.2.2]octa-2,5-diene) + $\Delta_f H°$(2-methylidenebicyclo[2.2.2]octane) − $\Delta_f H°$(bicyclo[2.2.2]octane) = $34 - 2 + 23$ = 55 kcal mol^{-1}.

Contrary to barrelene, 7-methylidenebicyclo[2.2.2]octa-2,5-diene is not destabilized electronically. Since allenes have the same relative stability as isomeric alkynes ($\Delta_f H°$(allene) = 45.5 kcal mol^{-1}, $\Delta_f H°$(propyne) = 44.2 kcal mol^{-1}) the heat of their cycloadditions is similar to those of alkynes.

Problem 1.18 Interpret the relative electron affinities of HO$^{\cdot}$, HS$^{\cdot}$, and HSe$^{\cdot}$. Interpret the relative electron affinities of halide radicals. What makes the trends observed?

Answer

Table 1.A.13 gives −EA(X$^{\cdot}$) = −78.4, −83.4, −77.6, and −70.5 kcal mol^{-1} for F$^{\cdot}$, Cl$^{\cdot}$, Br$^{\cdot}$, and I$^{\cdot}$, respectively. −EA(X$^{\cdot}$) = $\Delta_f H°$(X$^-$) − $\Delta_f H°$(X$^{\cdot}$). The trend observed for Cl$^{\cdot}$, Br$^{\cdot}$, and I$^{\cdot}$ is consistent with the fact that Cl is more electronegative than Br, and Br is more electronegative than I. One thus would have expected for fluorine atom −EA(F$^{\cdot}$) < −EA(Cl$^{\cdot}$), which is not the case. In the gas phase anions are stabilized by the electronegativity of the atom that bears the negative charge and by the polarizability of this atom. Polarizability is the property that chemical species and atoms have to delocalize the charge, and thus stabilize the anion or cation. This is done by the formation of induced dipole, π-conjugation and hyperconjugation (σ/π-interactions) for polyatomic species. The larger the atoms that constitute the charged species under examination, the larger their number, and the larger the number of bonds, the more the anion or cation is stabilized (electrostatic theory of substitutent effects on charge species: dipole/charge stabilization or destabilization: $V_C = \pm q\mu|\cos\theta|/\varepsilon r^2$; stabilization due to the induced dipole $V_I = -q^2\alpha/2\varepsilon r^4$; q, the charge; ε, the dielectric constant of the medium; θ, the angle that the dipole μ makes with the electrostatic field line; α, the polarizability; r, the distance between the substituent and the charged center). For monoatomic anions and cations, their polarizability is the ability of all electrons circulating about the nucleus to follow deformed trajectories and thus to generate induced dipoles that stabilize the charge by charge/dipole interactions. Since fluoride anion is smaller (has fewer electrons) than chloride anion, its negative charge cannot be stabilized by polarizability as well as in chloride anion. There is a competition between stabilization of the charge by electronegativity (attractive interaction between the charge and the nucleus) and formation of induced dipoles due to electron trajectory deformations that stabilize the charge.

Table 1.A.13 gives −EA(X$^{\cdot}$) = −42, −52.7, and −51 kcal mol^{-1} for HO$^{\cdot}$, HS$^{\cdot}$, and HSe$^{\cdot}$, respectively. Based on the greater electronegativity for oxygen than for sulfur one would have expected −EA(HO$^{\cdot}$) < −EA(HS$^{\cdot}$), which is not the case. Stabilization by electronegativity (attraction of the negative charge of the anion by the nucleus) competes with the stabilization of the anion by its polarizability. Thus the larger the atom (Se > S > O), the more the anion is stabilized by polarizability. The data for HS$^{\cdot}$ and HSe$^{\cdot}$ are consistent with the greater electronegativity of S compared with that of Se.

Problem 1.19 Why is the hydride anion more stable than hydrogen radical in the gas phase?

Answer

Table 1.A.13 gives IE(H$^{\cdot}$) = 313.6 kcal mol^{-1}, $\Delta_f H°$(H$^+$) = 365.2 kcal mol^{-1}, $\Delta_f H°$(H$^{\cdot}$) = ½$DH°$(H$^{\cdot}$/H$^{\cdot}$) = 52.1 kcal mol^{-1}, and $\Delta_f H°$(H$^-$) = 34.7 kcal mol^{-1}.

Hydride anion enjoys stabilization with respect to hydrogen radical due to the filling of its valence shell (doublet of electrons instead a single valence electron).

Problem 1.20 Interpret the differences in electron affinities between alkyl, alkenyl, and alkynyl radicals.

Answer

Tables 1.A.8 and 1.A.13 give $DH°(\text{Et}^{\cdot}/\text{H}^{\cdot}) = 100.7$, $DH°(\text{CH}_2\text{=CH}^{\cdot}/\text{H}^{\cdot}) = 110.6$, and $DH°(\text{CH≡C}^{\cdot}/\text{H}^{\cdot}) = 111.9\ \text{kcal mol}^{-1}$.

Table 1.A.13 gives $DH°(\text{Et}^-/\text{H}^+) = 420.1$, $DH°(\text{CH}_2\text{=CH}^-/\text{H}^+) = 406$, and $DH°(\text{CH≡C}^-/\text{H}^+) = 377\ \text{kcal mol}^{-1}$.

Table 1.A.14 gives $DH°(\text{Et}^+/\text{H}^-) = 270$, $DH°(\text{CH}_2\text{=CH}^+/\text{H}^-) = 291\ \text{kcal mol}^{-1}$.

The latter data demonstrate that the positive charge is better supported by an sp^3-hydrized carbon center than by an sp^2-hybridized carbon center. The greater the s-character the more the positive charge that approaches the protons of the nucleus, and the less stable is the carbocation. The situation is reversed for carbanions. The greater the s-character of the doubly occupied orbital of the anion, the more these two electrons are stabilized by attraction by the protons of the nucleus. Carbon-centered radicals are electron-deficient species since they lack one electron to fulfill the octet rule. One thus expects a parallelism between the relative stabilities of sextet carbocations (carbenium ions) and their radical analogs. Carbenium ions such as alkyl cations are stabilized by hyperconjugation. In the case of ethyl cation the positive charge is delocalized by three C—H bonds (Mulliken hyperconjugation, see Section 4.8.1). In vinyl cation, only one C—H bond can hyperconjugate with the cationic center, whereas in ethynyl cation hyperconjugation is not possible. Hyperconjugation is evidenced in radicals by the fine structure of their electron spin resonance (ESR) spectra that shows spin/spin coupling between the electron and the nearby C–H protons.

For simple alkyl cations their relative stabilities (as given by $DH°(\text{R}^+/\text{H}^-)$) follow the sequence: methyl cation < primary < secondary < tertiary cations. The more β σ(C—H) bonds are available to interact (hyperconjugate) with the empty $2p^{(+)}$ orbital localized at the electro-deficient carbon center, the more stable is the carbocation. A similar, but very much attenuated trend is observed for the corresponding alkyl radicals as given by $DH°(\text{R}^{\cdot}/\text{H}^{\cdot})$. This is has been attributed also to hyperconjugation (interaction between 2p and β σ(C—H) bonds) as suggested by the ESR spectra of the radicals that demonstrate delocalization of the spin onto their β H atoms. However, steric destabilizing interactions in the precursor (R–H) that are reduced in the radical R· might be the origin of this stability trend (see Section 4.8.5).

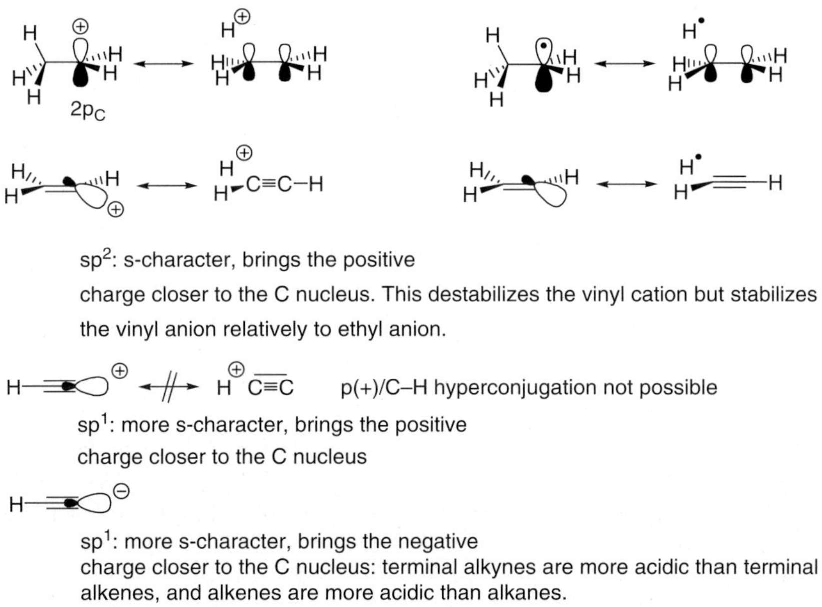

Problem 1.21 Why is there no linear relationship for the homolytic bond dissociation enthalpies $DH°(\text{Me}^{\cdot}/\text{Me}^{\cdot}) = 89\ \text{kcal mol}^{-1}$, $DH°(t\text{-Bu}^{\cdot}/\text{Me}^{\cdot}) = 81.8\ \text{kcal mol}^{-1}$ and $DH°(t\text{-Bu}^{\cdot}/t\text{-Bu}^{\cdot}) = 68.8\ \text{kcal mol}^{-1}$?

Answer

Alkane *t*-Bu–*t*-Bu is destabilized by front strain (F-strain) and back strain (B-strain) (steric repulsion between the methyl groups, see Section 2.5.1), thus rendering its homolysis less endothermic than expected ($DH°(t\text{-Bu}^{\cdot}/t\text{-Bu}^{\cdot}) = 89 - 2(7.2) = 74.6\ \text{kcal mol}^{-1}$ instead of the observed $DH°(t\text{-Bu}^{\cdot}/t\text{-Bu}^{\cdot}) = 68.8\ \text{kcal mol}^{-1}$) on the basis of the relative stability of the radical fragments formed in the homolysis. Both F- and B-strains are relieved in the two *t*-butyl radicals.

Problem 1.22 Why is the homolytic dissociation enthalpy of HF higher than that of HCl?

Answer

H—F bond is stronger than H—Cl bond because of the greater electronegativity difference between H/F pair of atoms than between H/Cl pair of atoms and shorter H—F bond than H—Cl bond.

Problem 1.23 Why are oxygen-centered radicals more reactive than analogous carbon-centered radicals?

Answer

Oxygen-centered radicals such as R–O and Ph–O$^{\cdot}$ are more reactive than alkyl radicals such as R–$CH_2^{\cdot}$ and Ph–CH_2 because they form alcohols (R–O–H) and phenols (Ph–O–H) in which the O—H bonds are significantly stronger that the corresponding C—H bonds in the corresponding hydrocarbons (R–CH_2–H, Ph–CH_2–H). This is a manifestation of the electronegativity difference, which is larger between atom pair H/O than between atomic pair H/C.

Problem 1.24 Give an interpretation for the better stabilizing effect of 4-amino group than of 4-CN and 4-nitro group in 4-substituted phenoxyl radicals.

Answer

Table 1.A.11 gives $DH^\circ(\mathrm{PhO^{\cdot}/H^{\cdot}}) = 90.4$, $DH^\circ(\mathrm{4\text{-}H_2N\text{–}C_6H_4O^{\cdot}/H^{\cdot}}) = 77.3$, $DH^\circ(\mathrm{4\text{-}O_2N\text{–}C_6H_4O^{\cdot}/H^{\cdot}}) = 94.7$ and $DH^\circ(\mathrm{4\text{-}NC\text{–}C_6H_4O^{\cdot}/H^{\cdot}}) = 94.2\ \mathrm{kcal\,mol^{-1}}$ (in DMSO). Phenoxy radical is electrophilic (electron-poor) because of its tendency to take an electron from its neighborhood, a manisfestation of the electronegativity of oxygen that is greater than that of carbon. Thus, the oxy radical pulls on the π-electrons of the phenyl ring, making it electron poor at position C(4). As a consequence, a 4-amino substituent, which is a strong electron-releasing group, better stabilizes the phenoxy radical than any electron-withdrawing 4-substituents such as CN and NO_2.

Problem 1.25 Compare the N—H homolytic bond dissociation enthalpies of amines (Table 1.A.11) and explain the non-additivity of phenyl substitution on the stability of nitrogen-centered radical.

Answer

Table 1.A.11 gives $DH^\circ(\mathrm{H_2N^{\cdot}/H^{\cdot}}) = 108\ \mathrm{kcal\,mol^{-1}}$ (gas phase) and $DH^\circ(\mathrm{PhNH^{\cdot}/H^{\cdot}}) = 88\ \mathrm{kcal\,mol^{-1}}$ (gas phase), indicating stabilizing effect of $20\ \mathrm{kcal\,mol^{-1}}$ by a phenyl substituent on the nitrogen-centered radical. This is significantly higher than for the phenyl substituent effect ($14.5\ \mathrm{kcal\,mol^{-1}}$) measured for the corresponding carbon-centered radical (methyl radical) as given (Table 1.A.7) by comparing $DH^\circ(\mathrm{H_3C^{\cdot}/H^{\cdot}}) = 104.2\ \mathrm{kcal\,mol^{-1}}$ (gas phase) and $DH^\circ(\mathrm{PhCH_2^{\cdot}/H^{\cdot}}) = 89.7\ \mathrm{kcal\,mol^{-1}}$ (gas phase).

Table 1.A.11 gives also $DH^\circ(\mathrm{PhNH^{\cdot}/H^{\cdot}}) = 92\ \mathrm{kcal\,mol^{-1}}$ and $DH^\circ(\mathrm{Ph_2N^{\cdot}/H^{\cdot}}) = 87.5\ \mathrm{kcal\,mol^{-1}}$ in DMSO solution. A comparison of the two latter values shows a much weaker phenyl substituent effect of $4.5\ \mathrm{kcal\,mol^{-1}}$ on the stabilization of the phenylaminyl radical. In the case of related carbon-centered radicals, $DH^\circ(\mathrm{PhCH_2^{\cdot}/H^{\cdot}}) = 89.7\ \mathrm{kcal\,mol^{-1}}$ (gas phase) and $DH^\circ(\mathrm{Ph_2CH^{\cdot}/H^{\cdot}}) = 81.4\ \mathrm{kcal\,mol^{-1}}$ (gas phase) are measured. This gives a phenyl substituent stabilizing effect of $8.3\ \mathrm{kcal\,mol^{-1}}$ on the benzyl radical, almost twice as much as the phenyl substituent effect on the stabilization of PhNH$^{\cdot}$. The difference between the nitrogen and carbon-centered radical arises from the fact that nitrogen is more electronegative than carbon. The electron demand of radical $H_2N^{\cdot}$ is higher than that of $H_3C^{\cdot}$. As a consequence, the electron-rich and highly polarizable phenyl substituent responds more when substituting $H_2N^{\cdot}$ instead of $H_3C^{\cdot}$. The spin is more delocalized in PhNH$^{\cdot}$ than in Ph$CH_2^{\cdot}$. This represents a first hypothesis to explain the smaller phenyl stabilizing effect on PhNH$^{\cdot}$ than on Ph$CH_2^{\cdot}$. Once the spin is delocalized by the phenyl substituent a second phenyl substituent cannot stabilize as much as the first one. A second hypothesis is to invoke the non-planarity of the two phenyl rings in $Ph_2CH^{\cdot}$ and $Ph_2N^{\cdot}$. Because of gauche interactions between the two geminal phenyl rings in these radicals, conjugation between the radicals and the π-systems is reduced compared with Ph$CH_2^{\cdot}$ and PhNH$^{\cdot}$, respectively. Because of the shorter C—N bond (between Ph and N) compared with the C—C bond (between Ph and $CH_2^{\cdot}$) the two phenyl rings in $Ph_2N^{\cdot}$ are closer to each other than in $Ph_2CH^{\cdot}$. Consequently, deviation from planarity might be larger in $Ph_2N^{\cdot}$ than in $Ph_2CH^{\cdot}$. The above discussion ignores the error on the phenyl substituent effect considered. If it should be ± 2–$3\ \mathrm{kcal\,mol^{-1}}$, more hypotheses should be examined!

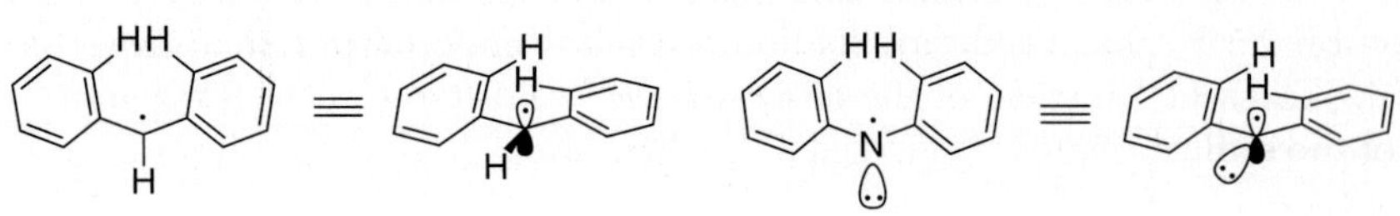

Problem 1.26 Calculate the standard gas phase heat of reaction of the following additions:

a) $R^{\cdot} + CO \leftrightarrows RCO^{\cdot}$ for R=Ph, Me, *t*-Bu
b) $Et^{\cdot} + MeCOMe \leftrightarrows EtC(Me)_2C–O^{\cdot}$
c) $Et^{\cdot}$ + acetone $\leftrightarrows$ $EtO–C^{(\cdot)}Me_2$
d) $Et^{\cdot}$ + cyclopentene $\leftrightarrows$ 2-ethylcyclopentyl$^{\cdot}$

Answer

Heats of formation of radicals are taken from Table 1.A.7.

a) $\Delta_r H°(Ph^{\cdot} + CO \leftrightarrows PhCO^{\cdot}) = 26\ (\pm?) - 81\ (\pm2.0) - (-26.4 \pm 0.04) = -28.8 \pm 3$.
$\Delta_r H°(Me^{\cdot} + CO \leftrightarrows MeCO^{\cdot}) = -2.9(\pm0.7) - 34.8(\pm0.2) - (-26.4 \pm 0.04) = -10.2 \pm 1$.
$\Delta_r H°(t\text{-}Bu^{\cdot} + CO \leftrightarrows t\text{-}BuCO^{\cdot}) = \Delta_f H°(t\text{-}BuCO^{\cdot}) - 11.0(\pm0.7) - (-26.4 \pm 0.04) = X$.
Since $\Delta_f H°(t\text{-}BuCO^{\cdot})$ is not given in the table, we can estimate $\Delta_r H°(t\text{-}Bu^{\cdot} + CO \leftrightarrows t\text{-}BuCO^{\cdot}) = \Delta_r H°(Me^{\cdot} + CO \leftrightarrows MeCO^{\cdot})$ + correction for the stability difference between *t*-Bu$^{\cdot}$ and Me$^{\cdot}$ that is given by the comparison of $DH°(Me^{\cdot}/H^{\cdot})$ = 104.7 with $DH°(t\text{-}Bu^{\cdot}/H^{\cdot})$ = 95.2 kcal mol^{-1}, i.e.: 9.5 ± 2.0 kcal mol^{-1}, taking ±1.0 kcal mol^{-1} for the average standard error on the homolytic bond dissociation enthalpies given in Table 1.A.7. Assuming that the *t*-butyl group of *t*-BuCO$^{\cdot}$ radical does not stabilize the radical more than the methyl group in MeCO$^{\cdot}$ (which must be the case since $DH°(MeCH_2^{\cdot}/H^{\cdot})$ = 100.7 and $DH°(PhCH_2^{\cdot}/H^{\cdot})$ = 89.7, but $DH°(MeCO^{\cdot}/H^{\cdot})$ = 86 and $DH°(PhCO^{\cdot}/H^{\cdot})$ = 87 kcal mol^{-1}, Table 1.A.7), one estimates $\Delta_r H°(t\text{-}Bu^{\cdot} + CO \leftrightarrows t\text{-}BuCO^{\cdot}) = -10.2(\pm1.0) + 9.5(\pm2.0) = -0.7 \pm 3.0$ kcal mol^{-1}. This reaction is not exothermic enough to compensate for the entropy cost (condensation). Pivalyl radical (*t*-BuCO$^{\cdot}$) might be formed as an intermediate by addition of *t*-butyl radical to carbon monoxide. This intermediate might give a stable product such as *t*-BuCO–*t*-Bu by reaction with a *t*-butyl radical, or form pivaldehyde (*t*-BuCHO) by hydrogen atom transfer from *t*-Bu$^{\cdot}$.
Disproportionation reaction: *t*-BuCO$^{\cdot}$ + *t*-Bu$^{\cdot}$ $\leftrightarrows$ *t*-BuCHO + isobutylene
But the latter reactions compete with a more frequent reaction that is the disproportionation 2 *t*-Bu$^{\cdot}$ $\leftrightarrows$ isobutylene + isobutane.
At room temperature, the concentration of *t*-BuCO$^{\cdot}$ remains always much lower than that of *t*-Bu$^{\cdot}$.

b) Since $\Delta_f H°(Me_2(Et)C–O^{\cdot})$ is not given in the tables, one assumes
$\Delta_r H°(Et^{\cdot} + Me_2CO \leftrightarrows Me_2(Et)CO^{\cdot}) = \Delta_r H°(Me^{\cdot} + Me_2CO \leftrightarrows Me_3CO^{\cdot})$ + correction for the difference in relative stability between methyl and ethyl radicals, which is given on comparing $DH°(Me^{\cdot}/H^{\cdot})$ = 104.7 with $DH°(Et^{\cdot}/H^{\cdot})$ = 100.7 kcal mol^{-1}, i.e. 4.0 ± 2.0 kcal mol^{-1}, taking ±1.0 kcal mol^{-1} for the average standard error on the homolytic bond dissociation enthalpies given in Table 1.A.7. This implies that the relative stability of the alkyloxy radicals is the same irrespective of the size of the alkyl group. This is indeed the case as $DH°(t\text{-}BuO^{\cdot}/H^{\cdot}) = DH°(MeO^{\cdot}/H^{\cdot})$ = 104–105 kcal mol^{-1} (Table 1.A.13).
One finds the following data:
$\Delta_f H°(Me^{\cdot}) = 34.8 \pm 0.2$ (Table 1.A.7), $\Delta_f H°(Me_2CO) = -52.23 \pm 0.14$ kcal mol^{-1} (Table 1.A.4).
$\Delta_f H°(Me_3C–O^{\cdot})$ can be calculated from $DH°(t\text{-}BuO^{\cdot}/H^{\cdot})$ = 105 kcal mol^{-1} (Table 1.A.11):
$DH°(t\text{-}BuO^{\cdot}/H^{\cdot}) = 105 \pm 1.0 = \Delta_f H°(t\text{-}BuO^{\cdot}) + \Delta_f H°(H^{\cdot}) - \Delta_f H°(t\text{-}BuOH) = \Delta_f H°(t\text{-}BuO^{\cdot}) + 52.1(\pm0.001) - (-74.72 \pm 0.21)$, which leads to $\Delta_f H°(t\text{-}BuO^{\cdot}) = -21.8 \pm 2.0$ (Table 1.A.13 gives $\Delta_f H°(t\text{-}BuO^{\cdot}) = -21.0$) This gives $\Delta_r H°(Me^{\cdot} + Me_2CO \leftrightarrows Me_3CO^{\cdot}) = -21.8(\pm2) - 34.8(\pm0.2) - (-52.23 \pm 0.14) = -4.4 \pm 2.4$ kcal mol^{-1}.
Thus, $\Delta_r H°(Et^{\cdot} + Me_2CO \leftrightarrows Me_2(Et)CO^{\cdot}) = -4.4 \pm (2.4) + 4.0 \pm (2.0) = -0.4 \pm 4.4$ kcal mol^{-1}.
Alternatively, one can estimate $\Delta_f H°(Me_2(Et)C–O^{\cdot})$ from $\Delta_f H°(Me_3C–O^{\cdot})$ by considering the difference $\Delta_f H°(Me_2(Et)C–OH) = -78.7$ and $\Delta_f H°(Me_3C–OH) = -74.72 \pm 0.21$ kcal mol^{-1} (Table 1.A.4) of -4.0 ± 0.5 kcal mol^{-1}. This gives $\Delta_f H°(Me_2(Et)C–O^{\cdot}) = \Delta_f H°(Me_3C–O^{\cdot}) - 4.0 \pm 0.5$ kcal mol^{-1} $= -21.8(\pm2.0) - 4.0(\pm0.5) = -25.8 \pm 2.5$ kcal mol^{-1}. Thus, $\Delta_r H°(Et^{\cdot} + Me_2CO \leftrightarrows Me_2(Et)CO^{\cdot}) = -25.8(\pm2.5) - 28.4(\pm0.5) - (-52.23 \pm 0.14) = -2.0 \pm 3.2$ kcal mol^{-1}.
The more the alkyl radical is stable, the least exothermic is its addition to the carbon center of a ketone. This is the case because the oxygen-centered radicals that result from these reactions have the same relative stabilities since they are not affected (Table 1.A.13) by the different tertiary alkyl groups. Indeed, Table 1.A.13 gives $DH°(MeO^{\cdot}/H^{\cdot}) = DH°(t\text{-}BuO^{\cdot}/H^{\cdot}) = DH°(Me_2(Et)CO^{\cdot}/H^{\cdot})$ = 104–105 kcal mol^{-1}.

c) $\Delta_r H°(Et^{\cdot} + Me_2CO \leftrightarrows EtO–(Me)_2C^{\cdot}) = \Delta_f H°(EtO–(Me)_2C^{\cdot}) - \Delta_f H°(Et^{\cdot}) - \Delta_f H°(MeCOMe) = \Delta_f H°(EtO–(Me)_2C^{\cdot}) - 28.4(\pm0.5) - (52.23 \pm 0.14)$.

$\Delta_f H°$(EtO–(Me)$_2$C$^•$) is estimated from $\Delta_f H°$(MeOCH$_2$$^•$) = −3 kcal mol^{-1} (Table 1.A.7) by considering the exchange of a methoxy group for an ethoxy group (contribution X) and by converting the primary alkyl radical into a dimethylalkyl radical: contribution Y.

X = $\Delta_f H°$(MeOEt) − $\Delta_f H°$(MeOMe) = −51.73(±0.16) − (−44.0(±0.12)) = −7.7 ± 0.3 kcal mol^{-1} (Table 1.A.4). Y = $\Delta_f H°$(Et(Me$_2$)C$^•$) − $\Delta_f H°$(EtCH$_2$$^•$) = (6.7 ± 0.7) − (23.5 ± 0.5) = −16.8 ± 1.2 kcal mol^{-1} (Table 1.A.7). One obtains $\Delta_f H°$(EtO–(Me)$_2$C$^•$) = $\Delta_f H°$(MeOCH$_2$$^•$) = −3.0 + X + Y = −3.0 − 7.7(±0.3) − 16.8(±1.2) = −27.5 ± 2.5 kcal mol^{-1}.

Thus, $\Delta_r H°$(Et$^•$ + Me$_2$CO ⇆ EtO–(Me)$_2$C$^•$) = $\Delta_f H°$(EtO–(Me)$_2$C$^•$) − $\Delta_f H°$(Et$^•$) − $\Delta_f H°$(MeCOMe) = −27.5(±2.5) − 28.4(±0.5) − (52.23 ± 0.14) = −3.7 ± 3.1 kcal mol^{-1}

d) $\Delta_r H°$(Et$^•$ + cyclopentene ⇆ 2-ethylcyclopent-1-yl$^•$) = $\Delta_f H°$(2-Et-cyclopentyl$^•$) − $\Delta_f H°$(Et$^•$) − $\Delta_f H°$(cyclopentene) = $\Delta_f H°$(2-Et-cyclopentyl$^•$) − 28.4(±0.5) − (8.5 ±?).

$\Delta_f H°$(2-Et-cyclopentyl$^•$) = $\Delta_f H°$(cyclopentyl$^•$) + difference $\Delta_f H°$(3-Et-pentane) − $\Delta_f H°$(pentane) = 24.3(±?) + W(Table 1.A.7).

W = $\Delta_f H°$(3-Et-pentane) − $\Delta_f H°$(pentane) = −45.34(±0.28) − (35.0 ± 0.14) = −10.3 ± 0.4 kcal mol^{-1} (Table 1.A.2). One obtains $\Delta_f H°$(2-Et-cyclopentyl$^•$) = 24.3 − 10.3 = 14.0 (±?) kcal mol^{-1}. This leads to $\Delta_r H°$(Et$^•$ + cyclopentene ⇆ 2-ethylcyclopent-1-yl$^•$) = 14.0 − 28.4 − 8.5 = −22.9 (±?) kcal mol^{-1}.

The addition of alkyl radicals to alkenes is much more exothermic than the addition of the same alkyl radicals to ketones. This is due mostly to the greater C=O bond energies than C=C bond energies. The heats of alkyl radical additions to alkenes and carbonyl compounds parallel those of the corresponding hydrocarbations (Section 1.7.1).

Problem 1.27 What is the preferred regioisomeric adduct of the following equilibria?

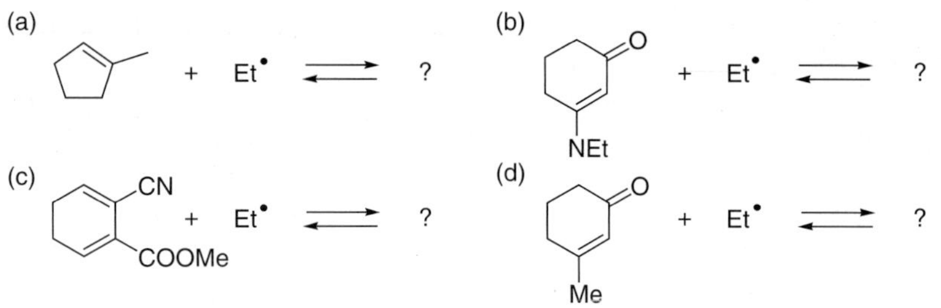

Answer

Under thermodynamic control (fully reversible reactions) the most stable regioisomeric adduct-radical should be preferred.

a) The two radicals **A**$^•$ and **B**$^•$ can form concurrently:

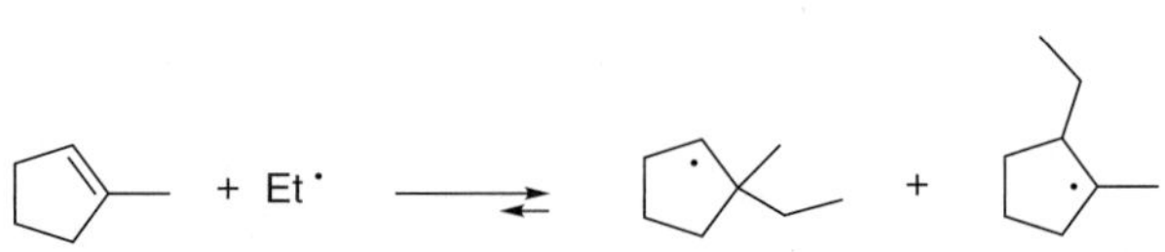

A$^•$ **B**$^•$

A$^•$ is a secondary alkyl radical with a quaternary carbon center (doubly branched alkane); **B**$^•$ is a tertiary alkyl radical with two monobranchings (branched hydrocarbons are more stable than their unbranched isomers; Section 1.6.4). One needs to consider the difference in standard heats of formation of the corresponding alkanes A–H and B–H and the corresponding homolytic bond dissociation enthalpies $DH°$(A$^•$/H$^•$) and $DH°$(B$^•$/H$^•$):

$\Delta_r H°$(**A**$^•$ ⇆ **B**$^•$) = $\Delta_f H°$(**B**$^•$) − $\Delta_f H°$(**A**$^•$) = $DH°$(B$^•$/H$^•$) − $DH°$(A$^•$/H$^•$) − $\Delta_f H°$(A–H) + $\Delta_f H°$(B–H) = $\Delta_f H°$(B$^•$) + $\Delta_f H°$(H$^•$) − $\Delta_f H°$(B–H) − $\Delta_f H°$(A$^•$) − $\Delta_f H°$(H$^•$) + $\Delta_f H°$(A–H) − $\Delta_f H°$(A–H) + $\Delta_f H°$(B–H).

Since $\Delta_f H°$(A–H) and $\Delta_f H°$(B–H) are not available $\Delta_f H°$(A–H) − $\Delta_f H°$(B–H) can be estimated to be similar to the difference $\Delta_f H°$(3,3-dimethylpentane) − $\Delta_f H°$(2,3-dimethylpentane) = −48.17 (±0.22) − (−47.62 ± 0.30) = −0.55 ± 0.52 kcal mol^{-1}.

$DH°$(A$^•$/H$^•$) and $DH°$(B$^•$/H$^•$) are not available. $DH°$(cyclopentyl$^•$/H$^•$) = 94.8 kcal mol^{-1} (Table 1.A.7) but $DH°$(1-alkylcyclopentyl$^•$/H$^•$) is not available, which is required to estimate the difference in

C—H bond energies between secondary and tertiary C—H bonds in alkylcyclopentanes. We thus need to turn to simpler model compounds for which this difference is available. Table 1.A.7 gives $DH°(\text{2-methylbut-2-yl}^{\cdot}/\text{H}^{\cdot}) = 95.3\,\text{kcal mol}^{-1}$ and $DH°(\text{pent-2-yl}^{\cdot}/\text{H}^{\cdot}) = 99.2\,\text{kcal mol}^{-1}$, a difference of c. $4\,\text{kcal mol}^{-1}$ in favor of the tertiary C—H bond. The same difference is found with the comparison $DH°(\text{2-methylprop-2-yl}^{\cdot}/\text{H}^{\cdot}) = 95.2\,\text{kcal mol}^{-1}$ and $DH°(n\text{-but-2-yl}^{\cdot}/\text{H}^{\cdot}) = 99.1\,\text{kcal mol}^{-1}$. Accepting standard deviations of $\pm 1.0\,\text{kcal mol}^{-1}$ on published $DH°$ values, one obtains:
$\Delta_r H°(\mathbf{A}^{\cdot} \leftrightarrows \mathbf{B}^{\cdot}) = DH°(\text{B}^{\cdot}/\text{H}^{\cdot}) - DH°(\text{A}^{\cdot}/\text{H}^{\cdot}) - \Delta_f H°(\text{A–H}) + \Delta_f H°(\text{B–H}) = -4(\pm 2.0) + 0.55(\pm 0.52) = -3.5 \pm 2.5\,\text{kcal mol}^{-1}$.
In this case the greater stability of the tertiary radical **B**˙ with respect to isomeric secondary radical **A**˙ overwhelms the relative small difference in stability of the alkyl skeletons due to changes in chain branching. Both radical **A**˙ and **B**˙ do not share any symmetry axis of rotation and have similar conformational flexibility. They are expected to have the same entropy, thus:
$\Delta_r G°(\mathbf{A}^{\cdot} \leftrightarrows \mathbf{B}^{\cdot}) = \Delta_r H°(\mathbf{A}^{\cdot} \leftrightarrows \mathbf{B}^{\cdot})$. Radical **B**˙ forms to a greater extent than radical **A**˙.

b) Ethyl radical adds to the C=C double bond, not to the C=O double bond (see Problem 1.26). Radicals **D**˙ and **E**˙ can form concurrently.

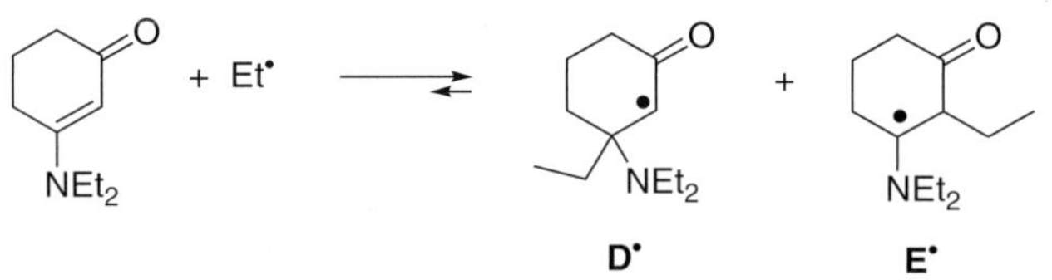

$\Delta_r H°(\mathbf{D}^{\cdot} \leftrightarrows \mathbf{E}^{\cdot}) = \Delta_f H°(\mathbf{E}^{\cdot}) - \Delta_f H°(\mathbf{D})^{\cdot} = DH°(\mathbf{E}^{\cdot}/\text{H}^{\cdot}) - DH°(\mathbf{D}^{\cdot}/\text{H}^{\cdot}) - \Delta_f H°(\mathbf{D}\text{–H}) + \Delta_f H°(\mathbf{E}\text{–H})$.
The corresponding hydrocarbons **D**–H and **E**–H are isomers in which **D**–H contains a tertiary alkylamine whereas **E**–H contains a secondary alkylamine. The stability difference $\Delta_f H°(\mathbf{D}\text{–H}) - \Delta_f H°(\mathbf{E}\text{–H})$ can be taken, to a first approximation, similar to $\Delta_f H°(\text{Me}_3\text{C–NH}_2) - \Delta_f H°(\text{MeCH(NH}_2\text{)CH}_2\text{Me}) = -28.8(\pm 0.2) - (-23.6 \pm 0.13) = -5.2 \pm 0.33\,\text{kcal mol}^{-1}$ (Table 1.A.4).
$DH°(\mathbf{D}^{\cdot}/\text{H}^{\cdot})$ can be estimated from $DH°(\text{MeCOCH}_2{}^{\cdot}/\text{H}^{\cdot}) = 98 \pm 2\,\text{kcal mol}^{-1}$ (Table 1.A.9) and by considering the fact that **D**˙ is a secondary, not a primary alkyl radical. Thus, $DH°(\mathbf{D}^{\cdot}/\text{H}^{\cdot}) = 98 - 2 = 96 \pm 3\,\text{kcal mol}^{-1}$.
$DH°(\mathbf{E}^{\cdot}/\text{H}^{\cdot})$ can be estimated from $DH°(\text{Me}_2\text{NCH}_2{}^{\cdot}/\text{H}^{\cdot}) = 83.8 \pm 1.0\,\text{kcal mol}^{-1}$ (Tables 1.A.7 and 1.A.9) and corrected for the stability difference between a primary and a secondary substituted alkyl radical (c. $4\,\text{kcal mol}^{-1}$). Thus, $DH°(\mathbf{E}^{\cdot}/\text{H}^{\cdot}) = 83.8\ (\pm 1.0) - 4 = 79.8 \pm 1.0\,\text{kcal mol}^{-1}$.
This leads to $\Delta_r H°(\mathbf{D}^{\cdot} \leftrightarrows \mathbf{E}^{\cdot}) = \Delta_f H°(\mathbf{E}^{\cdot}) - \Delta_f H°(\mathbf{D})^{\cdot} = DH°(\mathbf{E}^{\cdot}/\text{H}^{\cdot}) - DH°(\mathbf{D}^{\cdot}/\text{H}^{\cdot}) - \Delta_f H°(\mathbf{D}\text{–H}) + \Delta_f H°(\mathbf{E}\text{–H}) = 83.8 - 96 + 5.2 \approx -7\,\text{kcal mol}^{-1}$. Thus, radical **E**˙ is highly preferred over radical **D**˙ (two isomers with similar entropies).

c) Ethyl radical adds to the alkene moieties forming the isomeric tertiary-allylic radicals **F**˙ and **G**˙. Isomers **H**˙ and **I**˙ do not form as they are secondary alkyl radicals, not stabilized by allylic conjugation. Both radicals **F**˙ and **G**˙ have the same skeleton, the same degree of chain branching. They can be interconverted by exchange of their COOEt and CN groups.

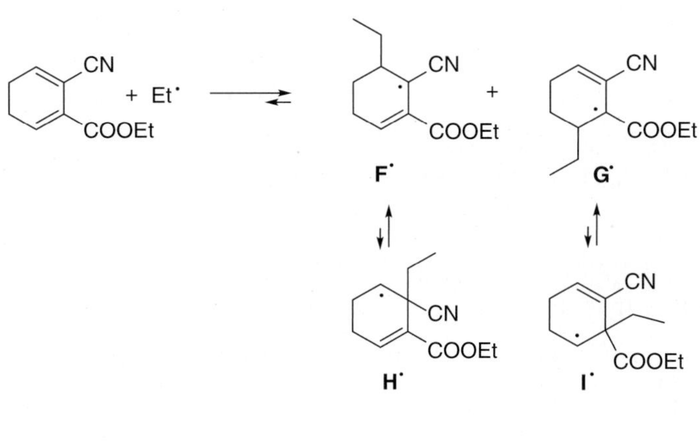

Table 1.A.9 gives the substituent effects on the relative stabilities of primary radicals (MeCOOC stabilizes by 9.7 ± 2.0 kcal mol^{-1}, and N≡C by 12 ± 2.7 kcal mol^{-1}). Assuming they are similar for tertiary allylic radicals, one predicts that radical **H**˙ will form to a greater extent than radical **G**˙.

d) Ethyl radical adds to 3-methylcyclohex-2-enone giving the regioisomeric radicals **J**˙ and **K**˙. The former is a secondary alkyl radical stabilized by 7 ± 2 kcal mol^{-1} (Table 1.A.9) through conjugation; the latter is a tertiary alkyl radical not stabilized by carbonyl conjugation, but intrinsically 2–4 kcal mol^{-1} more stable than a secondary radical. As the difference in stability between an alkane containing a quaternary carbon center and an isomer containing two tertiary carbon centers is very small (see Problem 1.27a), one can predict that radical **J**˙ will form to a greater extent than radical **K**˙.

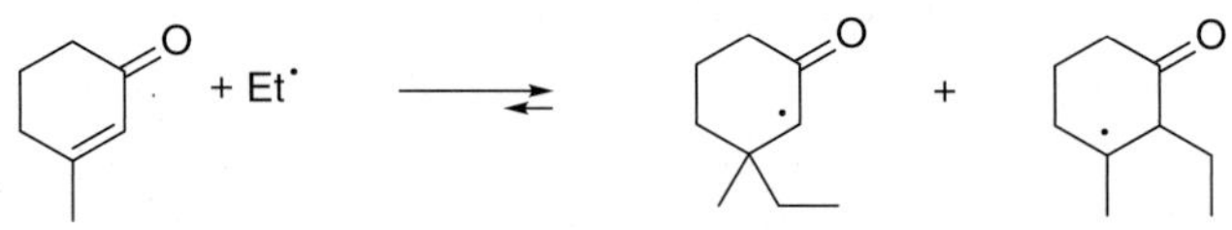

J˙ K˙

Problem 1.28 If you had to propose good radical scavenging agents, which compounds listed in Table 1.A.11 would you choose?

Answer

Those compounds with the smallest homolytic bond dissociation enthalpies such as the conjugate base of hydroquinone (O—H bond: 73 kcal mol^{-1}) and (t-Bu)$_2$NO—H (O—H bond: 68.2 kcal mol^{-1}).

Problem 1.29 Compare the gas phase heterolytic bond dissociation energies ROH ⇆ RO$^-$ + H$^+$ for methanol, ethanol, isopropanol, and *tert*-butanol. Is the trend the same in solution?

Answer

Table 1.A.13 gives $DH°$(MeO$^-$/H$^+$) = 380.5, $DH°$(EtO$^-$/H$^+$) = 377.5, $DH°$(i-PrO$^-$/H$^+$) = 375, and $DH°$(t-BuO$^-$/H$^+$) = 374.5 kcal mol^{-1}. The gas phase acidities of alcohols increases with the size of the alkyl group. The larger the alkyl group, the larger its polarizability, the more it stabilizes the alcoholate anion.

Bordwell's tables give pK_a values of 16.0, 18.8, and 19.9 for EtOH, i-PrOH, and t-BuOH in water (Table 1.A.24), and 29.0, 29.8, 30.3, and 32.2 for MeOH, EtOH, i-PrOH, and t-BuOH in DMSO (Table 1.A.25), respectively. In solution, the acidity of alcohols decreases with the size of the alkyl group. This arises from solvation that overwhelms the intrinsic stability of the alcoholate anions. The smaller the alcoholate anion, the better it is stabilized by interaction with the solvent molecules (charge/dipole interactions depend on the distance separating the charge from the solvent dipoles). In t-BuO$^-$ the solvent molecules cannot get close to the oxide anion as in MeO$^-$. There is a steric effect to solvation in large alcoholate anions.

Problem 1.30 Diazotization of primary alkyl amines with NaNO$_2$/HCl/H$_2$O at 0 °C leads to mixtures of alcohols, chlorides, and alkenes with the evolution of N$_2$. The same reaction with aniline and other aromatic primary amines generates at 0 °C persistent diazonium salts that decompose with N$_2$ evolution on heating above 60 °C. Why is there this difference in behavior between the diazonium salts resulting from alkyl and aryl primary amines?

Answer

Cyclopentyl–NH$_2$ + NaNO$_2$ + 2HCl ⇆ c-C$_5$H$_9$–N$_2$$^+$ + Cl$^-$ + NaCl + 2H$_2$O

At 20 °C: c-C$_5$H$_9$–N$_2$$^+$ + Cl$^-$ + H$_2$O → c-C$_5$H$_6$–Cl + c-C$_5$H$_9$–OH + cyclopentene + HCl + N$_2$

Phenyl–NH$_2$ + NaNO$_2$ + 2HCl ⇆ Ph–N$_2$$^+$/Cl$^-$ + NaCl + 2H$_2$O

At 20 °C, PhN$_2$$^+$/Cl$^-$ is stable.

Table 1.A.14 gives $DH°$(c-pentyl$^+$/H$^-$) = 249.8 ≪ $DH°$(Ph$^+$/H$^-$) = 287 kcal mol^{-1}. Thus, the dissociation of c-pentyl-N$_2$$^+$ into c-pentyl cation and N$_2$ is much easier than dissociation of Ph–N$_2$$^+$ into phenyl cation + N$_2$.

Problem 1.31 Explain the difference in heterolytic bond dissociation enthalpies $DH°$(cyclopent-2-enyl$^+$/H$^-$), $DH°$(cyclopentadienyl$^+$/H$^-$) and $DH°$(tropylium$^+$/H$^-$).

Answer
Table 1.A.14 gives $DH°$(cylopent-2-enyl$^+$/H$^-$) = 225, $DH°$(cylopentadienyl$^+$/H$^-$) = 255.6 and $DH°$ (cyloheptatrienyl$^+$/H$^-$) = $DH°$(tropylium cation/H$^-$) = 194 kcal mol^{-1}. The smaller the hydride affinity, the more stable the carbenium ion. Cyclopent-2-enyl cation is a 1,3-dialkylallyl cation stabilized by allylic conjugation. Cyclopentadienyl cation would be stabilized twice by allylic conjugation, but it realizes a Hückel π-system with four electrons (Section 4.5.8) that is antiaromatic, thus destabilized electronically. In contrast, tropylium cation (Section 4.5.13) is a six electron Hückel system that is stabilized by aromaticity.

Problem 1.32 Ethyl cation and methylsilicenium ion are both primary cations. Explain the data of Table 1.A.16, in particular $\Delta_r H°(Et^+ + MeSiH_3 \leftrightarrows MeSiH_2{}^+ + EtH) = -20$ kcal mol^{-1}. Why is this value not closer to zero?

Answer
In this equilibrium one exchanges one Si—H bond for a C—H bond. The latter is c. 14 kcal mol^{-1} more stable than the former as given by the comparison $DH°(CH_3{}^{\cdot}/H^{\cdot}) = 104.7$ kcal mol^{-1} with $DH°(SiH_3{}^{\cdot}/H^{\cdot}) = 90.3$ kcal mol^{-1} [18, 19]. This does not account totally for the 20 kcal mol^{-1} of exothermicity in favor of the silicenium cation. One needs to consider also the higher polarizability of the Si atom compared with that of the C atom. The Si atom has more electrons circulating about its nucleus than the C atom. The Si atom is thus more polarizable than the C atom. The positive charge in $MeSiH_2{}^+$ is better supported by the Si atom than by the carbon atom in $MeCH_2{}^+$.

Problem 1.33 Compare the gas phase hydride affinities of the following primary alkyl cations: Et$^+$, n-Pr$^+$ and n-Bu$^+$. Why are they not the same?

Answer
Table 1.A.14 gives $DH°(Me{-}CH_2{}^+/H^-) = 270$, $DH°(Me{-}CH_2{-}CH_2{}^+/H^-) = 267$, and $DH°(Me{-}CH_2{-}CH_2{-}CH_2{}^+/H^-) = 265$ kcal mol^{-1}.

In the absence of solvation, the larger the alkyl substituent, the larger its polarizability, the better it stabilizes the primary carbenium ion.

Problem 1.34 Calculate the heat of cyclopropanation of ethylene with CH_2: (methylene).

Answer
$\Delta_r H°(CH_2{:} + CH_2{=}CH_2 \leftrightarrows c\text{-}C_3H_6) = 12.74(\pm 0.14) - 62.9(\pm 0.6) - 12.54(\pm 0.1) = -62.7 \pm 0.85$ kcal mol^{-1}

Problem 1.35 Calculate the heat of the addition of trimethylenemethane to ethylene giving methylenecyclopentane.

Answer
$\Delta_r H°((CH_2)_3C + CH_2{=}CH_2 \leftrightarrows$ methylidenecyclopentane$) = 2.4 - 62 - 12.5 =$ c. -72 kcal mol^{-1}. The standard heat of formation of trimethylenemethane is estimated in the following way: $\Delta_f H°$(trimethylenemethane) = $\Delta_f H°$(isobutylene) + $2DH°$(2-methylallyl$^{\cdot}$/H$^{\cdot}$) – $DH°$(H$^{\cdot}$/H$^{\cdot}$) = $-4.3 + 2(85.1) - 104.4 \approx 62$ kcal mol^{-1}.

Problem 1.36 What is the major product of cyclodimerization of o-xylylene?

Answer
o-Xylylene is a highly reactive diene that searches to recover the aromatic stabilization of the benzene system. It can do that by entering into (4+2)- (Diels–Alder reactions) and (4+4)-cycloadditions. The cyclodimerization leading to the highest exothermicity is the [4+4]-cycloaddition forming dibenzocycloocta-1,5-diene.

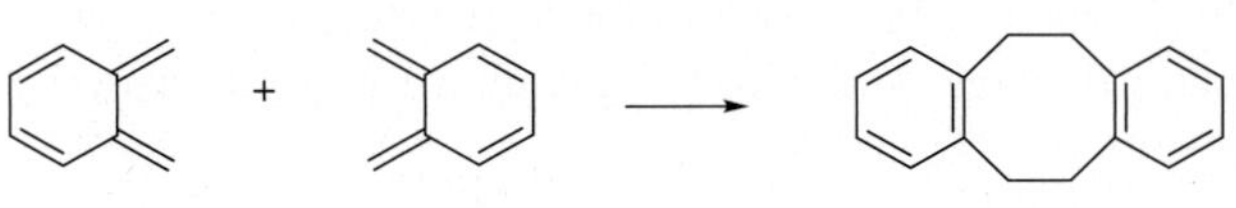

Problem 1.37 Estimate the equilibrium constant at 25 °C for phenol $\leftrightarrows$ cyclohex-2,4-dien-1-one [20, 21].

Answer

$\Delta_r H°$(phenol ⇆ cyclohex-2,4-dien-1-one) = gain of stability for enol ⇆ ketone ($-11\ kcal\ mol^{-1}$; Section 1.12.3), plus gain in stability for the π-conjugation in cyclohexa-2,4-dienone ($-5.4\ kcal\ mol^{-1}$; see below) and loss of aromaticity of phenol ($36.2\ kcal\ mol^{-1}$; see below) = $19.8\ kcal\ mol^{-1}$.

The following equilibrium gives a first estimate of the π-conjugation stabilization effect in cyclohex-2-enone ($-3.3\ kcal\ mol^{-1}$):

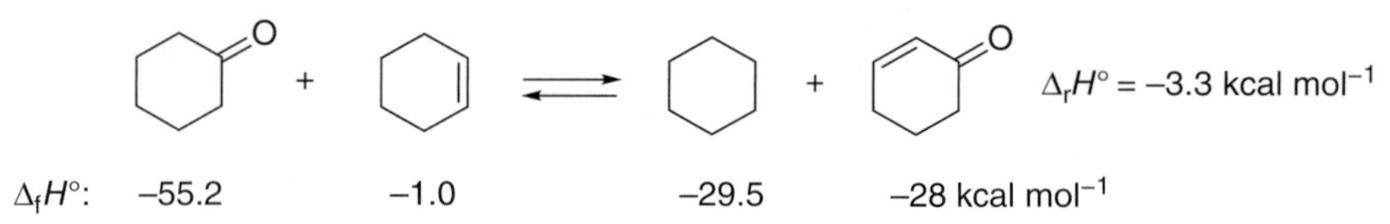

The above equilibrium compares the heats of hydrogenation of cyclohex-2-enone into cyclohexanone with the heat of hydrogenation of cyclohexene into cyclohexane. It ignores the differential hyperconjugation effects ($\sigma(CH_2)/\pi(C{=}O)$ and ($\sigma(CH_2)/\pi(C{=}C)$ interactions) between products and reactants. In the products one counts two such interactions and four in the reactants. The difference of c. $2\ kcal\ mol^{-1}$ in heat of formation of a terminal olefin (1-alkylethylene) with its 1,2-diakylethylene isomers could be used as the stabilization introduced by one ($\sigma(CH_2)/\pi(C{=}C)$) interaction. With the above equilibrium, one ignores also the differential ring strain effects between products and reactants.

Making the same simplifications (ignoring differential hyperconjugation effects and ring strain effects) the heat of the next equilibrium ($-2.1\ kcal\ mol^{-1}$) can be considered as the π-conjugation effect in cyclohexa-1,3-diene.

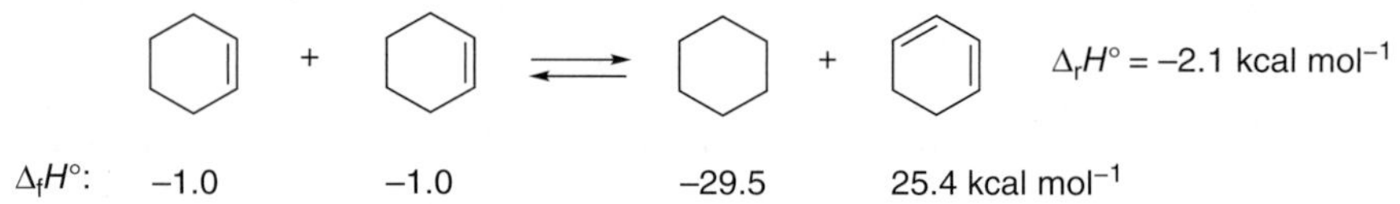

This leads to an estimate of $-3.3 - 2.1 = -5.4\ kcal\ mol^{-1}$ for the stabilization of cyclohexa-2,4-dien-1-one due to π-conjugation.

Stabilization by aromaticity in phenol is taken to be the same as that in benzene for which the following equilibrium gives a first estimate ($-36.2\ kcal\ mol^{-1}$, ignores differential hyperconjugation and ring strain effects):

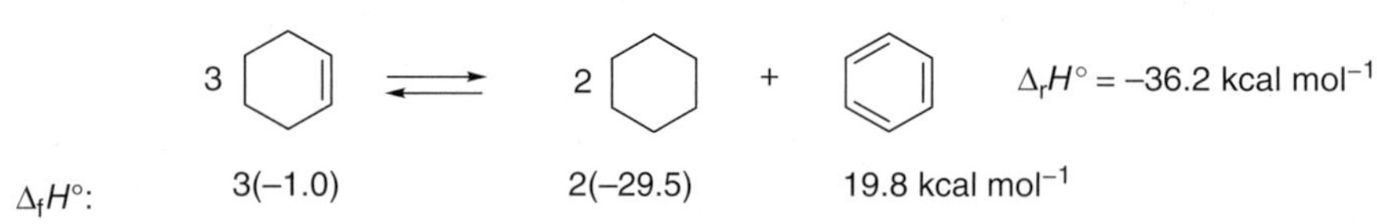

One estimates $\Delta_r G°$(phenol ⇆ cyclohex-2,4-dien-1-one) = $\Delta_r H°$(phenol ⇆ cyclohex-2,4-dien-1-one) − $T\Delta_r S° = 19\,800 - 298(\Delta_r S°)\ cal\ mol^{-1}$.

Phenol has one C_2-axis whereas cyclohexadienone does not have any rotation axis. Thus $\Delta_r S°$(phenol ⇆ cyclohex-2,4-dien-1-one) = $R\ln 2 = 1.987\ln 2 = 1.38$ eu. This gives

$\Delta_r G°$(phenol ⇆ cyclohex-2,4-dien-1-one) = $19\,800 - 298(1.38) = 19\,390\ cal\ mol^{-1}$.

This leads to $\log K = -(\Delta_r G°/RT)\cdot\log e = (-19\,390/1.987 \times 298)\cdot 0.4343 = -14.22$, and $K = 0.6 \times 10^{-14}$.

One can also use at 25 °C $\log K = \Delta_r G°$(phenol ⇆ cyclohex-2,4-dien-1-one)/1.36 = −14.25 (see equation (1.8)).

Our estimated value of K(phenol ⇆ cyclohex-2,4-dien-1-one, gas phase) can be compared with the calculated (CBS-QB3) value $\kappa = 7.15 \times 10^{-14}$ reported by Zhu and Bozzelli, and with the gas phase (flash photolysis) experimental value $K = 10^{-13\pm1}$ [21].

Problem 1.38 Can the Diels–Alder addition of methylidenemalonodinitrile with 1-phenylbutadiene occur through a diradical intermediate? Which one?

Answer

The most stable diradical resulting from the condensation of 1-phenylbutadiene with methylidenemalonodinitrile possesses an allyl radical enjoying stabilization by the phenyl substituent and an alkyl radical doubly substituted by the two cyano substituents. Any other diradicals would not be as stable.

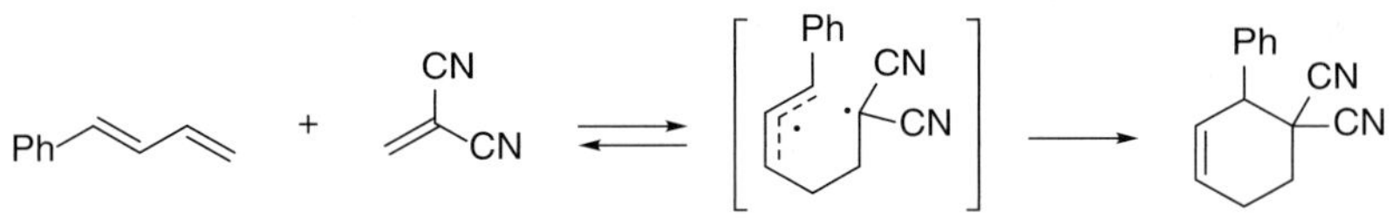

A B D P

$\Delta_r H°$(**A** + **B** ⇆ **D**) can be deduced from $\Delta_r H°$(butadiene + ethylene ⇆ hex-2-en-1,6-diyl diradical) = 32.1 kcal mol^{-1} by considering π-conjugation stabilization of cycloaddents **A** and **B** introduced by their substituent with respect to butadiene and ethylene, and the phenyl and cyano substituent effect on the relative stability of the radicals they substitute. $\Delta_f H°$(hex-2-en-1,6-diyl diradical) = $\Delta_f H°$(hex-1-ene) + $DH°$(but-3-en-2-yl$^•$/H$^•$) + $DH°$(but-1-yl$^•$/H$^•$) − $DH°$(H$^•$/H$^•$) = −10.2 + 83.9 + 101.1 − 104.2 = 70.6 kcal mol^{-1}.

$\Delta_f H°$: 26.0 12.5 70.6 −10.2 kcal mol^{-1}

$DH°$(CH$_2$=CH–C(Me)H$^•$/H$^•$) = 83.9 kcal mol^{-1}
$DH°$(n-Bu$^•$/H$^•$) = 101.1 kcal mol^{-1}
$DH°$(H$^•$/H$^•$) = 104.2 kcal mol^{-1} (Table 1.A.7)

The stabilizing π-conjugation effect (−2.1 kcal mol^{-1}) of the 1-phenyl substituent on butadiene can be evaluated by the standard heat of the following equilibrium, which compares the standard heats of hydrogenation of styrene and of propene.

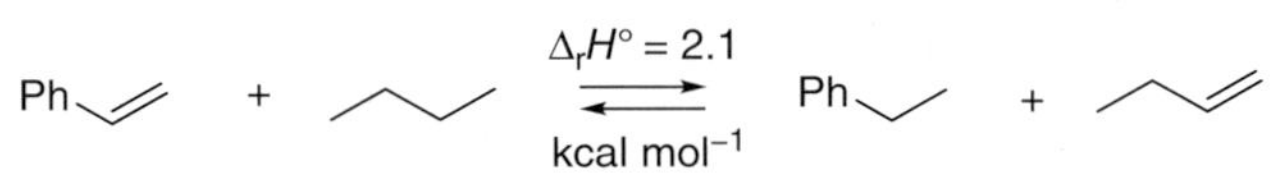

$\Delta_f H°$: 35.1 −30.0 7.1 0.1 kcal mol^{-1}

With the next equilibrium one compares the heats of hydrogenation of the C=C double bond of but-1-ene with that of acrylonitrile. One finds that π-conjugation in acrylonitrile does not stabilize the system; in fact, the available data indicate a destabilization of c. 2 kcal mol^{-1} in acrylonitrile. This arises from the stabilizing σ(CH$_2$)/π(CN) hyperconjugative interaction in propionitrile, which is not present in acrylonitrile. This is not compensated by the stabilizing σ(CH$_3$)/π(C=C) hyperconjugation in propene, an interaction which is absent in propane.

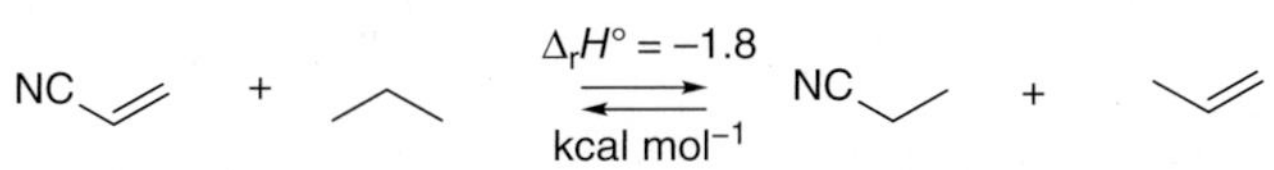

$\Delta_f H°$: 44 −25.0 12.3 4.9 kcal mol^{-1}

Thus, it appears that the stabilizing effect introduced on the diene by the 1-phenyl substituent is compensated somewhat by a destabilization effect of the 1-cyano substituent effect in the dienophiles. The 1,1-dicyano substitution introduces an electrostatic repulsion effect due to the two geminal cyano dipoles, a repulsion effect that is probably similar to that in the diradical and cycloadduct. In the above analysis, one ignores the differential σ(CH$_2$)/π(C=C) hyperconjugation effects. The stabilization effects introduced by the phenyl group on diradical **D** must be a part of the phenyl substitution stabilizing effect on methyl radical (11 kcal mol^{-1}, Table 1.A.9) as the diradical **D** is already stabilized by allylic conjugation. One chooses the phenyl stabilization effect on **D** to be c. 6 kcal mol^{-1} as the spin is delocalized on two carbon centers in the allyl radical.

Table 1.A.10 gives $DH°(\text{Me(NC)}_2\text{C}^{\cdot}/\text{H}^{\cdot})$ = 84.6 kcal mol^{-1} for homolysis in DMSO to be compared with $DH°(t\text{-Bu}^{\cdot}/\text{H}^{\cdot})$ = 95.7 kcal mol^{-1} in DMSO and 95.2 kcal mol^{-1} in the gas phase. The stabilization effect of methyl radical and in the gas phase by the cyano substituent effect amounts to −12 kcal mol^{-1}. Thus the difference $DH°/\text{Et}^{\cdot}/\text{H}^{\cdot}$, (gas) – $DH°(\text{Me(NC)}_2\text{C}^{\cdot}/\text{H}^{\cdot}$, DMSO) = 100.7 – 84.6 = 16 kcal mol^{-1} looks reasonable for the stabilizing effect of the two geminal cyano groups in diradical **D**. Therefore, a first estimate to the reaction heat $\Delta_r H°$(**A** + **B** ⇆ **D**) = 32.1 – 6 – 16 = 10.1 kcal mol^{-1} is obtained. If one considers now the reaction entropy, $\Delta_r S°$ (**A** + **B** ⇆ **D**) = $\Delta_r S°$(butadiene + ethylene ⇆ cyclohexene) – liberation of the rotation of three σ(C—C) bonds in the diradical **D** (see Section 2.11) = $S°$ (cyclohexene) – $S°$(butadiene) – $S°$(ethylene) + 15 eu = 74.2 – 66.6 – 52.5 + 15 = −30 eu, which corresponds to an entropy cost of $-T\Delta_r S° \approx 10$ kcal mol^{-1} at room temperature.

This gives $\Delta_r G°$(**A** + **B** ⇆ **D**) ≈ 20 kcal mol^{-1}. In Chapter 5, we shall show that **D,** which is only 20 kcal mol^{-1} above **A** + **B,** can be an intermediate of the Diels–Alder reaction **A** + **B** at room temperature.

Problem 1.39 Why are the yields of Diels–Alder additions usually better when adding a small amount of a radical scavenging agent such as 1,4-dihydroxybenzene (parahydroquinone) or 4-methyl-2,6-bis(*tert*-butyl)phenol?

Answer

Because diradicals resulting from the condensation of two cycloaddends are formed concurrently with the cycloadducts that may or may not follow one-step, concerted mechanisms, addition of radical scavenging agents such as hydroquinone and 4-methyl-2,6-bis(tertiobutyl)phenol reduces the diradicals and retards the polymerization of the cycloaddents.

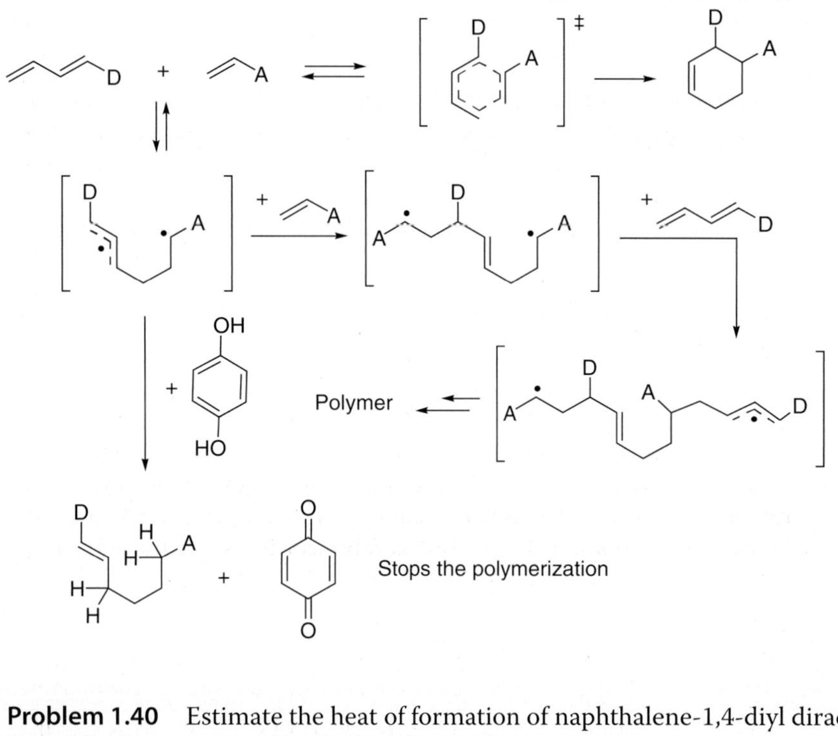

Problem 1.40 Estimate the heat of formation of naphthalene-1,4-diyl diradical and compare it with the experimental values proposed by Roth and coworkers.

Answer

$\Delta_f H°$(naphthalene-1,4-diyl diradical, est.) = $\Delta_f H°$(naphthalene) + 2$DH°(\text{Ph}^{\cdot}/\text{H}^{\cdot})$ – $DH°(\text{H}^{\cdot}/\text{H}^{\cdot})$ = 36.0(±2.0) + 2(113.3 ± 1.0) – 104.2 = 158.4 ± 4.0 kcal mol^{-1}. Squires and coworkers have measured $DH°$(naphth-1-yl$^{\cdot}$/H$^{\cdot}$) = 113.4 ± 5.2 kcal mol^{-1} similar to the C—H bond energy of benzene [22].

Roth and coworkers measured $\Delta_f H°$(naphthalene-1,4-diyl diradical, exp.) = 152.9 ± 1.4 kcal mol^{-1} [23].

Problem 1.41 Captopril® is one of the early discovered inhibitors of angiotensin converting enzymes and is an antihypertensive agent. It can be prepared according to the method presented below [25].

a) Why is the sulfur moiety of $CH_3C(S)OH$ adding to methacrylic acid and not the oxygen moiety?
b) Why does one add hydroquinone to this reaction mixture?
c) The second step of the procedure is an amidification. What is the role of dicyclohexylcarbodiimide (DCCI) added to this reaction mixture?
d) The third step implies the acidic hydrolysis of the *t*-butyl ester. What is the mechanism of this reaction?

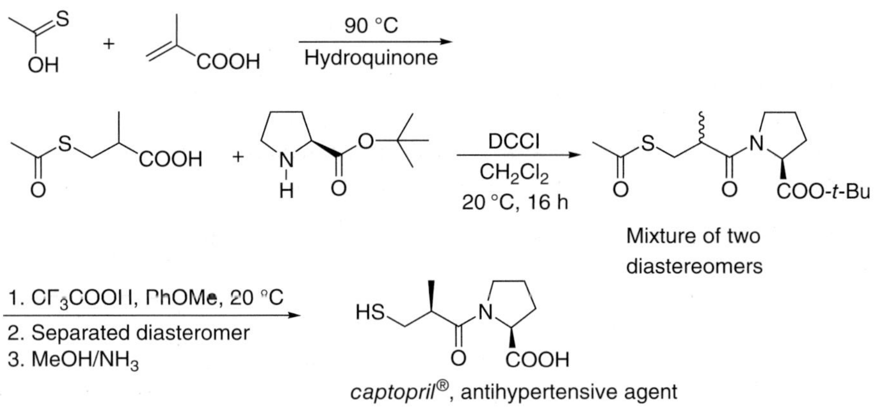

Answer

a) As per Newman–Kwart rearrangement, which converts phenols to thiophenols via their O- and S-thiocarbamates (Problem 1.9), thionocarboxylic esters (R—(C=S)—OR′) are less stable than their isomeric thioesters (R—(C=O)—SR′). Furthermore, the addition of an alcohol to an alkene and the addition of a carboxylic acid to an alkene are less exothermic than the addition of a thiol to the same alkene (Section 1.6.1, equilibria (1.48)).
b) One adds hydroquinone to trap radicals and diradicals that could be formed concurrently with the nucleophilic conjugate addition of the thionoacetic acid to methacrylic acid. In the absence of radical scavenger (hydroquinone) radical-induced polymerization of methacrylic acid could compete and this would lower the yield of the desired reaction.
c) DCCI is a carbodiimide (dicyclohexylcarbodiimide) that reacts with water, coproduct of the amidification, forming a highly stable urea derivative (*N*,*N*′-dicyclohexylurea). One thus shifts the equilibrium amine + carboxylic acid ⇆ carboxamide + water in favor of the amide, the desired product.
d) The acid-catalyzed hydrolyses of *tertio*-butyl esters are usually E_1-eliminations or S_N1-substitutions. The conjugate acid of the ester (RCOO–*t*-Bu + HX ⇆ $RC^{(+)}$(OH)–O–*t*-Bu + X^-)) undergoes heterolysis into *t*-Bu^+ + X^- + RCOOH (first order rate law, depends on the concentration of the ester and the acidic catalyst HX, not on the concentration of base or nucleophile). Then *t*-Bu^+ combines with X^- or the solvent to give isobutylene + HX (E_1-elimination) or eventually, at low temperature, *t*-Bu–X (S_N1 substitution).

Problem 1.42 Estimate the equilibrium constant of the isomerization butane ⇆ isobutane (2-methylpropane) at 300 K, at 600 K.

Answer

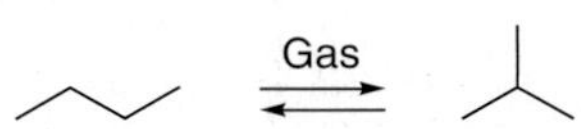

$\Delta_f H°$: −30.04 | −32.07 kcal mol^{-1} | $\Delta_r H° = -2.0$ kcal mol^{-1}

$S°$: 74.1 | 70.4 eu | $\Delta_r S° = -3.7$ kcal mol^{-1}

The branched hydrocarbon, 2-methylpropane (isobutane), is more stable than the linear hydrocarbon, butane, and its entropy is smaller than that of butane. With one methyl group more than butane, it thus has more local symmetry (for a freely rotating CH_3 group, the entropy is reduced by $-R\ln 3$; see Eq. (1.35)). Furthermore, isobutane has a C_3 axis, whereas butane has a C_2 axis. Thus the entropy of isobutane is further reduced by $R\ln 3 + R\ln 2$. Therefore, $\Delta_r S°$(butane ⇆ isobutane) $\cong -2.18 - 2.18 + 1.38 = -2.98$ eu. Experimentally, $\Delta_r S°$(butane ⇆ isobutane) $= -3.7$ eu; this is in fact more negative than the estimate based on variation on symmetry and local symmetry between product and reactant. Isobutane is more compact than butane, thus its three moments of inertia must be smaller than those of butane. From Eq. (1.35) one predicts that entropy decreases as the inertia moments diminish. This explains the experimental value $\Delta_r S°$(butane ⇆ isobutane) $= -3.7$ eu vs. the estimated value of -2.98 eu. At $T = 300$ K, one calculates $\Delta_r G^{300\text{K}} \cong \Delta_r H° - 300(\Delta_r S°) = -2000 - 300(-3.7) = -2000 + 1110 = -890$ cal mol^{-1}. This gives $\ln K^{300\text{K}} = -\Delta_r G^{300\text{K}}/1.99 \times 300 = 1.490$, or $K^{300\text{K}} = e^{1.490} = 4.4371$, which corresponds to 81.5% isobutane and 18.5% butane in equilibrium at 300 K. For the reaction at $T = 600$ K, one can use the standard heats of formation and entropies. This is reasonable as the error that will be made on $\Delta_r H^{600\text{K}}$ will be compensated more or less by the error made on $T\Delta_r S^{600\text{K}}$. Thus $\Delta_r G^{600\text{K}} \cong -2000 - 600(-3.7) = -2000 + 2220 = +220$ cal mol^{-1}. This gives $\ln K^{600\text{K}} = -220/1.99 \times 600 = -0.184$, or $K^{600\text{K}} = e^{-0.184} = 0.832$, which corresponds to 45.4% of isobutane and 54.6% of butane at equilibrium at 600 K.

Problem 1.43 Methyl isopropyl ketone (MIPK: 3-methylbutan-2-one) is an efficient high-octane (>100) oxygenate gasoline additive, without many of the undesirable effects of the widely used methyl *tert*-butyl ether. One of the methods of preparation of MIPK involves the rearrangement of pivalaldehyde (2,2-dimethylpropanal) catalyzed by strong acids [26]. Evaluate the equilibrium constant for this rearrangement at 25 °C using gas phase standard heats of formation for MIPK and *t*-BuCHO.

Answer

Table 1.A.4 gives $\Delta_f H°$(3-methylbutan-2-one) $= -62.76 \pm 0.21$ kcal mol^{-1}. It does not gives

$\Delta_f H°$(2,2-dimethylpropanal), but gives $\Delta_f H°$(2-methylpropanal) $= -51.57 \pm 0.37$ kcal mol^{-1} and $\Delta_f H°$(3,3-dimethylbutan-2-one) $= -69.47 \pm 0.21$ kcal mol^{-1}. Thus, 3-methyl substitution of 3-methylbutan-2-one amounts to $-69.47 - (-62.76) \approx -6.7$ kcal mol^{-1}. If one admits that the same substitution effect can be retained for the 2-methyl substitution of 2-methylpropanal, one estimates $\Delta_f H°$(2,2-dimethylpropanal,est.) $= -51.57 - 6.7 \approx -58.3$ kcal mol^{-1}. This gives $\Delta_r H°$(2,2-dimethylpropanal ⇆ 3-methylbutan-2-one) $= -62.76 - (-58.3) \approx -4.5$ kcal mol^{-1}.

To a first approximation, the entropy variation of this reaction must be close to zero as it is an isomerization with no blockage of liberalization of single bond rotation. There are three methyl groups in both the reactant and product. *t*-BuCHO has a local C_3 axis (*t*-Bu group), which is not present in MIPK. Thus, the overall symmetry of MIPK is lower than that of *t*-BuCHO, which contributes $R\ln 3 = 2.18$ eu to the entropy of reaction (reaction entropy increases because the product is less symmetrical than the reactant):

$\Delta_r S°$(2,2-dimethylpropanal ⇆ 3-methylbutan-2-one) $= 2.18$ cal mol^{-1} K^{-1}

Accordingly, $\Delta_r G°$(2,2-dimethylpropanal ⇆ 3-methylbutan-2-one) $= -4500 - 298(2.18) = 5156$ cal mol^{-1}. This gives $K = \exp.(-\Delta_r G°/RT) = \exp.(5156/1.987 \times 298) \approx 6050$ at 25 °C.

A possible mechanism for the acid-catalyzed isomerization of *t*-BuCHO into MIPK is given below. It involves protonation of the aldehyde follows by two successive Wagner–Meerwein rearrangements (Section 5.5.1) that are the 1,2-migrations ((1,2)-sigmatropic shifts) of a methyl group and of an hydride.

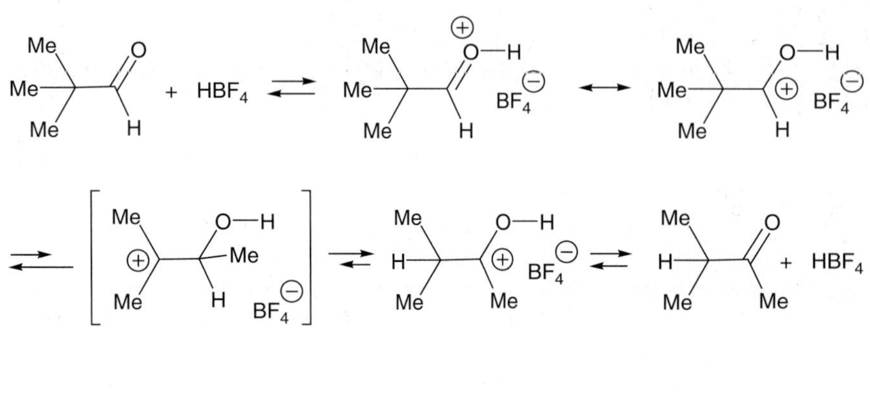

Problem 1.44 Complex **A** equilibrates with complexes **B, C,** and **D** via intermediate **I**. At equilibrium the relative proportions of **A**/**B**/**C**/**D** are 2.7 : 2 : 2 : 1. What are **B, C, D,** and **I**? Why are the relative proportions not 1 : 2 : 2 : 1?

Answer

In the absence of equilibrium isotopic effect (EIE) the proportion of **A**/**B**/**C**/**D** should be 1.0 : 2.0 : 2.0 : 1.0 (statistical distribution of the H atom between one Rh—H bond, two *ortho* C—H, two *meta* C—H, and one *para* C—H bonds of the phenyl group. At 25 °C in THF-d_6 the measured proportion are 2.7 : 2.0 : 2.0 : 1.0, which demonstrates that the heavy (deuterium) isotope prefers the stronger C—H bonds of the phenyl group to the weaker metal–H bond [26].

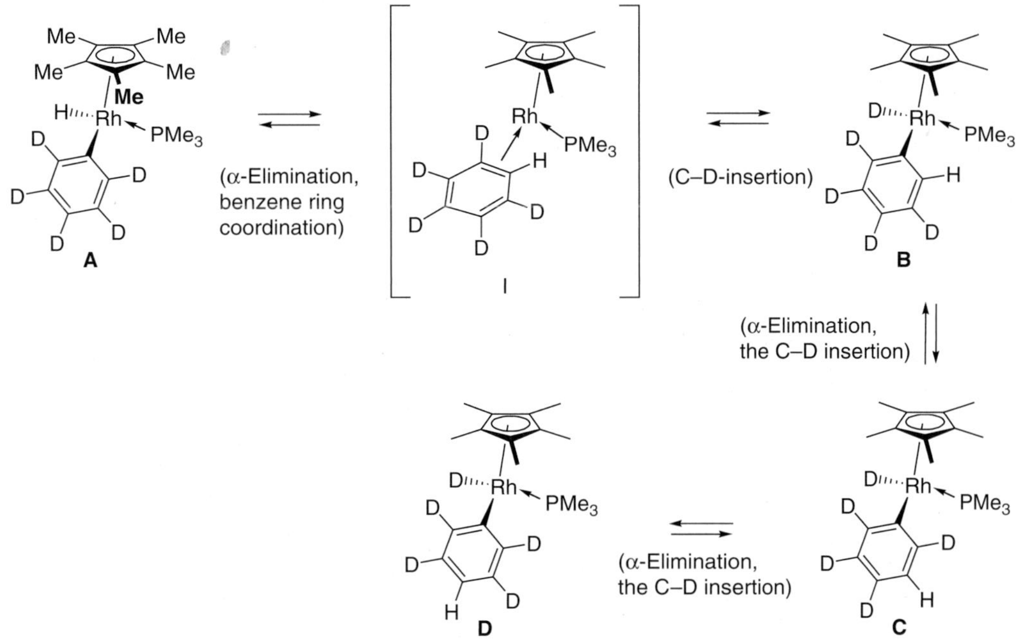

References

1 Peng, S., Qing, F.L., Li, Y.Q., and Hu, C.M. (2000). Palladium(0)/copper(I)-cocatalyzed cross-coupling of the zinc reagent of ethyl 3-bromo-3,3-difluoropropionate with aryl (alkenyl) halides: an efficient stereoselective synthesis of beta-fluoro-alpha,beta-unsaturated esters. *J. Org. Chem.* 65 (3): 694–700.

2 Kwart, H. and Evans, E.R. (1966). Vapor phase rearrangement of thioncarbonates and thioncarbamates. *J. Org. Chem.* 31 (2): 410–413.

3 Newman, M.S. and Karnes, H.A. (1966). Conversion of phenols to thiophenols via dialkylthiocarbamates. *J. Org. Chem.* 31 (12): 3980–3984.

4 Lloyd-Jones, G.C., Moseley, J.D., and Renny, J.S. (2008). Mechanism and application of the Newman–Kwart O→S rearrangement of *O*-aryl thiocarbamates. *Synthesis* 5: 661–689.

5 Dolbier, W.R., Alty, A.C., and Phanstiel, O. (1987). Kinetic and stereochemical effect of a fluorine substituent on the Cope and the homodienyl 1,5-hydrogen shift rearrangements. *J. Am. Chem. Soc.* 109 (10): 3046–3050.

6 Young, A., Lesmana, D., Dai, D.-J., and Wu, H.-S. (2013). *Bull. Chem. Ract. Eng. Catal.* 8: 1–13.

7 Christensen, J.M., Medford, A.J., Studt, F., and Jensen, A.D. (2014). High pressure CO hydrogenation over bimetallic Pt-Co catalysts. *Catal. Lett.* 144: 777–778.

8 Yu, B., Diao, Z.-F., Guo, C.-X. et al. (2013). Carboxylation of terminal alkynes at ambient CO_2 pressure in ethylene carbonate. *Green Chem.* 15: 2401–2407.

9 Ostapowicz, T.G., Hölscher, M., and Leitner, W. (2012). Catalytic hydrocarboxylation of olefins with CO_2 and H_2. A DFT computational analysis. *Eur. J. Inorg. Chem.* 5632–5641. and ref. cited therein.

10 Ding, L., Li, Y., and Huang, J. (2001). Polymerization of vinylidene carbonate as well as aminolysis and hydrolysis of polyvinylene carbonate. *Eur. Polymer J.* 37 (12): 2453–2459.
11 Paddock, R. L. and Nguyen, S: T. (2001). Chemical CO_2 fixation Cr(III) salen complexes as highly efficient catalysts for the coupling of CO_2 and epoxides. *J. Am. Chem. Soc.* 123 (46): 11498–11499.
12 Himbert, G. and Henn, L. (1982). Intramolecular Diels–Alder reaction of allenecarboxanilides. *Angew. Chem. Int. Ed. Engl.* 21 (8): 620–620.
13 Henn, L., Himbert, G., Diehl, K., and Kaftory, M. (1986). Cycloadditions. Part 6. Intramolecular Diels–Alder reaction of allenecarboxanilides – variation of the residue geminal to the carboxamide group. *Chem. Ber. Recl.* 119 (6): 1953–1963.
14 Diehl, K., Himbert, G., and Henn, L. (1986). Cycloadditions Part 7. Intramolecular Diels–Alder reactions of allenecarboxanilides – variation of the substituents in the para-position of the aniline nucleus. *Chem. Ber. Recl.* 119 (8): 2430–2443.
15 Schmidt, Y., Lam, J.K., Pham, H.V. et al. (2013). Studies on the Himbert Intramolecular arene/allene Diels–Alder cycloaddition. Mechanistic studies and expansion of scope to all-carbon tethers. *J. Am. Chem. Soc.* 135 (19): 7339–7348.
16 Himbert, G., Diehl, K. and Schlindwein, H.-J. (1989). Cycloadditionen 16. Einfluss von Alkyl und Phenylgruppen in der allenischen w-Position auf die intramolekulare Diels-Alder-Reaktionen bein Allencarboxamiden. *Chem. Ber.* 122: 1691–1699.
17 Cheng, G., He, X., Rian, L., Chen, J., Li, C., Jia, X., Li, J. (2015). Ugi/Himbert arene/allene Diels-Alder cycloaddition to synthesize strained polycyclic skeleton. *J. Org. Chem.* 80: 11100–11107.
18 Grev, R. S. and Schaefer, H. F. III, (1992). Thermochemistry of CH_n, SiH_n ($n = 0–4$), and the cation SiH^+, $SiH_2{}^+$, and $SiH_3{}^+$: A converged quantum mechanical approach. *J. Chem. Phys.* 97 (11): 8389–8406.
19 Ding, L. and Marshall, P. (1992). Does alkyl substitution affect the silicon-hydrogen bond strength in silane? Kinetic studies of the reactions of atomic chlorine and bromine with trimethylsilane and an ab initio investigation. *J. Am. Chem. Soc.* 114 (14): 5754–5758.
20 Capponi, M., Gut, I., and Wirz, J. (1986). The phenol reversible 2,4-cyclohexadienone equilibrium in aqueous-solution. *Angew. Chem. Int. Ed. Engl.* 25 (4): 344–345.
21 Zhu, L. and Bozzelli, J.W. (2003). Kinetics and thermochemistry for the gas-phase keto-enol tautomerism of phenol ↔ 2,4-cyclohexadienone. *J. Phys. Chem. A* 107 (19): 3696–3703.
22 Lardin, H.A., Squires, R.R., and Wenthold, P.C. (2001). Determination of the electron affinities of alpha- and beta-naphthyl radicals using kinetic method with full entropy analysis. The C—H bond dissociation energies of naphthalene. *J. Mass Spectrom.* 36 (6): 607–615.
23 Roth, W.R., Wasser, T., Zimmermann, H. et al. (1996). *Liebigs Ann. Chem.* (10) 1691–1695.
24 Ondetti, M.A., Rubin, B., and Cushman, D.W. (1977). Design of specific inhibitors of angiotensin-converting enzyme: new class of orally active antihypertensive agents. *Science* 196 (4288): 441–444.
25 Olah, G.A., Mathew, T., Marinez, E.R. et al. (2001). Acid-catalyzed isomerization of pivalaldehyde to methyl isopropyl ketone via a reactive protosolvated carboxonium ion intermediate. *J. Am. Chem. Soc.* 123 (47): 11556–11561.
26 Jones, W. D. and Feher, F. J. (1986). Isotope effects in arene C—H bond activation by $[(C_5Me_5)Rh(PMe_3)]$.. *Science* 108 (16): 4814–4819.

2

Additivity Rules for Thermodynamical Parameters and Deviations

Answers to Problems

Problem 2.1 Evaluate the entropy variations of the following reactions at 25 °C.

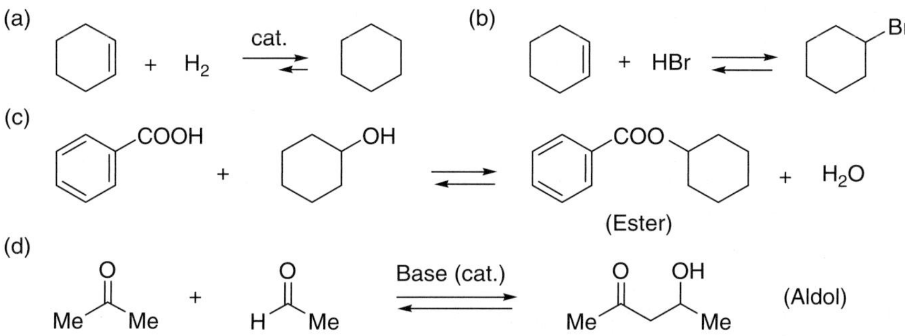

Answer

Using data given by the NIST Chemistry WebBook one obtains $\Delta_r S°$(cyclohexene + H_2 ⇄ cyclohexane) = $\Delta_r S°$(a) = 71.3 − 74.2 − 31.2 = −34.1 eu. By considering the entropy of translation only, which is expected to contribute the most to the entropies of molecules in the gas phase (S_{trans} = 6.86log Mr(g) + 11.44 log$T$ − 2.31 eu, Eq. (1.31)), one estimates $\Delta_r S°(a)_{est} = \Delta_r S°_{trans}$ (cyclohexene + H_2 ⇄ cyclohexane) = 6.86(log 84.162 − log 82.146 − log 2.016) − 11.44 log 298 + 2.31 = −28 eu. The symmetry numbers of cyclohexene, H_2, and cyclohexane are $\sigma = 2$, 2, and 6, respectively. One expects that the entropy of the reaction is more negative than −28 eu because of the greater symmetry of the product with respect to the reactants. The corrections to be appled should also consider the inertia moments of the compounds entering in the reaction.

(a) $\Delta_r S°(b)_{est} = \Delta_r S°_{trans}$(cyclohexene + HBr ⇄ cyclohexyl bromide) = 6.86(log 163.054 − log 82.148 − log 80.908) − 11.44 log 298 + 2.31 eu = −37 eu.

(b) $\Delta_r S°(c)_{est} = \Delta_r S°_{trans}$(benzoic acid + cyclohexanol ⇄ cyclohexyl benzoate + water) = 6.86(log 204.271 + log 18.016 − log 122.125 − log 100.162) eu = −3.6 eu.

(c) $\Delta_r S°(d)_{est} = \Delta_r S°_{trans}$(acetone + ethanal ⇄ 4-hydroxypentan-2-one) = 6.86(log 102.135 − log 58.081 − log 44.054) − 11.44 log 298 + 2.31 eu = −35.6 eu.

(d) Table 1.A.4 gives $S°$(acetone) = 70.5 eu and $S°$(ethanal) = 63.2 eu. It does not give $S°$(4-hydroxypentan-2-one). Table 1.A.2 gives $S°$(butane) = 74.1 eu, $S°$(pentane) = 83.1 eu, and $S°$(n-hexane) = 92.8 eu. Insertion of a CH_2 group in an alkane increases the entropy by c. 10 eu (see Section 2.4). Table 1.A.2 gives $S°$(2-methylpropane) = 70.4 eu and $S°$(2-methylpropene) = 70.2 eu, which is similar to $S°$(acetone) = 70.5 eu. Similarly, one observes that $S°$(propene) = 63.8 eu and $S°$(propane) = 64.5 eu are nearly the same as $S°$(ethanal) = 63.2 eu. Thus, one can assume $S°$(4-hydroxypenta-2-one) = $S°$(2,4-dimethylpentane), which is not given, but we estimate it to be $S°$(2,3-dimethylbutane) + 10 eu = 87.4 + 10 = 97.4 eu. This leads to another estimate for the aldol reaction: $\Delta_r S°(d)_{est}$ = 97.4 − 70.5 − 63.2 = −36.3 eu.

Problem 2.2 An enzyme **E** of molecular mass 120 000 activates a substrate **A** of molecular mass 200 making a complex **E·A** of molecular mass 120 200. Estimate the entropy variation of equilibrium **E** + **A** ⇄ **E·A** assuming

Organic Chemistry: Theory, Reactivity and Mechanisms in Modern Synthesis, Workbook, First Edition. Pierre Vogel and Kendall N. Houk.

that the flexibilities of **E** and **A** do not change in the complex **E·A**. If the same enzyme releases one, two, three, or four molecules of water during the formation of complex **E·A**, what are the entropy variations of these equilibria?

Answer

Using $S°_{trans} = 6.86 \log M_r(g) + 11.44 \log T - 2.31$ eu (Eq. (1.31)) one estimates $\Delta_r S°(\mathbf{E} + \mathbf{A} \rightleftarrows \mathbf{EA}) \cong \Delta_r S°_{trans}(\mathbf{E} + \mathbf{A} \rightleftarrows \mathbf{EA}) = 6.86(\log 120\,200 - \log 120\,000 - \log 200) - 11.44 \log 298 + 2.31 = -41.8$ eu.

For equilibrium $\mathbf{E} + \mathbf{A} \rightleftarrows \mathbf{EA} + H_2O$ one estimates $\Delta_r S°_{trans} = 6.86(\log 120\,182 + \log 18.016 - \log 120\,000 - \log 200)$ eu $= -7.2$ eu.

For equilibrium $\mathbf{E} + \mathbf{A} \rightleftarrows \mathbf{EA} + 2H_2O$ one estimates $\Delta_r S°_{trans} = 6.86(\log 120\,164 + 2 \log 18.016 - \log 120\,000 - \log 200) + 11.44 \log 298 - 2.31 = 27.4$ eu.

For equilibrium $\mathbf{E} + \mathbf{A} \rightleftarrows \mathbf{EA} + 3H_2O$ one estimates $\Delta_r S°_{trans} = 6.86(\log 120\,146 + 3 \log 18.016 - \log 120\,000 - \log 200) + 2(11.44 \log 298 - 2.31) = 61.9$ eu.

For equilibrium $\mathbf{E} + \mathbf{A} \rightleftarrows \mathbf{EA} + 4\ H_2O$ one estimates $\Delta_r S°_{trans} = 6.86(\log 120\,128 + 4 \log 18.016 - \log 120\,000 - \log 200) + 3(11.44 \log 298 - 2.31) = 96.7$ eu.

In the latter case, the enzyme is an entropy trap that can pay at least with -28.8 kcal mol^{-1} ($= -T\Delta_r S° = -298(96.7) = -28\,816$ cal mol^{-1}) for the binding and deformation of substrate **A** in the **EA** complex.

Problem 2.3 Using data of Table 1.A.4, determine the group equivalents for the molecular group of thiols.

Answer

Thiols:

$$\Delta_f H°(\text{Me—SH}) = [\text{C—(S)(H)}_3] + [\text{S—(C)(H)}] = -5.5 \text{ kcal mol}^{-1} \quad (1)$$

$$\Delta_f H°(\text{Et—SH}) = [\text{C—(C)(H)}_3] + [\text{C—(S)(C)(H)}_2] + [\text{S—(C)(H)}] = -11.0 \text{ kcal mol}^{-1} \quad (2)$$

$$\Delta_f H°(n\text{—Pr—SH}) = [\text{C—(C)(H)}_3] + [\text{C—(C)(H)}_2] + [\text{C—(S)(C)(H)}_2] + [\text{S—(C)(H)}] = -16.2 \text{ kcal mol}^{-1} \quad (3)$$

$$\Delta_f H°(n\text{—Bu—SH}) = [\text{C—(C)(H)}_3] + 2[\text{C—(C)(H)}_2] + [\text{C—(S)(C)(H)}_2] + [\text{S—(C)(H)}] = -21.0 \text{ kcal mol}^{-1} \quad (4)$$

We have five unknowns, the groups equivalents defined above, and four equations.

Let us define $x = [\text{S—(C)(H)}]$, $y = [\text{C—(S)(C)(H)}_2]$, $z = [\text{C—(C)(H)}_2]$, $w = [\text{C—(C)(H)}_3]$, and $u = [\text{C—(S)(H)}_3]$; the following equations can be written:

$$x + u = -5.5 \quad (1)$$

$$x + y + w = -11.0 \quad (2)$$

$$x + y + z + w = -16.2 \quad (3)$$

$$x + y + 2z + w = -21.0 \quad (4)$$

Comparison of Eqs. (2) and (3) gives a first value for $z = -5.2$ kcal mol^{-1}.

Comparison of Eqs. (3) and (4) gives a second value for $z = -4.8$ kcal mol^{-1}.

The average of these two values $z_m = -5.0$ kcal mol^{-1}, which is nearly the same value found for this group equivalent with alkanes. One can thus use $w = -10.0$ kcal mol^{-1}, the equivalent of the methyl group of alkanes. This leaves us with

$$x + u = -5.5 \quad (1)$$

$$x + y = -1.0 \quad (2)$$

$$x + y = -1.2 \quad (3)$$

and

$$x + y = -1.0 \quad (4)$$

Thus, we have now three unknowns (x, y, u) and two equations.

We are forced to define the two group equivalents:

$$x + y = [\text{C—(S)(C)(H)}_2] + [\text{S—(C)(H)}] = -1.0\ \text{kcal mol}^{-1}\ (\text{CH}_2\text{—SH in alkanethiols})$$

and

$$x + u = [\text{C—(S)(H)}_3] + [\text{S—(C)(H)}] = -5.5\ \text{kcal mol}^{-1}\ (\text{MeSH, unique group}).$$

Table 1.A.4 gives

$$\Delta_f H°(\text{HS—(CH}_2)_2\text{—SH}) = -2.3\ \text{kcal mol}^{-1} = 2[\text{C—(S)(C)(H)}_2] + [\text{S—(C)(H)}].$$

This gives

$$x + y = [\text{C—(S)(C)(H)}_2] + [\text{S—(C)(H)}] = -1.15\ \text{kcal mol}^{-1}, \text{a value similar to that obtained above.}$$

If one assumes $u = [\text{C—(S)(H)}_3] = [\text{C—(C)(H)}_3] = -10.0\ \text{kcal mol}^{-1}$ (reasonable since carbon and sulfur atoms have similar electonegativities), one can define

$$x = [\text{S—(C)(H)}] = -5.5 + 10.0 = 4.5\ \text{kcal mol}^{-1}.$$

This gives $y = [\text{C—(S)(C)(H)}_2] = -1.0 + 5.5 = -4.5\ \text{kcal mol}^{-1}$, a value similar to the group equivalent $z = [\text{C—(C)}_2(\text{H})_2]$, the CH_2 group in alkanes.

Table 1.A.4 also gives

$$\Delta_f H°(i\text{–Pr—SH}) = 2[\text{C—(C)(H)}_3] + [\text{C—(S)(C)}_2(\text{H})] + [\text{S—(C)(H)}] = -18.2\ \text{kcal mol}^{-1} \quad (5)$$

and

$$\Delta_f H°(i\text{–Bu—SH}) = 2[\text{C—(C)(H)}_3] + [\text{C—(C)(H)}_2] + [\text{C—(S)(C)}_2(\text{H})] + [\text{S—(C)(H)}] = -23.2\ \text{kcal mol}^{-1} \quad (6)$$

which add another unknown $v = [\text{C—(S)(C)}_2(\text{H})]$. Using standard increments for the groups in alkanes, one obtains

$[\text{C—(S)(C)}_2(\text{H})] + [\text{S—(C)(H)}] = 1.8\ \text{kcal mol}^{-1}$ with both Eqs. (5) and (6).

Using $[\text{S—(C)(H)}] = 4.5\ \text{kcal mol}^{-1}$ proposed above, one calculates $[\text{C—(S)(C)}_2(\text{H})] = 1.8 - 4.5 = -2.7\ \text{kcal mol}^{-1}$, to be compared with $[\text{C—(C)}_3(\text{H})] = -1.75\ \text{kcal mol}^{-1}$ obtained for mono-branched alkanes (compare with increment values proposed by Benson and coworkers; Table 2.A.1).

The following group equivalents are given in Table 2.A.1:

$$[\text{C—(C)}_2(\text{H})_2] = -4.92, [\text{C—(C)}_3(\text{H})] = -1.75, [\text{C—(C)}_4] = -0.06\ \text{kcal mol}^{-1}$$

$$[\text{C—(S)(C)(H)}_2] = -5.65, [\text{C—(S)(C)}_2(\text{H})] = -2.64, [\text{C—(S)(C)}_3] = -0.55\ \text{kcal mol}^{-1}$$

$$[\text{C—(N)(C)(H)}_2] = -6.6, [\text{C—(N)(C)}_2(\text{H})] = -5.2, [\text{C—(N)(C)}_3] = -3.2\ \text{kcal mol}^{-1}$$

$$[\text{C—(O)(C)(H)}_2] = -8.00, [\text{C—(O)(C)}_2(\text{H})] = -7.03, [\text{C—(O)(C)}_3] = -6.66\ \text{kcal mol}^{-1}$$

These data show the effect of electronegativity difference between carbon and the heteroelements X. Substitution of an alkyl group by X = S, N, and O moiety (alcohol, ether, carboxylate, etc.) renders the C—X bonds more polar than the corresponding C—C bond. According to Pauling, the dipolar limiting structure $C^{(+)}\ X^{(-)}$ stabilizes the C—X bonds with respect to the C—C bond, which shows up in the standard heats of formation of the polar compounds and which is expressed by the values taken by the group equivalents.

Table 1.A.4 gives $\Delta_f H°(\text{Ph—SH}) = 5[\text{C}_b\text{—(C)}_b(\text{H})] + [\text{C—(S)(C}_b)_2] + [\text{S—(C}_b)(\text{H})] = 26.7\ \text{kcal mol}^{-1}$

Using $5[\text{C}_b\text{—(C)}_b(\text{H})] = 5(3.3) = 16.5\ \text{kcal mol}^{-1}$ (Table 2.A.1), one obtains $[\text{C—(S)(C}_b)_2] + [\text{S—(C}_b)(\text{H})] = 26.7 - 16.5 = 10.2\ \text{kcal mol}^{-1}$

Problem 2.4 Calculate the equilibrium constant of isomerization (2.7) at 25 °C. At which temperature would this isomerization have to be carried out in order to obtain a yield of 80% in 2,2,3,3-tetramethylbutane by isomerization of *n*-octane?

Answer

Assuming that the conditions of an ideal gas under 1 atm apply, one can write at 298 K:

$$\Delta_r G° = \Delta_r H° - T\Delta_r S° = -4200 - 298(-18.5) = 1313\ \text{cal mol}^{-1} = -RT\ \ln K.$$

This gives $\ln K = -2.217433$ and $K^{298K} = 0.11$ ($K = 80/20 = 4$ for a yield [conversion] of 80%).

Since the error made on the heat of reaction on considering it to be the same at a temperature different from 298 K is compensated by the error made on the entropy that we keep constant also (because of the $-T\Delta_r S°$ term) we can use $\Delta_r G^T = \Delta_r G° = \Delta_r H° - T\Delta_r S° = -RT \ln K$ to calculate T with $K = 80/20 = 4$. Thus, $4200 - T(-18.5) = -RT \ln 4 = -1.987T(1.386)$ at T different from 298 K.

This gives $-4200 + 18.5T = -2.754T$ or $-4200 = -21.254T$. Therefore, $T = 197.6$ K or c. −75.5 °C.

This calculation shows that the direct conversion of *n*-octane into 2,2,3,3-tetramethylbutane, one of the best anti-knocking compounds for combustion engines, is not an easy route. Under equilibrium conditions (thermodynamic control) the process requires costly low temperatures even if a suitable catalyst should exist to catalyze the isomerization at such a low temperature. A more practical solution is to work under kinetically controlled conditions, for instance, one condenses isobutane with isobutylene that produces 2,2,4-trimethylpentane, also an good anti-knocking hydrocarbon (see Problem 2.5).

Problem 2.5 *n*-Heptane is the hydrocarbon with the highest knocking effect as engine fuel. 2,2,4-trimethylpentane is the fuel with the lowest knocking effect. A gasoline with 95% octane means that its knocking effect is the same as a 5 : 95 mixture of *n*-heptane and 2,2,4-trimethylpentane. Isomerize *n*-octane into 2,2,4-trimethylpentane in order to obtain a 5 : 95 mixture of octane and 2,2,4-trimethylpentane. At which temperature does the isomerization have to be carried out under 1 atm? *Hint*: use the $\Delta_f H°$(gas) values given in Table 1.A.2 and estimate the standard entropy of that isomerization.

Answer

Table 1.A.2 gives $\Delta_f H°$(*n*-octane) = −49.88 ± 0.16 kcal mol^{-1}; $\Delta_f H°$(2,2,4-dimethylpentane) = −53.57 ± 0.32 kcal mol^{-1}; $S°$(*n*-octane) = 111.6 eu; $S°$(2,2,4-dimethylpentane) is not given. One estimates it from $S°$(*n*-octane) correcting it for the increase in symmetry in 2,2,4-trimethylpentane, which contains three methyl groups and one *t*-butyl group more than *n*-octane. This leads to $S°(\text{2,2,4-dimethylpentane})_{est} = 111.6 + 3(-R \ln 3) - R \ln 3 = 111.6 - 8.73 = 102.87$ eu. Thus, $\Delta_r H°$(*n*-octane ⇄ 2,2,4-trimethylpentane) = −3.69 ± 0.48 kcal mol^{-1} and $\Delta_r S°$(*n*-octane ⇄ 2,2,4-trimethylpentane) = 102.87 − 111.6 = −8.73 eu. As in Problem 2.4 we assume that the isomerization occurs under the conditions of an ideal gas under 1 atm and, for the reasons invoked above, we can use $\Delta_r G^T = \Delta_r G° = \Delta_r H° - T\Delta_r S° = -RT \ln K$ at a temperature different from 298 K. This gives $-3690 - T(-8.73) = -1.987T \ln(95/5) = -5.85T$, or $-3690 = -14.58T$ leading to $T = 253.1$ K, or −20 °C.

Problem 2.6 Evaluate the strain energies of [2.2]paracyclophane (**A**), [2.2]paracyclophane-1-ene (**B**), and [2.2]paracyclophane-1,9-diene (**C**).

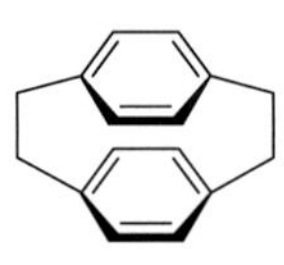

A
[2.2]Paracyclophane
$\Delta_f H°$(gas): 58.8 ± 0.8

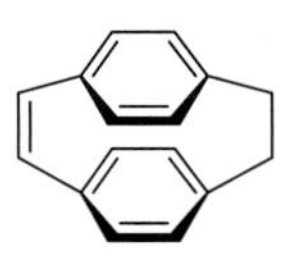

B
[2.2]Paracyclophane-1-ene
86.8 ± 1.0

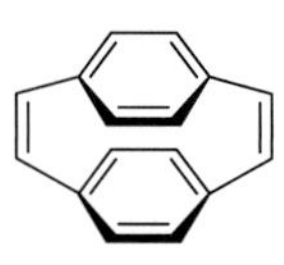

C
[2.2]Paracyclophane-1,9-diene
117.6 ± 1.2 kcal mol^{-1}

Answer

The strain energy of [2.2]paracyclophane (**A**), [2.2]paracyclophane-1-ene (**B**), and [2.2]paracyclophane-1,9-diene (**C**) can be defined as the difference between $\Delta_f H°$(gas) measured experimentally and the calculated sum of the Benson's group increments (Table 2.A.1). This leads to strain energies of 30.4 kcal mol^{-1} for **A**.

$$\Delta_f H°(\mathbf{A})_{est} = 2\Delta_f H°(p\text{–Me–}(C_6H_4\text{–(Me)} - 4[C\text{–}(C_b)(H)_3] + 4[C\text{–}(C_b)(C)(H)_2] = 2(4.3) - 4(-10.12) + 4(-5.18) = 28.36 \text{ kcal mol}^{-1}$$

$$\Delta_f H°(\mathbf{A})_{exp} - \Delta_f H°(\mathbf{A})_{est} = 58.8 - 28.4 = 30.4 \text{ kcal mol}^{-1}$$

In **B**, since the alkene moiety places its 2p electrons perpendicular to the π-systems of the aromatic ring there should not be any π-conjugative stabilization, not as in (*E*)-stilbene. The heats of hydrogenation of **B** → **A** amounts to $\Delta_f H°(\mathbf{A}) - \Delta_f H°(\mathbf{B}) = 58.8 - 86.8 = -28$ kcal mol^{-1}, which is the "norm" for 1,2-diallyl alkenes; thus, **B** has the same strain energy than **A**.

The heat of hydrogenation of **C** into **B** amounts to $\Delta_f H°(\mathbf{B}) - \Delta_f H°(\mathbf{C}) = 86.6 - 117.6 = -30.8$ kcal mol^{-1}, thus c. 3 kcal mol^{-1} more negative than the heat of hydrogenation of **B** → **C**. Thus, the strain energy of **C** must be that of **B** + 3 kcal mol^{-1} = 33.4 kcal mol^{-1}.

Problem 2.7 Hexoses (aldoses with six carbon centers, sugars) in water exist as equilibria of pyranoses and furanoses. Which hexose should show the highest pyranose content?

Answer

The hexose for which the α- and β-pyranose forms place a maximun number of substituents (2,3,4-hydroxy, 4-hydroxymethyl group) in equatorial positions should be the most stable one, and thus, favor the pyranose forms with respect to the furanose and open forms. This is the case with glucose for which the largest proportion of pyranoses is observed among all hexoses. In water glucose is made mostly of a 2 : 1 mixture of β- and α-glucopyranose. The most stable form has its anomeric OH (hydroxyl group at C(1)) in an equatorial position rather than in an axial position, contrary to expectation. Indeed, conformational anomeric effect in hemiacetals should favor the hydroxyl group in a gauche position with respect to the ring alkoxy group. In the case of glucose the water molecules (solvent) circulate about the β-anomer much better than around the α-anomer. This is a dynamic effect that renders β-glucopyranose slightly more stable than its α-anomer in water. With methyl glucopyranosides, as expected from the conformational anomeric effect, the α-anomer (with axial 1-methoxy group) is more stable than methyl β-glucopyranoside (with equatorial 1-methoxy group) (see Figure 2.9 and Table 2.A.3).

Problem 2.8 Estimate the standard heats of formation of 1,2-, 1,3-, and 1,4-dimethylcyclohexane isomers. Compare with the experimental values given below.

(a) (b) (c) (d) (e) (f)

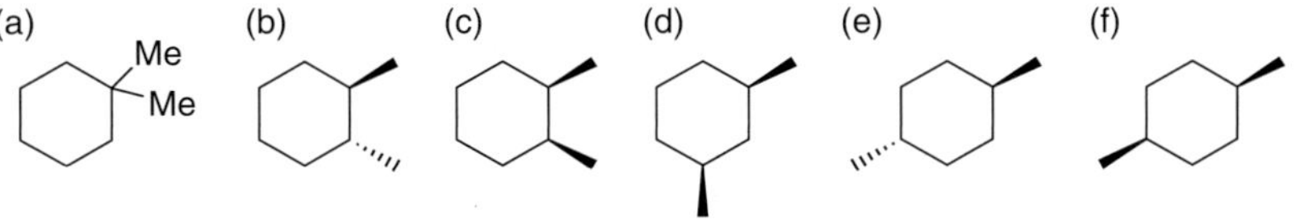

−43.2 −42.2 −41.1 −44.1 −44.1 −42.2 kcal mol^{-1}

Answer

We use the simplified group equivalents [C—(C)(H)$_3$] = −10.0 kcal mol^{-1} (instead of −10.12; Table 2.A.1), [C–(C)$_2$(H)$_2$] = −5.0 kcal mol^{-1} (instead of −4.92), [C—(C)$_3$(H)] = −1.8 kcal mol^{-1} (instead of −1.75), and [C–(C)$_4$] – 0.0 (instead of −0.06) and the A value = 1.6 kcal mol^{-1} for axial methyl group given in Table 2.A.4. Other gauche interactions will be considered such 0.7 kcal mol^{-1} for two vicinal methyl groups.

(a) $\Delta_f H°$(1,1-dimethylcyclohexane) = $\Delta_f H°$(cyclohexane) − [C—(C)$_2$(H)$_2$] + [C—(C)$_4$] + 2[C—(C)(H)$_3$] + A value for one axial methyl group = −29.5 − (−5.0) − 0.0 + 2(−10) + 1.6 = −42.9 kcal mol^{-1} (exp. −43.2 kcal mol^{-1}).

(b) $\Delta_f H°$(*trans*-1,4-dimethylcyclohexane) = $\Delta_f H°$(cyclohexane) − 2[C–(C)$_2$(H)$_2$] + 2[C–(C)$_3$(H)] + 2[C–(C)(H)$_3$] + no corrections for gauche effects = −29.5 − 2(−5.0) + 2(−1.8) + 2(−10) = −43.1 kcal mol^{-1} (exp. −44.1 kcal mol^{-1}). Using the increments of Table 2.A.1, one estimates $\Delta_f H°$(*trans*-1,4-dimethylcyclohexane) = $\Delta_f H°$(cyclohexane) − 2[C–(C)$_2$(H)$_2$] + [C–(C)$_3$(H)] + 2[C–(C)(H)$_3$] + no corrections for gauche effects = −29.5 − 2(−4.92) + 2(−1.75) + 2(−10.12) = −43.3 kcal mol^{-1}.

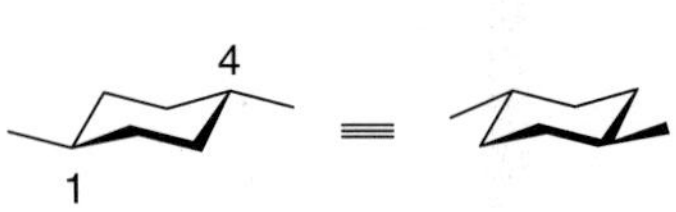

(c) $\Delta_f H°$(*cis*-1,3-dimethylcyclohexane) = $\Delta_f H°$(cyclohexane) – 2[C–(C)$_2$(H)$_2$] + 2[C–(C)$_3$(H)] + 2[C–(C)(H)$_3$] + no corrections for gauche effects = −29.5 – 2(−5.0) + 2(−1.8) + 2(−10) = −43.1 kcal mol^{-1} (exp. −44.1 kcal mol^{-1}).

(d) $\Delta_f H°$(*cis*-1,4-dimethylcyclohexane) = $\Delta_f H°$(cyclohexane) – 2[C–(C)$_2$(H)$_2$] + 2[C–(C)$_3$(H)] + 2[C–(C)(H)$_3$] + A value for one methyl group = −43.1 + 1.6 = −41.5 kcal mol^{-1} (exp. −42.2 kcal mol^{-1}).

Gauche interactions responsible for the A value of 1.6 kcal mol^{-1} of the methyl group

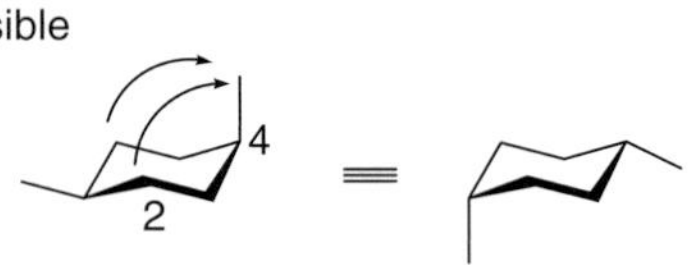

Newman projection along C(3)–C(4):

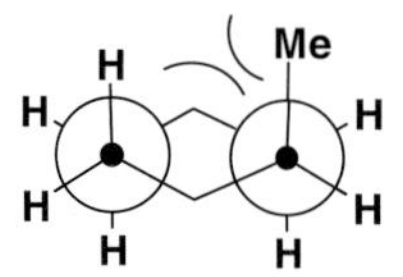

(e) $\Delta_f H°$(*cis*-1,2-dimethylcyclohexane) = $\Delta_f H°$(cyclohexane) – 2[C–(C)$_2$(H)$_2$] + 2[C–(C)$_3$(H)] + 2[C–(C)(H)$_3$] + A value for one methyl group + gauche effect between two vicinal methyl groups = −43.1 + 1.6 + 0.7 = −40.8 kcal mol^{-1} (exp. −42.2 kcal mol^{-1}).

(f) $\Delta_f H°$(*trans*-1,2-dimethylcyclohexane) = $\Delta_f H°$(cyclohexane) – 2[C–(C)$_2$(H)$_2$] + 2[C–(C)$_3$(H)] + 2[C–(C)(H)$_3$] + gauche effect between two vicinal methyl groups = −43.1 + 0.7 = −42.4 kcal mol^{-1} (exp. −41.1 kcal mol^{-1}).

If one considers the experimental standard deviation of 0.2 kcal mol^{-1} given for $\Delta_f H°$(cyclohexane) and those given for the heats of formation of dimethylcyclohexanes under study, our estimated heats of formation are not far off the experimental data. The reported experimental heats of formation of the latter present one surprise, however. Since all methyl groups are in equatorial position in *trans*-1,2-dimethylcyclohexane it is surprising that this compound is reported to be less stable than its *cis*-isomer for which one of the methyl group must occupy an axial position. If the data reported should be confirmed by more experiments, the results would imply that the gauche interaction between two equatorial vicinal methyl groups is more important than that between two vicinal methyl groups, one occupying an equatorial position and the other an axial position. In fact, one should wonder whether the data for *cis*- and *trans*-1,2-dimethylcyclohexane have not been interchanged!

Problem 2.9 In 3β-cholestanol ($C_{27}H_{48}O$), the A/B, B/C, and C/D ring junctions are all *trans*. The two methyl groups at the ring junctions are in axial positions. Estimate the heat of formation of this natural compound.

Answer

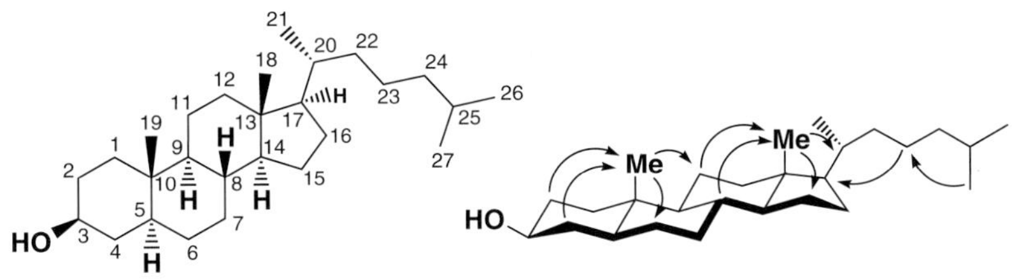

3β-Cholestanol

Newman projections

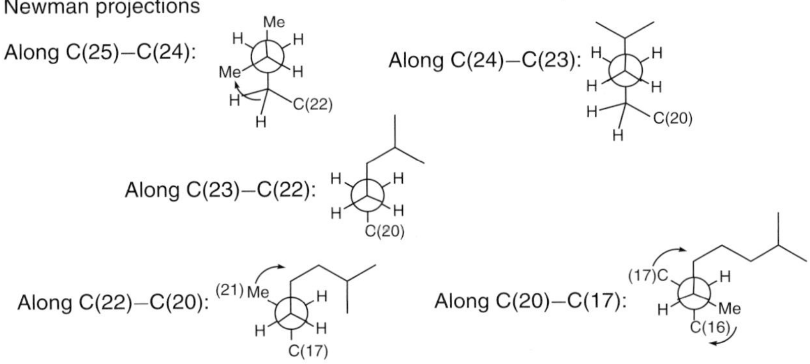

$$5\,[\mathrm{C{-}(C)(H)_3}] = 5(-10.12) = -50.6\ \mathrm{kcal\ mol^{-1}}\ (-50.0)$$

$$12\,[\mathrm{C{-}(C)_2(H)_2}] = 12(-4.92) = -59.04\ (-60.0)$$

$$7\,[\mathrm{C{-}(C)_3(H)}] = 7(-1.75) = -12.25\ (-12.6)$$

$$2\,[\mathrm{C{-}(C)_4}] = 2(-0.06) = -0.12\ (0.0)$$

$$[\mathrm{C{-}(O)(C)_2(H)}] + [\mathrm{O{-}(C)(H)}] = \Delta_f H^\circ(\text{pentan}-3-\text{ol}) - 2[\mathrm{C{-}(C)(H)_3}] - 2[\mathrm{C{-}(C)_2(H)_2}]$$
$$= -75.8 - 2(-10.12 - 4.92) = -75.8 + 30.08 = -45.7\ \mathrm{kcal\ mol^{-1}}$$

Corrections for 10 gauche interations: $10(0.7) = 7.0\ \mathrm{kcal\ mol^{-1}}$, and ring strain (Table 2.A.3) of the cyclopentane moiety: $6.3\ \mathrm{kcal\ mol^{-1}}$.

This leads to $\Delta_f H^\circ(\text{cholestanol, gas phase})_{est} = -154.4\ \mathrm{kcal\ mol^{-1}}$.

Using the simplified increments for alkanes, one obtains $-155.0\ \mathrm{kcal\ mol^{-1}}$.

Problem 2.10 On heating cyclohexanol on acidic alumina in the gas phase below 300 °C, a rapid elimination of water occurs with formation of cyclohexene. Above 350 °C 3.5% of methylcyclopentenes (mostly 1-methylcyclopentene, some 4-methylcyclopentene) are formed also. At 450 °C, up to 19.5% of methylcyclopentenes are present. Explain these results.

Answer

Elimination of water from simple alcohols is usually exergonic above 100 °C (Section 1.6.1). For instance, $\Delta_r H^\circ(i\text{-Pr—OH} \rightleftarrows \text{propene} + H_2O) = 12.2\ \mathrm{kcal\ mol^{-1}}$. This fragmentation has a positive entropy of 35 eu, which gives $-\Delta_r S^{373K} T = -13\ \mathrm{kcal\ mol^{-1}}$. In this case $\Delta_r H^\circ(\text{cyclohexanol} \rightleftarrows \text{cyclohexene} + H_2O) = 12.2\ \mathrm{kcal\ mol^{-1}}$ is estimated. The standard heat of formation of cyclohexanol is not given in Table 1.A.4. We estimate it from the heat of formation of cyclohexane and the oxidation of pentane into pentan-3-ol. $\Delta_f H^\circ(\text{pentan-3-ol}) - \Delta_f H^\circ(\text{pentane}) = -75.8 - (-35.08) = -40.7\ \mathrm{kcal\ mol^{-1}}$. Thus, $\Delta_f H^\circ(\text{cyclohexanol})_{est} = \Delta_f H^\circ(\text{cyclohexane}) - 40.7 = -29.5 - 40.7 = -70.2\ \mathrm{kcal\ mol^{-1}}$. This leads to $\Delta_r H^\circ(\text{cyclohexanol} \rightleftarrows \text{cyclohexene} + H_2O) = \Delta_f H^\circ(\text{cyclohexene}) + \Delta_f H^\circ(H_2O) - \Delta_f H^\circ(\text{cyclohexanol}) = -1.03 - 57.8 - (-70.2) = -11.4\ \mathrm{kcal\ mol^{-1}}$.

In the presence of a protic acid cyclohexanol equilibrates with its conjugate acid the corresponding alkylhydroxonium ion. The latter undergoes either an E_2 (concerted reaction, not a fragmentative process in the transition state) or an E_1 elimination (fragmentative process in the transition state) of water producing cyclohexene (see Figure 3.22, Section 3.9.3). Since fragmentation is favored at high temperature competition between the E_1 and E_2 eliminations increases with temperature. The higher the temperature the greater the concentration of cyclohexyl cation intermediate. This leaves a chance to this cation to undergo typical rearrangements of secondary carbenium ions before proton elimination [1]. Alternatively, cyclohexene is protonated to give a cyclohexyl cation that equilibrates with 1-methylcyclopentyl cation intermediate, a tertiary carbenium ion more stable than cyclohexyl cation, which is a secondary carbenium ion. 1-Methylcyclopentyl cation eliminates a proton to give 1-methylcyclopentene. It can also undergo hydride migrations that produce secondary methylcyclopentyl cations that, in turn, eliminate a proton to produce 3- and 4-methylcyclopentene.

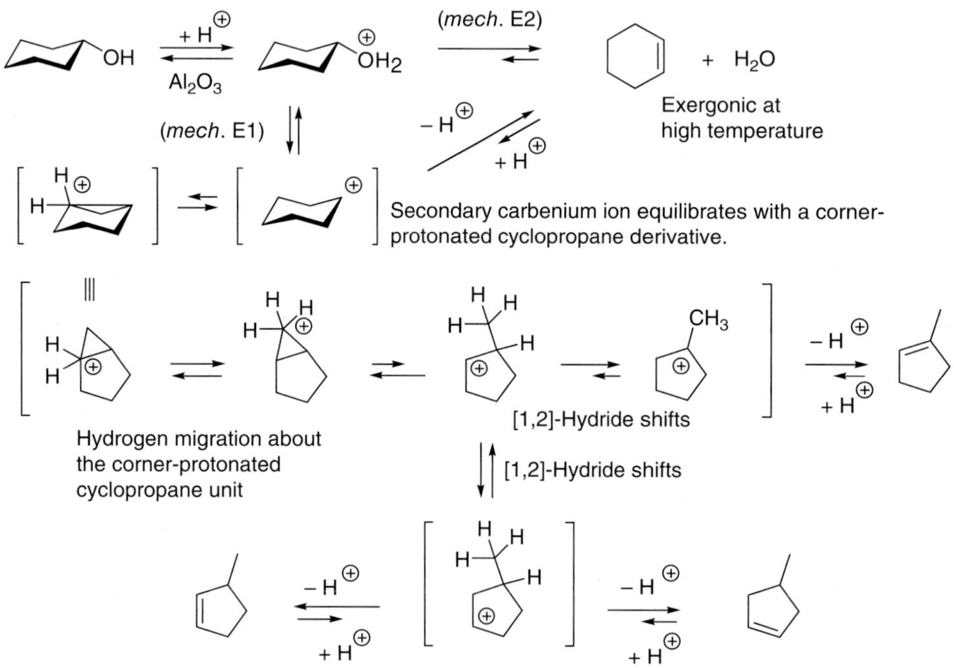

Problem 2.11 What are the products of the following reactions, for $n = 3, 7, 9$?

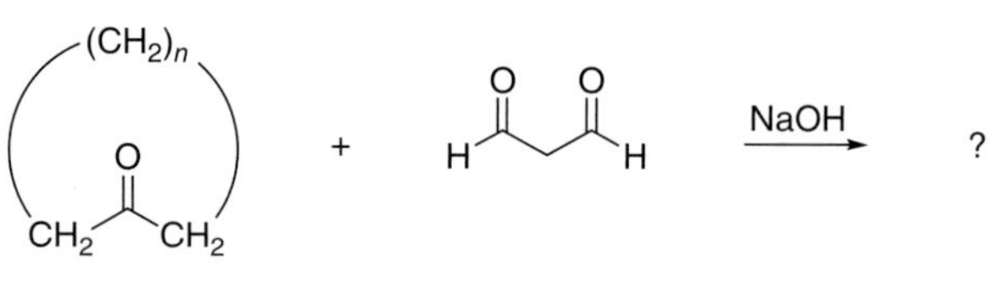

Answer

The enolate of the cycloalkanone undergoes a first nucleophilic addition to the biscarbaldehyde, which gives an aldol that eliminates an equivalent of water (crotonalization). Then, an intramolecular crotonalization occurs that generates the corresponding cyclohexa-2,5-dien-3-one. The latter equilibrates with the corresponding phenol only if $n > 3$.

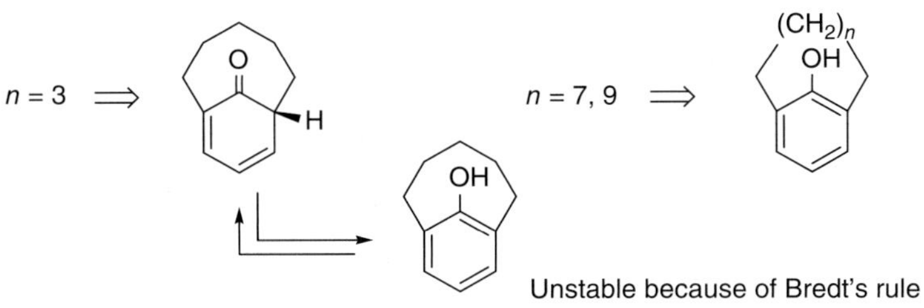

Problem 2.12 What is the product of the following hydrolysis?

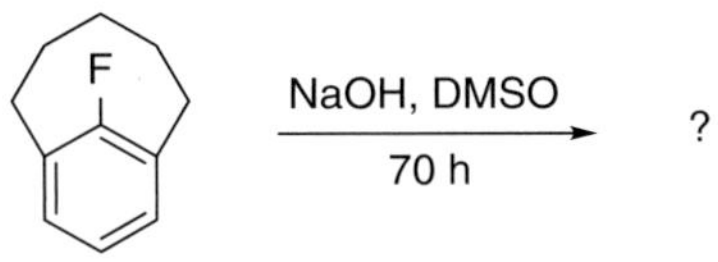

Answer
Nucleophilic addition of HO^- to the fluoroarene gives a Meisenheimer intermediate that eliminates F^- ($S_N$arom by add/elim mechanism) producing the corresponding phenol. In this case, because of severe ring strain (double Bredt's dialkene), the phenol equilibrates with the more stable cyclohexa-2,4-dien-1-one derivative:

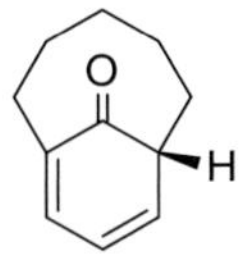

Problem 2.13 What is the most stable conformation of propene? Estimate the energy barrier to the methyl group rotation.

Answer
Applying the banana bond model for alkenes one predicts the staggered conformer **A** to be more stable than the eclipsed conformer **B** by about 2–3 kcal mol^{-1}, as in the case of ethane.

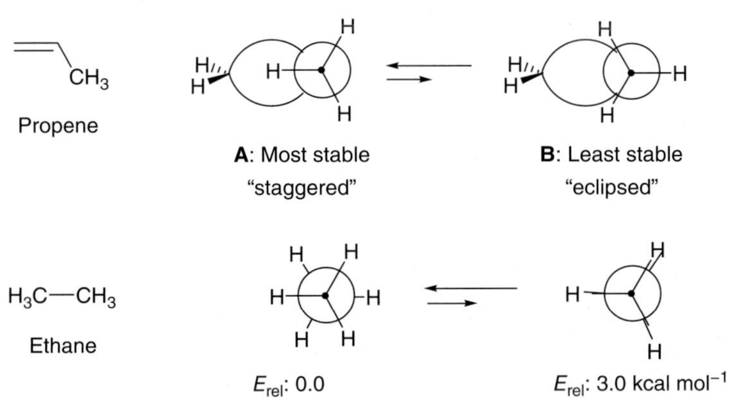

Problem 2.14 Predict the favored products (diastereoselectivity) of the ene-reactions (Section 5.7) of 3,4-dimethylpent-3-en-2-ol with singlet oxygen ($^1O_2$), *N*-phenyltriazolinedione (PTAD), and 4-nitrosobenzene (PhN=O).

Answer

Because of 1,3-allylic strain, allylic alcohol **A** adopts the preferred conformation shown. Lateral control by the hydroxyl group by hydrogen bridging with the enophile lead to the preferred *threo*-products of ene-reactions **B**. The formation of the *erythro* isomeric products **C** implies that the methyl group at the allylic position to interact severely (^{1,3}A strain) with another methyl group of **A**.

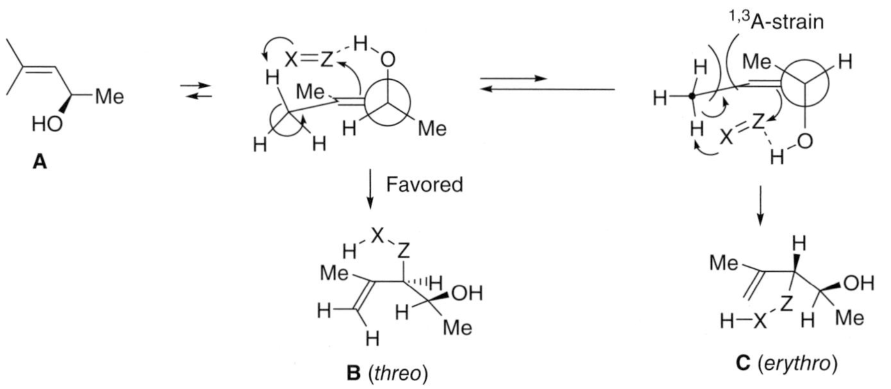

Experimentally: $^1O_2(CCl_4)$: 93 : 7
PTAD ($CHCl_3$): 87 : 13
ArNO ($CHCl_3$): >95 : 5

Problem 2.15 What is the most stable conformations of 2,3-dimethylbutadiene?

Answer

Applying the banana bond model for alkenes one predicts that both the *s-trans* and *s-cis* conformers are unstable, the most stable conformer being the *gauche* conformer that minimizes steric repulsions between the methyl substituents themselves and between them and the terminal methylidene moieties.

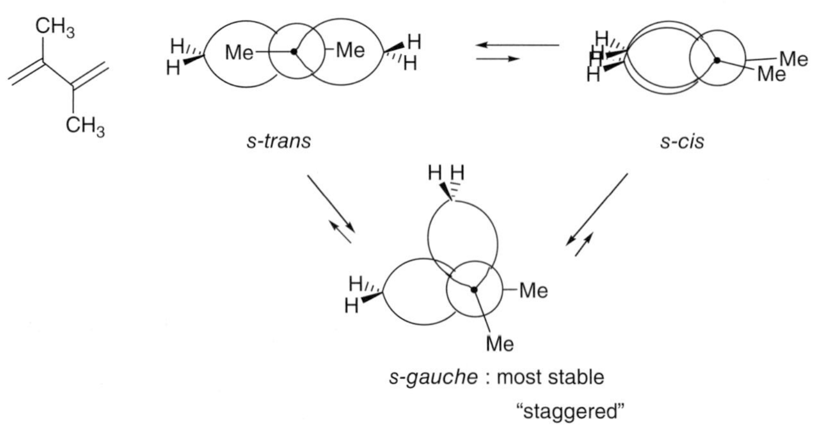

Problem 2.16 What experiments do you propose to carry out in order to prove that a given 2-substituted-3-methylbutadiene prefers a *s-gauche* (non-planar) conformation?

Answer

If the 2-substituted-3-methylbutadiene (A-**D**) with achiral 2-substituent A is not planar, the compound is chiral. The diene is an equilibrium (atropisomerism) between two enantiomeric forms (*R*)-A-**D** and (*S*)-A-**D**. If the energy barrier for the isomerization of the two enantiomeric forms is high enough one can resolve the two enantiomers (*R*)-A-**D** and (*S*)-A-**D** by chiral chromatography or by another method (Sections 1.14.3–1.14.5 and 3.7) and verify that each isolated enantiomer rotates polarized light or displays a dichroic effect (circular dichroism spectrum) of opposite sign. If the chromatographic separation and the chiroptical properties of the separated enantiomers can be carried out at 0 °C and all the operations take about 1 or 2 hours, the barrier to enantiomerization should not be lower than 21.5 kcal mol^{-1} (see Section 3.3, Fig. 3.12). If the barrier is too low, racemization will compete with the speed by which the enantiomer separation can be carried out. In such a case, one option we have to prove that the diene is chiral is to exchange the achiral 2-substituent A (e.g. Me_2CH) by a chiral substituent R* (e.g. $Me(CH_2F)CH$, Me(Et)CH, or $CD_3(CH_3)CH$) which can be either (*R*)-R or (*S*)-R. For a diene moiety that is a fast equilibrium of gauche atropoisomers (*R,R*)-R-**D** ⇄ (*R,S*)-R-**D** is a mixture of two diastereomers for which the equilibrium constant may differ from unity. The mixture is associated with a free energy hypersurface with two different minima and the equilibrium constant might depend upon temperature ($\Delta_r G = -RT\ln K$). Thus the chiroptical (and other physical or spectroscopic) properties of this mixture will depend upon temperature. Should the diene moiety be planar (either *s-cis* or *s-trans*-rotamer), one is in the presence of a single species (R)-R-**D**. The system belongs to a free energy hypersurface with a single minimum. The chiroptical (and other physical or spectroscopic) properties do not depend upon temperature.

If the chiral substituent R is not enantiomerically enriched, for a planar diene, on has a racemic mixture of (*R*)-R-**D** and (*S*)-R-**D**, which does not show any chiroptical properties. If the diene moiety adopts two gauche conformations which equilibrates rapidly, one is in the presence of a mixture of four diastereomers that are (*R,R*)-R-**D** ⇄ (*R,S*)-R-**D** and (*S,S*)-R-**D** ⇄ (*S,R*)-R-**D**. Diastereomers (*R,R*)-R-**D** and (*S,S*)-R-**D** are enantiomers and their proportion is 1 : 1. Diastereomers (*R,S*)-R-**D** and (*S,R*)-R-**D** are enantiomers and their proportion is also 1 : 1. The proportion of diastereomers (*R,R*)-R-**D** and (*S,R*)-R-**D** may differ form 1 : 1 and depend upon temperature. The same for their enantiomeric forms (*S,S*)-R-**D** and (*R,S*)-R-**D**. Fractional crystallization or achiral chromatography might separate two different racemic mixtures made of (*R,R*)-R-**D** and (*S,S*)-R-**D**, for one of them, and (*R,S*)-R-**D** and (*S,R*)-R-**D** for the other.

In the case of fast equilibrating atropisomers in solution distinction between the two pairs [(*R,R*)-R-**D** + (*S,S*)-R-**D**] and [(*R,S*)-R-**D** + (*S,R*)-R-**D**] is possible by IR spectroscopy, which has a time scale of the order 10^{-16} seconds. This requires that at least one vibration frequency of [(*R,R*)-R-**D** + (S,S)-R-**D**] is different from the corresponding vibration frequency in [(*R,S*)-R-**D** + (*S,R*)-R-**D**]. The time scale of NMR is much longer, of the order of 1 second. If the rate of isomerization of the atropisomers is >1 s^{-1} (half-life time c. 1 second; see Section 3.2.1, Table 3.1) the NMR signals of one given nuclei (e.g. ^{1}H, ^{13}C, ^{19}F) in both diastereomers are averaged. If the two pairs of diastereomers (*R,R*)-R-**D** ⇄ (*R,S*)-R-**D** and (S,S)-R-**D** ⇄ (*S,R*)-R-**D** have the same stability the proportion [(*R,R*)-R-**D** + (S,S)-R-**D**]/[(*R,S*)-R-**D** + (*S,R*)-R-**D**] is 1 : 1 = *K*. If they should not have the same stability the proportion $K \neq 1:1$. The equilibrium constant *K* depends on temperature because of the Gibbs law $\Delta_r G^T = -RT \ln K$ (Section 1.2). As a consequence the position of the averaged signals observed in the NMR spectra have chemicals shifts that vary with temperature. The method is not unambiguous as solvent effects can affect the chemical shifts and this as a function of temperature also. Furthermore, if the equilibrium constant is much smaller than unity or much larger than unity it does not vary significantly with operational temperature (−130 to 100 °C). It is for *K* values that do not differ very much from unity that the method will be the most sensitive, provided that at least one of the nuclei has chemical shifts different enough in the two diastereomers under study.

Another method is to prepare enantiomerically enriched dienes, e.g. (*R,R*)-R-**D** ⇄ (*R,S*)-R-**D** (in which the substituent (*R*)-R is enantiomerically enriched). If the substituent R does not have a chromophore (does not absorb UV light as does the diene moiety; see Section 6.2, Table 6.1) (*R,R*)-R-**D** and (*R,S*)-R-**D** will have chiroptical properties with opposite signs. Their circular dichroism (CD) spectra will nearly cancel if their proportion is $K = 1:1$. In contrast, if $K \neq 1$ the CD spectrum of (*R,R*)-R-**D** ⇄ (*R,S*)-R-**D** will be either positive ($\Delta\varepsilon < 0$) or negative ($\Delta\varepsilon < 0$) and the dichroic effect will increase on lowering temperature, if $\Delta_r G^T$ is constant (Section 1.14). For instance, if $K^{298K} = 1.10$ at 25 °C, $K^{198K} = 1.15$ at −75 °C. Thus, on lowering the temperature form 25 to −75 °C a 50% increase of $|\Delta\varepsilon|$ will be observed in the CD spectrum. For an application of the method, see Ref. [2].

Problem 2.17 Estimate the energy of π-conjugation of *p*-benzoquinone.

Answer

Table 1.A.4 gives $\Delta_f H°$(*p*-benzoquinone) = −29.4 kcal mol^{-1}, $\Delta_f H°$(cyclohexa-1,4-dione) = −79.5 kcal mol^{-1}.

Table 1.A.2 gives $\Delta_f H°$(cyclohexa-1,4-diene) = 24 kcal mol^{-1}, $\Delta_f H°$(cyclohexene) = −1 kcal mol^{-1}, and $\Delta_f H°$(cyclohexane) = −29.5 kcal mol^{-1}. If strain energy in *p*-benzoquinone can be taken as the same as that in cyclohexa-1,4-diene, and that the strain energy of cyclohexa-1,4-dione can be taken as the same as that of cyclohexane (0 kcal mol^{-1}), the comparison of the heat of double hydrogenation $\Delta_r H°$(*p*-benzoquinone → cyclohexa-1,4-dione) − 79.5 − (−29.4) = −50.1 kcal mol^{-1} with $\Delta_r H°$(cyclohexa-1,4-diene → cyclohexane) = −29.5 − (24) = −53.5 kcal mol^{-1} suggests a stabilization by π-conjugation in *p*-benzoquinone that amounts to c. 3.5 kcal mol^{-1}.

Problem 2.18 Do your expect an increase or a decrease in the stability difference between *s-cis* and *s-trans* conformers of acryl chloride and acryl fluoride compared with propenal?

Answer

On exchanging the CHO group in acrolein (propenal) for a COCl or a COF group one introduces steric repulsions that are larger in the *s-trans* form than in *the s-cis* form.

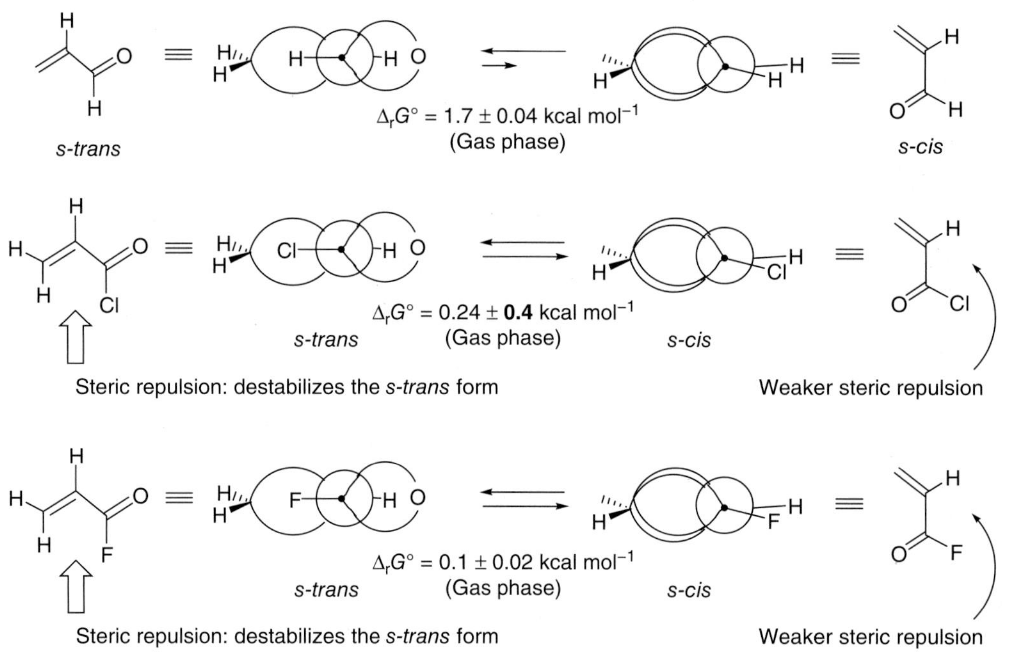

Problem 2.19 Can one obtain a product of addition of DCl to naphthalene? At which temperature can we expect a Gibbs energy for that equilibrium to be zero? Can one observe a product of addition of DCl to anthracene? If yes, at which temperature do you propose to run the reaction using one molar concentration of the reactants? Compare with the bromination of naphthalene and of anthracene.

Answer

The NIST Chemistry WebBook gives $\Delta_f H°$(naphthalene) = 36.0 ± 2.0 kcal mol^{-1}, $\Delta_r H°$(naphthalene + 2H$_2$ → benzocyclohexene, gas phase) = −29.8 ± 2.3 kcal mol^{-1}. This leads to $\Delta_f H°$(benzocyclohexene = 1,2,3,4-tetrahydronapthalene) ≈ 6.2 kcal mol^{-1}. The NIST Chemistry WebBook gives also $\Delta_r H°$(1,2-dihydronaphthalene + H$_2$ → benzocyclohexene, in AcOH, 302 K) = −24.1 ± 0.2 kcal mol^{-1}, and $\Delta_r H°$(1,4-dihydronaphthalene + H$_2$ → benzocyclohexene, in AcOH, 302 K) = −27.1 ± 0.1 kcal mol^{-1}. This allows one to estimate $\Delta_f H°$(1,2-dihydronaphthalene) = $\Delta_f H°$(benzocyclohexene) + 24.1 = 6.2 + 24.1 ≈ 30.3 kcal mol^{-1}, and $\Delta_f H°$(1,4-dihydronaphthalene) = $\Delta_f H°$(benzocyclohexene) + 27.1 = 6.2 + 27.1 ≈ 33.3 kcal mol^{-1}. This gives the heat of hydrogenation of naphthalene to give 1,2-dihydronaphthalene or 1,4-dihydronaphthalene to be c. −5.7 and −2.7 kcal mol^{-1}, respectively, which must be compared with the heat of hydrogenation of cyclohexene into cyclohexane that amounts to $-28.5\,\text{kcal mol}^{-1}$. In these hydrogenations one destroys part of the aromaticity of naphthalene. This part can be estimated to $-5.7-(-28.5)\approx 22.8\,\text{kcal mol}^{-1}$ for the formation of 1,2-dihydronapthalene and to $-2.7-(28.5)\approx 25.8\,\text{kcal mol}^{-1}$ for the formation of 1,4-dihydronapthalene. The heat of the addition of HCl to cyclohexene to give cyclohexyl chloride is $\Delta_r H°$(cyclohexene + HCl → cyclohexyl chloride, gas phase) = $\Delta_f H°$(cyclohexyl chloride) − $\Delta_f H°$(cyclohexene) − $\Delta_f H°$(HCl) = (−39.8 (Table 1.A.4)) − (−1.0 (Table 1.A.2)) − (−22.1 (Table 1.A.1)) = $-39.8+1.0+22.1\approx -16.7\,\text{kcal mol}^{-1}$. One can thus estimate the heat of addition of HCl to naphthalene producing 2-chloro-1,2-dihydronaphthalene to amount to $-16.7+22.8\approx 6.1\,\text{kcal mol}^{-1}$, an endothermic reaction that cannot occur at any temperature because of the negative entropy of condensation ($-T\Delta_r S$ term always positive). For the addition of HCl to naphthalene giving 1-chloro-1,4-dihydronaphtalene, the exothermicity of the reaction amounts to $-16.7+25.8\approx 9.1\,\text{kcal mol}^{-1}$. Thus, the Gibbs energy for the addition of HCl to naphthalene to give 2-chloro-1,2-dihydronaphthalene or/and 1-chloro-1,4-dihydronapthalene is always positive. There is no temperature at which it can become zero ($K=1$). Using DCl instead of HCl, D/H exchange will occur in naphthalene. As with other electrophiles, naphthalene can add it generating benzocyclohexadienyl cation intermediates that can exchange H for D.

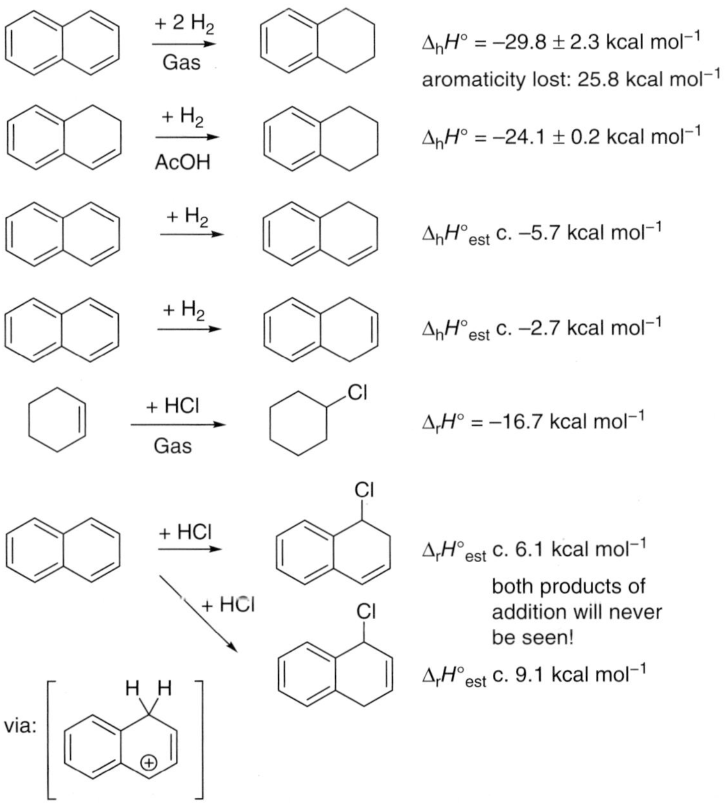

Table 2.A.2 gives $\Delta_r H°$(anthracene + H_2 → 1,9-dihydroanthracene, gas phase) = $38.2-55.2\approx -17\,\text{kcal mol}^{-1}$ to be compared with $\Delta_r H°$(cyclohexene + H_2 → cyclohexane, gas phase) = $-28.5\,\text{kcal mol}^{-1}$. Thus, one destroys a part of the aromatricity of anthracene in this hydrogenation that amounts to $-17-(-28.5)=11.5\,\text{kcal mol}^{-1}$. One estimates $\Delta_r H°$(anthracene + HCl ⇄ 9-chloro-9,10-dihydroanthracene, gas phase) = $-16.7+11.5=-5.2$ kcal mol^{-1}, an endothermic reaction that can be observed at a sufficiently low temperature because of the $-T\Delta_r S$ term that is always positive, but decreases on lowering the temperature. The major contribution to the entropy of this condensation arises from the variation of entropy of translation between products and reactants. Using Eq. (1.30), one can write $\Delta_r S_{\text{trans}}$(anthracene + HCl ⇄ 9-chloro-9,10-dihydroanthracene) = 2.9 ln[190.7/(36.6)(154.2)] − 4.97ln 298 + 2.31 eu = −35.8 eu at 25 °C. Anthracene has two C_2 axes of symmetry that are not present in the adduct with HCl. This gives an increase of entropy of reaction of R ln 4 = 2.7 eu (Eq. (1.33)). The adduct is more flexible than anthracene, which might also contribute to some positive entropy of reaction. We

retain $\Delta_r S^{298K}$ (anthracene + HCl ⇄ 9-chloro-9,10-dihydroanthracene) = −33 eu. Assuming that the heat of reaction and reaction entropy estimated at 25 °C can be used at a different temperature, one can write $\Delta_r G^T = \Delta_r H° - T\Delta_r S°$. The temperature at which $K = 1$ ($\Delta_r G^T = 0$) is thus given by $\Delta_r H° = T\Delta_r S°$, or by $T = \Delta_r H°/\Delta_r S° = 5200/33 = 158$ K or −115 °C.

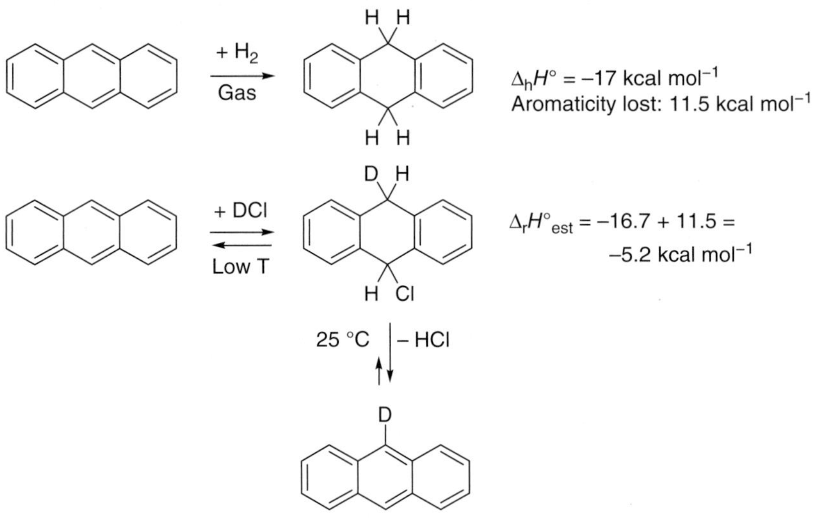

At 25 °C anthracene adds Br_2 to give isolable 9,10-dibromo-9,10-dihydroanthracene. A subsequent facile elimination of HBr gives the 9-bromoanthracene corresponding to the "classical" product of electrophilic aromatic substitution [3]. The product of addition forms as the estimated heat of this reaction in the gas phase amounts to c. −10.2 kcal mol^{-1}, enough to pay for the entropy cost of the condensation at room temperature. In the case of the bromination of naphthalene, the product of addition forms only in a small amount in competition with the "classical" product of substitution at temperatures below 25 °C. In this case, the estimated heat of the addition is c. 1.1 kcal mol^{-1} at 25 °C in the gas phase.

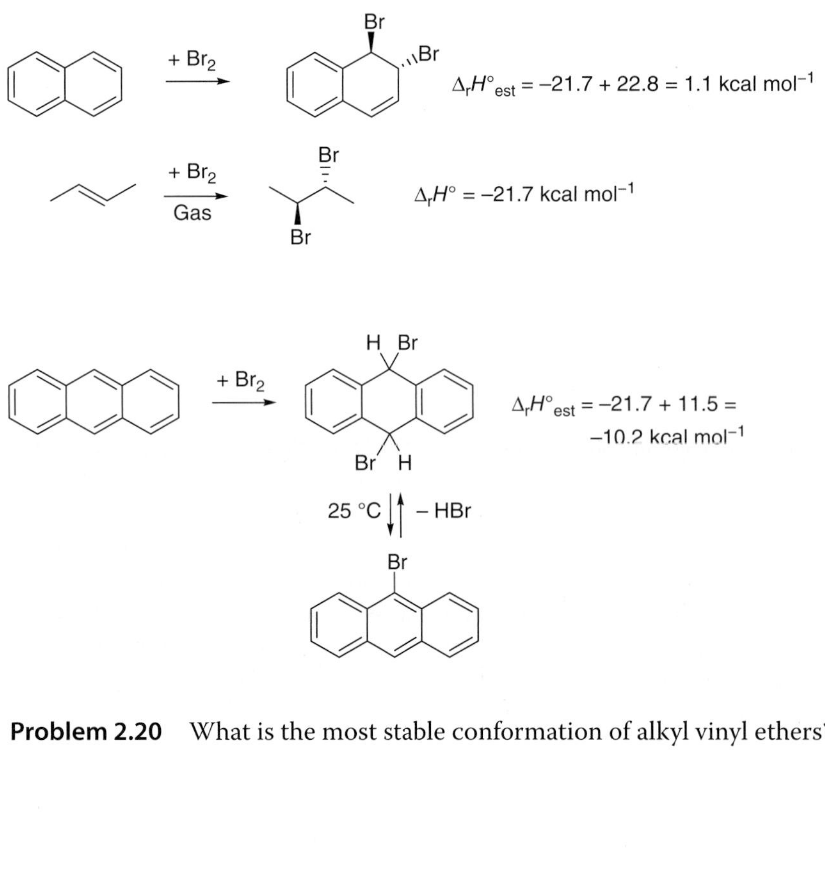

Problem 2.20 What is the most stable conformation of alkyl vinyl ethers?

Answer

Methyl vinyl ether favors the syn conformer to avoid electrostatic repulsion involving the lone pairs on O and the electrons in double bond, but larger groups give the anti conformer to avoid alkyl – cis-H steric repulsions; thus the *anti* conformer is expected to be preferred.

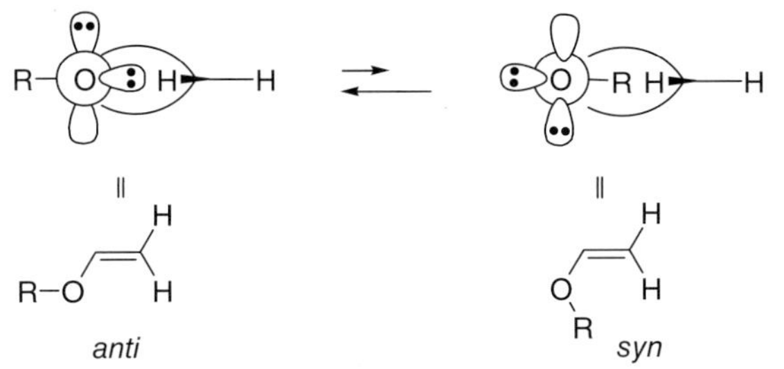

Problem 2.21 Predict whether (*E*)- or (*Z*)-planar conformers are preferred for planar amides.

Answer

In *N*-substituted formamides (*Z*)-conformers are preferred because they benefit from electrostatic stabilization due to partial charges on the oxygen center of the amide and the *N*-substituents. This stabilizing effect competes with the destabilizing gauche interactions between these two groups. Thus, the larger the *N*-substituent, the more the minor (*E*)-conformer is populated. In the case of acetamides, gauche interactions between the methyl group and the *N*-substituent reinforce the population of (*Z*)-conformers.

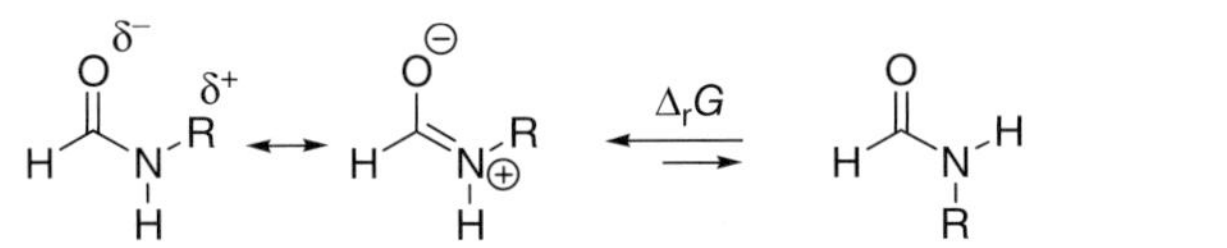

R = Me $\Delta_r G$ (solution, 21–22 °C): 1.22 kcal mol^{-1}		11.1% (*E*)
Et	0.93	16.8
Vinyl	0.42	32.7
n-Pr	0.93	17
i-Pr	0.83	19.5
c-Pr	0.35	35.3
c-Bu	0.65	24.6
c-Pent	0.81	19.9
Phenyl	0.14	44.1
t-Bu	0.17	42.6

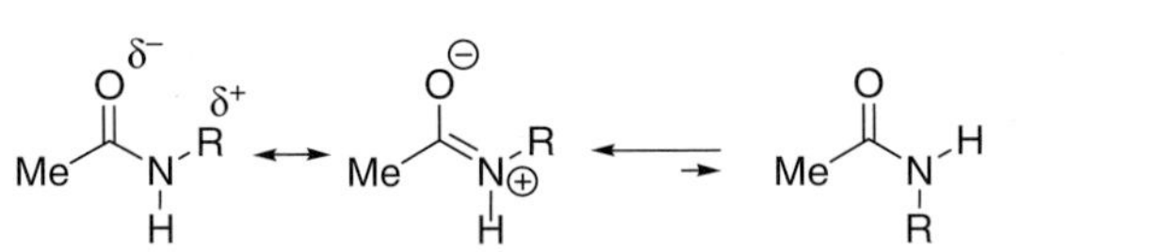

R = Me: 2.3% of (*E*) in $CDCl_3$; 7% of (*E*) in H_2O

Problem 2.22 Evaluate the Gibbs energy of the Diels–Alder reactions of 1*H*-pyrrole, furan, and cyclopentadiene with maleic anhydride giving the corresponding *exo* adducts at 300 K. Compare these values with those evaluated for the corresponding Michael additions.

Answer

Because of its aromaticity (c. 16.5 kcal mol^{-1}, given by the comparison of the heat of double hydrogenation of pyrrole and of cyclopentadiene) the Diels–Alder additions of pyrrole are much less exothermic than those of cyclopentadiene. One estimates to −8.7 kcal mol^{-1} the exothermicity of the Diels–Alder reaction of pyrrole to maleic anhydride, using ethylene as a model dienophile. This is not enough to compensate for the cost of entropy at room temperature (c. 10 kcal mol^{-1}), and thus pyrrole prefers to undergo a Michael addition (see below). Correction for the loss of possible π/π-conjugation stabilization in maleic anhydride is not considered because published standard heat of hydrogenation of maleic anhydride shows that such stabilization does not exist. The heat of the Diels–Alder reaction of pyrrole with maleic anhydride might be less than −8.7 kcal mol^{-1} because of destabilizing gauche interactions between the *exo*-succinic anhydride moiety and the rest of the bicyclic cycloadduct.

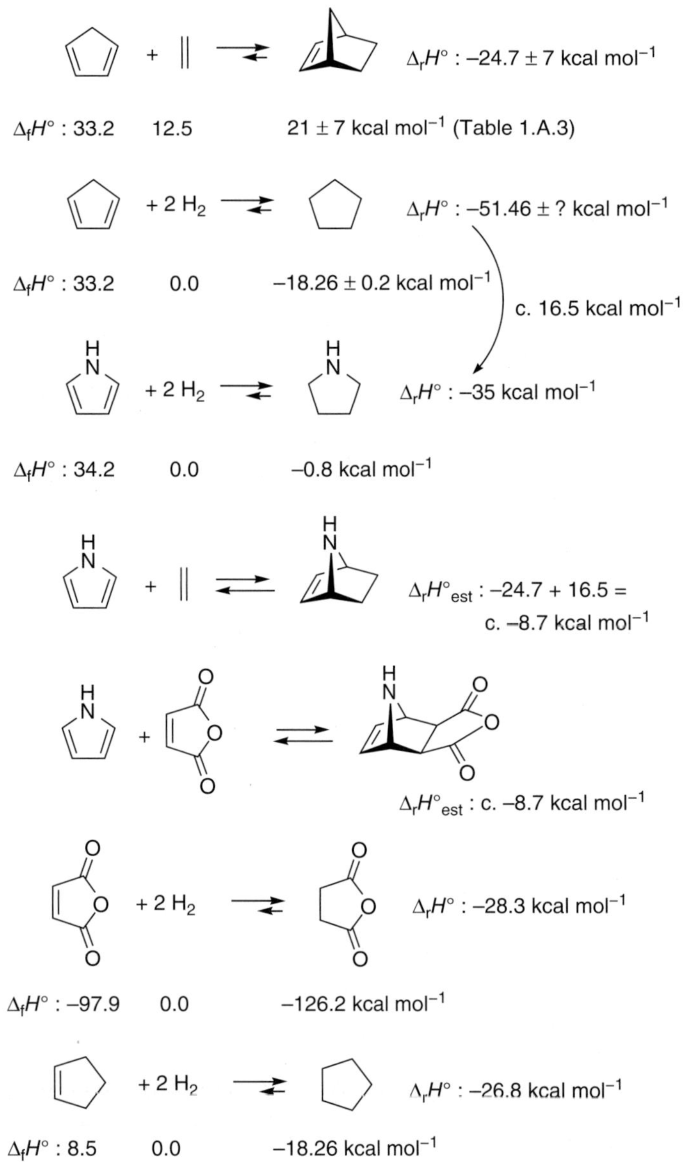

One estimates the heat of Michael addition of pyrrole to maleic anhydride (pyrrole + MA) to amount to c. −20 kcal mol^{-1}. This is done in the following way. Hydrocarbation of (*Z*)-but-2-ene by (*Z*)-but-2-ene is exothermic by -20.44 ± 1.3 kal mol^{-1}. Table 1.A.2 gives $\Delta_f H°$((*E*)-but-2-ene) but not $\Delta_f H°$((*E*)-3,4-dimethylhex-2-ene). The latter is derived from that of 3,4-dimethylhexane (Table 1.A.2) and corrected for the heat of hydrogenation of 2-methylpent-2-ene.

Stabilization of furan by aromaticity is less than that of pyrrole. It amounts to c. -14 kcal mol^{-1} on comparing the heat of double hydrogenation of furan (c. -37.4 kcal mol^{-1}) with that of cyclopentadiene (c. -51.5 kcal mol^{-1}). Thus, the heat of the Diels–Alder reaction of furan with maleic anhydride is estimated as c.-10.7 kcal mol^{-1}, which compensates more or less for the cost of the entropy of condensation at room temperature. Thus, the equilibrium constant of this reaction must be closed to unity at 25 °C, as found experimentally in solution ($K^{298K} = 0.275$, Scheme 2.1).

The heat of the Michael additions of furan and cyclopentadiene to maleic anhydride must also amount to c. -20 kcal mol^{-1}.

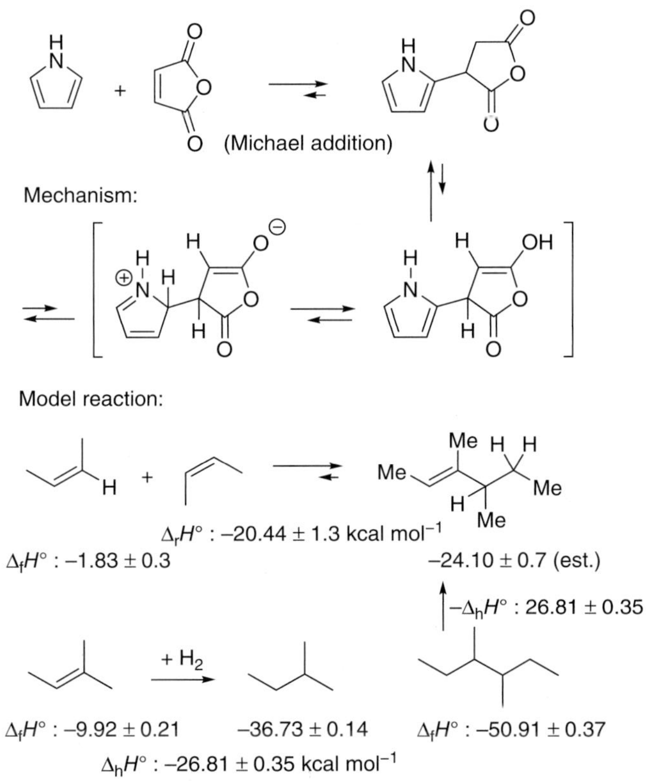

Problem 2.23 From the heats of formation of dibenzofuran (19.9 kcal mol^{-1}); diphenyl ether (10.6 kcal mol^{-1}), dibenzocyclopentadiene (42.2 kcal mol^{-1}), and diphenylmethane (36.8 kcal mol^{-1}) evaluate the extra aromaticity introduced by the furan ring of dibenzofuran. Compare with the aromaticity of anthracene.

Answer

The standard heats of the following deshydrogenation reactions suggest that the formation of dibenzofuran is stabilized only slightly by aromaticity. If one considers the experimental errors given with standard heats of formation of the compounds entering in these equilibria, one can state that there is no significant difference in electronic stabilization in dibenzofuran compared with dibenzocyclopentadiene. This contrasts with 9,19-dihydroanthracene for which a stabilization of c. -11.5 kcal mol^{-1} is gained upon double annulation of benzene by benzo rings (see Problem 2.19).

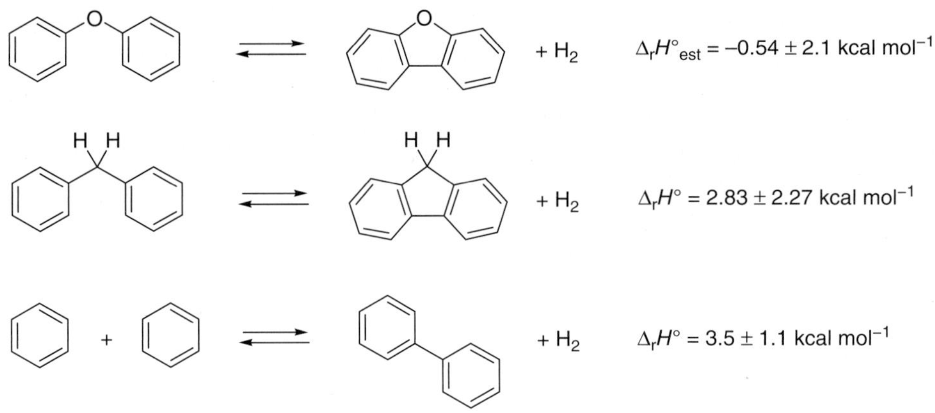

Problem 2.24 On treating 2,3,4,6-tetra-*O*-acetyl-*α*-D-glucopyranosyl bromide (**A**) with Bu_3SnH and AIBN ($Me_2C(CN)N{=}N(CN)Me_2$) in benzene at 80 °C, 2,3,4,6-tetra-*O*-acetyl-1,5-anhydro-D-glucitol (**B**) is formed concurrently with 1,3,4,6-tetra-*O*-acetyl-2-deoxy-*α*-D-arabinohexose (**C**). The proportion of these two products of reduction depends on the concentration of Bu_3SnH. With low concentration of Bu_3SnH **C** is the major product. Explain.

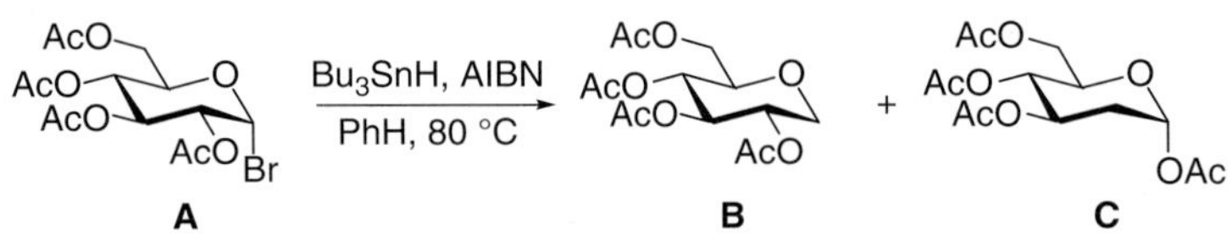

Answer

On heating AIBN undergoes decomposition into N_2 and 2-cyanoisopropyl radical. The latter is reduced by tributyltin hydride into isobutyronitrile via intermolecular hydrogen atom transfer, generating tributyltin radical. Because of the high affinity of metals for halogens in general, in this case, of tin for bromine, tributyltin radical abstracts selectively a bromine radical from **A** giving the glucosyl radical intermediate **R**$^\bullet$. In the presence of a high concentration of tributyltin hydride this alkyl radical is reduced into **B** (second order rate law – the rate of reaction depends on the concentration of tributyltin hydride) with generation of tributyltin radical that starts a chain reaction on reacting with another molecule of glycosyl bromide **A**. In the presence of a low concentration of tributyltin hydride, radical **R**$^\bullet$ has the time to undergo a rearrangement (first order rate law – the rate of the reaction does not depend on the concentration of tributyltin hydride) into the isomeric radical **Q**$^\bullet$, which is more stable than radical **R**$^\bullet$ because of the enthalpic anomeric effect realized by the *α*-substitution of the pyranosyl ether by an acetoxy group. This rearrangement is a [1,2]-shift of the 2-acetoxy group. Once formed, radical **Q**$^\bullet$ is reduced by tributyltin hydride into **C**.

Radical reaction initiation

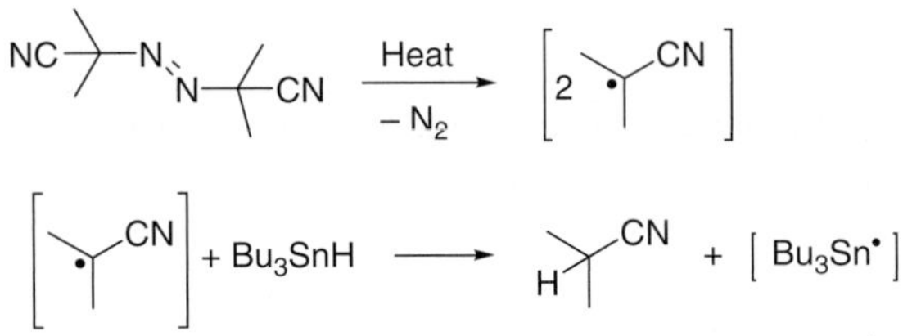

Propagation steps:

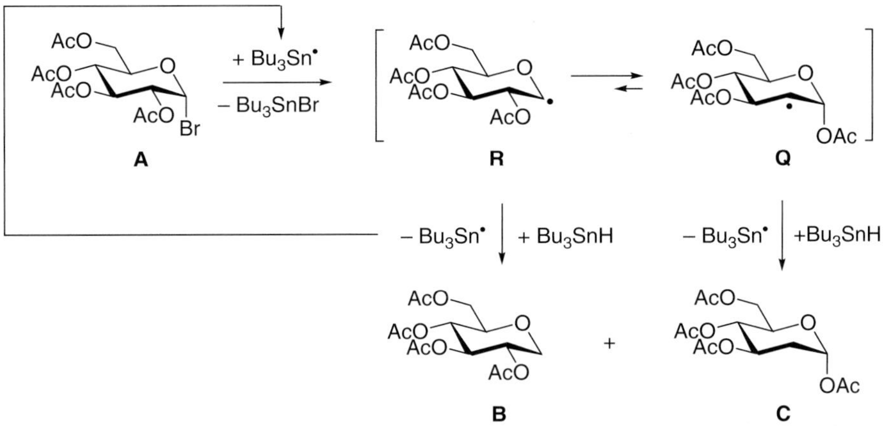

Problem 2.25 Bicyclo[2.1.0]pent-2-ene reacts with cyclopentadiene at 25 °C giving a Diels–Alder cycloadduct 3000 times as fast as bicyclo[2.2.0]hex-2-ene. Why?

Answer
The standard heat of hydrogenation of bicyclo[2.1.0]pent-2-ene amounts to c. −41.9 kcal mol^{-1} whereas that of bicyclo[2.2.0]hex-2-ene is c. −31.2 kcal mol^{-1} (Section 2.8). The Diels–Alder reaction of cyclopentadiene to bicyclo[2.1.0]pent-2-ene is expected to be, as for the hydrogenation, c. 10 kcal mol^{-1} more exothermic than that of cyclopentadiene to bicyclo[2.2.0]hex-2-ene. Since these Diels–Alder reactions are concerted (Section 5.3.8) one-step reactions, part of the exothermicity difference (Bell–Evans–Polanyi theory, Dimroth principle: $\Delta^{\ddagger}H = \alpha\Delta_r H + \beta$) will show up in the transition states of these reactions that follow the same mechanism and are affected similarly by geometric and steric factors.

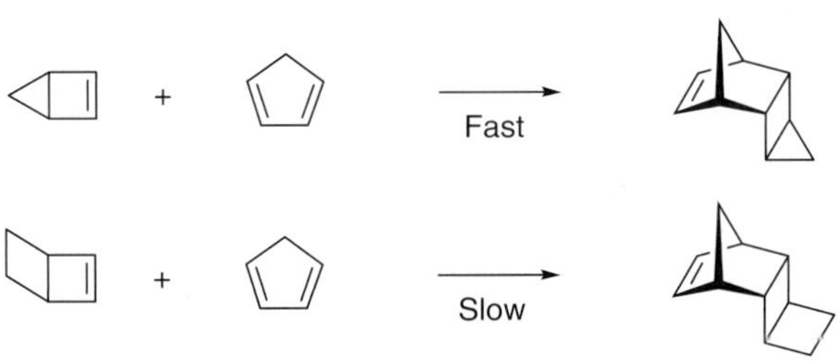

Problem 2.26 Using heat group increments (Table 2.A.1) estimate the standard heats of formation of CH_3COCHO and of $CH_3COCOCH_3$ and compare your estimates with the experimental data given in Table 1.A.4. Explain the deviations. Evaluate the heat of formation of penta-2,4-dione and compare it with the experimental value given in Table 1.A.4. Explain the deviation.

Answer
Table 1.A.4 gives

$$\Delta_f H^\circ(\text{MeCOCHO}) = -64.8 \pm 1.2\ \text{kcal mol}^{-1},$$

$$\Delta_f H^\circ(\text{MeCOCOMe}) = -78.1\ \text{kcal mol}^{-1}$$

$$\Delta_f H^\circ(\text{MeCOCH}_2\text{CCOMe}) = -91.87 \pm 0.31\ \text{kcal mol}^{-1}.$$

Using the Benson's group equivalents (Table 2.A.1) one estimates

$$\Delta_f H°_{est}(\text{MeCOCHO}) = [\text{C–(C)(H)}_3] + [\text{CO–(C)}_2] + [\text{CO–(C)(H)}] = -10.12 - 31.66 - 29.1$$
$$= -70.88\ \text{kcal mol}^{-1};$$

$$\Delta_f H°_{est}(\text{MeCOCOMe}) = 2[\text{C–(C)(H)}_3] + 2[\text{CO–(C)}_2] = -20.24 - 63.32 = -83.56\ \text{kcal mol}^{-1};$$

$$\Delta_f H°_{est}(\text{MeCOCH}_2\text{COMe}) = 2[\text{C–(C)(H)}_3] + 2[\text{CO–(C)}_2] + [\text{C–(C)}_2\text{(H)}_2] = -88.48\ \text{kcal mol}^{-1}.$$

The α-dicarbonyl compounds are destabilized by c. 6 kcal mol^{-1} because of electrostatic repulsions between the two carbonyl groups. In the case of the β-diketone, a stabilization is found (c. 3 kcal mol^{-1}) that can be explained by invoking the formation of MeCH(OH)=CH—COMe, an enol stabilized by conjugation with a COMe group and by chelation (intramolecular O—H$\cdots$O=C bridging).

Problem 2.27 Which aldehyde given below will lead to the highest equilibrium constant of aldol formation with an enolizable ketone? Aldehydes: acetaldehyde, formaldehyde, benzaldehyde, ethyl glyoxylate (CHOCOOEt).

Answer

The following standard heats of hydrogenation are given (Table 1.A.4): $\Delta_h H°(\text{H}_2\text{C=O} \rightarrow \text{CH}_3\text{OH}) = -22.3$ kcal mol^{-1}, $\Delta_h H°(\text{MeCH=O} \rightarrow \text{EtOH}) = -15.2$ kcal mol^{-1}, $\Delta_h H°(\text{PhCH=O} \rightarrow \text{BnOH}) = -15.2$ kcal mol^{-1}. That of ethyl glyoxalate is not given, but is expected to be more exothermic than that of ethanal because of electrostatic repulsion between the ester and carbaldehyde groups. The NIST Chemistry WebBook gives $\Delta_h H°(\text{MeCOCHO} \rightarrow \text{MeCH(OH)CH}_2\text{OH}) = -37.9$ kcal mol^{-1}, which is more exothermic by c. $-9.7(-37.98 + 13 + 15.2)$ kcal mol^{-1} than the hydrogenation of acetone (c. -13 kcal mol^{-1}) + hydrogenation of ethanal (-15.2 kcal mol^{-1}). One estimates $\Delta_h H°(\text{MeCOCHO} \rightarrow \text{MeCOCH}_2\text{OH}) = -37.9 + 13 = -24.9$ kcal mol^{-1}. In the case of a reversible reaction such as aldol formation the equilibrium constant K_{aldol} in favor of the aldol (adduct) will increase according to the sequence:

$$\text{MeCHO} \approx \text{PhCHO} < \text{H}_2\text{CO} < \text{EtCOOCHO}.$$

Problem 2.28 Triply branched octanes are usually obtained by the "alkyl process" that cracks high boiling alkanes into butane, isobutane, isobutene, but-1-ene, and (*E*)- and (*Z*)-but-2-ene. This mixture is then reacted over an acidic catalyst producing a high proportion of triply-branched octanes. What are they? Estimate the heats and entropy variations of these reactions. Are they reversible below 100 °C? Explain your answer. Propose mechanisms for these reactions using H_2SO_4 as catalyst.

Answer

There are four triply branched isomeric octanes: 2,2,4-trimethylpentane, 2,2,3-trimethylpentane, 2,3,3-trimethylpentane, and 2,3,4-trimethylpentane. Under acidic conditions but-1-ene ($\Delta_f H° = 0.15 \pm 0.16$ kcal mol^{-1}; $S° = 73.8$ eu), (*Z*)-but-2-ene ($\Delta_f H° = -1.83 \pm 0.30$ kcal mol^{-1}, $S° = 72.1$ eu), and (*E*)-but-2-ene ($\Delta_f H° = -2.58 \pm 0.24$ kcal mol^{-1}, $S° = 70.9$ eu) are equilibrated with isobutylene ($\Delta_f H° = -4.29 \pm 0.26$ kcal mol^{-1}, $S° = 70.2$ eu). At 25 °C, isobutylene is the major C_4 alkene. Using the data of Table 1.A.2 one calculates $\Delta_r G°$((*E*)-but-2-ene $\rightleftarrows$ isobutylene) $= -1.71(\pm 0.5) - 298(-0.0007) = -1.5 \pm 0.5$ kcal mol^{-1}, which leads to $5.4 < K^{298\text{K}}$((*E*)-but-2-ene $\rightleftarrows$ isobutylene) < 29. The mechanism of the acid-catalyzed isomerization of these alkenes implies their protonation with the formation of carbenium ion intermediates that can be rearranged as shown below. [4]

(a)

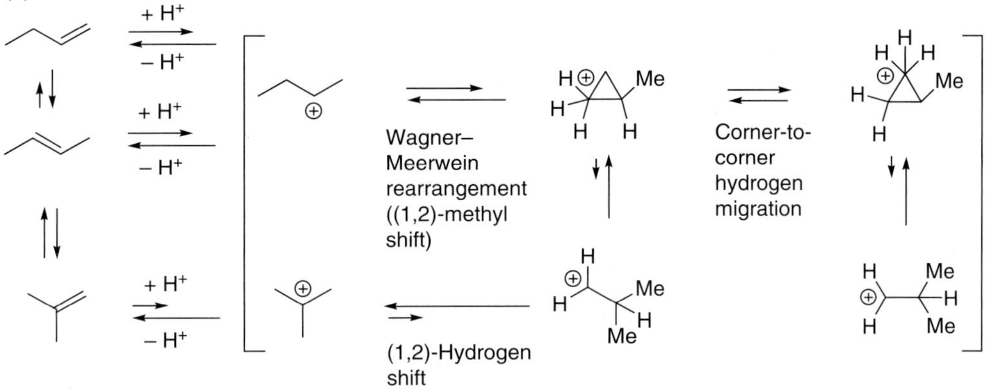

Tert-butyl cation is the most stable C_4-alkyl cation isomer. Thus, its concentration is much higher than that of isomeric secondary but-2-yl cation, and primary but-1-yl, 2-methylprop-1-yl cations, as expected from the hydride affinities in the gas phase (Table 1.A.14): $DH°(t\text{-Bu}^+/\text{H}^-) = 233\ \text{kcal mol}^{-1}$, $DH°(\text{but-2-yl}^+/\text{H}^-) = 248\ \text{kcal mol}^{-1}$, $DH°(\text{but-1-yl}^+/\text{H}^-) = 265\ \text{kcal mol}^{-1}$. *Tert*-butyl cation is capable of abstracting an hydride from butane generating but-2-yl cation + isobutane ($\Delta_r H°(t\text{-Bu}^+ + n\text{-Bu–H} \rightleftarrows \text{but-2-yl}^+ + t\text{-Bu–H}) = 15\ \text{kcal mol}^{-1}$). But-2-yl cation undergoes a double Wagner–Meerwein rearrangement ((1,2)-methyl shift followed by a (1,2)-hydrogen shift) into the more stable *t*-butyl cation. The latter is then reduced (hydride transfer) by *n*-butane or isobutane into isobutane.

(b)

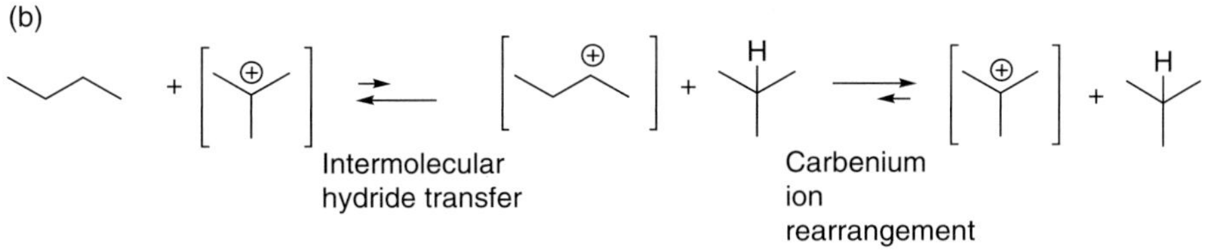

Hydrocarbation of isobutylene by isobutane, the major C_4-compound of the reaction mixture, generates 2,2,4-trimethylpentane. The mechanism of this reaction involves the addition of *t*-butyl cation onto isobutylene giving the tertiary carbenium ion **A**. Hydride transfer from isobutane to **A** gives 2,2,4-trimethylpentane and *t*-butyl cation, which starts another addition to isobutylene. This hydrocarbation is exothermic by $-17.2 \pm 0.75\ \text{kcal mol}^{-1}$ using data from Table 1.A.2. The NIST Chemistry WebBook gives $S°$(2,3,4-trimethylpentane) = 102.1 ± 0.3 eu and Table 1.A.2 gives $S°$(isobutane) = 70.4 eu andSo(isobutene) = 70.2 eu. If one considers the fact that 2,2,4-trimethylpentane has an extra *tert*-butyl group (C_3-axis of symmetry) its entropy amounts to $S°$(2,2,4-trimethylpentane) = $S°$(2,3,4-trimethylpentane) – $R \ln 3$ = 102.1 – 2.18 ≈ 100 eu.

Thus, $S°$(isobutane + isobutene ⇄ 2,2,4-trimethylpentane) ≈ –40.6 eu. At 100 °C the $-T\Delta_r S^T$ term amounts to –373(–40.6) = 15 144 cal mol^{-1} = 15.1 kcal mol^{-1}, which is compensated by the exothermicity (–17.2 ± 0.75 kcal mol^{-1}) of the reaction. Thus, the hydrocarbation is exergonic by –2.1 ± 0.75 kcal mol^{-1} at 100 °C. This leads to $6.2 < K^{373\,\text{K}} < 38.8$. If there is a moment during the reaction for which [isobutane] = [isobutylene] = [2,2,4-trimethypentane] = 1 mol dm^{-3}, the rate of formation of 2,2,4-trimethylpentane = k forward[isobutene][isobutylene] ($K = k$ forward/k reverse) is between 6 and 39 times as large as the rate of its fragmentation into isobutane + isobulylene. At low concentrations of isobutane and isobutylene and high concentration of 2,2,4-trimethylpentane, the fragmentation will compete with the hydrocarbation. Irrespective of its exergonicity any reaction must be considered as reversible *a priori*. It is the reaction conditions (concentration and relative concentration of reactants and products) and the time available to observe the reverse reaction that permit us to decide whether the reaction has reached equilibrium or not.

One may think that t-butyl cation will abstract an hydride from isobutylene and be equilibrated with methallyl cation + isobutane. Data from Tables 1.A.2 and 1.A.14 for a reaction in the gas phase predict it to be endothermic by 16.3 kcal mol^{-1}, although an allyl cation is generated at the expense of an alkyl cation! $\Delta_r H°(t\text{-Bu}^+ + \text{isobutylene} \rightleftarrows \text{2-methylallyl}^+ + t\text{-Bu–H}) = -32.1 + 211 - 167 + 4.3 \approx 16.3$ kcal mol^{-1} (hydride affinity of methallyl cation is higher than the hydride affinity of t-butyl cation. Table 1.A.14 gives $DH°(t\text{-Bu}^+/\text{H}^-) = 233$ kcal mol^{-1} and $DH°(\text{2-methylallyl}^+/\text{H}^-) = 248$ kcal mol^{-1}.

The entropy variation of this hydride transfer is close to unity; thus this reaction is endergonic and does not compete efficiently with the addition of t-butyl cation to isobutylene. The latter reaction is exergonic, even at 100 °C.

Cation **A** can equilibrate with the less stable secondary carbenium ion **B**, which also provides 2,2,4-trimethylpentane by hydride transfer from isobutane. Alternatively, cation **B** can undergo a Wagner–Meerwein rearrangement into secondary cation **C**. Reduction of the latter by isobutane furnishes 2,2,3-trimethylpentane. Note that cation **C** may result from the addition of t-butyl cation to either (E) or (Z)-but-2-ene. Alternatively, cation **C** can undergo a (1,2)-shift of hydrogen with formation of the more stable tertiary cation **D**. The latter carbenium ion undergoes a nearly degenerate ($K \approx 1$) Wagner–Meerwein rearrangement into tertiary cation **E**, the reduction of which by isobutane provides 2,3,3-trimethylpentane and t-butyl cation. Cation **B,** which results from the endothermic (c. 15 kcal mol^{-1}) rearrangement of cation **A** can also undergo Wagner–Meerwein rearrangements, for instance, into tertiary cation **F**. Reduction of cation **F** by isobutane gives 2,3,4-trimethylpentane + t-butyl cation.

Rearrangement of tertiary cation **A** into secondary cation **B** is endothermic by c. 15 kcal mol^{-1} (see Table 1.A.14). If it does not compete with the process that reduces cation **A** into 2,2,4-trimethylpentane, the formation of the latter triply branched octane is the product of kinetic control in the acid-catalyzed hydrocarbation of isobutylene by isobutane. If the monomolecular rearrangement **A** ⇄ **B** (first order rate law, does not depend upon the concentration of isobutane) competes with the hydrogen transfer (second order rate law, depends upon the concentration of isobutane), the other isomeric triply branched octane will be formed and be equilibrated with 2,2,4-trimethylpentane, as outlined below. In Section 3.3 we show that a reaction with an activation free energy of c. 27 kcal mol^{-1} will reach 99% conversion within about one hour at 100 °C. Thus, the chances for 2,2,4-trimethylpentane to be equilibrated with the other triply branched octane below 100 °C depends on the concentration of isobutane and the concentration of t-butyl cation. The concentration of t-butyl cation, and of all other tertiary carbenium ion intermediates, depends on the concentration of the protic source, the catalyst.

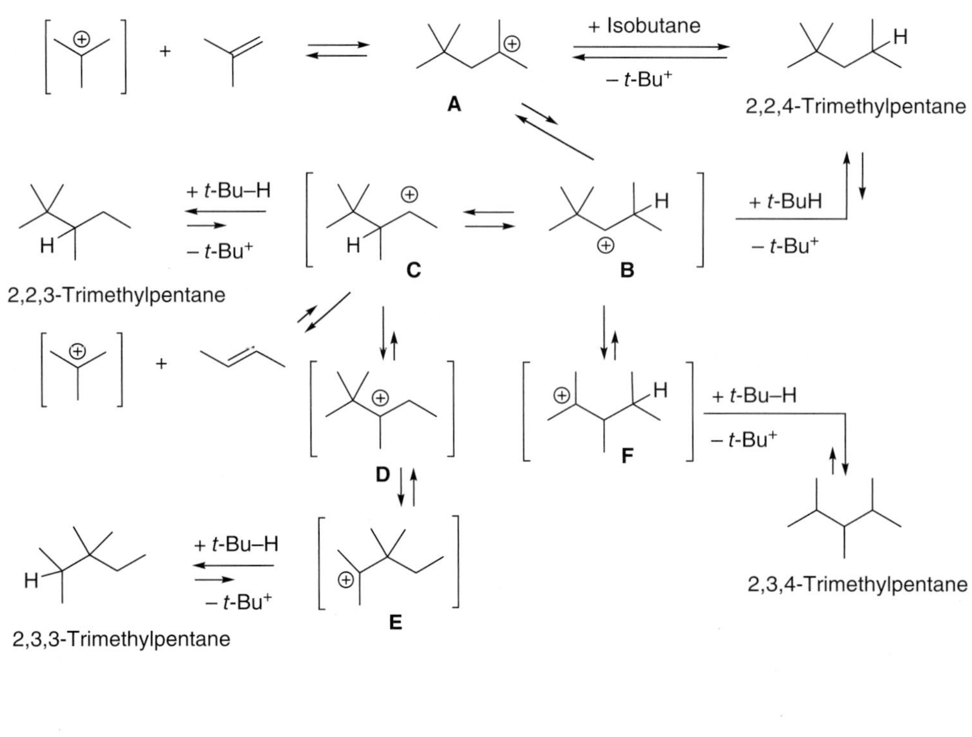

⇄

The NIST Chemistry WebBook gives

$$\Delta_f H°(2,2,4\text{–trimethylpentane}) = -53.57 \pm 0.32 \text{ kcal mol}^{-1}$$

$$\Delta_f H°(2,2,3\text{–trimethylpentane}) = -52.61 \pm 0.32 \text{ kcal mol}^{-1}$$

$$\Delta_f H°(2,3,3\text{–trimethylpentane}) = -51.73 \pm 0.33 \text{ kcal mol}^{-1}$$

$$\Delta_f H°(2,3,4\text{–trimethylpentane}) = -51.93 \pm 0.40 \text{ kcal mol}^{-1}.$$

The following heats of reaction are obtained:

$$\Delta_r H°(2,2,4\text{–trimethylpentane} \rightleftarrows 2,2,3\text{–trimethylpentane}) = 0.96 \pm 0.64 \text{ kcal mol}^{-1},$$

$$\Delta_r H°(2,2,4\text{–trimethylpentane} \rightleftarrows 2,3,3\text{–trimethylpentane}) = 1.84 \pm 0.65 \text{ kcal mol}^{-1},$$

$$\Delta_r H°(2,2,4\text{–trimethylpentane} \rightleftarrows 2,3,4\text{–trimethylpentane}) = 1.64 \pm 0.72 \text{ kcal mol}^{-1}.$$

All four triply branched octanes have five Me groups. The entropy of 2,2,4- and 2,2,3-trimethylpentane must be the same but smaller than that of 2,3,3- and 2,3,4-trimethypentane by $-R \ln 3$. The two first octanes have a *tert*-butyl group (local C_3-axis of symmetry) that is not present in the other isomers. Thus the variation of entropies of the above equilibria are

$$\Delta_r S°(2,2,4\text{–trimethylpentane} \rightleftarrows 2,2,3\text{–trimethylpentane}) \cong 0 \text{ eu}$$

$$\Delta_r S°(2,2,4\text{–trimethylpentane} \rightleftarrows 2,3,3\text{–trimethylpentane}) \cong R \ln 3 = 2.18 \text{ eu}$$

$$\Delta_r S°(2,2,4\text{–trimethylpentane} \rightleftarrows 2,3,4\text{–trimethylpentane}) \cong 2.18 \text{ eu}.$$

This gives

$$\Delta_r G^{373K}(2,2,4\text{–trimethylpentane} \rightleftarrows 2,2,3\text{–trimethylpentane}) = 0.96(\pm 0.64) - 373(0)$$
$$= 0.96 \pm 0.64 \text{ kcal mol}^{-1},$$
$$\Delta_r G^{373K}(2,2,4\text{–trimethylpentane} \rightleftarrows 2,3,3\text{–trimethylpentane}) = 1.84(\pm 0.65) - 373(2.18)$$
$$= 1.03 \pm 0.65 \text{ kcal mol}^{-1},$$
$$\Delta_r G^{373K}(2,2,4\text{–trimethylpentane} \rightleftarrows 2,3,4\text{–trimethylpentane}) = 1.64(\pm 0.72) - 373(2.18)$$
$$= 0.83 \pm 0.72 \text{ kcal mol}^{-1}.$$

One estimates the following equilibrium constant for the isomerization of triply branched octanes:

$$0.1 < K^{373K}(2,2,4\text{–trimethylpentane} \rightleftarrows 2,2,3\text{–trimethylpentane}) < 0.65$$

$$0.1 < K^{373K}(2,2,4\text{–trimethylpentane} \rightleftarrows 2,3,3\text{–trimethylpentane}) < 0.6$$

$$0.12 < K^{373K}(2,2,4\text{–trimethylpentane} \rightleftarrows 2,3,4\text{–trimethylpentane}) < 0.8$$

Under conditions of thermodynamic control (the relative stability of the products determines their proportion), 2,2,4-trimethylpentane remains the major triply branched octane formed at 100 °C.

Problem 2.29 Calculate the heat of the reaction that equilibrates 3-methylcyclohexanone and methanol with the corresponding hemiacetal knowing that $K = 0.4 \text{ dm}^3 \text{ mol}^{-1}$ at 25 °C and compare it with the values measured for the following equilibrium:

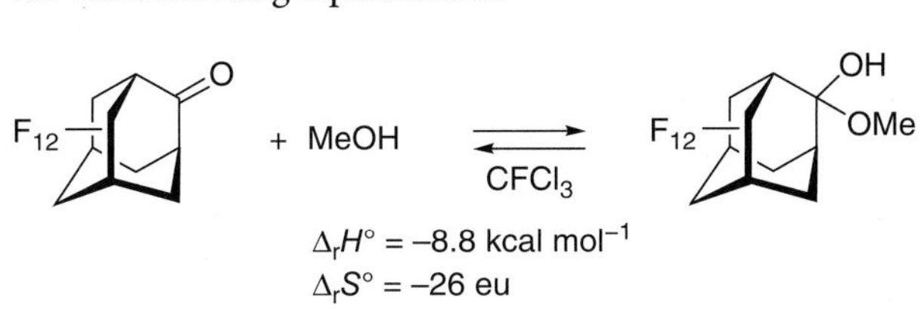

Answer

Table 1.A.4 gives $S°(\text{MeOH}) = 57.3$ eu. The standard entropy of 3-methylcyclohexanone is estimated from that of methylcyclohexane (82 eu), cyclohexane (71.3 eu), and cyclohexanone (80.2 eu) in the following way (application of additivity rules): $S°$(3-methylcyclohexanone) = $S°$(cyclohexanone) + $S°$(methylcyclohexane) $-S°$(cyclohexane) = 80.2 + 82 − 71.3 = 90.9 eu. The standard entropy of the hemiacetal is estimated in the following way using data from Table 1.A.2 and 1.A.4:

$$S°(\text{1-methoxy-3-methylcyclohexanol}) = S°(\text{methylcyclohexane}) + S°(\text{butan–2–ol}) \\ + S°(t\text{–Bu–OMe}) - S°(\text{butane}) - S°(\text{isobutane}) = 82 + 85.8 + 85.5 - 74.1 - 70.4 = 108.8 \text{ eu.}$$

This leads to $\Delta_r S°$(3-methylcyclohexanone + MeOH ⇄ 1-methoxy-3-methylcyclohexanol) = 108.8 − 90.9 − 57.3 = −39.4 eu.

Using Eq. (1.30) for the estimation of translation entropies one obtains

$\Delta_r S°_{\text{trans}}$(3-methylcyclohexanone + MeOH ⇄ 1-methoxy-3-methylcyclohexanol) = 2.98 ln[144.2/(112.2)(32)] − 4.97 ln(298) + 2.31 ≅ −35.6 eu. Both estimates are more negative than the entropy variation measured for the addition of methanol to perfluoroadamantanone (−26 eu). This can be interpreted in terms of a differential solvent effect between adduct and reactants. The adduct is less strongly solvated than reactants.

Using $\Delta_r S°$(3-methylcyclohexanone + MeOH ⇄ 1-methoxy-3-methylcyclohexanol) = 108.8 − 90.9 − 57.3 = −39.4 eu one calculates $\Delta_r H°$(3-methylcyclohexanone + MeOH ⇄ 1-methoxy-3-methylcyclohexanol) = $\Delta_r G° + T\Delta_r S° = -RT \ln K + T\Delta_r S° = -1.987 \times 298 \ln(0.4) + 298(-39.4) = 542 - 11\,741 = -11\,199$ cal mol^{-1} ≅ −11.2 kcal mol^{-1}. Using $\Delta_r S°_{\text{trans}}$(3-methylcyclohexanone + MeOH ⇄ 1-methoxy-3-methylcyclohexanol) = −35.6 eu one calculates $\Delta_r H°$(3-methylcyclohexanone + MeOH ⇄ 1-methoxy-3-methylcyclohexanol) = $\Delta_r G° + T\Delta_r S°$ ≅ −10.1 kcal mol^{-1}.

α-Substitution of a ketone by electron-withdrawing groups such as two fluoro substituents is expected to destabilize the carbonyl moiety and to increase the exothermicity of its 1,2-additions such as the formation of hemiacetals. This is not verified in this case, suggesting that the hemiacetal of perfluoroadamantanone is destabilized by steric factors such as gauche. In the hemiacetal of 3-methylcyclohexanone, either the OH group or the MeO group occupies an equatorial position. Electrostatic repulsion between the β-fluoro substituents and the OH and MeO groups in the hemiacetal of perfluoroadamantanone might also destabilize the latter adduct. Differential solvent effects could also render the formation of hemiacetal of perfluoroadamantanone less exothermic that the addition of methanol to 3-methycyclohexanone.

Problem 2.30 Hydration of acetone in water at 25 °C has a Gibbs energy of 3.74 kcal mol^{-1} whereas that of bicyclo[2.2.1]heptan-7-one gives $\Delta_r G° = -0.42$ kcal mol^{-1} ($\Delta_r H° = -7.34$ kcal mol^{-1}, $\Delta_r S° = -23.2$ eu. Explain why hydration is more complete with the bicyclic ketone than with acetone.

Answer

The 1,2-addition of water to bicyclo[2.2.1]heptan-7-one is more exothermic than the addition of water to acetone because of the ring strain about C(7) in the bicyclic system that is released partially upon the 1,2-addition.

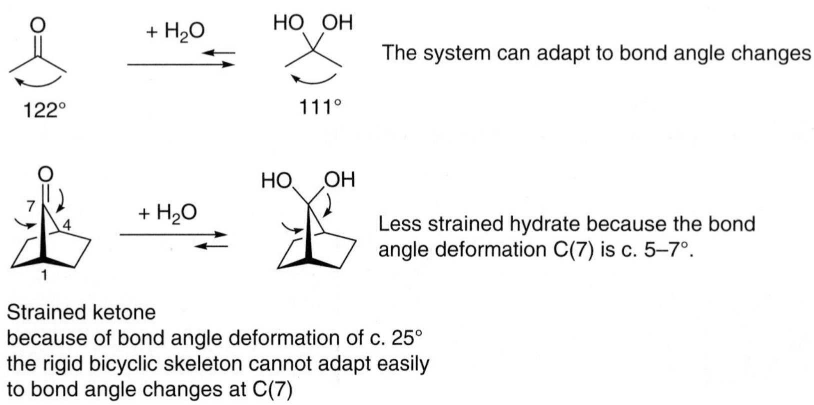

Problem 2.31 ZnI_2-catalyzed reaction of furan (excess) with 1-cyanovinyl acetate gives a mixture of 2-cyano-7-oxabicyclo[2.2.1]hept-5-en-2-yl acetates (**A**). Upon alkaline hydrolysis the corresponding cyanohydrines **B** are obtained. What can one add to the reaction mixture to generate 7-oxabicyclo[2.2.1]hept-5-en-2-one (**P**)?

Answer

Nucleophilic 1,2-additions of ketones are less exothermic than those of formaldehyde because the carbonyl dipole in ketones enjoys stabilization by the alkyl groups, which is not the case in formaldehyde. This is evidenced by the heats of hydrogenation, e.g. $\Delta_h H°(H_2C{=}O \rightarrow CH_3OH) \cong -22.3\,\text{kcal mol}^{-1}$ and $\Delta_h H°(Me_2C{=}O \rightarrow i\text{-Pr–OH}) \cong -13\,\text{kcal mol}^{-1}$. Thus, on adding formaline (aqueous solution of formaldehyde) cyanohydrines **B** eliminate HCN that adds to formaldehyde forming a more stable cyanohydrine **Q** and bicyclic ketone **P**.

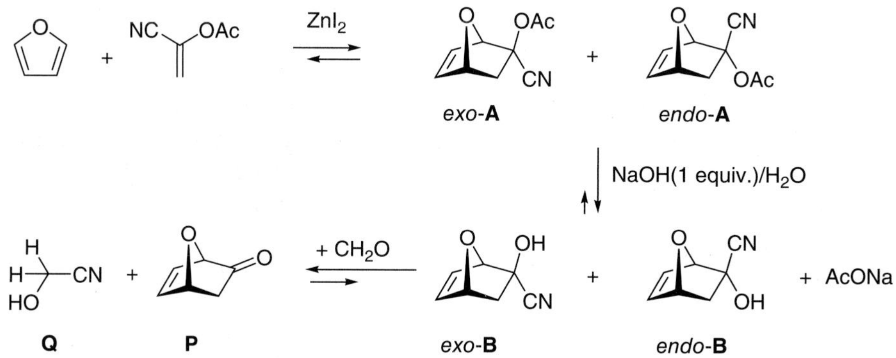

Problem 2.32 Negative enthalpy is associated with many carbohydrate + lectin bindings, while entropy plays a relatively minor role. The thermodynamic parameters for the binding of nojirimycin (**N**) and isofagomine (**I**), two competitive inhibitors of almond β-glucosidase, are given below. Give an interpretation of these data.

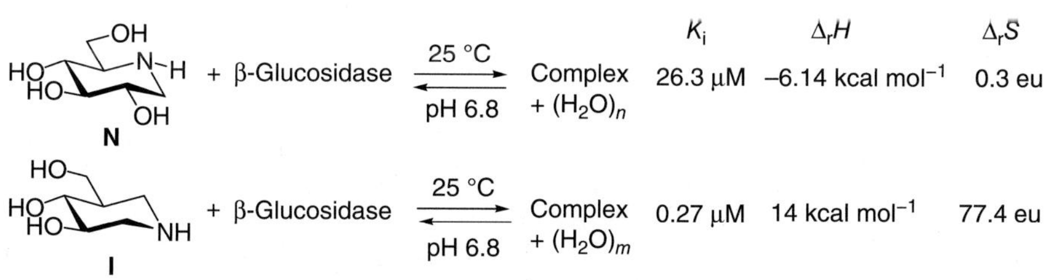

Answer

Binding of **N** with β-glucosidase is favorable in terms of bond (hydrogen bonding, electrostatic NH_2^+/COO^- interaction energies). One or two molecules of water ($n = 1,2$) are liberated during the complex formation. The binding of **I**, on the other hand, is very different: the enthalpy is positive, and there is a very large positive entropy. This means that the binding of **I** is unfavorable in terms of bond energies, but this is more than compensated for by the large entropy increase, probably due to a larger number of water molecules that are liberated ($m \gg 2$) and by gain of flexibility of the enzyme in the bond state with isofagomine.

Problem 2.33 Why is the oxidation of the yellow adduct of Figure 2.12 an easy process?

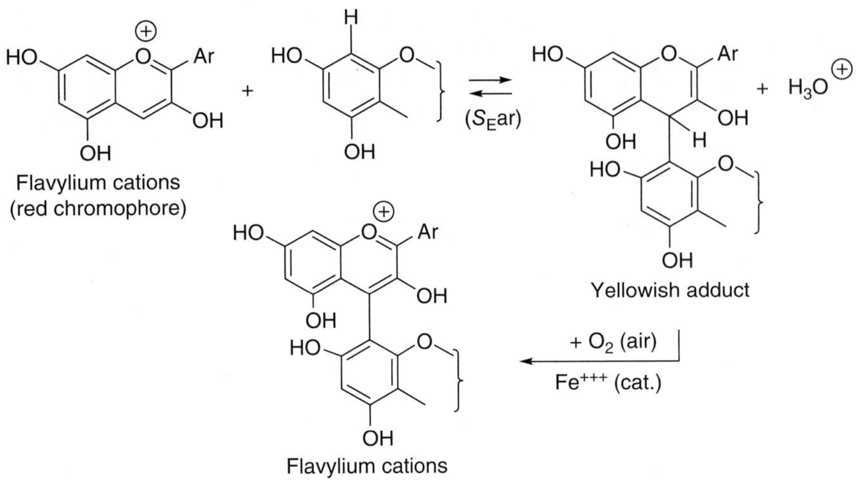

Answer

The yellowish adduct contains an allylic bisbenzylic C—H moiety, which is a relatively weak C—H bond able to transfer a hydrogen radical to the oxygen-centered radical of diradical, for instance. The radical so obtained reacts with O_2 giving a peroxide radical that can undergo various reactions providing finally the flavylium cation. One possible mechanism for the iron-catalyzed aerobic oxidation of hydrocarbons R—H is outlined below.

$$\mathrm{FeX_3} \rightleftarrows \mathrm{FeX_2} + \mathrm{X^{\cdot}}(\mathrm{X} = \mathrm{Cl, ROO, RO, etc.})$$

$$\mathrm{R{-}H} + \mathrm{X^{\cdot}} \rightleftarrows \mathrm{R^{\cdot}} + \mathrm{HX}$$

$$\mathrm{R^{\cdot}} + \mathrm{O_2} \rightleftarrows \mathrm{ROO^{\cdot}}$$

$$\mathrm{ROO^{\cdot}} + \mathrm{R{-}H} \rightleftarrows \mathrm{ROOH} + \mathrm{R^{\cdot}}$$

$$\mathrm{ROOH} + \mathrm{HX} \rightleftarrows \mathrm{R^+} + \mathrm{H_2O_2} + \mathrm{X^-}$$

$$\mathrm{H_2O_2} + 2\mathrm{FeX_2} \rightleftarrows 2\mathrm{Fe(OH)X_2}$$

Problem 2.34 Propose a mechanism for the elimination of AcOH from alkyl acetate involving diradical intermediates. Which are the diradicals?

Answer

$DH°(n\text{-Pr}^{\cdot}/\text{AcO}^{\cdot}) = \Delta_f H°(n\text{-Pr}^{\cdot})$ (Table 1.A.7) + $\Delta_f H°(\text{AcO}^{\cdot}) - \Delta_f H°(n\text{-PrOAc})$ [5] = 23.5 − 44.9 − (−111.5) ≅ 90.1 kcal mol^{-1}. Acetyloxy radical is not stable as it fragmentates into methyl radical and CO_2 with an exothermicity of −14.3 kcal mol^{-1} ($\Delta_f H°(\text{Me}^{\cdot}) = 34.8$ kcal mol^{-1} (Table 1.A.7), $\Delta_f H°(\text{CO}_2) = -94.0$ kcal mol^{-1} (Table 1.A.1)). Formally, hydrogen atom transfer from the β-position to the carbonyl group of the alkyl acetate generates diradical **D**. The latter is more stable by c. 14 kcal mol^{-1} with respect to the alkyl + acetyloxy radicals. Based on the estimated reaction enthalpy, isomerization **A** ⇄ **D** is easier than homolysis **A** ⇄ n-Pr$^{\cdot}$ + AcO$^{\cdot}$. However, entropy does not vary from **A** to **D** (isomerization) but it increases from **A** to MeCOO$^{\cdot}$ + n-Pr$^{\cdot}$ (fragmentation). Elimination of acetic acid from alkyl acetate requires high temperatures (>300 °C), which leaves diradical **D** with little chance to be an intermediate. Indeed, at 300 °C $\Delta_r G^T$(**A** ⇄ MeCOO$^{\cdot}$ + n-Pr$^{\cdot}$) = 90 − 573(0.030) = 72.8, and $\Delta_r G^T$(**A** ⇄ **D**) ≅ 76 − 573(0) = 76 kcal mol^{-1}.

$\Delta_f H°$(**D**) = $\Delta_f H°$(**B**) + $DH°(i\text{-Pr}^{\cdot}/\text{H}^{\cdot})$ + $DH°(\text{MeOCH}_2^{\cdot}/\text{H}^{\cdot})$ (Table 1.A.7) + radical stabilization by a hydroxy group (Table 1.A.9) + radical stability difference between tertiary and secondary alkyl radical (Table 1.A.7) − $DH°(\text{H}^{\cdot}/\text{H}^{\cdot})$ = $\Delta_f H°$(**B**) + 99.4 (±0.5) + 93.1 (±2.2) − 10 (±2.2) − 2 − 104.2 = $\Delta_f H°$(**B**) + 76.3 (±5) kcal mol^{-1}.

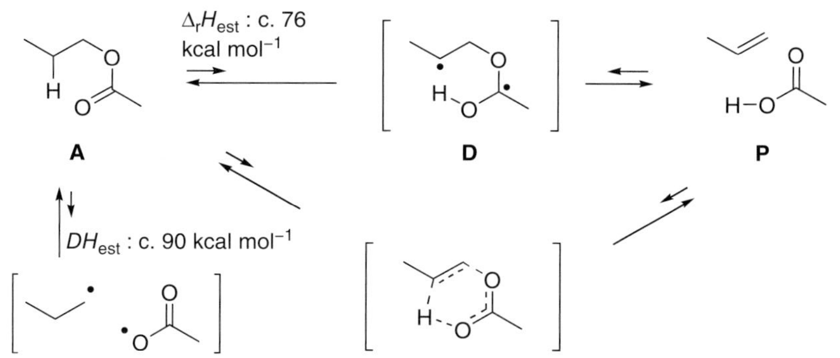

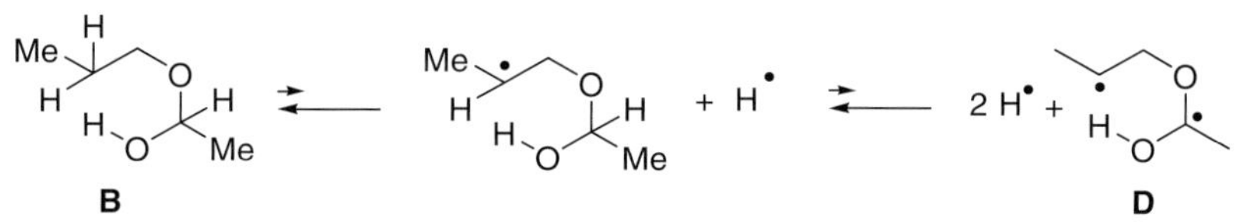

$\Delta_fH°(\mathbf{B}) = \Delta_fH°(n\text{-Pr–O–Et})$ (Table 1.A.4) + α-substitution by a MeO group (−41.5 kcal mol^{-1}) and exchange of the MeOH for an OH group (−5 kcal mol^{-1}) (see below) = −65 − 41.5 − 5 ≅ −111.5 kcal mol^{-1}. α-Substitution of an ether by a MeO group is estimated from $\Delta_fH°((\text{MeO})_2\text{CHMe}) - \Delta_fH°(\text{MeOCH}_2\text{Me})$ (Table 1.A.4) = −93.15 − (−51.73) ≅ −41.5 kcal mol^{-1}, and exchange of the MeO group for an HO group is given by $\Delta_fH°(i\text{-Pr–OH}) - \Delta_fH°(i\text{-Pr–OMe})$ (Table 1.A.4) = −65.2 − (−60.24) ≅ −5 kcal mol^{-1}. We assume that the enthalpic anomeric effect is the same for (MeO)C(OH) as for $(\text{MeO})_2\text{C}$. $\Delta_fH°(\mathbf{B})$ can also be estimated from $\Delta_fH°((\text{MeO})_2\text{CHMe})$ + exchange of MeO for a OH group (−5 kcal mol^{-1}) and exchange of a MeO group for an n-PrO group as given by $\Delta_fH°(n\text{-Pr–OEt}) - \Delta_fH°(\text{Me–OEt}) = -65.90 - (-51.7) = -13.3$ kcal mol^{-1}.

One obtains the estimate: $\Delta_fH°(\mathbf{D})_{est} = -111.5 + 76.3\,(\pm 5) = -35.2 \pm 5$ kcal mol^{-1}. This leads to $\Delta_rH°(\mathbf{A} \rightleftarrows \mathbf{D}) = \Delta_fH°(\mathbf{D})_{est.} - \Delta_fH°(\mathbf{A}) = -35.2 - (-111.5) = 76.3$ kcal mol^{-1}. This is c. 14 cal mol^{-1} less than the enthalpy of homolytic dissociation into acetyloxy and n-propyl radical. The homolysis is favored entropically (fragmentation) but not the formation of the diradical intermediate **D** for which a negative entropy of activation is expected since the process blocks freely rotating σ-bond in the transition state of the process equilibrating **A** with **D**. Thus, at high temperature, homolysis will compete with the formation of diradical **D** for thermodynamic and kinetic reasons. If the homolysis and the β-hydrogen atom transfer are concerted processes, elimination of AcOH from the alkyl acetate might be easier than predicted on the basis of non-concerted mechanisms involving the formation of either radical of diradical intermediates. This is the E_i elimination mechanism that is generally followed by low pressure and high temperature fragmentation of alkyl esters. The reaction can be seen as a retro-oxa-oxo-ene reaction (Section 5.7.3).

Problem 2.35 Estimate the heat of isomerization of [6]radialene (**A**) into tris(cyclobuteno)benzene (**B**) assuming a heat of isomerization of butadiene into cyclobutene of 11.3 kcal mol^{-1} and an ASE of −37 kcal mol^{-1} for tris(cyclobuteno)benzene (**B**). Can [6]radialene be isomerized into cyclododeca-1,5,9-triyne?

Answer

Using Benson's group equivalents (Table 2.A.1) one estimates the following heats of formation:

$\Delta_fH°([6]\text{radialene}) = 6[\text{C—(C}_d)_2(\text{H})_2] + 6[\text{C}_d\text{—(C}_d)(\text{C})_2] = 6(6.22 + 9.75) = 95.8$ kcal mol^{-1}, no ring strain, no stabilization energy due to π/π-conjugation since the diene moieties are not planar [6].

$\Delta_fH°(\text{tris(cyclobuteno)benzene}) = 6[\text{C}_b\text{—(C}_b)_2(\text{C})] + 6[\text{C—(C}_d)(\text{C})(\text{H})]$ + 3(ring strain of cyclobutene, Section 2.6.8) = 6(5.62 − 5.18) + 3(30) = 92.6 kcal mol^{-1}. If one considers the triple isomerization of [6]radialene into tris(cyclobuteno)benzene, $\Delta_fH°_{est}(\text{tris(cyclobuteno)benzene}) = \Delta_fH°([6]\text{radialene}) + (11.3) - 37 = 92.7$ kcal mol^{-1}!

$\Delta_f H°$(cyclododeca-1,5,9-triyne), no ring strain? $= 6[C_t\text{-}(C_t)(C)] + 6[C\text{-}(C_t)(C)(H)_2] + 6\,\text{kcal mol}^{-1} = 6(27.41 - 4.92) = 134.9\,\text{kcal mol}^{-1}$. If tris(cyclobuteno)benzene should not suffer from more ring strain than three times that of cyclobutene, it would be more stable than [6]radialene and cyclododeca-1,5,9-triyne. Non-annulated cyclobutene has H—C=C—H bond angles larger than 120°. When annulated with benzene the corresponding C—C=C—C bond angles cannot differ very much from 120°, which should impose a larger ring strain to tris(cyclobuteno)benzene than three times 30 kcal mol^{-1}. Obviously [6]radialene cannot be isomerized into cyclododeca-1,5,9-triyne under thermal conditions.

High vacuum pyrolysis of 1,3,5-(trichloromethyl)-2,4,6-trimethylbenzene at 800 °C gives hexamethylidenecyclohexane or [6]radialene [7]. On heating, cyclododeca-1,5,9-triyne is also isomerized into [6] radialene [8]. The process does not produce tris(cyclobuteno)benzene as an intermediate, but follows the mechanism outlined below [9].

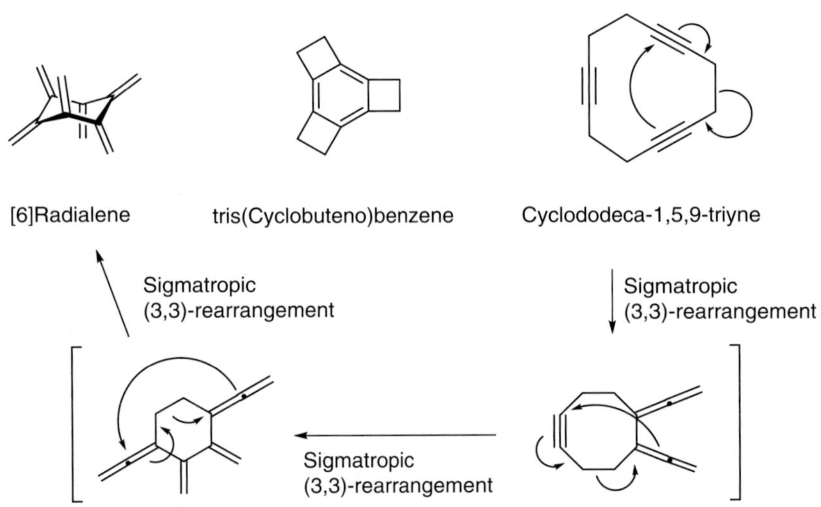

Problem 2.36 What is the product of flash high-vacuum pyrolysis of 2,3,6,7-tetrakis(chloromethyl)-1,4,5,8-tetramethylnaphthalene?

Answer

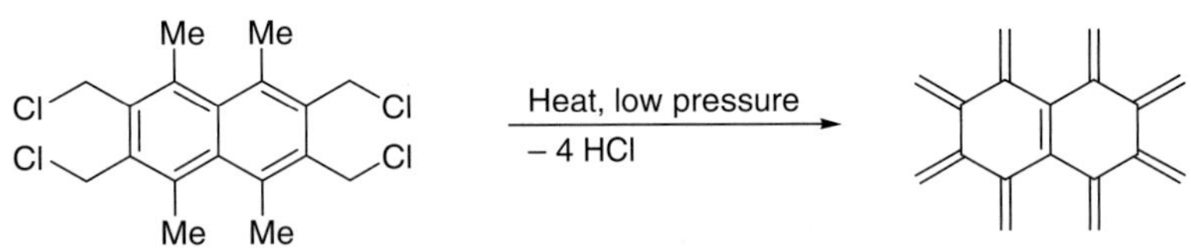

Problem 2.37 What is the major product of pyrolysis of 2,3-dimethylidene-7-oxabicyclo[2.2.1]heptane under high-vacuum. What are the favored products if the pyrolysis tube contains acidic chips?

Answer

The weakest bonds in **A** are the allylic σ(C—C) bonds (see Table 1.A.7). Thus, heating **A** under high vacuum leads to breakage of its C(1)—C(6) bond forming diradical **D**. The latter then fragmentates into ethylene and diradical **B,** which is stabilized by double allylic resonance and by formation of a carbonyl ylide **B′** (Section 5.3.17, Table 5.5). The latter then undergoes electrocyclic cyclization (Section 5.2.1) into furocyclobutene (**P**). Fragmentation of **A** into ethylene +**B** ↔ **B′** might be a concerted cycloreversion; Section 5.3).

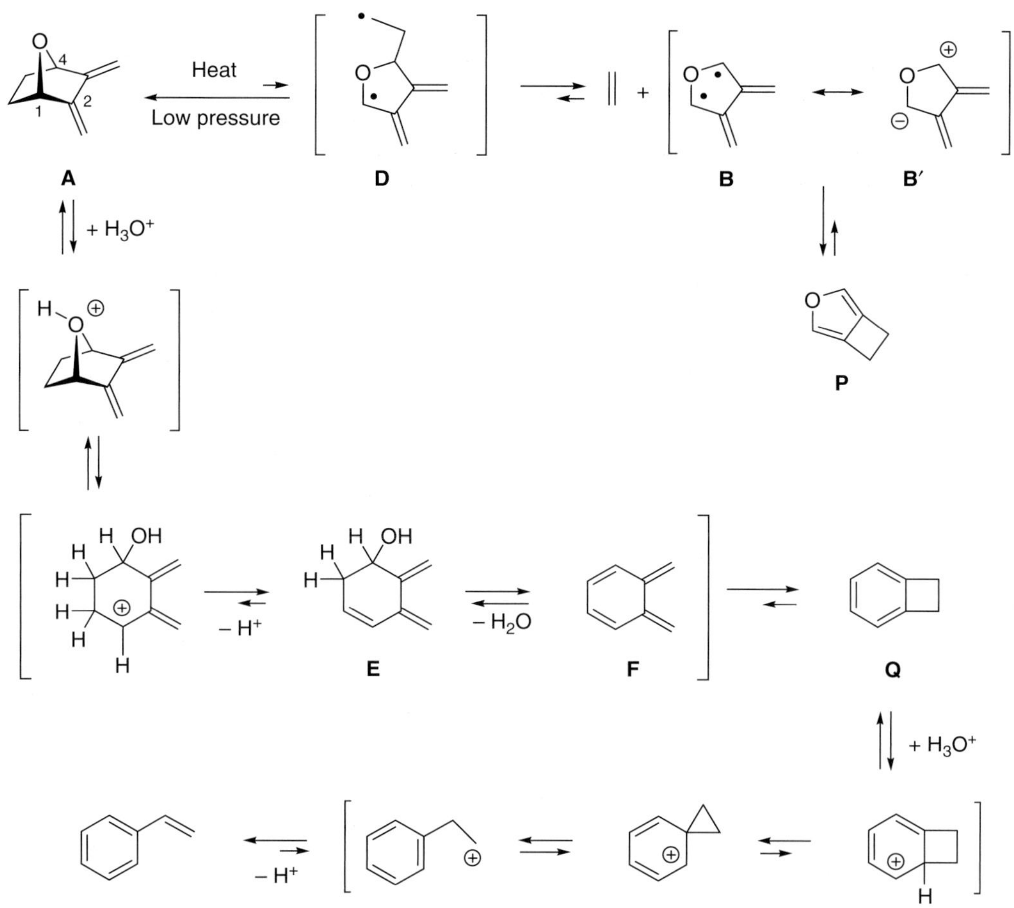

Under acidic conditions, protonation of **A** on its diene unit generates an allylic cation. Protonation of its ethereal moiety competes and generates an oxonium ion intermediate that undergoes allylic heterolysis. After deprotonation intermediate alcohol **E** is formed which eliminates one equivalent of water forming *ortho*-xyxylene (**F,** also named *ortho*-quinodimethane), a compound less stable that its isomer **Q** (benzocyclobutene) resulting from electrocyclic cyclization. As shown in Figure 2.14, benzocyclobutene (**Q**) is c. 5.3 kcal mol^{-1} more stable than *ortho*-xyxylene (**F**). Depending upon the residence time of **Q** in the acidic pyrolysis tube, it can also be protonated. This induces its rearrangement into styrene.

Problem 2.38 Pyrolysis of (2-methylfur-3-yl)methyl benzoate at 640 °C under 0.0001 Torr gives a diene that dimerizes in the condensed state at 0–20 °C. What is this diene? What is its dimer?

Answer

Flash pyrolysis of **A** induces 1,4-elimination of benzoic acid producing diene **P**.

When two unsaturated compounds collide, they might generate diradical intermediates. Usually the most stable diradicals form and evolve into products of cycloaddition, or fragmentate back into the cycloaddents. These processes may compete with other one-step concerted [π^2+π^2]-, [π^2+π^4]-, and [π^4+π^4]-cycloadditions (see Section 5.3). In the case of diene **P** its [π^4+π^4]-cyclodimerization leads to the most stable cycloadducts since the aromaticity of both furan rings is recovered in the two possible regioisomeric products **Q** and **W**. Alternative [π^2+π^2]- or [π^2+π^4]-cycloaddition would not produce cyclodimers as stable as **Q** and **W**, as illustrated with the hypothetical cyclodimerization of **P** into its **X** (Diels–Alder reaction). In **X** only one of the molecules **P** entering in the cyclodimerization is converted into a furan moiety, not two.

There are three possible diradicals **D1**, **D2**, and **D3** that result from the condensation of two molecules of **P** at 0–20 °C in the condensed state. For all of them the aromaticity of two furan units is realized in one of their resonance limiting structures. For diradical **D2** one finds a resonance limiting structure that shows twice the dienyl conjugation, and this stabilizes the diradical better than twice the allyl conjugation as found for **D1**. One cannot write any resonance limiting structure for **D1** that contains the dienyl conjugation. In the case of **D3**, one finds one limiting resonance structure realizing one dienyl conjugation and one allyl conjugation. This is less stabilizing than two dienyl conjugation as in **D2**. Thus, one predicts that the $[\pi^4+\pi^4]$-cyclodimer **Q** must be the major product formed under conditions of kinetic control (no possibility of equilibration of the $[\pi^4+\pi^4]$-cycloadducts at 0–20 °C because their formation is highly exothermic: c. $-60\,\text{kcal mol}^{-1}$). This is observed experimentally! In Section 5.3.4 (Scheme 5.21) we show that two molecules of butadiene equilibrate with octa-1,7-dien-3,6-diyl diradical with a free energy of c. 23.4 kcal mol^{-1} at room temperature. Because of the bis(dienyl) conjugation and the recovery of the aromaticity of two furan units, diradical **D2** equilibrates with **P** with a much smaller endergonicity.

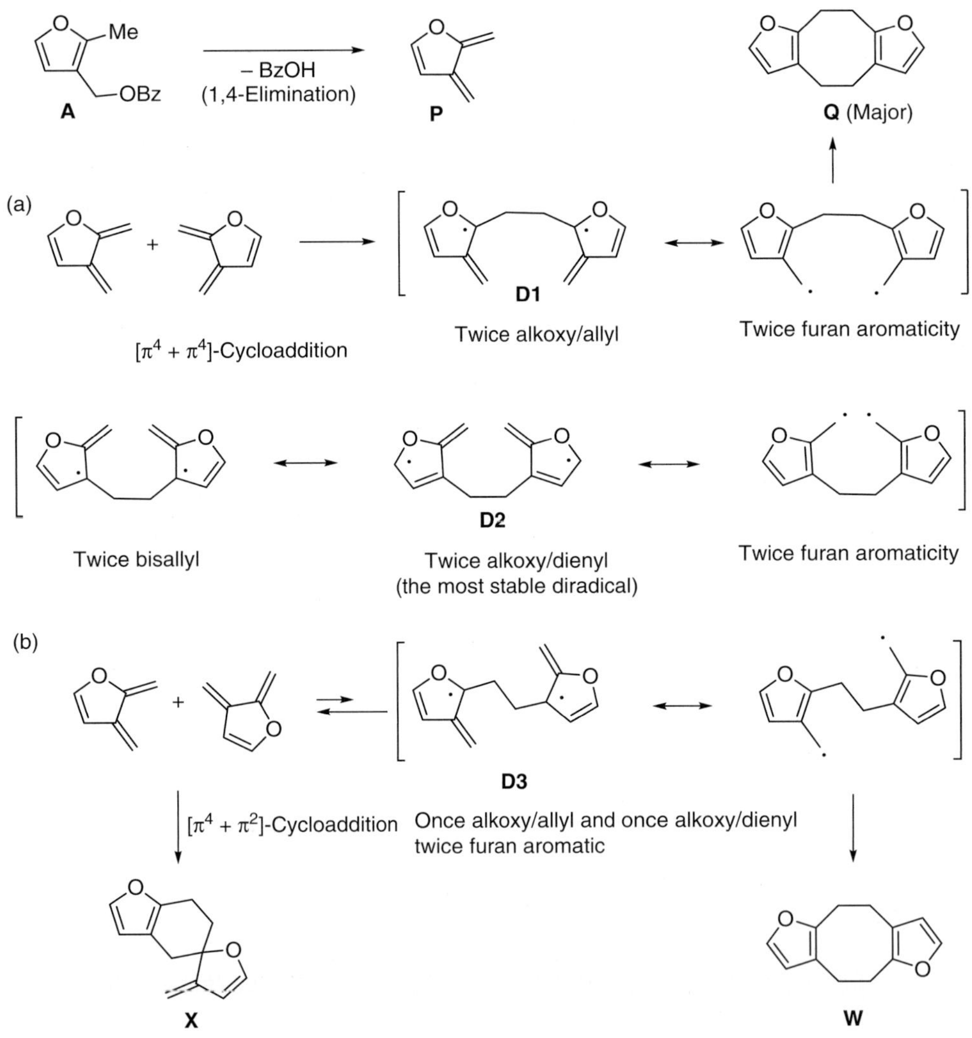

Problem 2.39 (*E*)-1-Methoxybutadiene and methyl vinyl ketone give a mixture of Diels–Alder adducts above 40 °C. The adducts eliminate MeOH on further heating. What conditions of concentration and temperature should be applied in order to avoid this elimination?

Answer

Formally all ethers can fragmentate on heating into alcohols + alkenes. In the case of the Diels–Alder adducts of (*E*)-1-methoxybutadiene to methyl vinyl ketone the adducts **A** can eliminate one equivalent of MeOH producing dienone **B**. The uncatalyzed reaction is much easier (lower activation free energy) than for simple ethers because of the acidity of the α-C—H group of the ketone, which is also, β to the ether moiety (the leaving group). A first model reaction for this fragmentation can be 3-methoxycyclohexene that eliminates methanol and produces cyclohexa-1,3-diene. The NIST Chemistry WebBook gives $\Delta_f H°$(1-methoxycyclohexene) = −38.5 kcal mol^{-1}. Considering 3 kcal mol^{-1} of stabilization because of n(O:)/π conjugation in 1-methoxycyclohexene one estimates $\Delta_f H°$(3-methoxycyclohexene) = −38.5 + 3 = −35.5 kcal mol^{-1}. This allows one to calculate $\Delta_r H°$(3-methoxycyclohexene ⇄ MeOH + cyclohexa-1,3-diene) = −49 + 25.4 (Table 1.A.4) − (−35.5) = 11.9 kcal mol^{-1}. If one considers the stabilization gained in **B** because of π/π-conjugation between the diene and ketone moieties, one obtains $\Delta_r H°_{est}$(**A** ⇄ **B** + MeOH) = 11.9 − 2 ≅ 10 kcal mol^{-1}. The variation of entropy for this elimination can be estimated from the calculated entropies of translation (Eq. (1.30)): $\Delta_r S°_{trans}$(**A** ⇄ **B** + MeOH) = 2.98 ln[(122.2)(32)/154.2] + 4.97 ln 298 − 2.31 = 9.63 + 28.31 − 2.31 = 35.6 eu. At 25 °C $\Delta_r G°_{est}$(**A** ⇄ **B** + MeOH) = 10 000 − 298 (35.6) = 32 cal mol^{-1} ≅ 0.03 kcal mol^{-1}. Thus, already at room temperature, and for one molar concentration of reactants, the elimination of methanol can compete thermodynamically with the formation of the cycloadduct **A**. If the uncatalyzed cycloaddition requires heating above 40 °C, the chances for the elimination **A** → **B** + MeOH to compete with the cycloaddition will increase. Thus, it is necessary to work under the highest possible concentration of cycloaddents, if possible without solvent. If a solvent is necessary, MeOH should be chosen to reduce the formation of **B** (mass law effect). Furthermore, one cannot wait the time for the reaction mixture to reach full conversion. The rate law of the uncatalyzed cycloaddition is second order and the rate law of the uncatalyzed elimination is first order. The longer the time of the reaction, the higher the proportion of products of elimination. Only acids catalyze the Diels–Alder reaction of methyl vinyl kekone. Thus, if an acid catalyst is found that accelerates the Diels–Alder reaction significantly, it should be run below room temperature. Under these conditions $\Delta_r G^T$(**A** ⇄ **B** + MeOH) > 0, the addition of MeOH to **B** is favored over the elimination (the $-T\Delta_r S^T$ term becomes less negative than at $T \geq 25$ °C). As a rule, when an elimination (fragmentation) competes with a cycloaddition (condensation) the lower the temperature of the reaction, the less the elimination will compete for thermodynamic reasons. The same rule is valid under kinetic control (see Section 3.3.1).

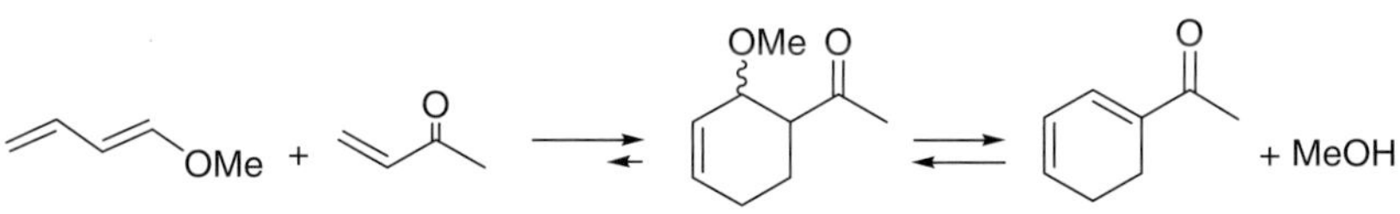

A **B**

Problem 2.40 Methyl isopropyl ketone (MIPK) is an efficient high-octane (>100) oxygenate gasoline additive (see Problem 1.43). One method of preparation of MIPK is the acid-catalyzed hydrocarbation of CO with isobutane that produces pivalaldehyde. The latter compound is isomerized under acidic conditions into MIPK. Propose a mechanism for the acid-catalyzed hydrocarbation of CO.

Answer

Olah and coworkers have found that isobutane undergoes formylation with CO in the presence of $HF{\cdot}BF_3$ to give pivalaldehyde, which is subsequently rearranged into MIPK. These reactions may involve dicationic intermediates [10].

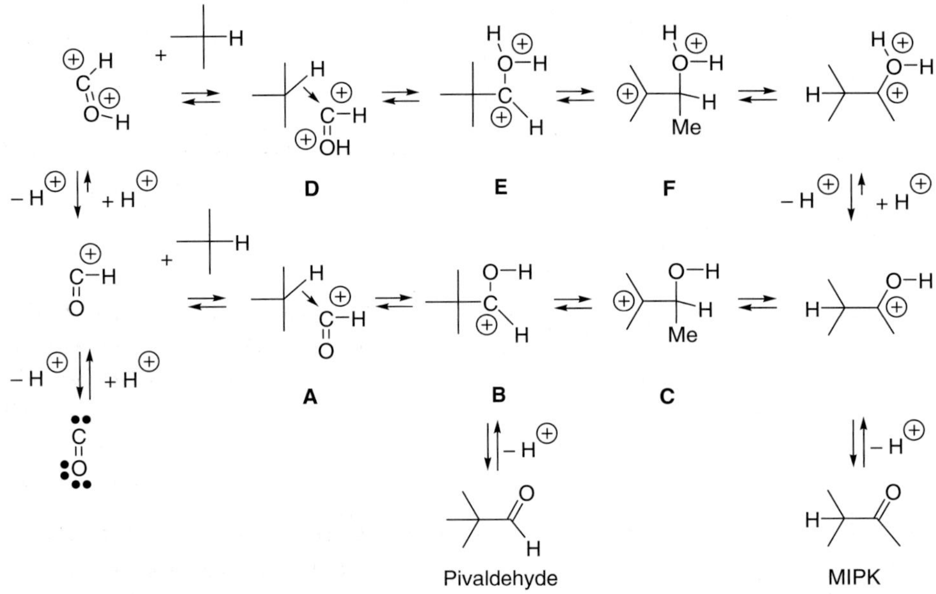

The acid-promoted carbonylation of isobutane is an example of hydrocarbation of an unsaturated compound (C=O) by an alkane (see Section 8.3 for more examples of hydrocarbations). The mechanism shown above for this reaction involves the C—H activation of the tertiary C—H bond of isobutane. CO can be seen as an oxocarbene and an electron-rich carbene since its 2p LUMO is partially filled by donation from one nonbonding 2p(n(O:)) electron pair: :C=O: ↔ – C≡O$^+$ (see Section 7.7.2). A carbene like $:CH_2$ (an unsaturated carbon species with a sextet of electrons) reacts with alkanes inserting into C—H bonds. The reaction is initiated by a LUMO(carbene)/HOMO(C–H) interaction (see Chapter 4) and leads to a H atom transfer ($R–H + :CH_2 \rightleftarrows \rightarrow R^{\cdot} + {}^{\cdot}CH_2–H \rightarrow R–CH_2–H$) or to a concerted insertion. Unsaturated transition metal complexes with 16 (or less) valence electrons can also insert into C—H bonds in reactions called oxidative additions: $ML_n + R–H \rightleftarrows R–ML_n–H$. These reactions are also initiated by LUMO(ML_n)/HOMO(C–H) interactions and may involve intermediates that are σ-complexes of type $R–H \rightarrow ML_n$ (Section 7.7.4). The electron-rich carbene C=O is made electron poor upon protonation and thus more prone to interact with the C—H bond than CO itself. Thus, the monocationic intermediate $HC^{(+)}{=}O$ (formyl cation) has a LUMO lower lying in energy than the LUMO of CO. This is even more the case in the dicationic species $HC^{(+)}{=}O^{(+)}H$ or: $C{=}O^{(2+)}H_2 \leftrightarrow :C^{(+)}—O^{(+)}H_2$ that are much better carbon electrophiles than: C=O: ↔ $^-$C≡O$^+$. It is proposed that intermediate **A** (σ-complex resulting from the interaction of the LUMO of the formyl cation with the C—H bond) forms first. The latter equilibrates with intermediate **B**, a relatively stable hydroxycarbenium ion (conjugate acid of pivaldehyde). Then, a (1,2)-sigmatropic shift (Section 5.5.1) of a methyl group (Wagner–Meerwein rearrangement) equilibrates **B** with the tertiary carbenium **C**. A (1,2)-hydride shift (see the pinacolic rearrangement) follows, giving the conjugate acid of MIPK. Olah and coworkers have proposed that dicationic intermediates **D**, **E**, and **F** might also be involved in the process.

Problem 2.41 Using gas phase standard heats of formation what base should be used to generate oxirane (ethylene oxide) by 1,3-elimination of HCl from 2-chloroethanol?

Answer

In the gas phase, displacement of EtCl + EtOH ⇄ EtOEt + HCl is nearly thermoneutral ($\Delta_r H° = +0.7$ kcal mol^{-1}) and has an entropy of reaction ($\Delta_r S°$) near zero since the number of reactants is the same as the number of products and their molecular masses are comparable.Thus, one estimates $\Delta_r G°$(EtCl + EtOH ⇄ EtOEt + HCl) ~ 0 kcal mol^{-1}, and thus $K^{298K} \sim 1$.

$$\text{EtCl} + \text{EtOH} \rightleftarrows \text{EtOEt} + \text{HCl} \quad \Delta_r H° = 0.7 \text{ kcal mol}^{-1}$$

Table 1.A.2. and 1.A.4 give:

$$\Delta_f H°(\text{EtCl}) = -28.8 \text{ kcal mol}^{-1}$$
$$\Delta_f H°(\text{EtOH}) = -56.2 \text{ kcal mol}^{-1}$$
$$\Delta_f H°(\text{Et}_2\text{O}) = -60.2 \text{ kcal mol}^{-1}$$
$$\Delta_f H°(\text{HCl}) = -22.1 \text{ kcal mol}^{-1}$$
$$\Delta_f H°(\text{oxirane}) = -12.6 \text{ kcal mol}^{-1}$$

The standard heat of formation of 2-chloroethanol is estimated from that of ethanol corrected for the chloro substitution that is taken as the difference in heats of formation between ethane and ethyl choride (chloroethane). Thus, $\Delta_f H°(\text{ClCH}_2\text{CH}_2\text{OH}) = \Delta_f H°(\text{CH}_3\text{CH}_2\text{OH}) + \Delta_f H°(\text{ClCH}_2\text{CH}_3) - \Delta_f H°(\text{CH}_3\text{CH}_3) = -56.2 - 26.8 + 20.2 = -62.8$ kcal mol^{-1}.

$$\text{Cl} - \text{CH}_2 - \text{CH}_2 - \text{OH} \rightleftarrows \text{oxirane} + \text{HCl} \quad \Delta_r H° = 28.1 \text{ kcal mol}^{-1}$$

The difference in heats of the two preceding reactions gives an estimate of the ring strain of oxirane, i.e. 28.1 – 0.7 = 27.4 kcal mol^{-1}. This is nearly the same as the ring strain of cyclopropane (Table 2.A.3).

The 1,3-elimination of HCl from 2-chloroethanol must have a positive entropy of reaction since two products are formed at the expense on one reactant. The variation of entropy of translation for this reaction is calculated (Eq. (1.31)) to be: $\Delta_r S°_{trans}$($\text{HOCH}_2\text{CH}_2\text{Cl}$ ⇄ oxirane + HCl) = 34.9 eu. One needs to consider the blockage of the rotation of one C—C bond (c. –5 eu) and one C—O bond (c. –5 eu) when going from the reactant to the product (oxirane) and the increase of symmetry (variation of entropy of rotation: $-R \ln 2$, C_2 axis of rotation in oxirane). Thus, one estimates $\Delta_r S°$($\text{HOCH}_2\text{CH}_2\text{Cl}$ ⇄ oxirane + HCl) = +34.9 – 10 – 1.36 eu = 23.5 eu. At 25 °C, one obtains $\Delta_r G°$($\text{HOCH}_2\text{CH}_2\text{Cl}$ ⇄ oxirane + HCl) = 28 100 – 298(23.5) = 28 100 – 7003 = 21 097 cal mol^{-1}. This reaction is highly endergonic at room temperature. In order to displace the equilibrium in favor of the oxirane we must neutralize HCl formed by a base that gives a salt with an exergonicity <–21.1 kcal mol^{-1}. For reaction in water,

$\Delta_r G°$(HCl + B: ⇄ BH^+Cl^-) < –21.1 kcal mol^{-1}, which corresponds to a $\Delta \text{p}K_a = 21.1/1.36 > 15.5$. Since HCl has a $\text{p}K_a = -7$ (H_2O, 25 °C), $\text{p}K_a(\text{BH}^+)$ must be larger than 15.5 – 7 = 8.5. NH_3 ($\text{p}K_a(\text{NH}_4^+) = 9.3$) could be sufficient. But this base cannot be used as it is a nucleophile that may react with the chloride and the oxirane leading to the formation of 2-hydroxyethyl amine, which then will react with 2-chloroethanol giving $\text{HOCH}_2\text{CH}_2\text{NHCH}_2\text{CH}_2\text{OH}$, and so forth. With Me_3N (and other tertiary amines; $\text{p}K_a(\text{Me}_3\text{NH}^+) = 9.8$) its reversible quaternization may occur with 2-chloroethanol competitively with elimination of HCl. Me_3N will not react readily with oxirane as the zwitterion $\text{Me}_3\text{N}^+\text{CH}_2\text{CH}_2\text{O}^-$ contains an alkoxide moiety that is much more basic than Me_3N. Under forcing conditions this zwitterion may induce the polymerization of oxirane, generating a polyethyleneglycol. Irrespective of the base used, the latter reactions may compete with the 1,3-elimimation of HCl. An alternative base is NaHCO_3. It reacts with HCl generating neutral NaCl and CO_2 that bubbles away from the reaction mixture, displacing the equilibrium in favor of oxirane and NaCl (Le Chatelier principle):

$$\text{ClCH}_2\text{CH}_2\text{OH} + \text{NaHCO}_3 \rightleftarrows \text{NaCl} + \text{H}_2\text{O} + \text{CO}_2 + \text{oxirane}.$$

The first step of the 1,3-elimination is the deprotonation of 2-chloroethanol into the corresponding alkoxide that undergoes rate-determining S_Ni reaction forming oxirane, as demonstrated for the reaction at 25 °C using NaOH as base [11].

References

1 Vogel, P. (1985). *Carbocation Chemistry*, Chapter 8.1, 330–349. Amsterdam: Elsevier.

2 Gabioud, R. and Vogel, P. (1984). Conformational-analysis of 2,3,5,6-tetrakis(methylene)bicyclo[2.2.2]octanes by circular-dichroism. *Tetrahedron Lett.* 25 (16): 1729–1732.

3 Kong, J., Galabov, B., Koleva, G. et al. (2011). The inherent competition between addition and substitution reactions of Br-2 with benzene and arenes. *Angew. Chem. Int. Ed.* 50 (30): 6809–6813.

4 Vogel, P. (1985). *Carbocation Chemistry*, Chapter 8, 323–372. Amsterdam: Elsevier.

5 *NIST Chemistry WebBook*

6 Shanshal, M. and Kubba, R.M. (2006). Bonding, electron densities and vibration frequencies of [3]-, [4]-, [5]- and [6]-radialenes; a B_3LYP density functional study. *Z. Naturforsch., A: Phys. Sci.* 61: 95–98.

7 Harruff, L.G., Brown, M., and Boelkelheide, V. (1978). Hexaradialene – Precursors and structure. *J. Am. Chem. Soc.* 100: 2893–2894.

8 Barbovich, A.J. and Vollhardt, K.P.C. (1976). 1,5,9-Cyclododecatriyne – Synthesis and conversion to intermediate 1,2-3,4-5,6-tricyclobutabenzene. *J. Am. Chem. Soc.* 98: 2667–2668.

9 Dower, W.V. and Vollhardt, K.P.C. (1982). Mechanism of the isomerization of 1,5,9-cyclododecatriyne to hexaradialene – 1,2-3,4-5,6-Tricyclobutabenzene is not an intermediate. *J. Am. Chem. Soc.* 104 (24): 6878–6879.

10 Olah, G.A., Surya Prakash, G.K., Mathew, T., and Marinez, E.R. (2000). Superacid-catalyzed selective formylation – rearrangement of isoalkanes with carbon monoxide to branched ketones. *Angew. Chem. Int. Ed.* 39 (14): 2547–2548.

11 Swain, C.G., Ketley, A.D., and Bader, R.F.W. (1959). The mechanism of ethylene oxide formation from 2-chloroethanol. *J. Am. Chem. Soc.* 81 (10): 2353–2359.

3

The Rates of Chemical Reactions*

Answers to Problems

Problem 3.1 Estimate the time for 99% conversion of **A** → **P**, an irreversible reaction following a first order rate law with $k = 10^{-3}$ s^{-1}. What is the half-life of this reaction?

Answer

Table 3.1 gives for a first order reaction (rate law: d[products]/dt = k^{I}[one of the reactants]), half-life $\tau_{1/2} = \ln 2/k^{I}$. In our case $\tau_{1/2} = \ln 2/k^{I} = 0.693/0.001$ = 693 seconds. For a simple reaction **A** → **P** with rate law $-$ d[**A**]/dt = d[**P**]/dt = k^{I}[**A**] we need to wait for $7\tau_{1/2}$ in order to reach 99% of the equilibrium. If the reaction is irreversible, we understand that $k_{reverse}$ is at least 1000 times slower than $k_{forward}$, or that K(**A** ⇄ **P**) > 1000. Thus 99% conversion is obtained after 4852 seconds ≅ 1 hour and 21 minutes.

Problem 3.2 Estimate the time for 99% conversion of reaction **A** + **B** → **P** following a first order rate law in [**A**] and a first order rate law in [**B**]. A 20-fold excess of **B** is used and $k = 10^{-3}$ dm^3 mol^{-1} s^{-1}. Estimate the half-life of this reaction.

Answer

As in the case of problem 3.1 one assumes reaction **A** + **B** → **P** to be irreversible, which implies that the equilibrium constant K(**A** + **B** ⇄ **P**) = [**P**]/[**A**][**B**] > 1000. With a 20-fold excess of reactant **B** one can consider that the concentration of **B** does not vary during the reaction. Thus, we are in the presence of a pseudo-first order rate law d[**P**]/dt = k′[**A**] with $k' = 0.001$ s^{-1}. If [**B**] = 1 mol dm^{-3}, and the initial concentration of **A** is 0.05 mold m^{-3}, the half-life $\tau_{1/2} = \ln 2/k' = 0.692/0.001$ = 693 seconds. We need to wait (Table 3.1) $7\tau_{1/2}$ = 4852 seconds or 1 hour and 21 minutes to reach 99% conversion of **A** into **P**.

If [**B**] = 1 mol dm^{-3} and the initial concentration of **A** is 0.1 mol dm^{-3}, $k' = 0.002$ s^{-1}. Thus, the reaction will reach 50% conversion after ln2/0.002 = 347 seconds, and 99% conversion after $7\tau_{1/2}$ = 2426 seconds.

If [**B**] = [**A**] = 2 mold m^{-3} $\tau_{1/2} = 1/0.002$ = 500 seconds, the time for 50% conversion of **A** and **B** into **P**. For 99% conversion one needs to wait for $99\tau_{1/2}$ ≅ 50 000 seconds, or 13 hours and 54 minutes.

If one considers that 1 mol of **A** and 1 mol of **B** + solvent are engaged in a 0.5 l reactor (initial [**A**] = initial [**B**] = 2 mold m^{-3}) and assuming that **P** and **A** + **B** can be separated readily and quantitatively from the mixture, 50% yield, or 0.5 mol of **P** will be obtained after 500 seconds from the above reaction mixture.

If one engages 1 mol of **A**, as above, and 20 mol of **B** + solvent in a 10 l reactor (initial [**B**] = 2 mold m^{-3}, initial [**A**] = 0.1 mold m^{-3}), 0.5 mol of **P** will be obtained after 347 seconds from this reaction mixture, which requires a reactor 20 times larger than that used above for a 1 : 1 mixture of both reactants **A** and **B**. If one places 0.5 mol of **A** and 10 mol of **B** in a 5 l reactor and waits for 99% conversion (13 hours and 54 minutes: overnight reaction), 0.99 × 0.5 = 0.495 mol of **P** will be obtained, nearly the same yield as above. The advantage in the latter case is that **A** will be consumed, but **B** and **P** have to be separated and purified.

Because side reactions that destroy **A**, **B**, and **P** may compete with the desired reaction, one might find that the yield of the desired product **P** decreases on waiting the time for a full conversion of the reactant not in excess.

*See end of the chapter for additional material.

Organic Chemistry: Theory, Reactivity and Mechanisms in Modern Synthesis, Workbook, First Edition. Pierre Vogel and Kendall N. Houk.

Problem 3.3 Consider the two parallel irreversible reactions **A** → **P** (rate constant k_1, $\Delta^{\ddagger}H_1 = 23\ \text{kcal mol}^{-1}$, $\Delta^{\ddagger}S_1 = -1\ \text{cal K}^{-1}\ \text{mol}^{-1}$) and **A** → **Q** (rate constant k_2, $\Delta^{\ddagger}H_2 = 17.4\ \text{kcal mol}^{-1}$, $\Delta^{\ddagger}S_2 = -20\ \text{cal K mol}^{-1}$) following first order rate laws. Estimate the temperature T_1 at which 99% conversion of **A** will be realized in two hours. What will the product ratio [**P**]/[**Q**] be at this temperature T_1? What will this product ratio be at temperature $T_i = T_1 - 50$ K, $T_1 - 30$ K, $T_1 + 30$ K and $T_1 + 50$ K? What are the Arrhenius pre-exponential factors A_1 and A_2 at temperature T_1 for these two reactions?

Answer

An irreversible reaction **A** → **P** (with $K(\mathbf{A} \rightleftarrows \mathbf{P}) = [\mathbf{P}]/[\mathbf{A}] > 1000$) with a first order rate law $-\text{d}[\mathbf{A}]/\text{d}t = \text{k}_1[\mathbf{A}]$ will reach 99% conversion after 7 half-lives (Table 3.1). In our case $7\tau_{1/2} = 7200$ seconds. Since $\tau_{1/2} = \ln 2/k_1$ one calculates $k_1 = \ln 2/\tau_{1/2} = 0.000\,674\,389\,3\ \text{s}^{-1}$. With $\Delta^{\ddagger}S_1 = -1$ eu, $\Delta^{\ddagger}G_1^{300\text{K}} = \Delta^{\ddagger}H_1 - T\Delta^{\ddagger}S_1 \cong 23\,300\ \text{cal mol}^{-1}$. From Figure 3.12, one can estimate that reaction **A** → **P** occurred between 25 and 60 °C, or 298–333 K.

The estimated temperature T_1 must give the same value of $\Delta^{\ddagger}G^{\text{T}}$ using $\Delta^{\ddagger}G_1^{\text{T}} = \Delta^{\ddagger}H_1 - T\Delta^{\ddagger}S_1$ and using Eq. (3.67) ($\Delta^{\ddagger}G^{\text{T}} = RT(23.76 + \ln T - \ln k)$. We do the calculations with the following trial temperatures:

At 310 K, $\Delta^{\ddagger}G_1^{310\text{K}} = \Delta^{\ddagger}H_1 - T\Delta^{\ddagger}S_1 = 23\,000–310(-1) = 23\,310\ \text{cal mol}^{-1}$ to compare with $\Delta^{\ddagger}G_1^{310\text{K}}(\text{est.}) = RT(23.76 + \ln T - \ln k) = 1.987 \times 310(23.76 + \ln 310 - \ln(0.000\,67)) = 22\,670\ \text{cal mol}^{-1}$.

This values is smaller than $\Delta^{\ddagger}G_1^{310\text{K}}$ by 640 cal mol^{-1}.

At 320 K, $\Delta^{\ddagger}G_1^{320\text{K}} = \Delta^{\ddagger}H - T\Delta^{\ddagger}S = 23\,000 - 320(-1) = 23\,320\ \text{cal mol}^{-1}$ to compare with $\Delta^{\ddagger}G_1^{320\text{K}}(\text{est.}) = RT(23.76 + \ln T - \ln k) = 1.987 \times 320(23.76 + \ln 320 - \ln(0.00067)) = 23\,422\ \text{cal mol}^{-1}$.

This value is larger than $\Delta^{\ddagger}G_1^{320\text{K}}$ by 102 cal mol^{-1}. By interpolation one estimates $T_1 = 320–10(102/752) = 318.6$ K.

At $T_1 = 319$ K, $\Delta^{\ddagger}G_2^{319\text{K}} = \Delta^{\ddagger}H_2 - T\Delta^{\ddagger}S_2 = 17\,400–319(-20) = 23\,780\ \text{cal mol}^{-1}$.

This gives the first order rate constant $k_2^{319\text{K}} = 2.085 \times 10^{10}T\ \exp(-\Delta^{\ddagger}G_2^{319\text{K}}/RT)$ (Eq. (3.60′)) = 2.085×10^{10} (319)exp.$(-23\,780/1.987 \times 319) = 3.3857 \times 10^{-4}\ \text{s}^{-1}$. This leads to $k_1/k_2 = [\mathbf{P}]/[\mathbf{Q}] = 6.743893/3.3857 \cong 1.99$ at 319 K.

Using $\Delta^{\ddagger}G_1^{319\text{K}} = 23\,000 + 319 = 23\,319\ \text{cal mol}^{-1}$ and $\Delta^{\ddagger}G_2^{319\text{K}} = 23\,780\ \text{cal mol}^{-1}$ ($\Delta\Delta^{\ddagger}G^{319\text{K}} = -461\ \text{cal mol}^{-1}$) and Eq. (3.68)

$k_1/k_2 = [\mathbf{P}]/[\mathbf{Q}] = \exp.(-\Delta\Delta^{\ddagger}G^{319\text{K}}/RT) = \exp.[-(-461)/1.987 \times 319] = 2.07$ at 319 K.

At $T_2 = 319 - 50 = 269$ K, $\Delta^{\ddagger}G_1^{269\text{K}} = \Delta^{\ddagger}H_1 - T\Delta^{\ddagger}S_1 = 23\,000 - 269(-1) = 23\,269\ \text{cal mol}^{-1}$, $\Delta^{\ddagger}G_2^{269\text{K}} = \Delta^{\ddagger}H_2 - T\Delta^{\ddagger}S_2 = 17\,400 - 269(-20) = 22\,780\ \text{cal mol}^{-1}$, which gives $\Delta\Delta^{\ddagger}G^{269\text{K}} = 489\ \text{cal mol}^{-1}$ and $k_1/k_2 = [\mathbf{P}]/[\mathbf{Q}] = \exp.(-\Delta\Delta^{\ddagger}G^{269\text{K}}/RT) = \exp.(-(489/1.987 \times 269) = 0.4$ at 269 K.

Note that lowering temperature does not increase selectivity in favor of **P** but reduces it!

At $T_3 = 319–30 = 289$ K, $\Delta^{\ddagger}G_1^{289\text{K}} = \Delta^{\ddagger}H_1 - T\Delta^{\ddagger}S_1 = 23\,000–289(-1) = 23\,289\ \text{cal mol}^{-1}$, $\Delta^{\ddagger}G_2^{289\text{K}} = \Delta^{\ddagger}H_2 - T\Delta^{\ddagger}S_2 = 17\,400–289(-20) = 23\,180\ \text{cal mol}^{-1}$, which gives $\Delta\Delta^{\ddagger}G^{289\text{K}} = 109\ \text{cal mol}^{-1}$ and $k_1/k_2 = [\mathbf{P}]/[\mathbf{Q}] = \exp.(-\Delta\Delta^{\ddagger}G^{269\text{K}}\ /RT) = \exp.(-(109/1.987 \times 289) = 0.83$ at 289 K.

At $T_4 = 319 + 30 = 349$ K, $\Delta^{\ddagger}G_1^{349\text{K}} = \Delta^{\ddagger}H_1 - T\Delta^{\ddagger}S_1 = 23\,000–349(-1) = 23\,349\ \text{cal mol}^{-1}$, $\Delta^{\ddagger}G_2^{349\text{K}} = \Delta^{\ddagger}H_2 - T\Delta^{\ddagger}S_2 = 17\,400–349(-20) = 24\,380\ \text{cal mol}^{-1}$, which gives $\Delta\Delta^{\ddagger}G^{349\text{K}} = -1031\ \text{cal mol}^{-1}$ and $k_1/k_2 = [\mathbf{P}]/[\mathbf{Q}] = \exp.(-\Delta\Delta^{\ddagger}G^{349\text{K}}\ /RT) = \exp.(-(-1031/1.987 \times 349) = 4.42$ at 349 K.

At $T_5 = 319 + 50 = 369$ K, $\Delta^{\ddagger}G_1^{369\text{K}} = \Delta^{\ddagger}H_1 - T\Delta^{\ddagger}S_1 = 23\,000 - 369(-1) = 23\,369\ \text{cal mol}^{-1}$, $\Delta^{\ddagger}G_2^{369\text{K}} = \Delta^{\ddagger}H_2 - T\Delta^{\ddagger}S_2 = 17\,400 - 369(-20) = 24\,780\ \text{cal mol}^{-1}$, which gives $\Delta\Delta^{\ddagger}G^{369\text{K}} = -1411\ \text{cal mol}^{-1}$ and $k_1/k_2 = [\mathbf{P}]/[\mathbf{Q}] = \exp.(-\Delta\Delta^{\ddagger}G^{369\text{K}}/RT) = \exp.(-(-1411/1.987 \times 369) = 6.85$ at 369 K.

In this case selectivity in favor of **P** increases with temperature. This illustrates the fact that reactions with negative activation entropies are less accelerated than reactions with positive activation entropies on heating.

The Arrhenius pre-exponential factors A are given by $A = 2.085 \times 10^{10}T\ \exp(\Delta^{\ddagger}S/\text{R})$ (Eq. (3.65)).

At $T_1 = 319$ K, $A_1(\mathbf{A} \rightarrow \mathbf{P}) = 2.085 \times 10^{10}(319)\exp.(-1/1.987) = 4.0 \times 10^{12}\ \text{s}^{-1}$, and $A_2(\mathbf{A} \rightarrow \mathbf{Q}) = 2.085 \times 10^{10}(319)\exp.(-20/1.987) = 2.8 \times 10^{8}\ \text{s}^{-1}$

Problem 3.4 At room temperature N-methyl-2-phenylpyrrolidine exits as an equilibrium of *trans* (**A**) and *cis* (**B**) isomers with $K = [\mathbf{A}]/[\mathbf{B}] = 17.0$. One measures $k_{\text{WH}} = 3 \times 10^{-3}\ \text{M}^{-1}\ \text{s}^{-1}$ for its quaternization with $^{13}CH_3I$ and a product ratio [**M**]/[**N**] = 1.72 is found. Calculate rate constants $k(\mathbf{A} \rightarrow \mathbf{M})$ and $k(\mathbf{B} \rightarrow \mathbf{N})$.

Answer

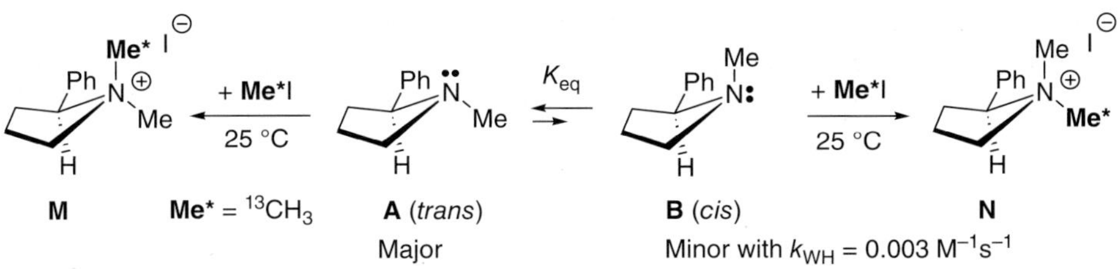

Reactants **A** and **B** are rapidly equilibrated at 25 °C with $K_{eq} = [\mathbf{A}]/[\mathbf{B}] = 17 = K$. They react with H_3C^{13}–I irreversibly giving products **M** and **N**, respectively, in a product ratio [**M**]/[**N**] = 1.72. Note that **M** and **N** have nearly the same stability. Under conditions of reversible quaternization reactions the product ratio should be nearly unity (not exactly unity because of the equilibrium ^{13}C isotopic effect).

A is quaternized more slowly than **B** since **M** is formed to a lesser extent relative to **N**. Thus, $k(\mathbf{A} \to \mathbf{M}) = k_q$ must be smaller than $k(\mathbf{B} \to \mathbf{N}) = k_p$.

Using Eq. (3.73′) one calculates

$k_p = k_{WH}(K+1)/(P+1) = 0.003[(17+1)]/(1.72+1)] = 0.0195\ M^{-1}\ s^{-1}$.

Using Eq. (3.74′) one calculates

$k_q = k_{WH}[(K+1)/K][P/(P+1)] = 0.003[(17+1)]/17]/[1.72/(1.72+1)] = 0.0020\ M^{-1}\ s^{-1}$.

If one had chosen $k(\mathbf{B} \to \mathbf{N}) = k_q$ and $k(\mathbf{A} \to \mathbf{M}) = k_p$ one would have obtained the former rate constant smaller than the latter, which is contrary to common sense.

Problem 3.5 The racemization of 1,3-dimethylallene (−)-**A** has activation parameters $\Delta^{\ddagger}H = 45.0$ kcal mol^{-1} and $\Delta^{\ddagger}S = 0.5$ eu (at 280 °C). Two mechanisms can be proposed for that isomerization. Hypothesis I: rotation about one double C=C bond via diradical **B**. Hypothesis II: equilibration with cyclopropylidene intermediates **C** and **D**. Applying thermochemical calculations, which of these two hypotheses is more consistent with the experimental data?

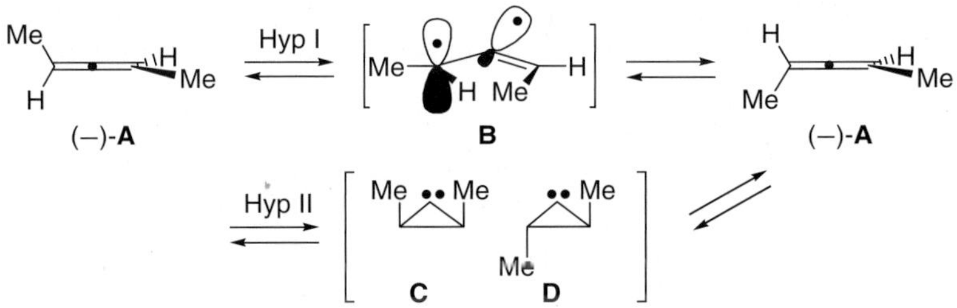

Answer

The standard heats of formation of diradical **B** and of carbene **C** or **D** can be estimated and compared to the heat of formation of (−)-**A**. Table 1.A.2 gives $\Delta_f H°(CH_2{=}C{=}CH_2) = 45.5$ and $\Delta_f H°(MeCH{=}C{=}CH_2) = 38.8$ kcal mol^{-1}. Thus, methyl substitution of allene leads to a change of −6.7 kcal mol^{-1}. This allows one to estimate $\Delta_f H°_{est}((-)\text{-}\mathbf{A}) = \Delta_f H°_{est}(MeCH{=}C{=}CHMe) = 45.5 + 2(-6.7) = 31.1$ kcal mol^{-1}. Diradical **B** is derived from the double C–H homolysis (*Z*)-pent-2-ene (see Section 1.12.5) for which the standard heat of formation is -7.0 ± 1.0 kcal mol^{-1} (Table 1.A.2).

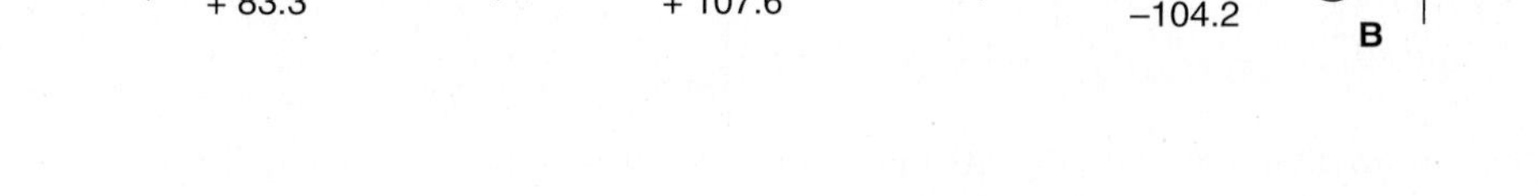

The C—H bond dissociation enthalpy $DH°(MeCH{=}CH{-}C^{\cdot}(Me)H/H^{\cdot}) = 81.6 \pm ?$ kcal mol^{-1} is given in Table 1.A.7. The two radical moieties in **B** are perpendicular; thus one can use $DH°(CH_2{=}CH^{\cdot}/H^{\cdot}) = 110.6 \pm ?$ kcal mol^{-1}. This value must be corrected for the vinyl radical stabilization due to methyl substitution; this leads to $DH°_{est}(CH_2{=}C^{\cdot}Me/H^{\cdot}) = 107.6$ kcal mol^{-1}. Thus, $\Delta_f H°_{est}(\mathbf{B}) = -7.0 + 81.6 + 107.6 - 104.2 = 78.0$ kcal mol^{-1}. This gives the heat of reaction $\Delta_r H°_{est}((-)\text{-}\mathbf{A} \rightleftarrows \mathbf{B}) = \Delta_f H°(\mathbf{B}) - \Delta_f H°((-)\text{-}\mathbf{A}) = 78.0 - 31.1 = 46.9$ kcal mol^{-1}. Considering errors (± 3 kcal mol^{-1}) that are associated with our estimate hypothesis I can be retained (rotation around C=C double bond with $\Delta^{\ddagger}H = 45.0$ kcal mol^{-1}) for the racemization of 1,3-dimethylallene.

The heat of formation of carbene **C** (or **D**) can be estimated as follows. The didedhydrogenation of propane into cyclopropane amounts to $\Delta_f H°$(cyclopropane) – $\Delta_f H°$(propane) = 12.7-(−25) = +37.7 kcal mol^{-1}. The heat of formation of *trans*-1,2-dimethylcyclopropane can thus be derived from $\Delta_f H°$(pentane) by adding 37.7 kcal mol^{-1}. This gives $\Delta_f H°$(*trans*-1,2-dimethylcyclopropane) = −35.1 + 37.7 = 2.6 kcal mol^{-1}. Homolysis of cyclopropane gives cyclopropyl radical, and a second homolysis gives triplet cyclopropylidene in which the two radicals ignore each other (a triplet 1,1-diradical with no interaction between the two unshared electrons). Table 1.A.7 gives $DH°(c\text{-}C_3H_5^{\cdot}/H^{\cdot}) = 106.3$ kcal mol^{-1}. Table 1.A.8 give $\Delta_r H°(Me^{\cdot} \rightarrow CH_2{:} + H^{\cdot}) = 110.3$ kcal mol^{-1}. Thus, a first estimate $\Delta_f H°_{est}$(triplet **C**) = 2.6 + 106.3 + 110.3 − 104.2 = 115 kcal mol^{-1}. Quantum calculation proposed $\Delta_f H°_{quant}$(singlet cyclopropylidene) = 120.7 kcal mol^{-1} and $\Delta_f H°_{quant}$(triplet cyclopropylidene) = 129.8 kcal mol^{-1}. Considering the 2,3-dimethyl substitution (correction of 2(−10.12 + 4.91 – 1.5) kcal mol^{-1}) one estimates $\Delta_f H°_{quant}$(singlet *trans*-2,3-dimethylclopropylidene) = 106.4 kcal mol^{-1} and $\Delta_f H°_{quant}$(triplet *trans*-2,3-dimethylcyclopropylidene) = 115.8 kcal mol^{-1}. The latter estimate is similar to that obtained above by the two successive C–H homolysis of *trans*-1,2-dimethylcyclopropane. One calculates $\Delta_f H°$(triplet **C**) – $\Delta_f H°$((−)-**A**) = 115–31.1 = 83.9 kcal mol^{-1}, a value much larger than $\Delta^{\ddagger}H((-)\text{-}\mathbf{A} \rightarrow (+)\text{-}\mathbf{A}) = 45.0$ kcal mol^{-1}. Using the estimate for the singlet *trans*-2,3-dimethylcyclopropylidene, one obtains $\Delta_f H°$(singlet **C**) – $\Delta_f H°$((−)-**A**) = 106 – 31.1 = 75 kcal mol^{-1}. This makes hypothesis II (equilibrium with cyclopropylidene species **C** or/and **D**) incompatible with the experimental data. For both mechanism I and II activation entropies are expected to be near zero, as observed. The thermal rearrangement of allene into methylacetylene has an experimental activation enthalpy of 60 to 64 kcal mol^{-1}. The process implies a first step that is a (1,2)-hydrogen shift forming the carbene vinylmethylene (CH$_2$=CH—CH:). The second step is an electrocyclic ring closure into cyclopropene. The third step isomerizes cyclopropene into methylvinyl carbene (MeCH=C:), which finally undergoes a (1,2)-hydrogen shift to generate methylacetylene. Quantum calculations suggest that cyclopropylidene is not an intermediate of the thermal rearrangement of allene into methylacetylene since $\Delta_r H(CH_2{=}C{=}CH_2 \rightleftarrows \text{cyclopropylidene}) = 62$ kcal mol^{-1} is calculated [1].

Problem 3.6 Give possible mechanisms for the following isomerizations and estimate their activation entropies.

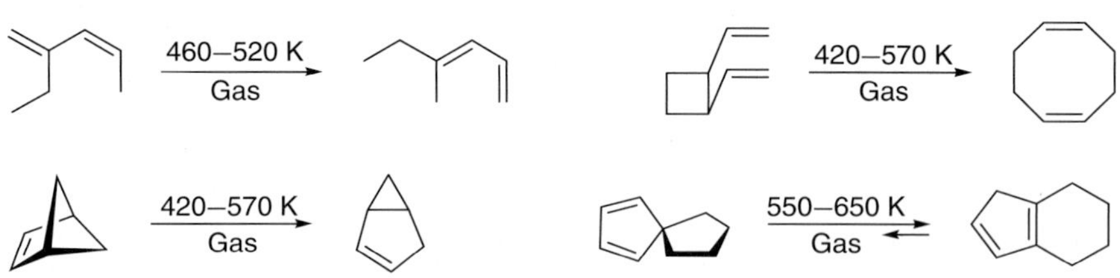

Answer

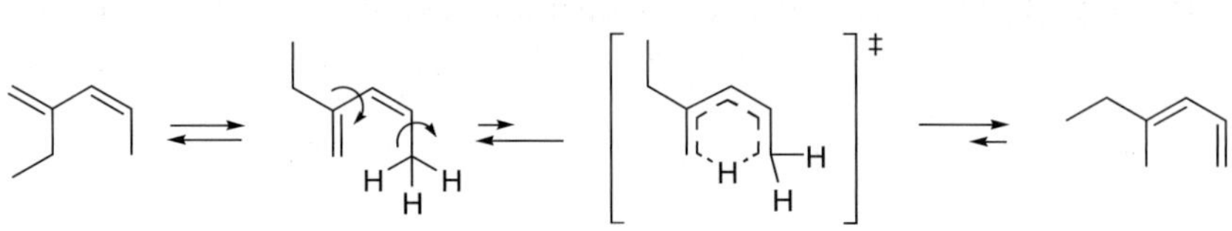

(i)

This is a concerted rearrangement involving a (1,5)-sigmatropic shift of hydrogen (Section 5.5.6). To reach the transition state of this reaction, two freely rotating σ(C—C) bonds must be blocked, which costs twice −5 eu. Thus, $\Delta^{\ddagger}S$ = c. −10 eu.

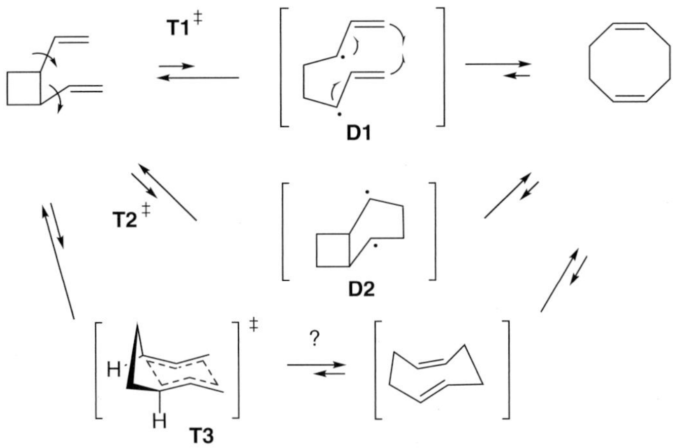

(ii)

This is a formal Cope rearrangement or a (3,3)-sigmatropic rearrangement (Section 5.5.9.6). The Cope rearrangement of alicyclic hexa-1,5-dienes follows concerted mechanisms adopting chair-shaped six-membered transition states. Such a concerted mechanism cannot be excluded in our case since the cyclobutane unit can adopt a puckered conformation as required by transition state **T3**. Nevertheless, **T3** is a bicyclo[4.2.0]octane system containing one partial (*E*)-C=C double bond, which is costly in terms of enthalpy (Bredt's rule). Such a mechanism implies the blockage of two σ(C—C) bonds and thus will have $\Delta^{\ddagger}S \cong -10$ eu. The same activation entropy is expected also for the Grob mechanism, which involves the formation of diradical **D2** in the rate-determining step. In the case of the Cope rearrangement of unsubstituted hexa-1,5-diene, which follows a concerted mechanism with an activation enthalpy of 37.2 kcal mol^{-1}), the Grob mechanism (formation of cyclohexa-1,4-diyl diradical intermediate) requires an activation enthalpy only 3 kcal mol^{-1} higher. If one considers the ring strain of transition **T3,** as discussed above, one must maintain the possibility of the Grob mechanism in the rearrangement of *cis*-1,2-divinylcyclobutane. The Cope rearrangement of hexa-1,5-diene does not occur through homolysis into two allyl radicals as it would imply an energy barrier higher than 56 kcal mol^{-1} (Figure 3.16). In the case of the rearrangement of *cis*-1,2-divinylcyclobutane the formation of diradical **D1** requires a much lower activation enthalpy because of the liberation of the cyclobutane ring strain (26.5 kcal mol^{-1}, Table 2.A.3) when going from the starting diene to **D1**, a diradical that is further stabilized by twice 2 kcal mol^{-1} because of dialkyl substitution with respect to two allyl radicals. Thus, $\Delta^{\ddagger}H$(*cis*-1,2-divinycyclobutane → **D1**) > 56 − 26.5 − 4 ≅ 25.5 kcal mol^{-1}. This mechanism is plausible in our case at $T > 420\,\text{K} = 147\,°\text{C}$ (see Figure 3.12).

The next reaction is a (1,3)-sigmatropic shift of an alkyl group. Irrespective of whether it is a concerted process or a two-step process involving a diradical intermediate, its activation entropy must be close to zero.

(iii)

The following rearrangement is a (1,5)-sigmatropic shift of an alkyl group. If the reaction is an one-step, concerted process, its entropy of activation should be close to zero since no degree of rotation is lost or gained in the transition state. However, if the reaction should follow a two-step mechanism with the formation of a diradical intermediate the activation entropy should be slightly positive, not because the alkyl chain can rotate freely in the diradical, but because of the C—C bond elongation in the transition state between the starting spiro-compound and the diradical.

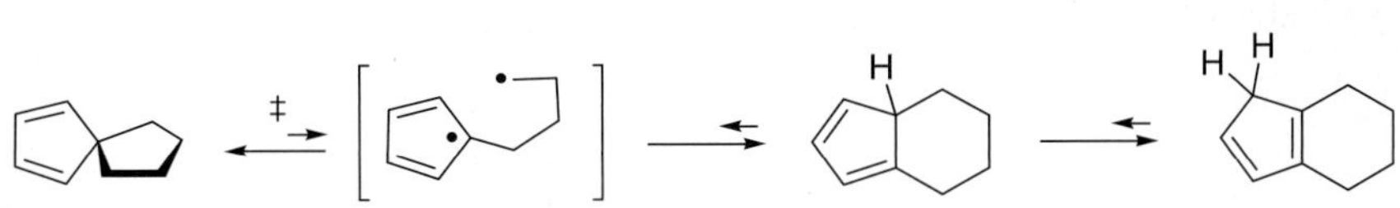

(iv)

Problem 3.7 What is the probable mechanism of the Claisen rearrangement of allyl vinyl ether into pent-4-enal for which $\Delta^{\ddagger}H = 25.7 \pm 0.3$ kcal mol^{-1}, $\Delta^{\ddagger}S = -14.1 \pm 0.7$ eu have been determined between 118 and 153 °C in deuterated benzene (see Section 5.5.9.2). Estimate the heat of this reaction.

Answer

The prototype of the Claisen rearrangement is a [3,3]-sigmatropic rearrangement. It follows a concerted mechanism in which the new C—C bond forms in concert with the breakage of the allylic C—O bond. This hypothesis is supported by the relatively low activation enthalpy and the negative activation entropy measured for this reaction. The reaction is exothermic by c. −16 kcal mol^{-1}. The standard heat of formation of allyl vinyl ether is derived from that of *n*-propyl vinyl ether, and that of pent-4-enal from the standard heat of formation of pentanal (Table 1.A.4, see below), considering an exothermicity of −30 kcal mol^{-1} for the hydrogenation of monoalkylalkenes.

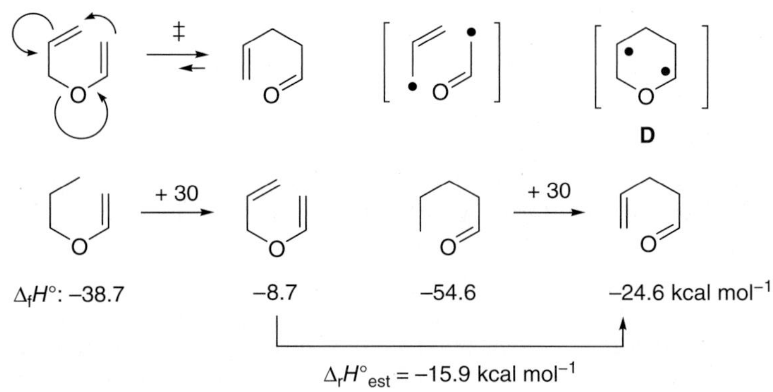

A dissociative mechanism involving the formation of allyl and vinyloxy radical as intermediate can be rejected as it would require a positive activation entropy and a much higher activation enthalpy. Indeed, using $DH°(\mathrm{O{=}CH{-}CH_2^{\bullet}/H^{\bullet}}) = 94.8 \pm 2$ kcal mol^{-1} (Table 1.A.9), $\Delta_fH°(\mathrm{MeCHO}) = -40.8 \pm 0.35$ kcal mol^{-1} (Table 1.A.4) and $\Delta_fH°(\mathrm{H^{\bullet}}) = 52.1 \pm 0.001$ kcal mol^{-1} (Table 1.A.7) one estimates $\Delta_fH°_{est}(\mathrm{O{=}CH{-}CH_2^{\bullet}}) = 1.9 \pm 2.4$ kcal mol^{-1}. Table 1.A.7 gives $\Delta_fH°(\mathrm{CH_2{=}CH{-}CH_2^{\bullet}}) = 40.9 \pm 0.7$ kcal mol^{-1}; this leads to $\Delta_rH°$(allyl vinyl ether ⇄ allyl radical + vinyloxy radical) = 40.9(±0.7) + 1.9(±2.4) − (−8.7(±?)) ≅ 51.5 ± 3.5 kcal mol^{-1}.

An associative mechanism generating diradical **D** as intermediate is consistent with the negative activation entropy measured. It is not possible considering the heat of formation of this diradical, which is given by $\Delta_fH°(\mathbf{D}) = \Delta_fH°(\text{tetrahydropyran}) + DH°(i\text{-Pr}^{\bullet}/\mathrm{H^{\bullet}}) + DH°(\mathrm{MeOCH_2^{\bullet}/H^{\bullet}})$ + correction for secondary alkoxyalkyl radical $-DH°(\mathrm{H^{\bullet}/H^{\bullet}})$ = −53.5(±0.25) (Table 1.A.4) + 99.4(±1.0) + 93.1(±1.0) − 2.0 − 104.2 (Table 1.A.7) ≅ 32.8 ± 2.3 kcal mol^{-1}. This leads to $\Delta_rH°$(allyl vinyl ether ⇄ diradical **D**) = 32.8(±2.3) − (−8.7) ≅ 41.5 ± 3.0 kcal mol^{-1}, which is significantly larger than the measured $\Delta^{\ddagger}H = 25.7 \pm 0.3$ kcal mol^{-1}.

Problem 3.8 What are the major products of the gas phase pyrolysis of allyl methyl amine and of diallyl ether?

Answer

A retro-ene reaction involving a concerted (1,5)-hydrogen shift is observed for the pyrolysis of allyl methyl amine in the gas phase at high temperature ($E_a = 43.4 \pm 1.6$ kcal mol^{-1}, $\log A = 11.4 \pm 0.6$). The reaction produces propene and methanimine. A similar reaction is observed for the pyrolysis of bis(allyl) ether ($E_a = 40.9 \pm 0.35$ kcal mol^{-1}, $\log A = 11.8 \pm 0.13$). A two-step process involving the formation of allyl and allyloxy radicals would be associated with a higher activation enthalpy: $DH°$(allyl-*O*-allyl ⇄ $\mathrm{CH_2C{=}CH{-}CH_2^{\bullet} + CH_2{=}CH{-}CH_2{-}O^{\bullet}}$) = 40.9(±0.7) + 19.5(±1.0) − (−10.2(±1.0)) ≅ 52.6 ± 3.0 kcal mol^{-1} ($\Delta^{\ddagger}H = E_a - RT$). The observation of logA <13 ($\Delta^{\ddagger}S < 0$) for both reactions confirms that they involve transition states that blocks the rotation of some sigma bonds, as required for concerted mechanisms in which the hydrogen atom transfer occurs in concert with the σ(C—N) and σ(C—O) bond breakages.

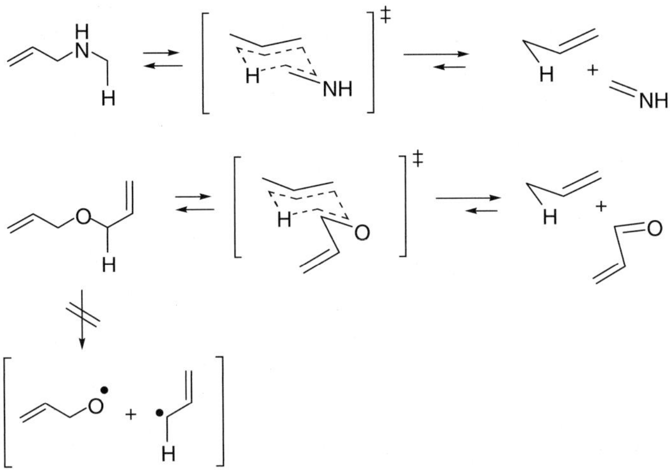

Problem 3.9 The ^{1}H-NMR spectrum of the dication obtained by dissolving 4-(2-methylpropyl)-2,6-dimethylocta-2,6-diol in magic acid and SO_2ClF at −120 °C shows the following signals at δ_H: 3.20 (6H), 2.88 (18H), 2.88 (1H), −0.78 ppm (1H). Propose a structure for this dication.

Answer

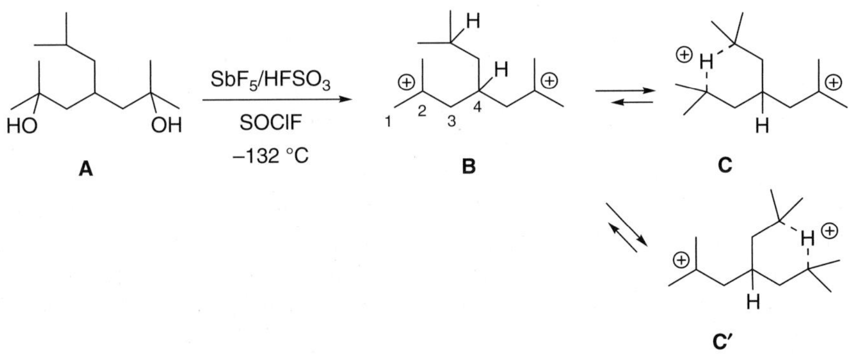

Diol **A** is doubly ionized into a tertiary alkyl dication **B**. The 2-methylpropyl group of this dication permits the formation of an intramolecular μ-hydrido species **C** in a reversible manner. At −120 °C dication **C** equilibrates in a fast process with dications **B** and **C′**. This renders all the six methyl groups equivalent (δ_H = 2.88 ppm) on the NMR time scale, as well as the three methylene groups (δ_H = 3.20 ppm). The signal with the negative chemical shift is typical of a μ-hydrido bridging hydrogen atom. The fact that δ_H = −0.78 ppm and not <−3 ppm as in dication **32** (4,4-bis(2-methypropyl)-2,6-dimethylhepta-2,6-diyl dication, δ_H = −4.89 ppm, Scheme 3.7) suggests that cation **B** has a similar stability as **C** and **C′** and contributes to the spectral properties of this dication. The signal at δ_H = 2.88 ppm is assigned to the proton at C(4) of dication **B** ⇄ **C** ⇄ **B** ⇄ **C′**. The significant contribution of bridged dications **C** and **C′** in equilibrium with **B** results from the Thorpe–Ingold effect (Section 3.5.1) introduced by the methyl substituents.

Problem 3.10 On treating nona-1,8-dien-5-one (**A**) with the molybdenum complex **C** (catalyst for metathesis of alkenes, Section 8.6.2) a polymer is obtained together with ethylene. Under similar conditions 4,4,6,6-tetramethylnona-1,8-dien-5-one generates 2,2,7,7-tetramethylcyclohept-4-en-1-one (**B**) in 95% yield. Explain the difference in behavior between **A** and **B**.

C:

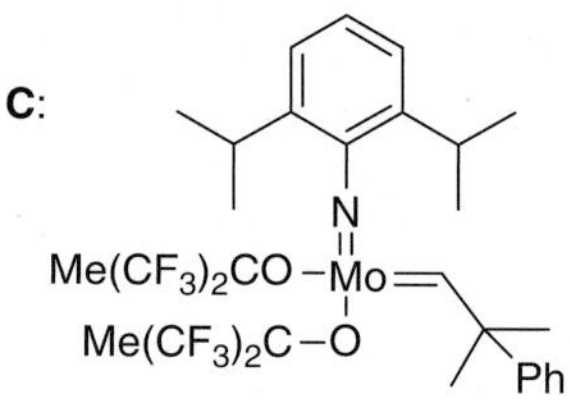

Answer

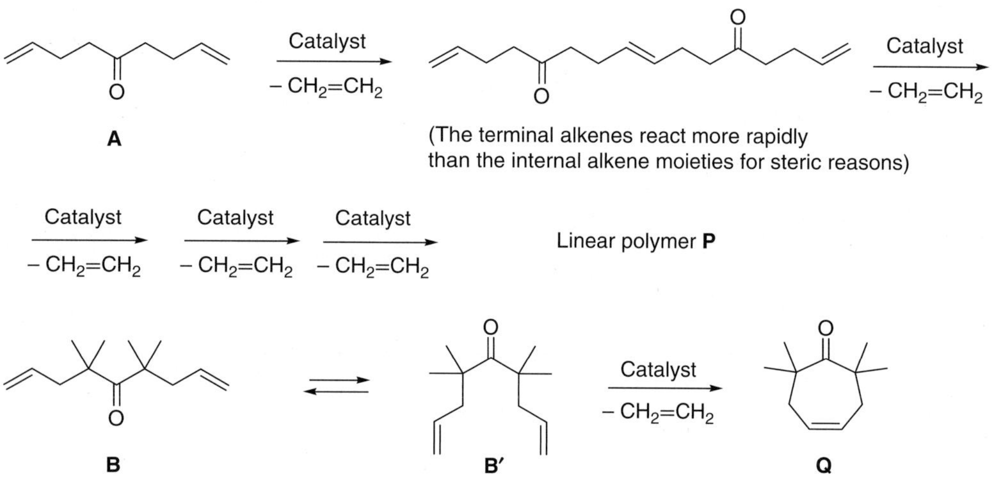

Because of the double *gem*-dimethyl substitution the pseudo cyclic conformer **B′** of dienone **B** is highly populated, thus permitting an intramolecular reaction between its two olefinic moieties, which competes with the intermolecular alkene metathesis reaction that would lead to a polymer. This leads to the production of 2,2,7,7-tetramethylcyclohept-4-enone (**Q**).

Problem 3.11 Explain why enone **21** (Scheme 3.6) refuses to equilibrate with **25** and **26**, whereas **D** equilibrates with **E** at 80 °C in benzene.

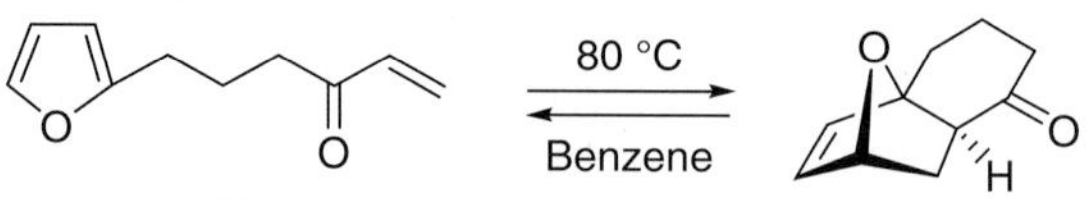

D E

Answer

The intramolecular Diels–Alder reaction **21** ⇄ **25** + **26** is not observed ($K^{393K} \ll 1$, $\Delta_r G^{393K} > 0$ kcal mol^{-1}) because it is not exothermic enough to compensate at 120 °C for the entropy cost (blockage of the rotation of 4 σ(C—C) bonds, $\Delta^{\ddagger}S$ c. −20 eu). In the case of reaction **D** ⇄ **E**, the reaction is more exothermic because a six-membered ring is formed in the cycloadduct **E**. In this case, $K^{353K} \geq 1$ and $\Delta_r G^{353K} \leq 0$ kcal mol^{-1}. Cycloadducts **25** and **26** are more strained than **E** because they contain an extra five-membered ring instead of a six-membered ring.

Problem 3.12 Ethylenetetracarbonitrile (tetracyanoethylene) adds to 1,1-dimethylbutadiene competitively in the [π^4+π^2] and [π^2+π^2] mode giving 3,3-dimethylcyclohex-4-ene-1,1,2,2-tetracarbonitrile and 3-(2-methylprop-1-enyl) cyclobutane-1,1,2,2-tetracarbonitrile, respectively. In toluene the activation volume of the [π^2+π^2]-cycloaddition is more negative than that of the [π^4+π^2]-cycloaddition. In butyronitrile reverse results are found. Explain.

Answer

The Diels–Alder reaction ([π^4s+π^2s]-cycloaddition) undergoes a transition state that is less polar than that of the [π^2+π^2]-cycloaddition. Electrostriction leads to more negative $\Delta^{\ddagger}V$ values for polar transition states. Electrostriction is the largest with the less polar solvent (PhMe) than with the more polar solvent (butyronitrile);

thus on increasing pressure, one increases the proportion of the $[\pi^2+\pi^2]$-cycloadduct more in toluene than in butyronitrile [2–6].

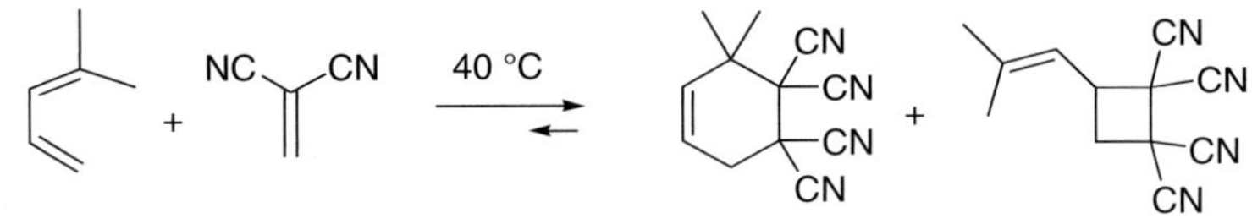

In toluene:	1 bar	3.4	:	1
	7 bars	2.3	:	1 $\Delta\Delta^{\ddagger}V = 3.5\ cm^3\ mol^{-1}$
In butyronitrile:	1 bar	0.15	:	1
	7 bars	0.43	:	1 $\Delta\Delta^{\ddagger}V = -4.0\ cm^3\ mol^{-1}$

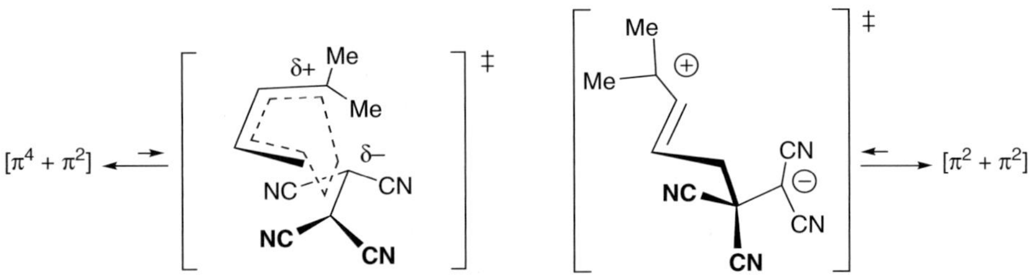

Problem 3.13 Explain the change of $\Delta^{\ddagger}V$ values from 16, 7.1 to $-17.9\ cm^3\ mol^{-1}$ for the hydrolysis of 2,4-dinitrophenyl 4-hydroxybenzoate at pH 8.0, 10.1, and 12.5 (at 29 °C).

Answer

At high pH values $[HO^-]$ is relatively high. It is the base that can deprotonate the phenol but also the nucleophile that can add to the carboxylic group. Whether the phenol is deprotonated or not there is enough HO^- anion to add to the carboxylate group in the slowest step of the reaction ($B_{Ac}2$ mechanism). This condensation is the rate-determining step, which is accompanied by a negative activation volume, $-17.9\ cm^3\ mol^{-1}$ in our case (Ar = 2,4-dinitrophenyl).

At pH = 12.5

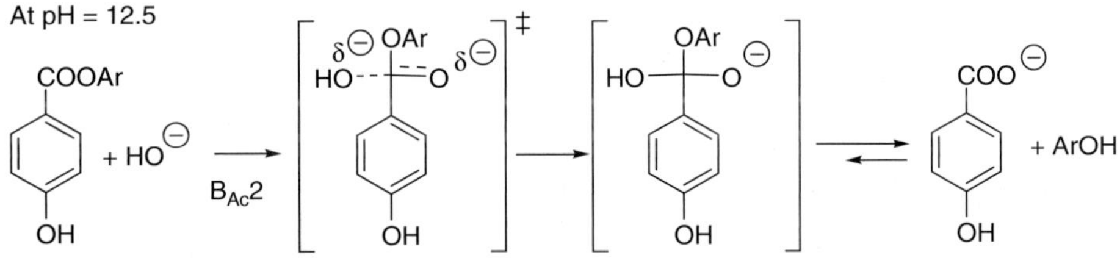

Relatively high concentration of hydroxide anion: the base and good nucleophile

(i)

At pH 8, $[HO^-]$ is very low; the base that deprotonates the phenol ($pK_a \cong 8$) is water. The phenolate intermediate eliminates the aryloxy anion, a good leaving group. This is the rate-determining step that corresponds to a fragmentation and that is accompanied by a positive activation volume of $16.5\ cm^3\ mol^{-1}$ in our case.

At pH = 8.0

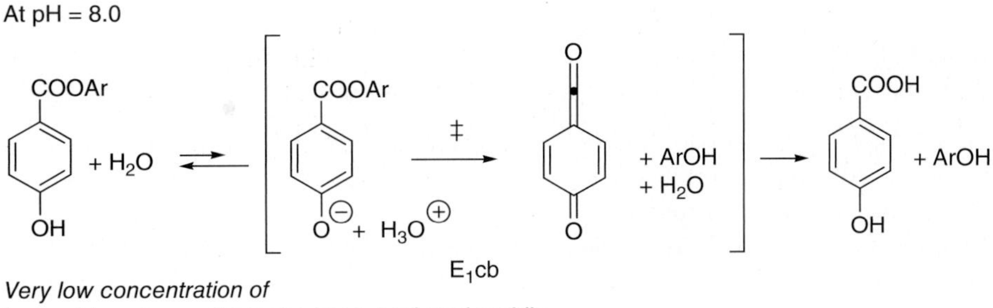

Very low concentration of hydroxide anion; water is the base, bad nucleophile

(ii)

Problem 3.14 Evaluate whether the Cope rearrangement of *rac*-(*E,E*)- and *meso*-(*E,E*)-1,3,4,6-tetraphenylhexa-1,5-diene can occur by a dissociative mechanism with the formation of 1,3-diphenylallyl radicals, considering the following activation parameters for their interconversion: $\Delta^{\ddagger}H = 30.7 \pm 0.2\,\text{kcal mol}^{-1}$, $\Delta^{\ddagger}S = 2.1 \pm 0.4$ eu, and $\Delta^{\ddagger}V = 13.5 \pm 0.1\,\text{cm}^3\,\text{mol}^{-1}$ between 77 and 115 °C. Resolution of *rac*-(*E,E*)-1,3,4,6-tetraphenylhexa-1,5-diene can be done by chiral chromatography. This allowed one to observe a fast racemization at 40–65 °C with $\Delta^{\ddagger}H = 21.3 \pm 0.1\,\text{kcal mol}^{-1}$, $\Delta^{\ddagger}S = -13.2 \pm 0.3$ eu, and $\Delta^{\ddagger}V = -7.4 \pm 0.4\,\text{cm}^3\,\text{mol}^{-1}$. What is the probable mechanism of this racemization?

Answer

Tetraphenyl substituted hexa-1,5-diene *meso*-**A** is equilibrated at temperature ranging from 77 to 115 °C with the racemic *threo*-**A** (1 : 1 mixture of (*S,S*)-**A** and (*R,R*)-**A**). The phenyl substituent effect on the relative stability of benzyl radical compared with methyl radical amounts to $DH°(\text{CH}_3^{\cdot}/\text{H}^{\cdot}) - DH°(\text{Ph–CH}_2^{\cdot}/\text{H}^{\cdot}) = 104.7(\pm 0.2) - 89.7(\pm 1.3) = 15.0 \pm 1.5\,\text{kcal mol}^{-1}$. The phenyl substituent effect on allyl radical must not be larger than $7.5\,\text{kcal mol}^{-1}$ as the spin is delocalized equally onto two carbon centers. Thus, the four phenyl groups of diene **A** are expected to make $DH°(\mathbf{B}^{\cdot}/\mathbf{B}^{\cdot}) = DH°(\text{allyl}^{\cdot}/\text{allyl}^{\cdot}) = 56.2–4\,(7.5) +$ correction for the stabilization of **A** by π-conjugation $= 26.2 + 4\,\text{kcal mol}^{-1} \approx 30.2\,\text{kcal mol}^{-1}$, a value similar to the enthalpy of activation $\Delta^{\ddagger}H = 30.7 \pm 0.2\,\text{kcal mol}^{-1}$ measured for the thermal isomerization *meso*-**A** ⇄ *threo*-**A**. Thus, this reaction might generate two radicals **B**$^{\cdot}$ in its rate-determining step. This hypothesis proposed to explain the non-Cope rearrangement is supported by the observation (ESR, electron spin resonance) of free radical **B**$^{\cdot}$ on heating *meso*-**A** in diphenyl ether to 251 °C (equilibrium **A** ⇄ 2**B**$^{\cdot}$).

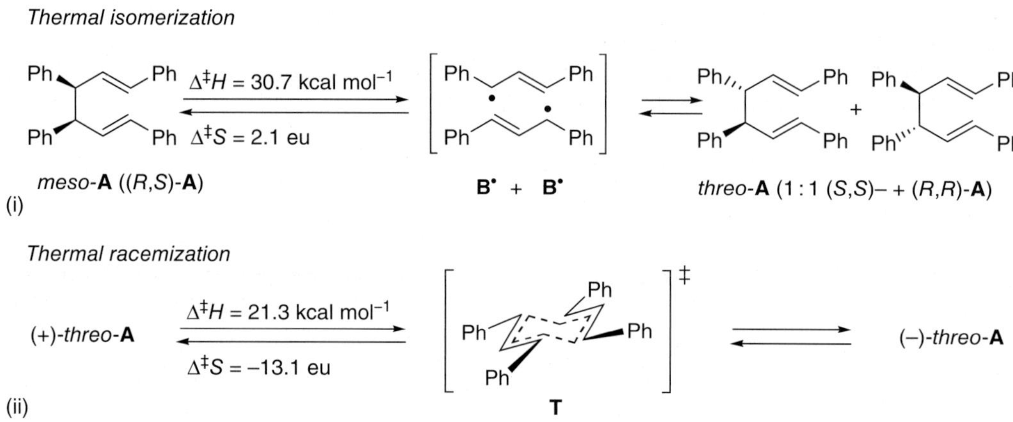

Enantiomerically enriched (+)-*threo*-**A** is racemized rapidly at temperature from 40 to 65 °C with activation parameters consistent with a concerted, one-step mechanism implying a C_s-symmetrical transition structure **T**.

Problem 3.15 What occurs to the polycyclic system **49** or **50** upon heating above 160 °C? Propose a synthesis for **49** (see Section 5.6.3).

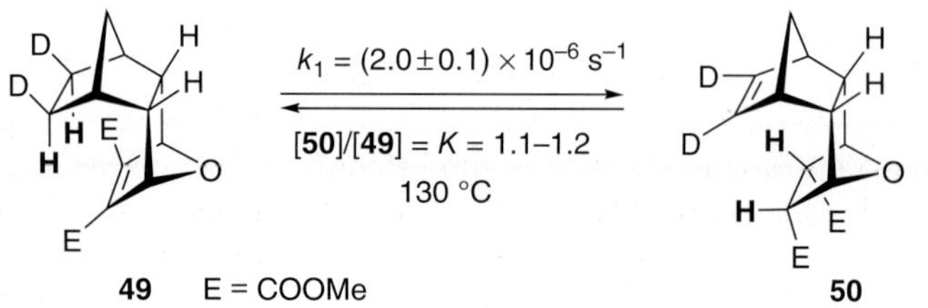

(3.91)

Answer

A retro-Diels–Alder reaction ([4+2]-cycloreversion) provides bicyclo[2.2.1]hept-2-ene (norbornene) and dimethyl furane-3,4-dicarboxylate. Alternative retro-Diels–Alder reaction furnishing cyclopentadiene and dimethyl 7-oxabicyclo[2.2.1]hept-5-ene-2,3-dicarboxylate is not observed. The latter fragmentation is more endothermic, thus slower than the former retro-Diels–Alder reaction. This arises from the aromaticity of the furan derivative that forms as product.

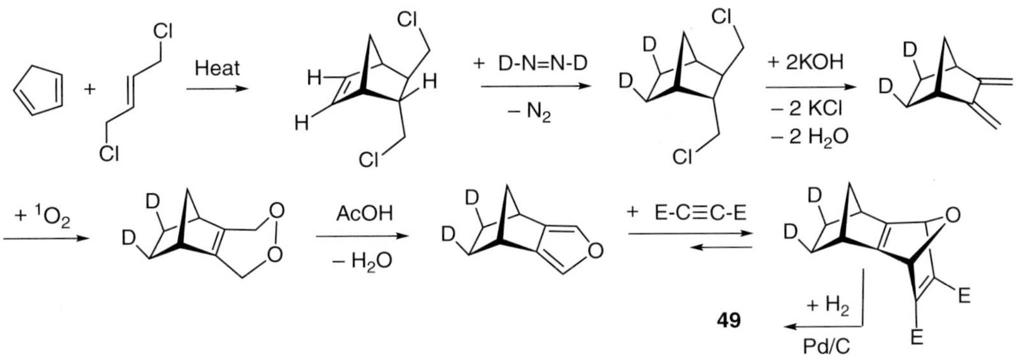

49

Problem 3.16 The degenerate Cope rearrangement of bullvalene that exchanges all the proton signals in its ^{1}H-NMR spectrum and all the ^{13}C signals in its ^{13}C-NMR spectrum at room temperature has activation parameters $\Delta^{\ddagger}H = 12.6 \pm 0.05$ kcal mol^{-1} and $\Delta^{\ddagger}S = 0.8 \pm 0.2$ eu (233–413 K). Why is this isomerization so much easier than the Cope rearrangement (3.82) (Section 3.4.3) or that of 1,4-dimethylidenecyclohexane for which $\Delta^{\ddagger}H = 39.0 \pm 1$ kcal mol^{-1}, $\Delta^{\ddagger}S = -13.8 \pm 2$ eu have been determined at 600 K?

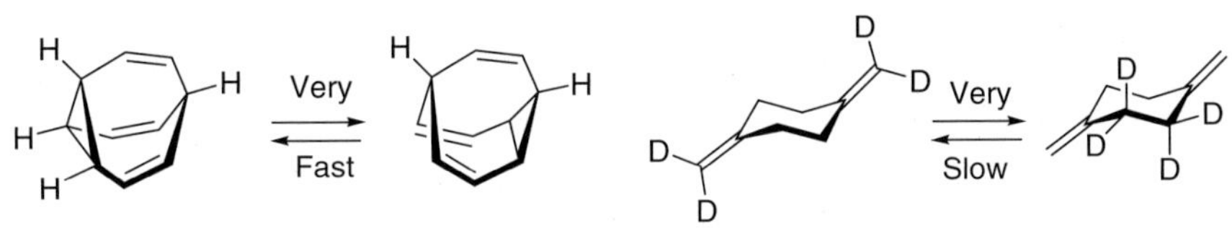

Bullvalene

Answer

Because of the polycyclic structure of bullvalene, the hexa-1,6-diene units it contains are forced to have their terminal carbon centers close to each other. This brings the hexa-1,6-diene units toward the transition state of their concerted Cope rearrangement. This is a case of compressed, or distorted, reactants along their reactive coordinates, which leads to a rate enhancement of the reaction.

Problem 3.17 Devise a synthetic procedure to prepare (−)- and (+)-penta-2,3-diene (see Problem 3.5).

Answer

In the absence of functional "handles" for resolution, chiral allenes of high enantiopurity can be obtained, in theory, by chiral chromatography, provided the chiral stationary phase can interact with the π-system of the allenes. This can be envisaged with columns having an immobile phase charged with transition metal complexes or transition Lewis acids coordinated to enantiomerically pure ligands (silica gel impregnated with salts of silver and iron permits to separate alkene isomers). An alternate method is to apply an enantioselective reaction to the racemic mixture of the allenes to realize a kinetic resolution. This has been proposed first by Caserio in 1968 [7] and by Moore and coworkers in 1973 who have applied enantioselective hydroboration [8]. During the early 1960s, Brown and coworkers introduced the use of optically active alkylboranes as hydroborating agents, which could effect the transformation of an appropriate alkene into an optically active alcohol after oxidative work-up [9, 10]. They also showed that it is possible to achieve a kinetic resolution of racemic alkenes [11].

One of the most useful reagents is the (−)-tetra-3-pinanyldiborane, (−)-$Pn_4B_2H_2$, a reagent also called (−)-tetraisopinocampheylborane. It is obtained by reacting 4 equiv. of (+)-α-pinene with 1 mol of diborane. Similarly, reagent (+)-$Pn_4B_2H_2$ is obtained by the reaction of (−)-α-pinene with diborane. In practice, the racemic allene is allowed to react with either (−)-$Pn_4B_2H_2$ or (+)-$Pn_4B_2H_2$ until more than 50% of the allene has reacted; then unreacted chiral allene is distilled off the reaction mixture. This gives 1,3-dimethylallene with moderate enantiopurity.

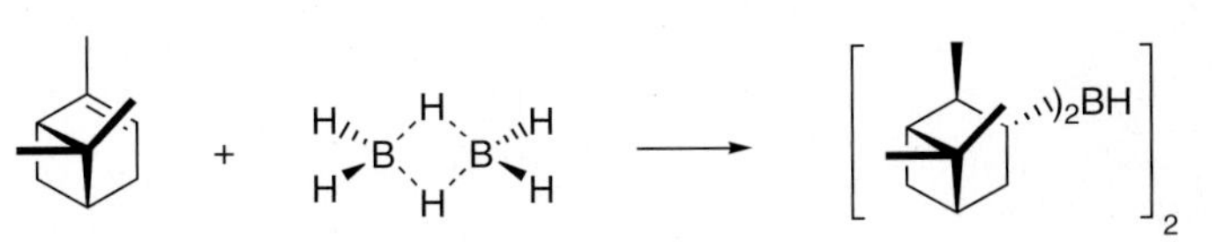

(+)-α-Pinene Diborane (−)-tetra-3-Pinanyldiborane

Corey and Borden in 1969 reported the synthesis of optically active 1,3-bis(*t*-butyl)allene from propargylic derivatives by hydride displacement as outlined below [12].

The method relies upon S_N2' substitution of propargylic esters. It furnishes the chiral allene with high enantiopurity, but suffers from competition by a S_N2 displacement that produces achiral alkynes. The competition depends to some extent upon the nature of the hydride reagent ($LiAlH_4$, AlH_3, $LiAl(OMe)H_3$, $LiAl(OMe)_3H$), but the formation of the acetylene cannot be suppressed.

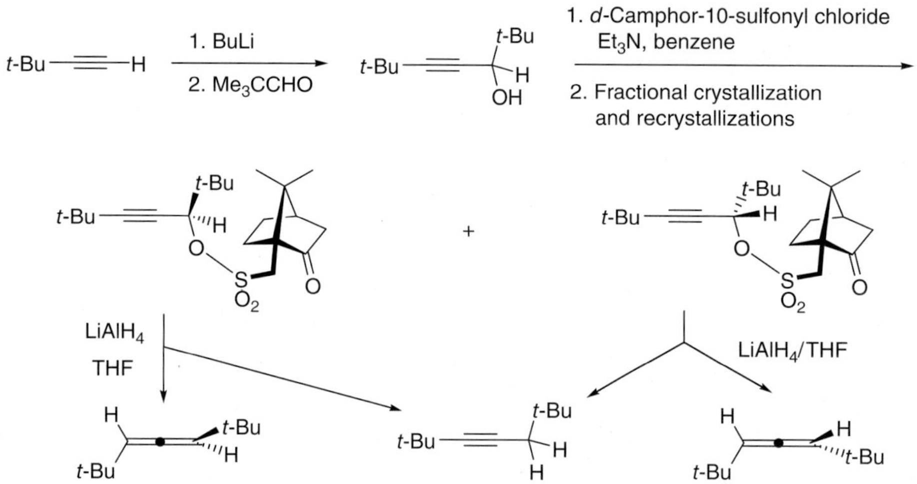

Crabbé and coworkers reported that propargylic acetates undergo chemo- and stereoselective S_N2' displacement reactions with dialkylcuprates affording allenes stereoselectively [13]. Diastereomeric carbamates derived from (*R*)-1-[1-naphthyl]ethyl isocyanate and racemic secondary propargylic alcohols are readily separated by multigram HPLC. The separated carbamates react with dialkylcuprates at low temperature in high yields and with 60–80% ee [14]. For instance,

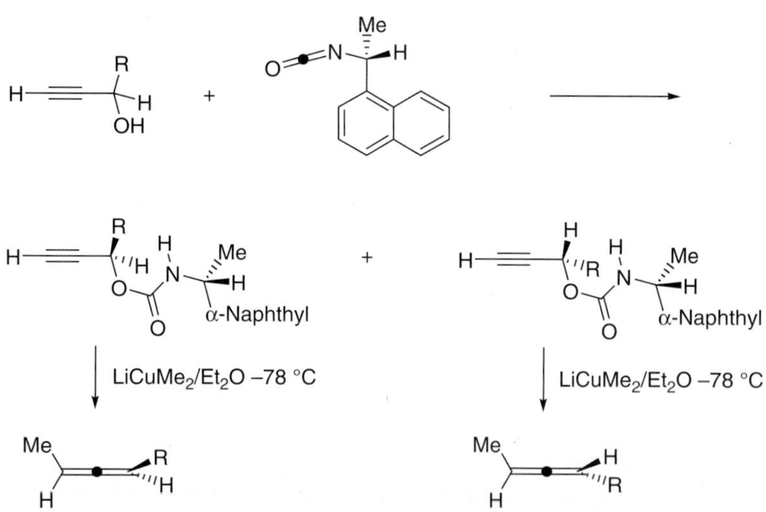

Improved stereoselectivity has been reported for the propargyl methanesulfonates and sulfinates displaced by organocopper(I) reagents of type [(RCuX)MgX·LiX] [15].

Problem 3.18 Explain the enantioselective conversion of (±)-**A** into **P**.

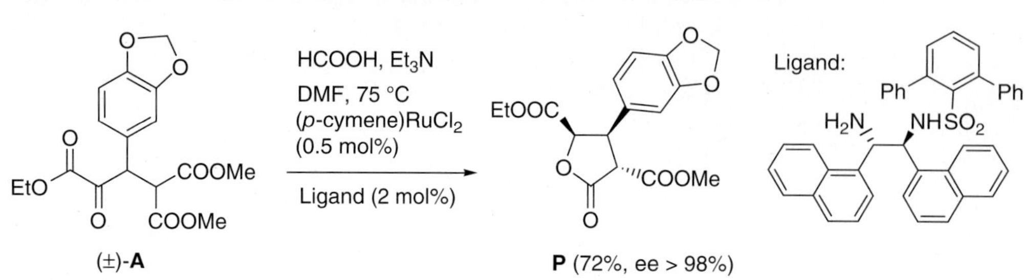

Answer

The hydrogenation catalyst is:

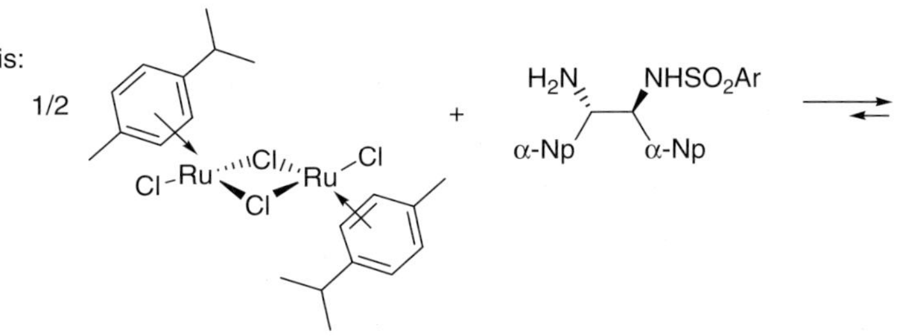

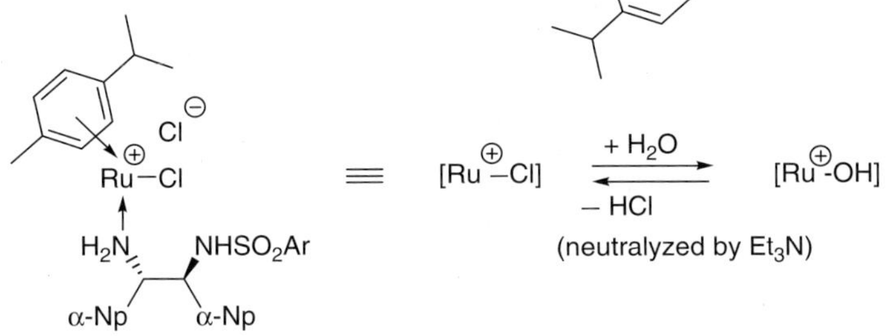

Hydrogenation of the ketone:

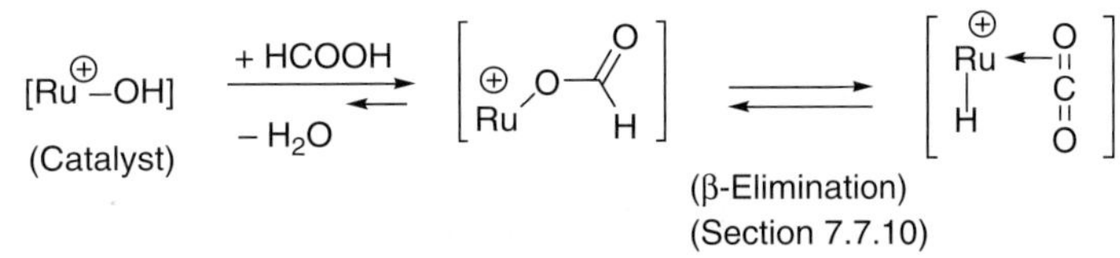

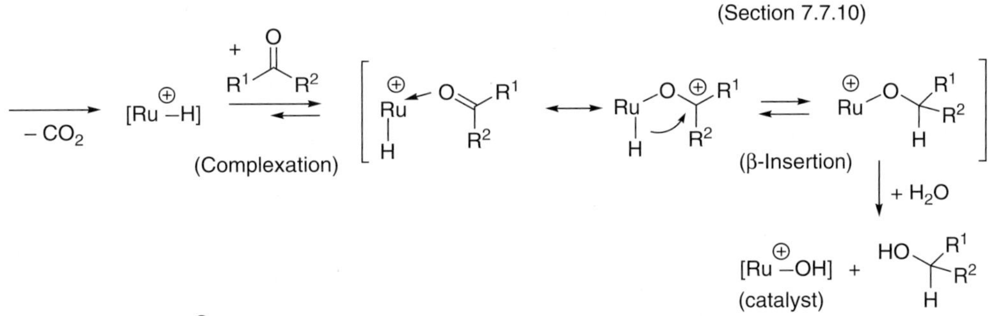

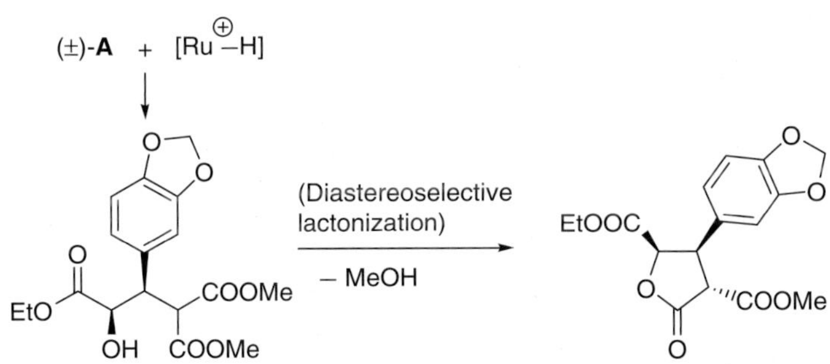

Epimerization of unreacted α-ketoester:

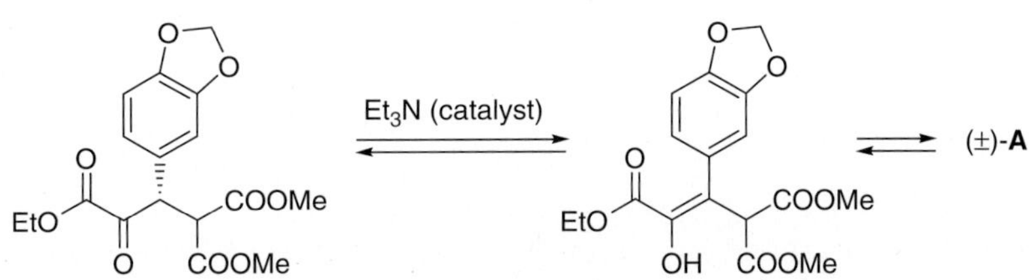

Problem 3.19 Give a possible mechanism for the following dynamic kinetic asymmetric transformation.

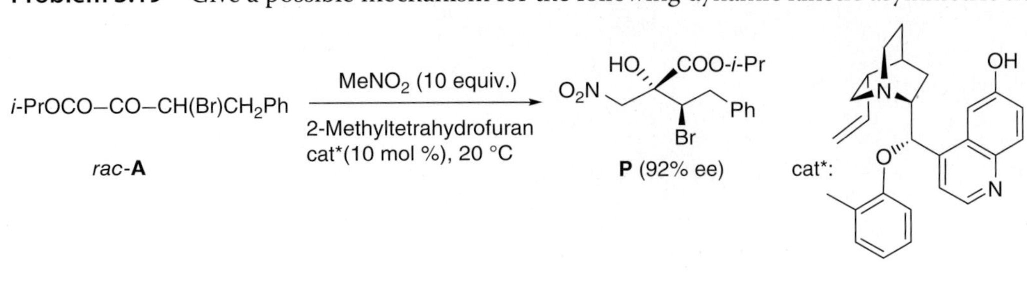

Answer

The catalyst cat* is basic. It deprotonates **A** and racemizes it more rapidly than both (*R*)-**A** and (*S*)-**A** add nitromethane coordinated to cat*(hydrogen-bridging with the phenolic group?). For steric reasons, (*R*)-**A** reacts faster with coordinated nitromethane (Henry reaction, analogous to aldol reaction) than (*S*)-**A**.

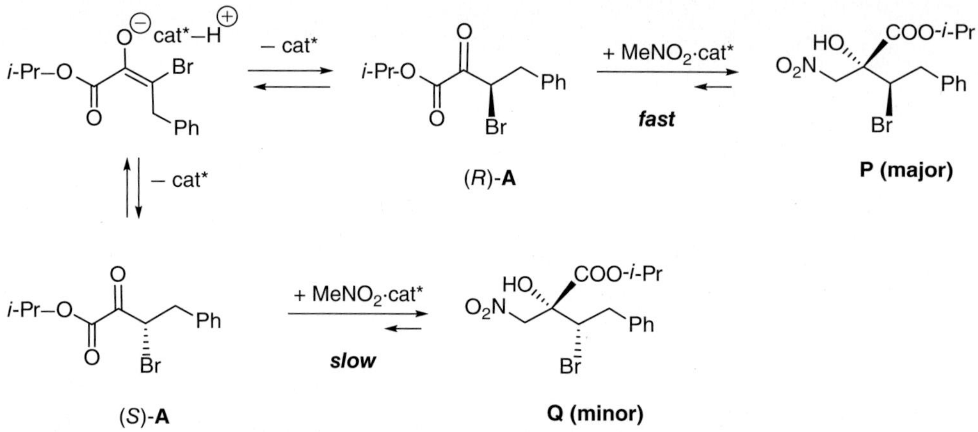

Problem 3.20 Acetophenone reacts with Br_2 in alkaline solutions with the rate law $d[PhCOMe]/dt = -k[PhCOMe][HO^-]$. The reaction shows important KIE k_D/k_H and k_T/k_H as shown with the following Arrhenius activation parameters for $PhCOCH_2T$: $E_a(T) = 15.35\,kcal\,mol^{-1}$, $A(T) = 10^{8.58}\ M^{-1}\ s^{-1}$; for $PhCOCD_3$: $E_a(D) = 14.6\,kcal\,mol^{-1}$, $A(D) = 10^{9.24}\ M^{-1}\ s^{-1}$, and for $PhCOCH_3$: $E_a(H) = 12.85\,kcal\,mol^{-1}$, $A(H) = 10^{8.82}\ M^{-1}\ s^{-1}$. What are the products of this reaction and what is the rate-determining step?

Answer

Br_2 in alkaline solution generates hypobromite anion BrO^- according to $Br_2 + 2MOH \rightleftarrows MBr + MOBr + H_2O$. It reacts with $PhCOCH_3$ according to $PhCOCH_3 + 3BrO^- \rightarrow PhCOO^- + CHBr_3 + 2HO^-$ (the haloform (bromoform in this case) reaction of methyl ketones. The rate-determining step of this reaction is the α-deprotonation of the ketone. This generates an enolate intermediate that reacts rapidly with Br_2, giving first the mono α-bromoketone. Then the process occurs two more times rapidly (increased acidity of α-bromoketone with respect to PhCOMe, and so forth).

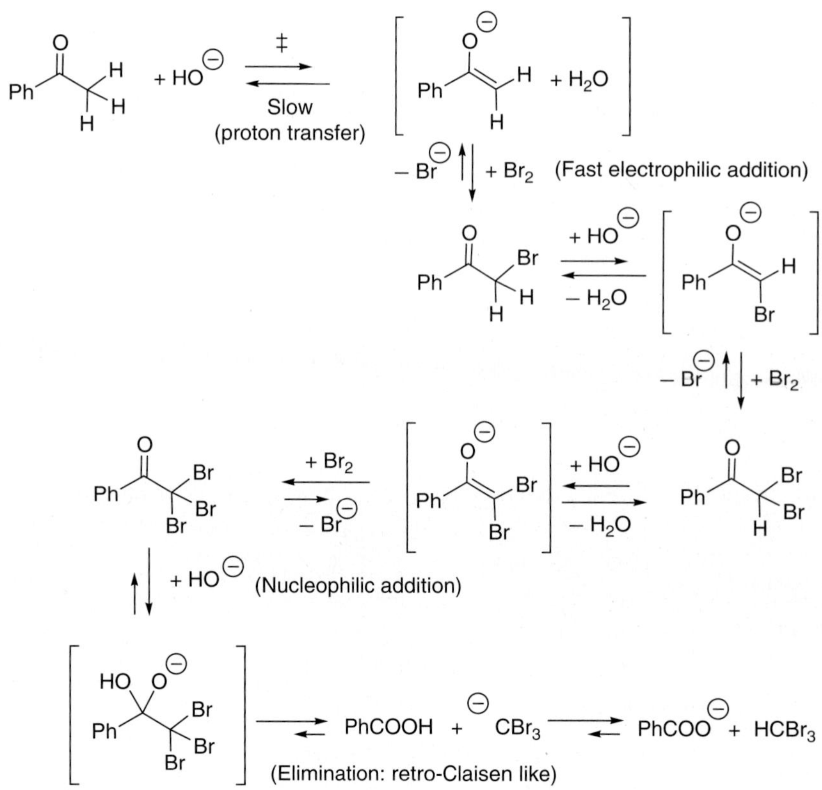

Problem 3.21 In MeCN 9-phenylxanthium tetrafluoroborate (**A**) adds isopropanol giving **B**. Using *i*-PrOD equilibrium isotope effect (EIE) $K_H/K_D = 2.67$ is measured at 62 °C. Explain this EIE. A slower reaction converts **A** into **P** + **Q**. At 62 °C, a KIE $k_H/k_D = 4.4$ is measured using *i*-PrOH and $Me_2C(D)$-OH. Give a possible mechanism for the reduction of **A** by isopropanol.

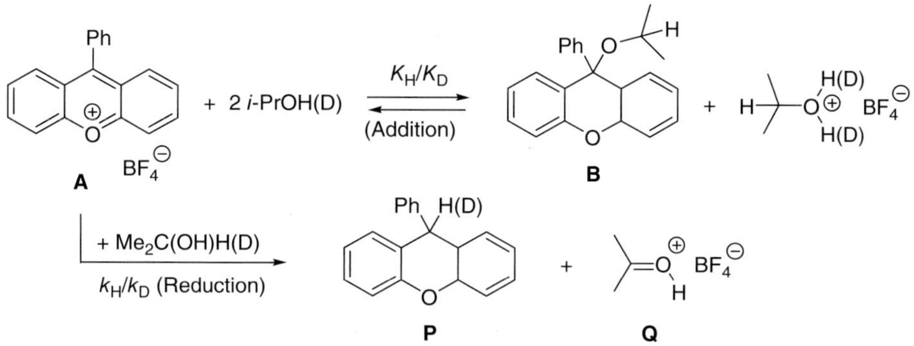

Answer

Reaction **A** + *i*-PrOH → **C** + MeCOMe is a model reaction for the biochemical reduction of NAD^+ into NADH: $NAD^+ + H^- \rightarrow$ NADH. The metabolism of primary alcohols in living organisms involves the hydride transfer from α-H of alcohols to C(4) of the pyridinium moiety of NAD^+ coenzyme.

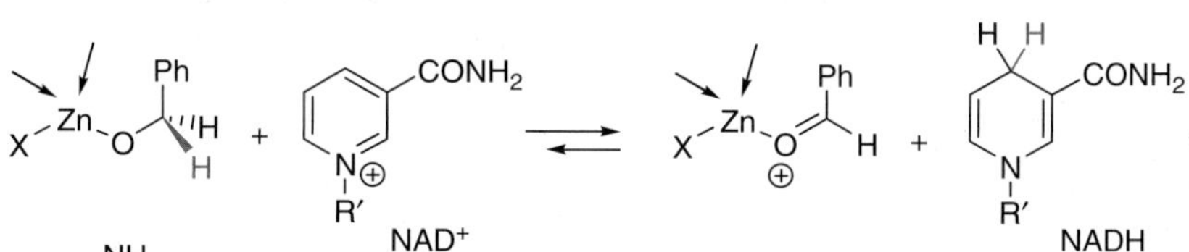

(3.120)

NAD+:

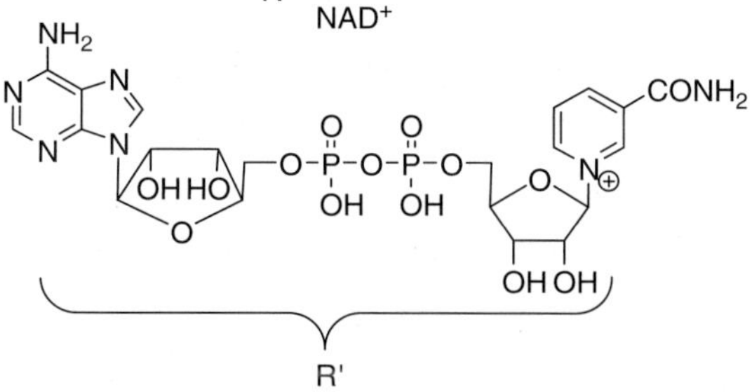

Equilibrium deuterium isotope effect:

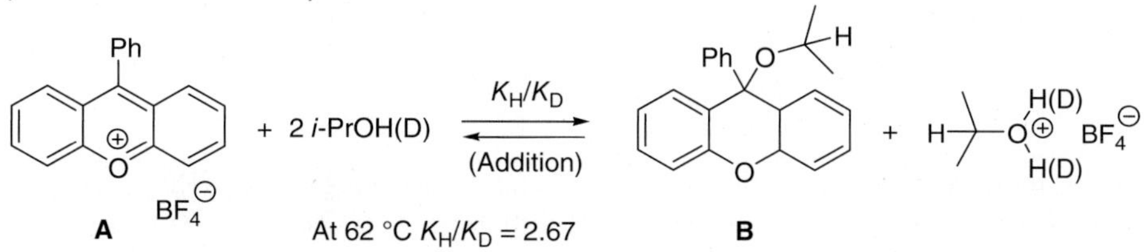

Deuterium prefers the strongest bond. The O–H(D) is stronger in *i*-Pr–O–H than in *i*-Pr–$O^{(+)}H_2(D_2)$; *i*-Pr–O–H(D) is a weaker acid than *i*-Pr–$O^{(+)}H_2(D_2)$.

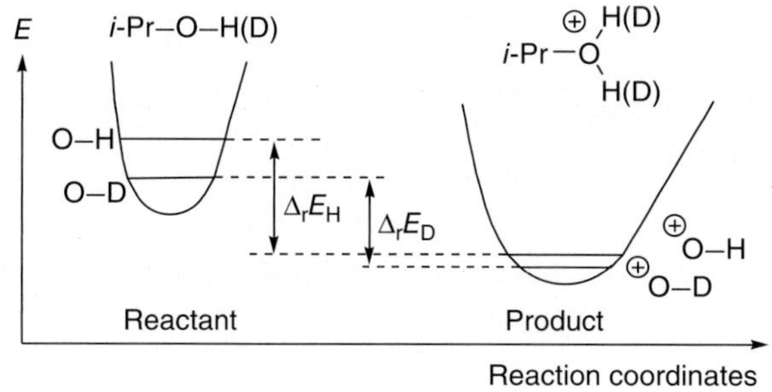

Assuming $\Delta H = \Delta E$

$\Delta_r G = \Delta_r H - T\Delta_r S$

If differential isotope effect on the entropy can be neglected:

$\Delta_r G \sim \Delta_r E$

Thus, $\Delta_r G_H \sim \Delta_r E_H$ and $\Delta_r G_D \sim \Delta_r E_D$

$\Longrightarrow K_H \sim \exp(-\Delta_r E_H/RT)$

$\Longrightarrow K_D \sim \exp(-\Delta_r E_D/RT)$

With $\Delta_r E_H < \Delta_r E_D \Longrightarrow K_H > K_D$

($\Delta_r E_H$ and $\Delta_r E_D < 0$ in our case)

Primary kinetic deuterium isotope effect in the reduction:

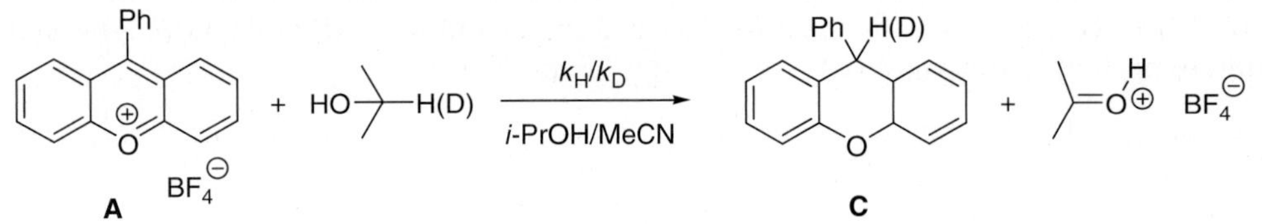

At 62 °C: $k_H/k_D = 4.4$ ⟹

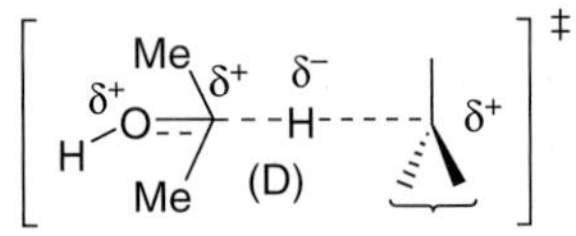

Direct hydride transfer: the C...H(D) bond in the transition state is much weaker than in the reactants.

Secondary kinetic deuterium isotope effects in the reduction:

(i)

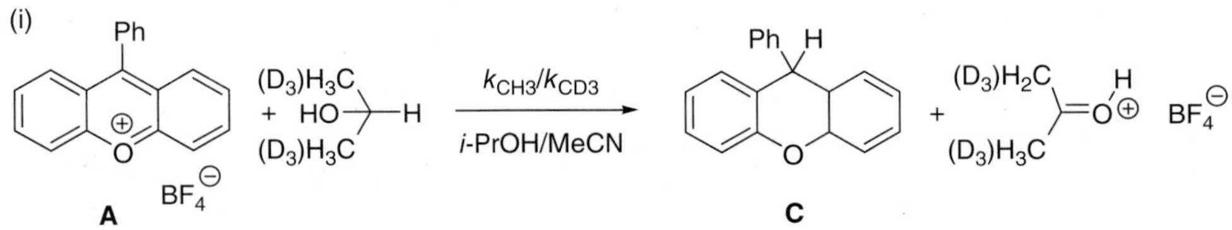

At 62 °C: $k_{CH3}/k_{CD3} = 1.1$ ⟹ Due the formation of carbenium ion character at C(2) of isopropanol in the transition state (see above), hyperconjugation weakens the β-C–H(D) bonds, which results in a normal KIE. This confirms that the hydrogen atom is transferred as an hydride (nucleofugal group).

(ii)

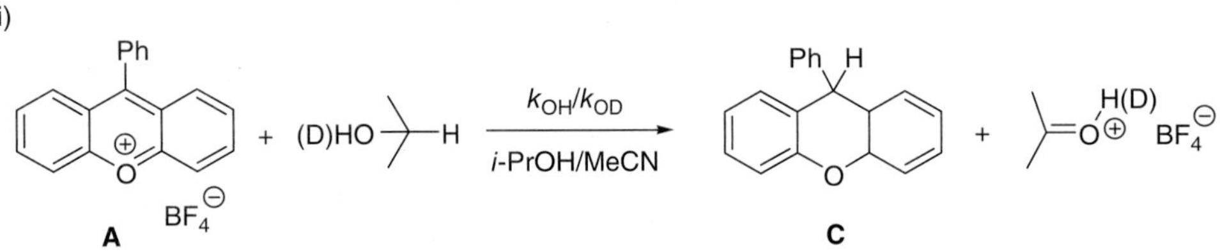

At 62 °C: $k_{OH}/k_{OD} = 0.39$ ⟹ *Inverse solvent kinetic isotope effect*

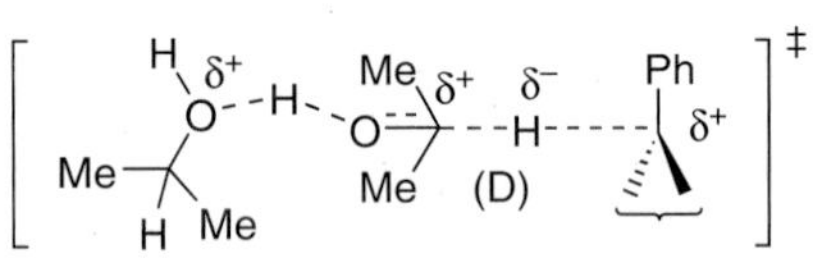

Problem 3.22 Base-induced elimination of HBr from bromocyclohexane gives cyclohexene. KIEs $k_{\alpha\text{-H}}/k_{\alpha\text{-D}} =$ 1.13–1.15 and $k_{\beta\text{-H}}/k_{\beta\text{-D}} = 1.15$–1.25 have been measured for this elimination. Propose a mechanism for this reaction.

Answer

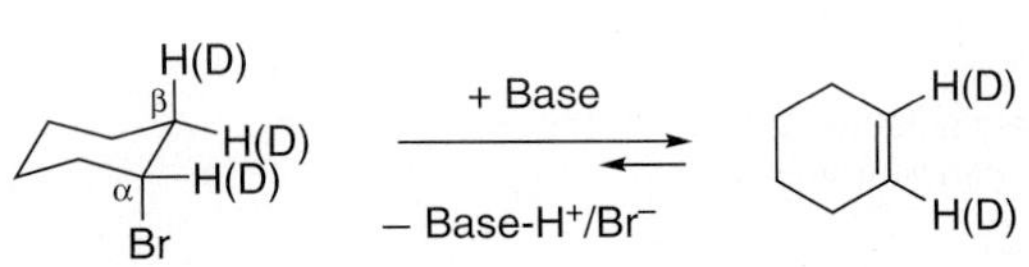

$k_{\alpha\text{-H}}/k_{\alpha\text{-D}} = 1.13 - 1.15$

$k_{\beta\text{-H}}/k_{\beta\text{-D}} = 1.15 - 1.25$

The KIEs are both consistent with either an E2C or an E1 mechanism, but not with an E2 or an E2H mechanism. Distinction between the E2C and E1 mechanisms can be done by measuring the rate law of the reaction. In the case of an E1 mechanism it depends only on the concentration of the substrate (alkyl bromide), whereas in the case of an E2C mechanism, the rate law depends upon both the concentration of the substrate and that of the base. For an E2, or an E2H mechanism, $k_{\beta\text{-H}}/k_{\beta\text{-D}} \gg 1.25$ should be observed since the β-C—H(D) bonds are weakened to a significant degree in the transition state compared with the substrate.

Problem 3.23 Eliminations of AcOH(D) from 3-(2-acetoxyprop-2-yl)-(1*H*)-indene and from its 1-deuterated analog, promoted by tertiary amines, show $k_{\beta\text{-H}}/k_{\beta\text{-D}} = 0.9–2.0$ in solvents ranging from *t*-BuOH to MeOH/H_2O. Using MeONa as base in MeOH $k_{\beta\text{-H}}/k_{\beta\text{-D}} = 7.6$. Propose mechanisms for these reactions.

Answer

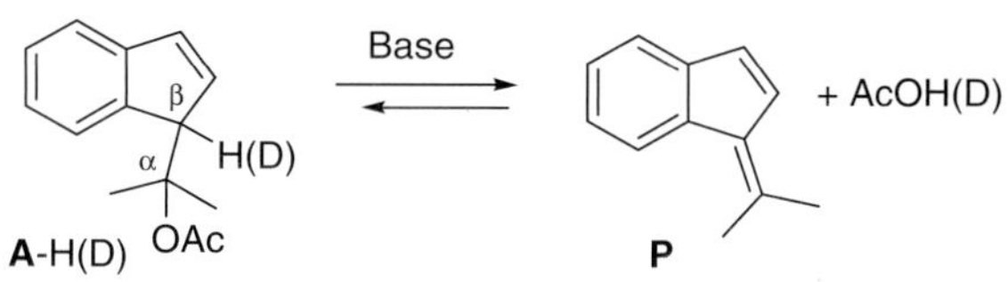

When the base is a tertiary amine, $k_{\beta\text{-H}}/k_{\beta\text{-D}} = 0.9–2.0$. At first glance, this is consistent with a spectrum of mechanism E2C – E1 – E2. When $k_{\beta\text{-H}}/k_{\beta\text{-D}} = 1.1–1.2$ the E1 mechanism is the most probable. In this case, the rate-determining step is the formation of the ion pair **B** (transition state TS1). The tertiary carbenium ion intermediate renders the β-C—H(D) bond weaker because of hyperconjugation. Ion pair **B** may form reversibly and the rate-determining step follows its formation (transition state TS2). Departure of the proton in the latter transition state is assisted by the base or/and solvent, thus leading to $k_{\beta\text{-H}}/k_{\beta\text{-D}} > 1.2$ (greater β-C—H(D) bond weakening in the transition state TS2 than in TS1). Because of steric hindrance between the large base (tertiary amine) and the tertiary substrate A-H(D) there is little chance for an E2C mechanism to intervene.

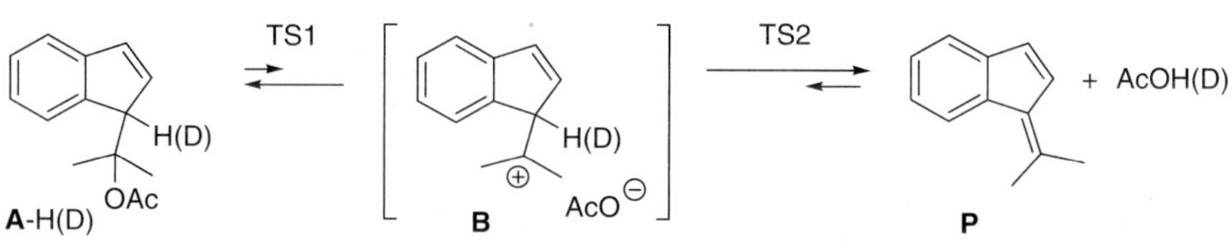

When the base is MeONa in MeOH, the small and unstable alkoxide (MeO^-) can approach the β-C—H(D) bond of **A**–H(D) and pull on it promoting either an E2H or an E1cb mechanism through transition state TS3. Both mechanisms imply large $k_{\beta\text{-H}}/k_{\beta\text{-D}}$ values, as observed (7.6 in our case).

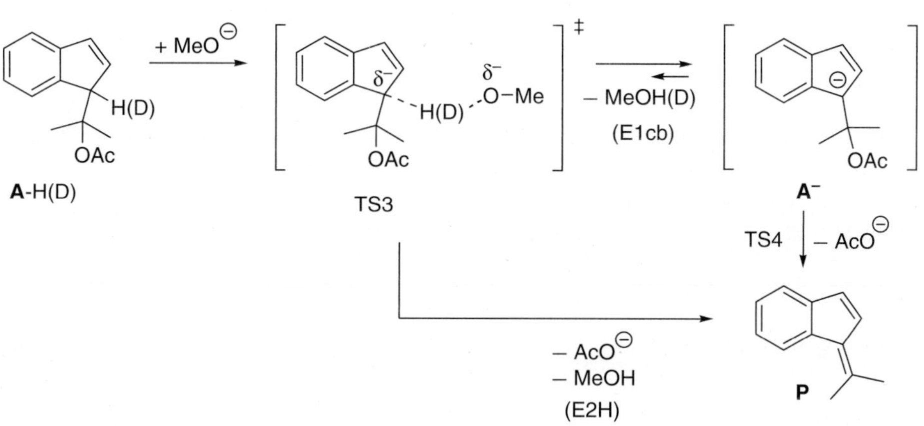

The chances for carbanion $\mathbf{A}^-$ (the conjugate base of substrate **A**–H(D)) to form as an intermediate is given by $pK_a(\mathbf{A}\text{-H}) - pK_a(\text{MeOH})$. It is quite high since $\mathbf{A}^-$ is stabilized by aromaticity such as the cyclopentadienyl anion

(Section 4.5.9: pK_a(indene) = 22.6, pK_a(MeOH) = 15.5 in water, Table 1.A.24). If the formation of **A**$^-$ should be faster than its conversion into alkene **P**, the rate-determining step of the E1cb elimination would be TS4.

In such a case a pre-equilibrium **A**–H(D) + MeONa ⇄ **A**$^-$ + Na$^+$ + MeOH(D) would exist with equilibrium constants K_H and K_D. EIE K_H/K_D should be smaller than unity as the C—H(D) bond in **A**–H(D) is weaker than the O—H(D) bond in MeOH(D). The relative concentration of **A**$^-$ for the elimination **A**–D + MeONa ⇄ **P** + AcONa + MeOD is expected to be slightly higher than for **A**–H + MeONa ⇄ **P** + AcONa + MeOH. This should lead to an inverse $k_H/k_D < 1$, which is not observed.

Problem 3.24 For the aminolysis of methyl ethyl (**A**) and di(isopropyl) chorothiophosphate (**B**) with 3-chloroaniline $k_H/k_D = 0.659 \pm 0.006$ and 1.21 ± 0.002, respectively, have been measured. Propose mechanisms for these reactions [16].

Answer

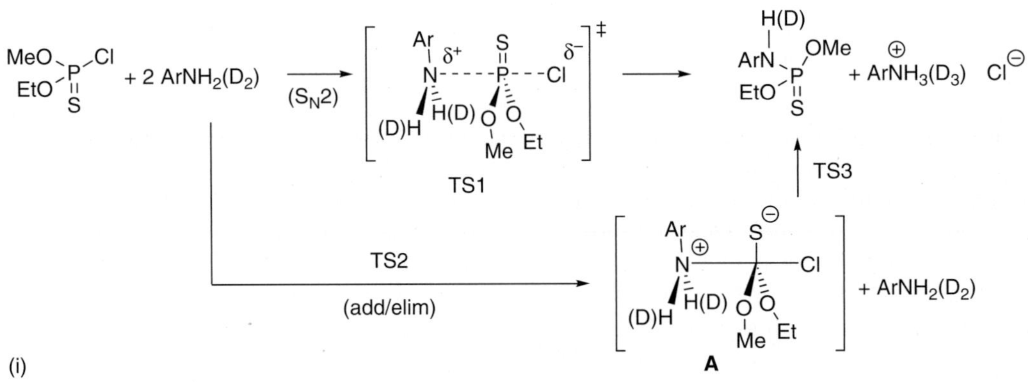

(i)

The observed inverse KIE $k_H/k_D = 0.66$ is consistent with transition state TS1 (see KIE for the Mentchukin reaction). The back-side attack is faster with the deuterated amine than with the non-deuterated amine for a steric reason, in part. This results from the fact that the N—D bond is shorter than the corresponding N—H bond. Both an S_N2 or an add/elim mechanism are consistent with the KIE reported. In the case of a add/elim mechanism the rate-determining step precedes (transition state TS2) or follows (transition state TS3) the formation of adduct **A**.

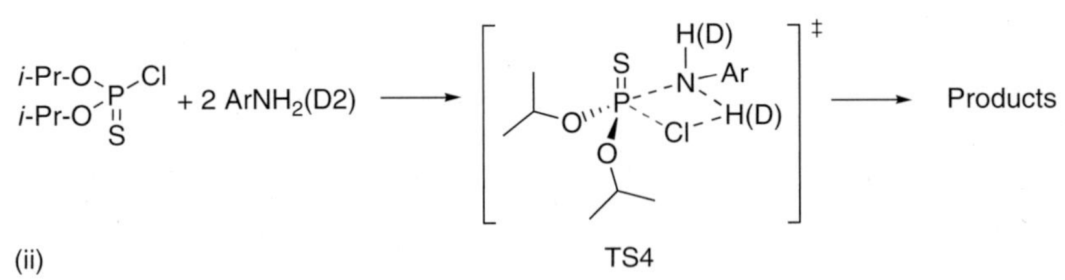

(ii)

The observed KIE $k_H/k_D = 1.21$ is consistent with the intervention (competition with the S_N2 or add/elim mechanism?) of a front-attack by the amine as shown with transition state TS4. In the latter mechanism the proton transfer from the amine is coupled with the displacement of the chloride anion. Note that

an alternative mechanism that would imply the formation of adduct **A** in a pre-equilibrium, followed by pseudo-rotation and a rate-determining step that implies the syn-elimination of HCl as shown with TS4, cannot be rejected on the basis of the KIE reported for this reaction.

Problem 3.25 Explain the following secondary KIE.

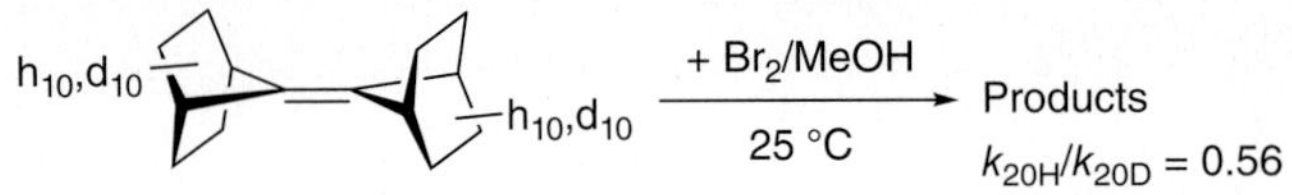

Answer

Bromine addition to a tetraalkyl substituted alkene **A** [17, 18].

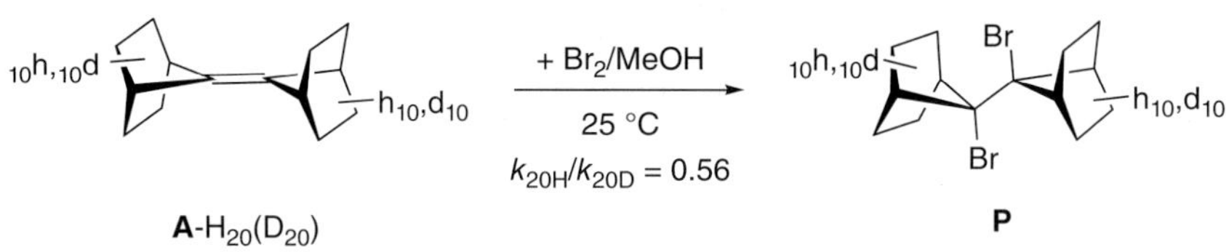

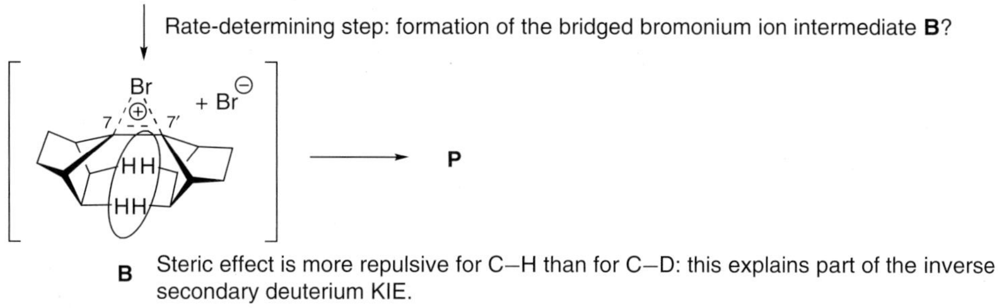

If a classical β-bromocarbenium ion **C** would form in the rate-determining step such steric effect would not be evidenced by k_{20H}/k_{20D} because the two norbornyl units can escape in **C** (rotation about the σ(C(7)—C(7′)) bond) geometries of type **B** in which four *exo*-H(D) are forced to collide.

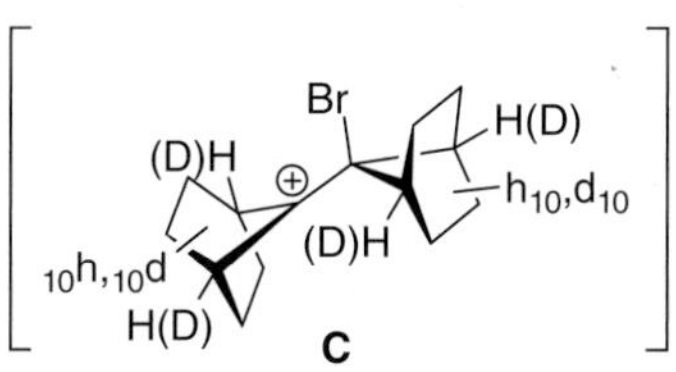

Since 2p(+)/σ(β-C–H(D)) hyperconjugation in carbenium ion **C** is forbidden for reasons of symmetry, $k_{\beta\text{-H}}/k_{\beta\text{-D}} \sim 1$ is expected, not a normal secondary KIE > 1. If one considers that 20 C—H and 20 C—D bonds are concerned the measured inverse $k_{20H}/k_{20D} = 0.56$ does not constitute a definitive proof that **B** rather than **C** is formed in the rate-determining step. The bromide anion cannot attack **B** in an *anti* fashion for steric reasons. Thus, cation **C** may form after the rate-determining step to allow quenching of the bromide anion to give **P**.

3.1 Additional Material

For reactions following concerted mechanisms during which one bond is formed in concert with the cleavage of another bond we refer to the Dimroth principle (the more exothermic a reaction, the faster it is) and to the Bell–Evans–Polanyi theory ($\Delta^{\ddagger}H = \alpha\Delta_r H^\circ + \beta$). We present here the latter theory and some of its corollaries.

3.1.1 The Bell–Evans–Polanyi Theory for Radical Exchanges

In the 1920s Brønsted observed that strongly exothermic acid- and base-catalyzed reactions have a low activation energy [19]. In 1933, Dimroth [20–24] showed for series of analogous reactions that a relationship exists between the rates and the heats of these reactions. For comparable reactions, the more exothermic the reaction, the faster it occurs. For instance, in the case of reaction (1) in which sodium atoms react with alkyl halides in the gas phase to generate the corresponding alkyl radicals and sodium halides, Evans and Polanyi [25, 26] found relationship (2) between the activation enthalpies and the heats of the reactions, where α and β are constants typical of the type of reactions under consideration [27]. For the exchange of proton in solution (acid/base equilibrium), the coefficient α varies between 0 and 1, and this is generally known as the Brønsted relationship [28, 29].

$$Na^{\cdot} + R - X \rightleftarrows NaX + R^{\cdot} \quad (1)$$

$$\Delta^{\ddagger}H = \alpha\Delta_r H^\circ + \beta \quad (2)$$

3.1.2 The Effect of Reactant Polarizability

For reaction (3) that exchange radicals,

$$Y^{\cdot} + R - X \rightleftarrows R^{\cdot} + X - Y \tag{3}$$

there are two reaction coordinates, the distance separating R· · ·X and that separating X· · ·Y. The reaction can follow several mechanisms. One of them is the dissociative mechanism in which homolysis $RX \rightarrow R^{\cdot} + X^{\cdot}$ occurs in a first step. In a second step, radical $X^{\cdot}$ combines with $Y^{\cdot}$ to produce XY (Figure 1). Alternatively, in an associative mechanism, radical $Y^{\cdot}$ adds to RX to generate an intermediate adduct $[Y{-}R{-}X]^{\cdot}$ that then undergoes dissociation into products $R^{\cdot} + XY$. For these two (dissociative and associative) mechanistic limits R—X bond breakage and X—Y bond formation are two processes that ignore each other (no assistance). One can imagine an ensemble of mechanisms in which the R—X bond breakage and X—Y bond formation occur in concert; the two processes assist each other. They are shown in Figure 1.

Figure 2 shows several features that make up a concerted reaction on a two-dimensional diagram. From left to right is given a Morse curve for the dissociation of RX to form $R^{\cdot}$ and $X^{\cdot}$. From right to left, a Morse curve for the dissociation of XY into $X^{\cdot} + Y^{\cdot}$ is shown. Evans and Polanyi described the reaction of $RX + Y^{\cdot} \rightarrow R^{\cdot} + XY$ as involving a superposition of these curves (in their original works these curves were taken as straight lines). In his theory developed initially for electron transfers, Marcus considered these curves to be parabolas [30, 31]. The reaction starts by dissociation of R–X, but at the transition state, switches over to follow the product potential energy to form the product. The upper limit to the activation barrier is the crossing of the reactant and product curves. The activation energy may actually be less than this if there are interactions between RX and $Y^{\cdot}$ that stabilize the crossing point. If $Y^{\cdot}$ is an electronegative radical, its electron affinity will be relatively high $(-EA(Y^{\cdot}))$. One can thus consider a charge transfer configuration in which $Y^{\cdot}$ has taken an electron from RX giving the ion pair $RX^{\cdot+} + Y^{-}$ (Figure 4). The energy $E_{CT}(R)$ of this charge-transfer configuration corresponds to the sum $EI(RX) + (-EA(Y^{\cdot}))$. Another charge transfer configuration might also intervene, which is the one implying an electron transfer from $Y^{\cdot}$ to RX giving ion pair $RX^{\cdot-} + Y^{+}$, the energy of which is $EI(Y^{\cdot}) + (-EA(RX))$.

The dissociation of the charge-transfer configuration, $RX^{\cdot+} + Y^{-}$, follows the energy potential curve (e) in Figure 3. Then, in a second step, X^{+} combines with Y^{-} to give products $R^{\cdot} + XY$, following the energy profile (f). For this dissociative mechanism, the charge-transfer configurations do not help the reaction to occur with a

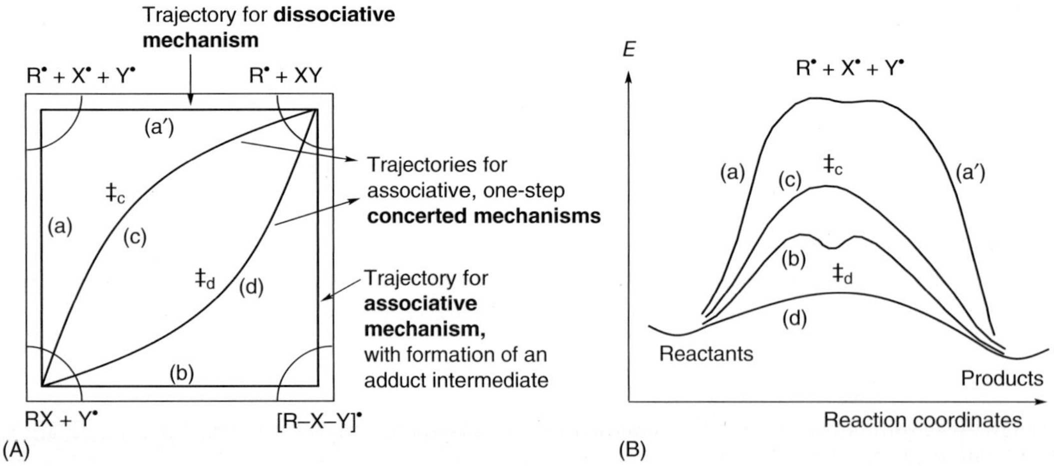

Figure 1 More O'Ferrall type of diagram for the reaction $R{-}X + Y^{\cdot} \rightleftarrows R^{\cdot} + X{-}Y$. (A) The potential energy is perpendicular to the two reaction coordinates: separation between X···Y (the *x*-axis) and between R···X (the *y*-axis); (B) the potential energy *E* as a function of the mechanism. (a)–(d) represent four possible paths for a radical substitution reaction. Path (a) corresponds to a dissociative mechanism involving unassisted homolysis of R–X into radicals $R^{\cdot}$ and $X^{\cdot}$ and the combination of $X^{\cdot}$ with $Y^{\cdot}$ into product X–Y. It is a two-step process. Paths (c) and (d) represent two possible concerted mechanisms in which the R—X bond breakage is assisted by the formation of the X—Y bond. They are single step processes with transition structures ‡c and ‡d, respectively. Path (c) and (d) are associative mechanisms: they combine both reactants in their transition states. Path (b) corresponds to another associative mechanism in which an adduct intermediate $[R{-}X{-}Y]^{\cdot}$ is formed. It is a two step process that can be seen as an addition/elimination mechanism.

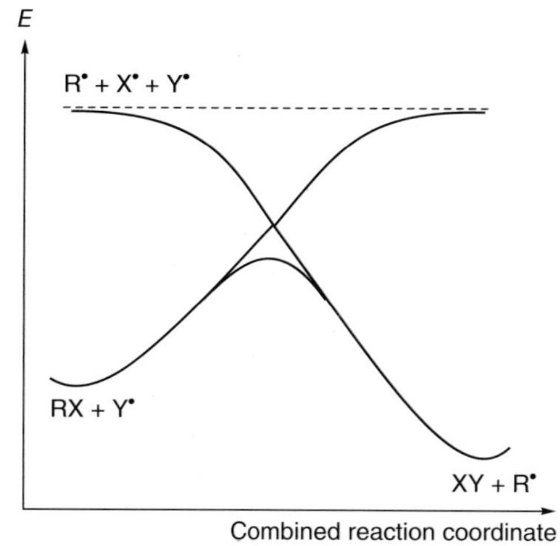

Figure 2 Two-dimensional representation of the Morse potential energies for dissociation mechanism and for the concerted reaction $RX + Y^{\cdot} \rightleftarrows XY + R^{\cdot}$.

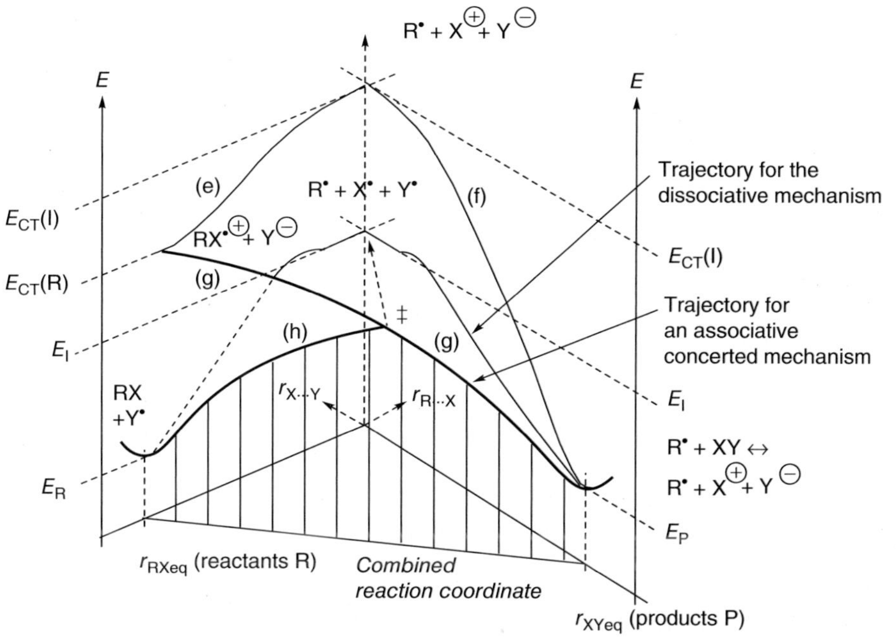

Figure 3 Representation of several reaction paths for reaction $RX + Y^{\cdot} \rightleftarrows R^{\cdot} + XY$. The charge transfer configuration of the reactants has an energy $E_{CT}(R) = IE(RX) + (-EA(Y^{\cdot}))$. For the dissociative, non-concerted mechanism, the energy barrier of the process is $\geq DH°(R^{\cdot}/X^{\cdot})$. The crossing point of the potential energy curves (h) and (g) represents the transition structure of the concerted reaction (bond breakage R···X occurs in concert with bond formation X···Y). This energy can be lower than the energy of radicals $R^{\cdot} + X^{\cdot} + Y^{\cdot}$.

lower activation enthalpy than that required for the homolysis of RX. Now considering the concerted reaction pathway (h) and the charge-transfer configuration, $RX^{\cdot+} + Y^-$, when Y^- approaches $RX^{\cdot+}$, the potential energy decreases as a Coulomb potential ($E \approx q^2/r$) as shown in (g). Before bond R—X in $RX^{\cdot+}$ is completely broken, there may be a transition structure $[Y^-RX^{\cdot+}]^{\ddagger}$ for which the potential energy E_i is below that of radicals $R^{\cdot} + X^{\cdot} + Y^{\cdot}$. From that point on, the system will evolve to products $R^{\cdot} + X^+Y^- \leftrightarrow R^{\cdot} + XY$, with a continual decrease in the energy to reach that of products $R^{\cdot} + XY$. The crossing point of potential energy curves (g) and (h) might lie below E_i, the energy of the radical intermediates formed in a non-concerted, dissociative mechanism. The same is represented in the two-dimensional diagram of Figure 4 [32].

Considering now the two reactions, $RX + Y^{\cdot} \rightarrow R^{\cdot} + XY$ and $RX + Z^{\cdot} \rightarrow R^{\cdot} + XZ$ that have the same exothermicity but different energies for their charge-transfer configurations $E_{CT}(RX + Y^{\cdot}) = IE(RX) + (-EA(Y^{\cdot})) < E'_{CT}$

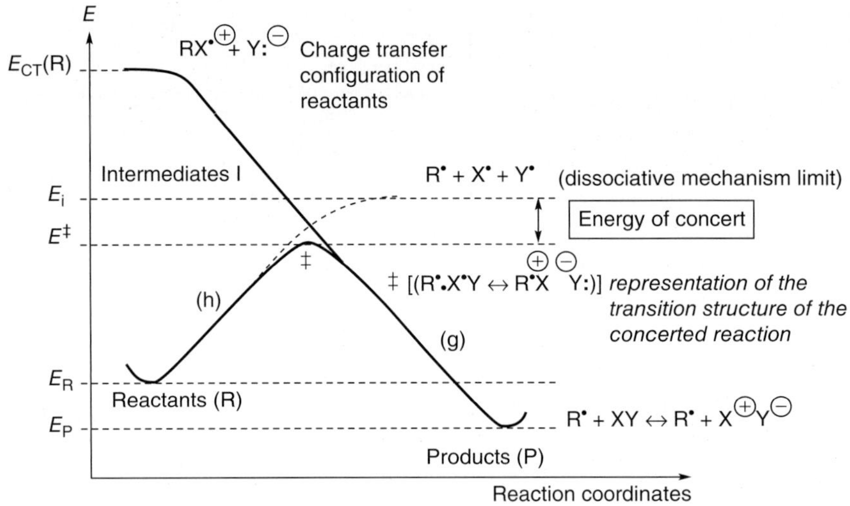

Figure 4 Bell–Evans–Polanyi treatment of the reaction $RX + Y^{\cdot} \rightleftarrows R^{\cdot} + XY$ showing the energy of concert = difference between the activation energy of the non-concerted, dissociative mechanism and that of the concerted mechanism ($E_i - E^{\ddagger}$).

$(RX + Z^{\cdot})$ = IE(RX) + (−EA($Z^{\cdot}$)), the crossing point of the dissociative curve (h) with the Coulombic potential curves (g) is at lower energy than for curves (g′) and (h). This analysis uses the Morse potential curve for (h) and Coulombic charge annihilation potential (g, g′) at the beginning of the reaction (when $RX^{\cdot+}$ approaches Y^- and Z^-). The degree of concert (or the energy of concert) is higher when the energy of the charge-transfer configuration of the reactants is lower. This occurs because the ease of electron transfer (as given by the sum of measurable thermodynamic values IE(RX) + (−EA($Y^{\cdot}$)) and *IE*(RX) + (−EA($Z^{\cdot}$)) stabilizes the transition state of the concerted process.

Based on thermodynamics α-H-atom abstraction from amino acid derivatives such as MeCONHCH(CH_2–CH_2–CH_2–CH_2–CH_3)$COOR^2$ should be easier (formation of a carbon-centered radical conjugated with the COOH group and further stabilized by the MeCONH substituent) than for any other positions of its side chain $(CH_2)_4$–Me (H-atom abstraction from β, γ, δ, and ε-positions generates secondary alkyl radicals, and from the terminal methyl group a primary alkyl radical is formed). With $Cl^{\cdot}$ and $HO^{\cdot}$ radicals the α-C—H bond is the least reactive. These two radicals are electrophilic and prefer the more nucleophilic C—H bonds of the alkyl side chain [33, 34]. The chemoselectivity does not arise from a steric effect since quantum mechanical calculations for the H-atom transfers from MeCONH($CH_2CH_2CH_2CH_2$Me)COOH to $H_3N{:}\rightarrow B^{(\cdot)}Me_2$ find the lowest energy barrier (8.3 kcal mol^{-1}) for the α-C—H bond than for the other C—H bonds (21–24 kcal mol^{-1}). Polar effects with the electrophilic radicals $Cl^{\cdot}$ and $HO^{\cdot}$ direct the H-atom abstraction away from the thermodynamically preferred α-C–H position, at which electron-withdrawing substituents have a deactivating effect in the transition structure. The opposite chemoselectivity is expected for the nucleophilic boron-centered radical $H_3N{:}\rightarrow B^{(\cdot)}Me_2$ [35].

3.1.3 The Effect of Exothermicity

The ideas introduced in the previous section can be illustrated by a comparison of two parallel reactions, $R^1X + Y^{\cdot} \rightleftarrows R^{1\cdot} + XY$ and $R^2X + Y^{\cdot} \rightleftarrows R^{2\cdot} + XY$ of Figure 5. The difference in exothermicity of the reactions is given by $\Delta\Delta E_r$, which corresponds to the difference of bond dissociation enthalpies $DH°(R^{2\cdot}/H^{\cdot}) - DH°(R^{1\cdot}/H^{\cdot})$. Assuming that the charge transfer configurations $E_{CT}(R^1X + Y^{\cdot})$ and $E_{CT}(R^2X + Y^{\cdot})$ have the same energy, $IE(R^1X) = IE(R^2X)$, then the more exothermic reaction $R^1X + Y^{\cdot} \rightleftarrows R^{1\cdot} + XY$ has the bond dissociation potential curve (h) and the bond formation potential curve (g) in Figure 6. The less exothermic reaction $R^2X + Y^{\cdot} \rightleftarrows R^{2\cdot} + XY$ has the bond dissociation potential curve (h′) and the bond formation potential curve (g′) in Figure 6. The crossing-point of curves (g) and (h) occurs at a lower energy than that of curves (g′) and (h′), showing that the more exothermic reaction has a lower activation energy than the less exothermic reaction, the activation energy difference being $\Delta\Delta E^{\ddagger} = \alpha\Delta\Delta E_r$ in this case (same steric factors, same reactant polarity, and polarizability).

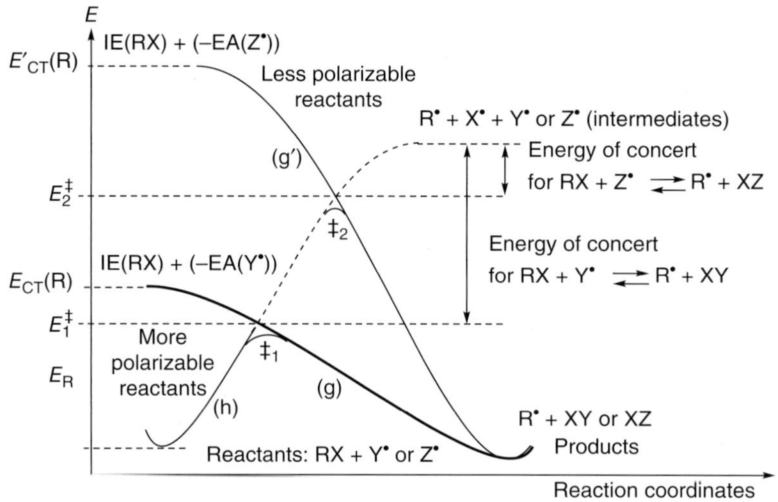

Figure 5 Effect of the reactant polarizability (ability to exchange an electron) on the energy of concert.

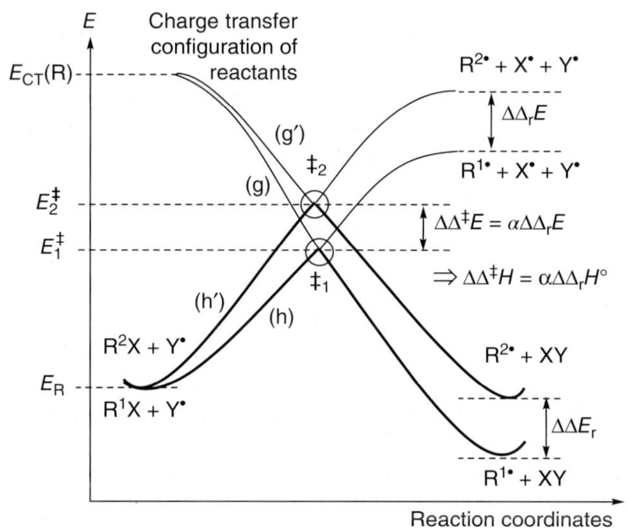

Figure 6 Bell–Evans–Polanyi demonstration of the Brønsted or Dimroth principle: $\Delta^{\ddagger}H = \alpha\Delta_r H + \beta$ for two concerted, parallel reactions implying reactants of same polarity and polarizability, but with different exothermicities (no differential steric effects for the two comparable reactions).

3.1.4 The Hammond–Leffler Postulate is a Theorem Demonstrated by the Bell–Evans–Polanyi Theory

The Hammond postulate [36] (also referred to as the Hammond–Leffler postulate [37]) states that the enthalpy and geometry of a transition structure (structure of the transition state) of a concerted one-step reaction resembles the least stable species of the energy hypersurface [38]. For an exothermic reaction, the transition structure resembles the reactants (starting materials); for an endothermic reaction, the transition structure resembles the product, generally unstable intermediates in a two-step reaction. A corollary of the Hammond postulate is that exothermic one-step reactions have early transition states whereas endothermic one-step reactions have late transition states. The Bell–Evans–Polanyi theory, which uses Morse potential energy curves for bond breaking and bond forming processes that operate in concert, demonstrates the above postulate (Figure 7). For exothermic reactions, the crossing-points between the charge-transfer configurations (g) and the bond breaking Morse curves (h) occur at the beginning of the reactions (early transition states), whereas for endothermic reactions, the crossing-points of curves (g) and (h) occur at a later stage, closer to the product coordinates. Reactions involving the formation of reactive intermediates in their rate-determining steps have structures for their transition states resembling these intermediates in terms of free energy and geometry.

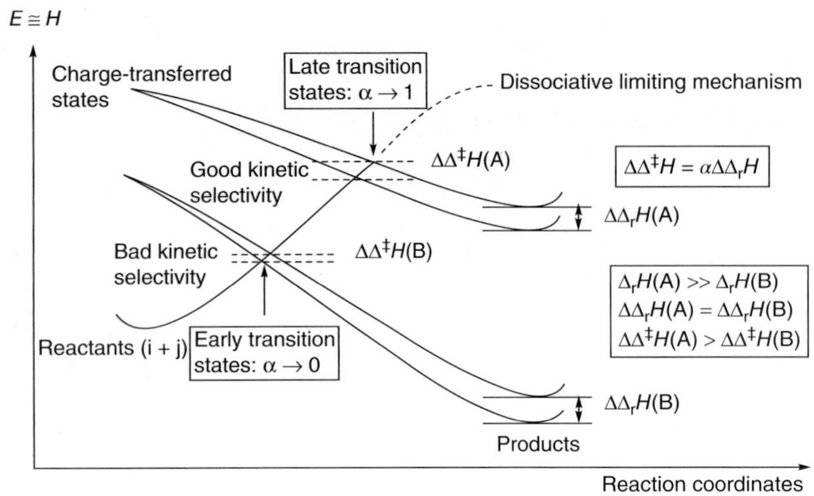

Figure 7 Exothermic reactions (B) have early transition states; endothermic reactions (A) have late transition states. Kinetic selectivity is better for endothermic ($\alpha \to 1$) than for exothermic ($\alpha \to 0$) concerted reactions.

3.1.5 The Selectivity/Reactivity Relationship

An important consequence of the theory is that the kinetic selectivity (taking $\Delta\Delta^{\ddagger}G \cong \Delta\Delta^{\ddagger}H$ in this case) of two concerted, parallel reactions is poor for exothermic reactions, and good for endothermic reactions (Figure 7). This has been verified experimentally for the radical chlorination and bromination of alkanes (Figures 8 and 9). For a critical discussion of the reactivity–selectivity principle, see [39].

Kinetic isotope effects (KIEs) represent one of the best experimental tools to characterize transition state of the rate-determining step of a reaction (Section 3.9). In the case of the Wohl–Ziegler bromination $Br^{\cdot}$ abstracts a benzylic or allylic H-atom in the rate-determining step of the reaction: $R{-}H + Br^{\cdot} \rightleftarrows R^{\cdot} + H{-}Br$. The heat of this reaction increases with the relative stability of radical $R^{\cdot}$, with the lowering of $DH°(R^{\cdot}/H^{\cdot})$. The transition state of the H-atom transfer represented by $[R\cdots H\cdots Br]^{\ddagger}$ is earlier the higher the exothermicity. Thus, the degree of C—H bond breaking in the transition state decreases with exothermicity. This is evidenced by the primary KIEs $k_H(R{-}H + Br^{\cdot})/k_D(R{-}D + Br^{\cdot})$ [40, 41]. An example is the Wohl–Ziegler reaction of *N*-bromosuccinimide (NBS) with toluene, ethylbenzene, and isopropylbenzene for which $k_H/k_D = 4.86$, 2.67, and 1.81, respectively, at 77 °C [42, 43]. The stability of the radical formed increases in the sequence $Bn^{\cdot} < PhC^{(\cdot)}(Me)H < PhC^{(\cdot)}Me_2$ (Table 1.A.7). For the photochemically initiated chlorination of CH_2D_2 a larger primary KIE $k_H/k_D = 12.1$ was measured at 0 °C, and the variation with temperature was given by $k_H/k_D = 1.09{\cdot}\exp.(1300/RT)$ [44, 45]. For the reactions of $Cl^{\cdot}$ with CH_4 and CD_4 giving $CH_3^{\cdot}$ ($\Delta_r H°(Me{-}H + Cl^{\cdot} \rightleftarrows Me^{\cdot} + HCl) = 1.5\,\text{kcal mol}^{-1}$)

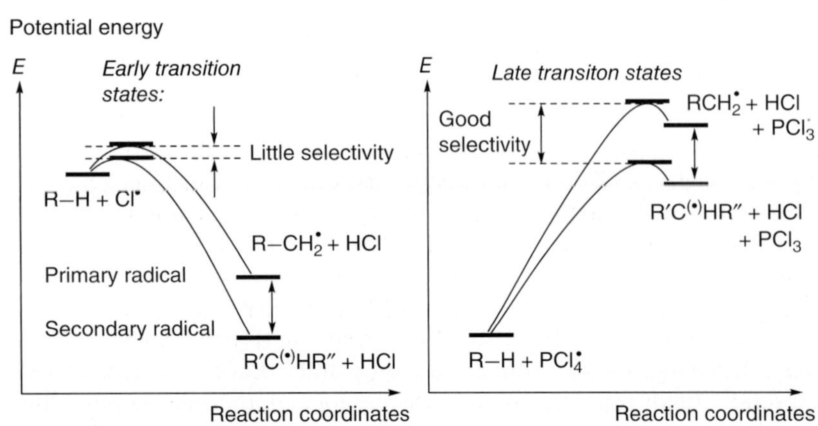

Figure 8 Direct radical chlorination of alkanes are not selective unless the chlorine atom is stabilized by "solvation" more than the alkyl radicals.

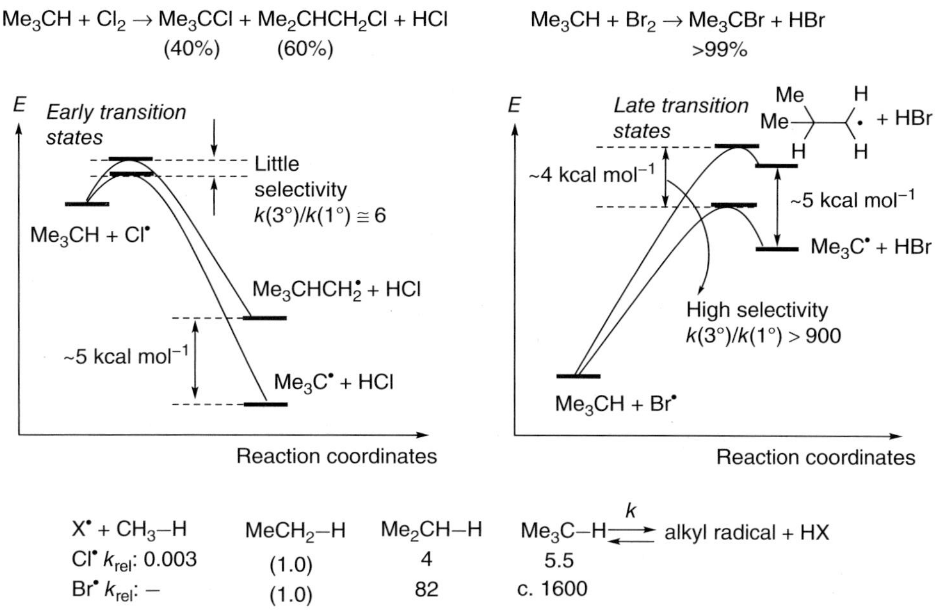

Figure 9 Direct radical brominations are more selective than direct radical chlorinations.

and $CD_3^{\cdot}$, respectively, $k_H/k_D = 2.2\cdot\exp.(1180/RT)$ has been reported [46]. The transition state of the $Cl^{\cdot}$ attack of methane can be represented as $[Me\cdots H\cdots Cl]^{\ddagger}$. It has weaker bonds than the transition state $[Bn\cdots H\cdots Br]^{\ddagger}$ of the benzylic bromination. For the highly exothermic reaction of $F^{\cdot}$ with CH_4 and CD_4 ($\Delta_r H°(Me–H + F^{\cdot} \rightleftarrows Me^{\cdot} + HF) = -31.7\ \text{kcal mol}^{-1}$) much smaller $k_H/k_D = (0.99 \pm 0.02)\exp.[(100 \pm 5)/RT\}$ has been measured between 183 K ($k_H/k_D = 1.72 \pm 0.05$) and 298 K ($k_H/k_D = 1.38 \pm 0.04$) [47]. This is in agreement with the selectivity/reactivity principle (Figure 7) and with a much tighter transition state $[Me\cdots H\cdots F]^{\ddagger}$ for the direct fluorination than transition state $[Me\cdots H\cdots Cl]^{\ddagger}$ of the direct chlorination [48].

References

1 Yoshimine, M., Pacansky, J., and Honjou, N. (1989). Ab initio studies of the C_3H_4 surface. 3. Thermal-isomerization. *J. Am. Chem. Soc.* 111 (12): 4198–4209.

2 Humski, K., Malojcic, R., Borcic, S., and Sunko, D.E. (1970). Thermodynamic and kinetic secondary isotope effects in Cope rearrangement. *J. Am. Chem. Soc.* 92 (22): 6534–6538.

3 Gajewski, J.J. and Conrad, N.D. (1979). Variable transition-state structure in 3,3-sigmatropic shifts from alpha-secondary deuterium-isotope effects. *J. Am. Chem. Soc.* 101 (22): 6693–6704.

4 Gajewski, J.J. and Conrad, N.D. (1979). Aliphatic claisen rearrangement transition-state structure from secondary alpha-deuterium isotope effects. *J. Am. Chem. Soc.* 101 (10): 2747–2748.

5 Houk, K.N., Gustafson, S.M., and Black, K.A. (1992). Theoretical secondary kinetic isotope effects and the interpretation of transition-state geometries. 1. The Cope rearrangement. *J. Am. Chem. Soc.* 114 (22): 8565–8572.

6 Quast, H., Seefelder, M., Becker, C. et al. (1999). Extension of Saunders' isotopic perturbation method as probe for the structures in solution of 2,4,6,8-substituted barbaralanes – NMR-spectroscopic evidence for the coexistence of localised and delocalised states. *Eur. J. Org. Chem.* (11): 2763–2779.

7 Waters, W.L. and Caserio, M. (1968). Resolution and absolute configuration of 1,3-dimethylallene. *Tetrahedron Lett.* 5233–5236.

8 Moore, W.R, Anderson, H.W., and Clark, S.D. (1973). Recovery of allenes from partial hydroboration of racemic allenes with (+)-tetra-3-pinanyldiborane – Model for hydroboration of olefins and allenes. *J. Am. Chem. Soc.* 95: 835–844.

9 Brown, H.C. and Zweifel, G. (1961). Hydroboration as a convenient procedure for asymmetric synthesis of alcohols of high optical purity. *J. Am. Chem. Soc.* 83: 486.

10 Brown, H.C., Ayyangar, N.R., and Zweifel, G. (1964). Hydroboration. XVIII. The reaction of diisopinocampheylborane with representative *cis*-acyclic, cyclic, and bicyclic olefins. A convenient synthesis of optically active alcohols and olefins of high optical purity and established configuration. *J. Am. Chem. Soc.* 86: 397–403.

11 Brown, H.C., Ayyangar, N.R., and Zweifel, G. (1964). Diisopinocampheylborane as a reagent for the production of optically active olefins and the establishment of their configuration. *J. Am. Chem. Soc.* 86: 4341–4342.

12 Borden, W.T. and Corey, E.J. (1969). A synthesis of optically active 1,3-di-*t*-butylallene from propargylic derivatives by hydride displacement. *Tetrahedron Lett.* (5): 313–316.

13 Luche, J.L., Barreiro, E., Dolla, J.M., and Crabbé, P. (1975). Formation of allenes by reaction of lithium diorganocurates on propargyl acetates – Mechanistic approach. *Tetrahedron Lett.* 4615–4618.

14 Pirkle, W.H. and Boeder, C.W. (1978). Preparation of simple chiral allenes – Reaction of propargylic carbamates with lithium dialkylcuprates. *J. Org. Chem.* 43: 1950–1952.

15 Elsevier, C.J. and Vermeer, P. (1989). Synthesis and stereochemistry of allenes. 3. Highly stereoselective synthesis of chiral alkylallenes by organocopper(I)-induced anti.1,3-sustitution of chiral propynyl esters. *J. Org. Chem.* 54: 3726–3730.

16 Barai, H.R., Hoque, M.E.U., and Lee, H.W. (2013). Kinetics and mechanism of anilinolyses of ethyl methyl, ethyl propyl and diisopropyl chlorothiophosphates in acetonitrile. *Bull. Korean Chem. Soc.* 34 (12): 3811–3816.

17 Nagorski, R.W., Slebockatilk, H., and Brown, R.S. (1994). Observation of an unusually large inverse secondary deuterium kinetic isotope effect for the reaction of a sterically congested olefin with Br_2. *J. Am. Chem. Soc.* 116 (1): 419–420.

18 Slebockatilk, H., Motallebi, S., Nagorski, R.W. et al. (1995). Electrophilic bromination of 7-norbornylidene-7′-norbornane – the observation of an unusually large inverse deuterium kinetic isotope effect. *J. Am. Chem. Soc.* 117 (34): 8769–8776.

19 Bronsted, J.N. (1928). Acid and basic catalysis. *Chem. Rev.* 5: 231–338.

20 Dimroth, O. (1933). Relationships between affinity and reactions speed. *Angew. Chem. Int. Ed. Engl.* 46: 571–576.

21 Marcus, R.A. (1964). Chemical + electrochemical electron-transfer theory. *Annu. Rev. Phys. Chem.* 15: 155–196.

22 Cohen, A.O. and Marcus, R.A.O. (1968). Slope of free energy plots in chemical kinetics. *J. Phys. Chem.* 72: 4249–4256.

23 Marcus, R.A. (1969). Unusual slopes of free energy plots in kinetics. *J. Am. Chem. Soc.* 91: 7224–7225.

24 Guthrie, J.P. (1996). Multidimensional Marcus theory: an analysis of concerted reactions. *J. Am. Chem. Soc.* 118: 12878–12885.

25 Evans, M.G. and Polanyi, M. (1936). Further considerations on the thermodynamics of chemical equilibria and reaction rates. *Trans. Faraday Soc.* 32: 1333–1359.

26 Evans, M.G. and Polanyi, M. (1938). Inertia and driving force of chemical reactions. *Trans. Faraday Soc.* 34: 11–24.

27 Dill, K.A. and Bromberg, S. (2003). *Molecular Driving Forces: Statistical Thermodynamics in Chemistry and Biology*, 1–666. New York, NY: Garland Science.

28 Bronsted, J.N. and Pedersen, K. (1924). The catalytic disintegration of nitramide and its physical-chemical relevance. *Z. Phys. Chem.* 108: 185–235.

29 Bell, R.P. (1936). The theory of reactions involving proton transfers. *Proc. R. Soc. London, Ser. A* 154: 414–429.

30 Marcus, R.A. (1956). On the theory of oxidation–reduction reactions involving electron transfer. I. *J. Chem. Phys.* 24: 966–978.

31 Marcus, R.A. (1956). Electrostatic free energy and other properties of states having nonequilibrium polarization. I. *J. Chem. Phys.* 24: 979–989.

32 Polanyi, J.C. (1967). Relation between electron affinity or ionisation potential, and reaction rate. *Chem. Phys. Lett.* 1: 421–423.

33 O'Reilly, R.J., Chan, B., Taylor, M.S. et al. (2011). Hydrogen abstraction by chlorine atom from amino acids: remarkable influence of polar effects on regioselectivity. *J. Am. Chem. Soc.* 133: 16553–16559.

34 Amos, R.I., Chan, B., Easton, C.J., and Radom, L. (2015). Hydrogen-atom abstraction from a model amino acid: dependence on the attacking radical. *J. Phys. Chem. B* 119: 783–788.

35 Chan, B., Easton, C.J., and Radom, L. (2015). Outcome-changing effect of polarity reversal in hydrogen-atom-abstraction reactions. *J. Phys. Chem. A* 119: 3843–3847.
36 Hammond, G.S. (1955). A correlation of reaction rates. *J. Am. Chem. Soc.* 77: 334–338.
37 Leffler, J.E. (1953). Parameters for the description of transition states. *Science* 117: 340–341.
38 Meany, J.E., Minderhout, V., and Pocker, Y. (2001). Application of Hammond's postulate – an activity for guided discovery learning in organic chemistry. *J. Chem. Educ.* 78: 204–207.
39 Mayr, H. and Ofial, A.R. (2006). The reactivity-selectivity principle: an imperishable myth in organic chemistry. *Angew. Chem. Int. Ed.* 45: 1844–1854.
40 Wiberg, K.B. (1955). The deuterium isotope effect. *Chem. Rev.* 55: 713–743.
41 Wiberg, K.B. and Slaugh, L.H. (1958). The deuterium isotope effect in the side chain halogenation of toluene. *J. Am. Chem. Soc.* 80: 3033–3039.
42 Pearson, R.E. and Martin, J.C. (1963). Identity of chain-carrying species in brominations with *N*-bromosuccinimide – selectivity of substituted *N*-bromosuccinimides toward substituted toluenes. *J. Am. Chem. Soc.* 85: 3142–3146.
43 Russell, G.A., Desmond, K.M., and Deboer, C. (1963). Mechanisms of benzylic bromination. *J. Am. Chem. Soc.* 85: 365–366.
44 Wiberg, K.B. and Motell, E.L. (1963). Kinetic isotope effect in photochlorination of methane. *Tetrahedron* 19: 2009–2023.
45 Motell, E.L., Boone, A.W., and Fink, W.H. (1978). Large kinetic isotope effects with unsymmetrical transitions-states. *Tetrahedron* 34: 1619–1626.
46 Strunin, V.P., Serdyuk, N.K., Chesnokov, E.N., and Panfilov, V.N. (1975). Kinetic isotope-effect in chlorination of methane. *React. Kinet. Catal. Lett.* 3: 97–104.
47 Persky, A. (2006). The temperature dependence of the kinetic isotope effect in the reaction of F atoms with CH_4 and CD_4. *Chem. Phys. Lett.* 430: 251–254.
48 Foon, R., Tait, K.B., and Reid, G.P. (1972). Kinetic hydrogen isotope-effect in fluorination of H_2, CH_4 and $CHCl_3$. *J. Chem. Soc., Faraday Trans. 1 F* 68: 1131–1138.

4

Molecular Orbital Theories

Answers to Problems

Problem 4.1 Olah and coworkers have prepared tetramethylcyclobutenediyl dication as a stable salt in solution by treating 3,4-dichloro-1,2,3,4-tetramethylcyclobutene in SbF_5/SO_2ClF at −75 °C. They also prepared the tetraphenyl substituted derivative in a similar way. Estimate the Hückel E_π energy of cyclobutenediyl dication. Is it stabilized by aromatic stabilization energy?

Answer

For planar cyclobutenediyl dication the Hückel $E_\pi = 2(\alpha + 2\beta) = 2\alpha + 4\beta$. Since the Hückel E_π(ethylene) + $2E_\pi(Me^+) = 2\alpha + 2\beta$, cyclobutenediyl dication has an aromatic stabilization corresponding to 2β. Quantum mechanical calculations suggest that cyclobutenediyl dication and its tetramethyl derivative adopt nonplanar geometries. The puckered geometries reduce charge/charge repulsive interactions [1, 2].

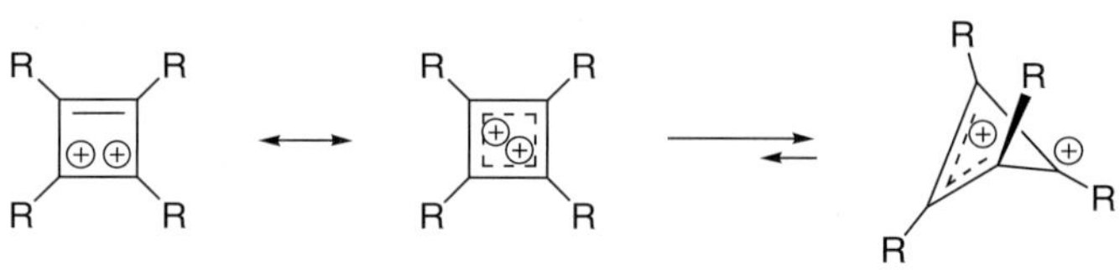

Problem 4.2 Estimate the hydride affinity of cyclohexadienyl cation in the gas phase using proton affinity of benzene (185 kcal mol^{-1}, Table 1.A.15).

Answer

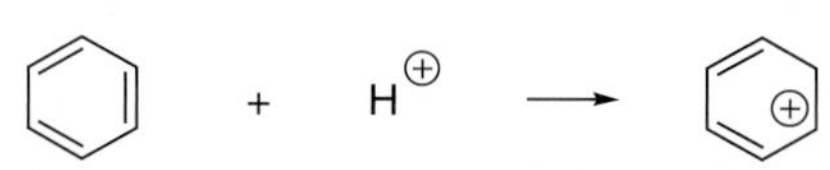

$\Delta_r H° = -185$ kcal mol^{-1}

$\Delta_f H° = 19.7$ (Table 1.2) 365.7 (Table 1.14) kcal mol^{-1}

This gives $\Delta_f H°$(cyclohexadienyl cation) = 200.4 kcal mol^{-1}

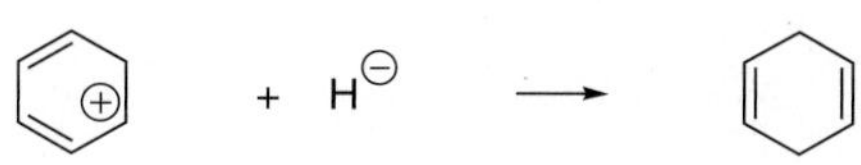

$\Delta_f H° = 200.4$ 34.7 24 kcal mol^{-1} (Table 1.2)

Hydride affinity of cyclohexadienyl cation $c\text{-}C_6H_7^+$: $DH°(c\text{-}C_6H_7^+/H^-) = \Delta_f H°(c\text{-}C_6H_7^+) + \Delta_f H°(H^-) - \Delta_f H°$ (cyclohexa-1,4-diene) = 200.4 + 34.7 − 24 = 211.1 kcal mol^{-1}.

Organic Chemistry: Theory, Reactivity and Mechanisms in Modern Synthesis, Workbook, First Edition. Pierre Vogel and Kendall N. Houk.

Problem 4.3 [4]Radialene (1,2,3,4-tetramethylidenecyclobutane: **A**) reacts with highly reactive dienophiles such as ethylenetetracarbonitrile (TCNE), 4-phenyl-1,2,4-triazolinedione (PTAD), or diethyl azodicarboxylate (DEAD) giving the corresponding monoadducts **B**. No bisadducts **C** can be detected under forcing conditions (large excess of dienophiles, prolonged reaction time, heating) [3]. Why?

Answer

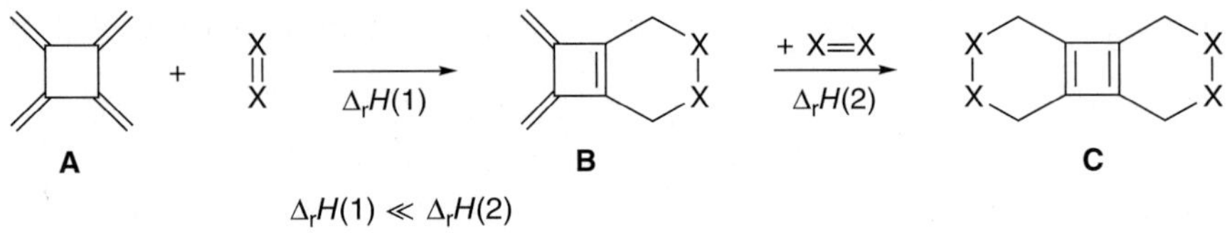

Double adducts **C** are destabilized by antiaromaticity as they are cyclobutadiene derivatives. As a consequence, the exothermicity of the cycloaddition **B** + X = X ⇆ **C** is much lower than that of **A** + X = X ⇆ **B**, and thus **C** cannot be formed in equilibrium with **B**.

The heat of hydrogenation of cyclobutene into cyclobutane is $-30.9\,\text{kcal mol}^{-1}$ (Table 1.A.2) and that of cyclobutadiene into cyclobutene is $-67.7 \pm 4\,\text{kcal mol}^{-1}$ (Section 4.5.6). Thus, the antiaromaticity of **C** must amount to c. $37\,\text{kcal mol}^{-1}$, which makes its formation impossible at 20 °C.

Problem 4.4 Acidity of 6,6-dimethylfulvene ($pK_a = 22.7$, DMSO) is higher than that of toluene ($pK_a \sim 43$, DMSO), diphenylmethane ($pK_a = 32.3$, DMSO), and triphenylmethane ($pK_a = 30.6$, DMSO, Table 1.A.25). Why?

Answer

Because of the dipolar character of 6,6-dimethylfulvene, its methyl groups are acidified like the α-C–H moieties in carbonyl compounds.

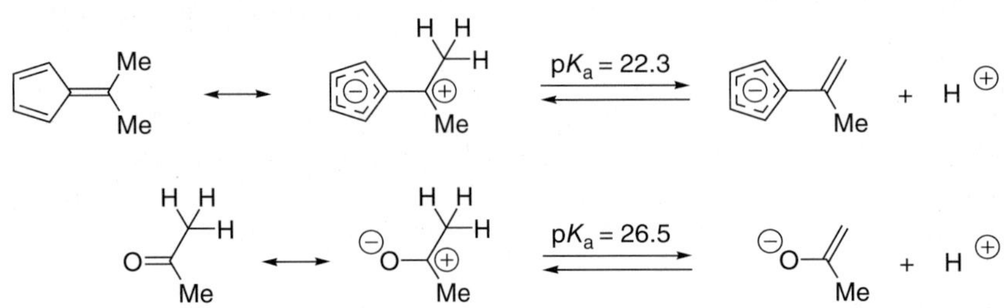

Furthermore, the conjugate base of 6,6-dimethylfulvene, isopropenylcyclopentadienyl anion, is stabilized by aromaticity, just as the enolate is stabilized by the electronegative oxygen atom.

Problem 4.5 Thiele prepared fulvenes by reaction of cyclopentadiene with aldehydes or ketones under base catalyzed conditions [4, 5]. They can be obtained in better yields by reaction of imines with cyclopentadiene [6]. Propose a mechanism for the following metathesis of substituted fulvenes catalyzed by *n*-butylamine [7].

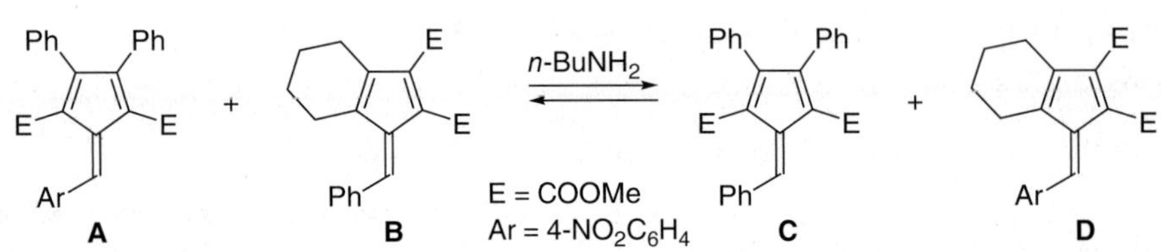

Answer

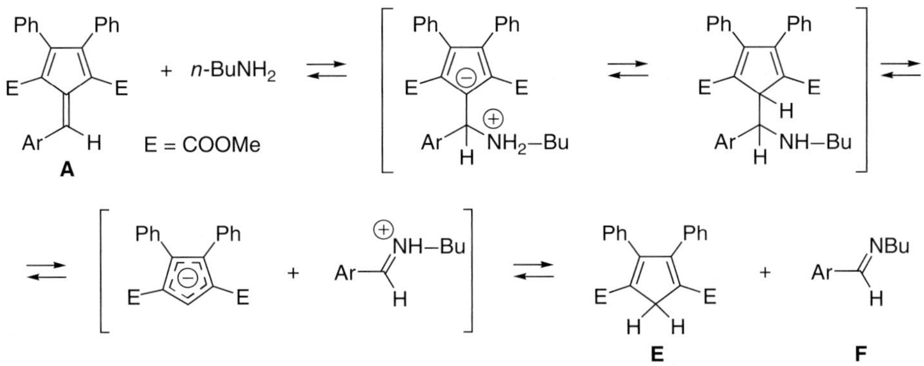

The same process occurs also with **B** leading to intermediates **G** + **H**.

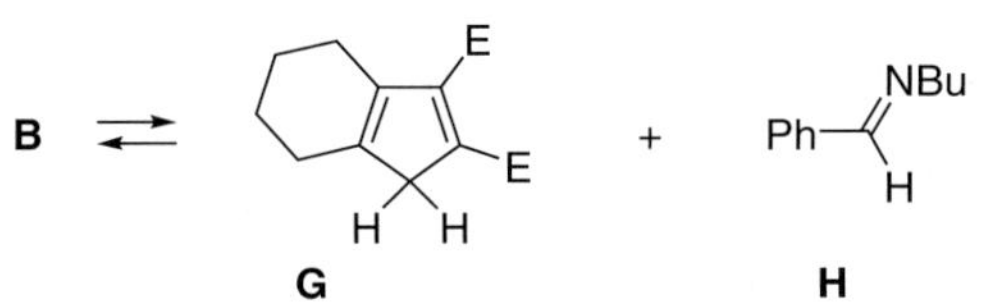

G and **H** react with intermediates **E** and **F** leading to the formation of fulvenes **C** and **D**. Fulvene formation from cyclopentadienes and imines is reversible. This is possible because of the high electrophilicity of fulvenes, making them resemble carbonyl compounds, and because of the related high stability of the cyclopentadienyl anion intermediates attributed to their aromaticity. In case **A** + **B** ⇆ **C** + **D**, extra-stabilization of the cyclopentadienyl anion is realized by the two methoxycarbonyl substituents (E = COOMe).

Problem 4.6 Write the Hückel secular determinant for [3]radialene (1,2,3-trimethylidenecyclopropane). Using PMO theory, do you expect this compound to enjoy significant electronic stabilization energy due to π-conjugation?

Answer

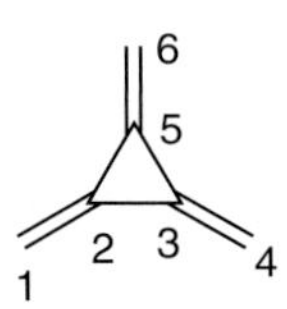

$$\begin{vmatrix} \alpha - E & \beta & 0 & 0 & 0 & 0 \\ \beta & \alpha - E & \beta & 0 & \beta & 0 \\ 0 & \beta & \alpha - E & \beta & \beta & 0 \\ 0 & 0 & \beta & \alpha - E & 0 & 0 \\ 0 & \beta & \beta & 0 & \alpha - E & \beta \\ 0 & 0 & 0 & 0 & \beta & \alpha - E \end{vmatrix} = 0$$

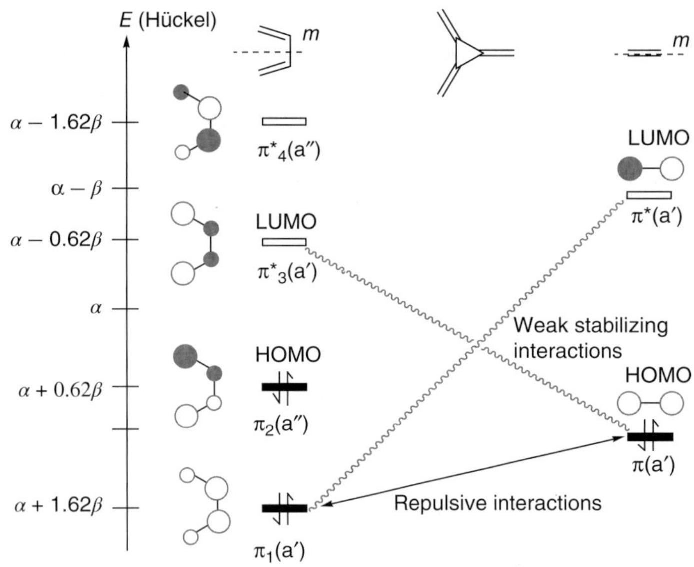

This PMO diagram summarizes the interactions between the π MOs of a *s-cis*-butadiene moiety and those of an ethylene unit approaching in a geometry that maintains a mirror plane of symmetry. The repulsive subHOMO(π_1(a′))/HOMO(ethylene, π) interaction is perhaps compensated by the stabilizing LUMO(diene)/HOMO(ethylene) interaction. Overlap between these two orbitals is relatively small (small coefficients at C(2), C(3) in HOMO(diene)), a situation different than that of fulvene. As found for fulvene, there is no LUMO(ethylene, π*)/HOMO(diene, π_2) interaction for reasons of symmetry. Thus, the PMO theory predicts [3]radialene to be less stabilized than fulvene due to π-conjugation.

Problem 4.7 Outline the synthesis of cyclooctatetraene presented by Willstätter. Propose another synthesis.

Answer

Willstätter's synthesis

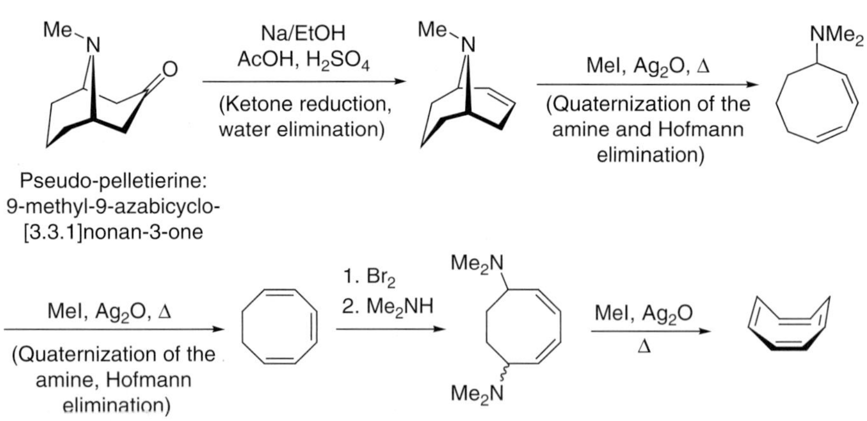

Reppe's synthesis, see Section 8.4.18

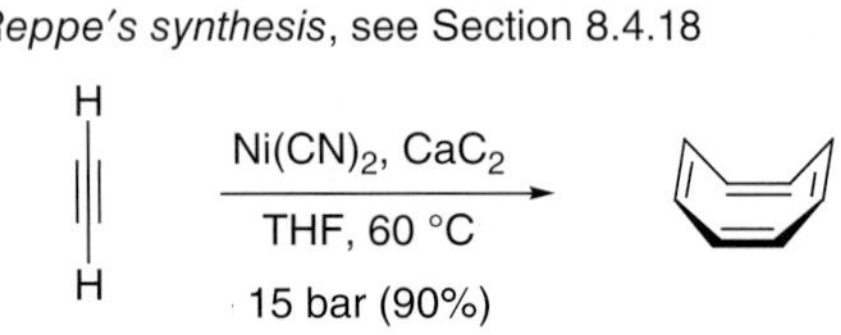

Problem 4.8 Write the Hückel secular determinant for methyl vinyl ether. Applying the PMO theory do you expect this compound to enjoy electronic stabilization due to n(O:)/π conjugation?

Answer

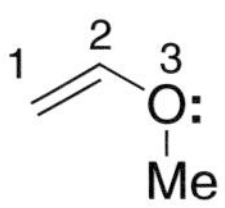

$$\begin{vmatrix} \alpha_C - E & \beta_{CC} & 0 \\ \beta_{CC} & \alpha_C - E & \beta_{CO} \\ 0 & \beta_{CO} & \alpha_O - E \end{vmatrix} = \begin{vmatrix} \alpha - E & \beta & 0 \\ \beta & \alpha - E & \beta \\ 0 & \beta & \alpha + \beta - E \end{vmatrix} = 0$$

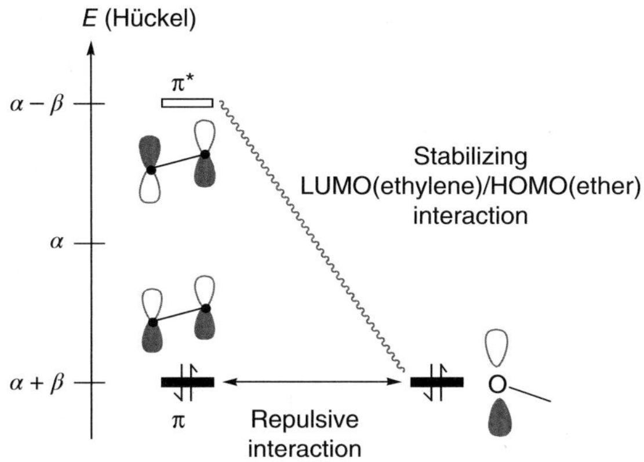

In this PMO diagram, alkenyl ethers enjoy some stabilization due to n(O:)/π conjugation because of the LUMO(ethylene, π*)/HOMO(ether) interaction. This stabilizing effect overcompensates the destabilizing, four-electron HOMO(ethylene, π)/HOMO(ether) interaction.

Problem 4.9 Write the Hückel determinant for oxirene (oxacyclopropene: *c*-C_2H_2O). Using the PMO theory do your predict this compound to enjoy electronic stabilization due to n(O:)/π conjugation?

Answer

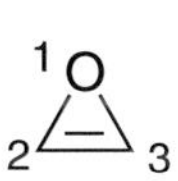

$$\begin{matrix} \alpha_O - E & \beta_{CO} & \beta_{CO} \\ \beta_{CO} & \alpha_C - E & \beta_{CC} \\ \beta_{CO} & \beta_{CC} & \alpha_C - E \end{matrix} = \begin{matrix} \alpha + \beta - E & b & \beta \\ \beta & \alpha - E & \beta \\ \beta & b & \alpha - E \end{matrix} = 0$$

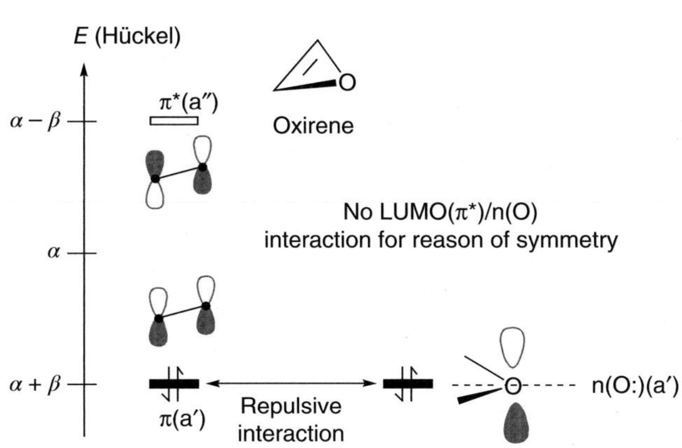

LUMO(ethylene, π^*) is antisymmetrical with respect to the mirror plane of symmetry of oxirene. Its overlap with 2p(O) is nil. There is no LUMO(π^*)/n(O) interaction for symmetry reasons. Oxirene is destabilized electronically, and it is isoelectronic with cyclopropenyl anion, an example of "antiaromatic" π-system.

Problem 4.10 What are the products of the reactions of dimethyl acetylenedicarboxylate with cyclopentadiene, furan, *N*-methylpyrrole, *N*-benzenesulfonylpyrrole? (Z = CH_2, O, NMe, NSO_2Ph. Section 2.7.8)

Answer

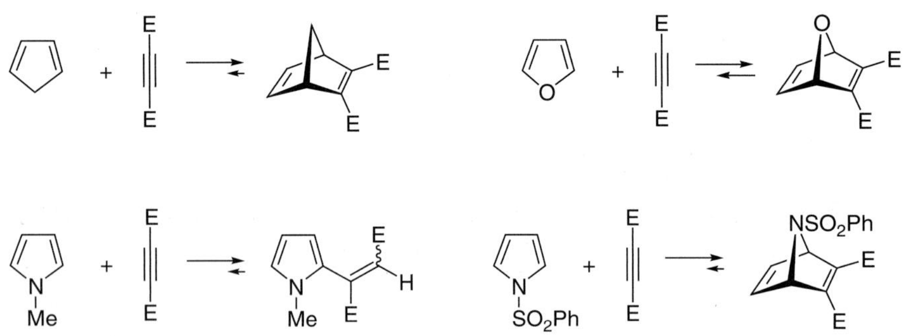

Problem 4.11 What are the products of direct bromination of furan in dimethylformamide? [8]

Answer

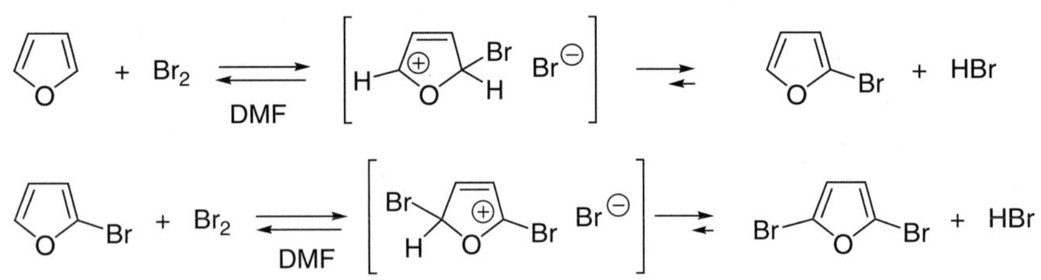

Problem 4.12 Which of 1-methylpyrrole and pyridine is the better nucleophile to react with MeI?

Answer

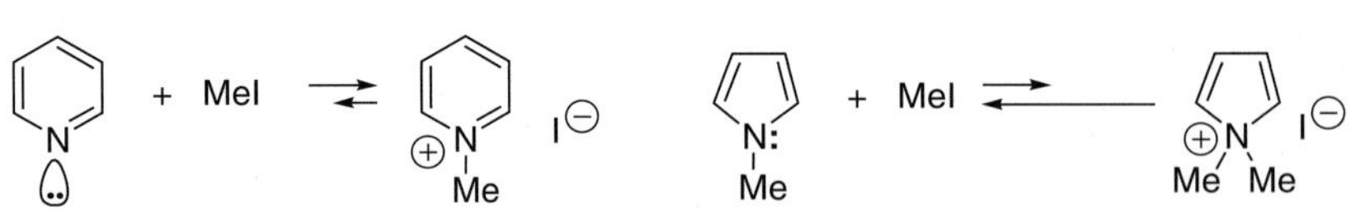

Pyridine reacts to give the corresponding quaternary ammonium salt. The nonbonding n(N:) electrons do not contribute to the aromaticity of the ring. On the contrary, the lone pair of pyrrole is needed for aromaticity and thus *N*-methylpyrrole is reluctant to react as a nucleophile. Quaternization of *N*-methylpyrrole with MeI is less exothermic than quaternization of pyridine.

Problem 4.13 Using data of Table 2.A.10 decide whether pyridine is more or less basic than pyrrole.

Answer

Table 1.24 gives:

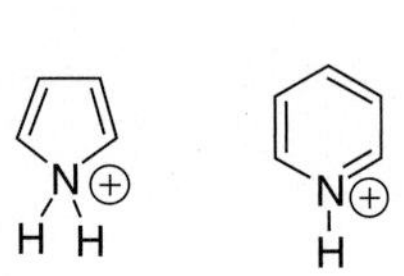

pK_a (H_2O): 0.0   5.2

The conjugate acid of pyrrole is a stronger acid than the conjugate acid of pyridine. Pyridine is a much stronger base than pyrrole. Pyrrole loses its n(N:) electron pair by protonation and thus loses its aromaticity. Pyridinium ion is aromatic, N-protonated pyrrole ion is not.

Problem 4.14 Why is pyrrole more acidic than piperidine?

Answer

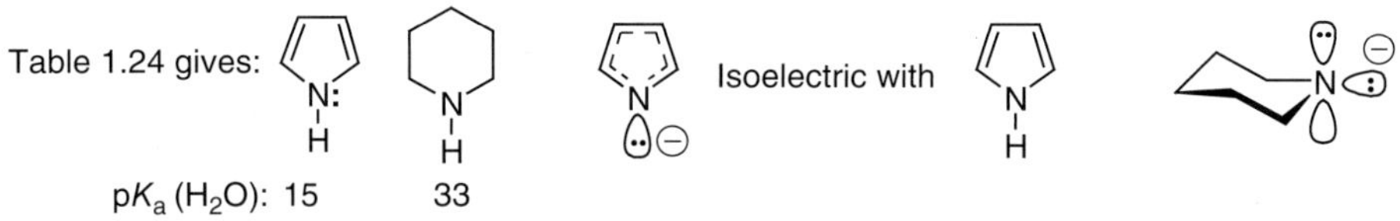

The conjugate base of pyrrole enjoys from a greater aromatic stabilization than pyrrole. In the former $\alpha(N^-)$ is higher than α(HN:), thus leading to a reduced energy gap between LUMO(diene) and HOMO(N^-). This is not possible with piperidine. Its conjugate base does not enjoy from any π-conjugation.

Problem 4.15 Why does the acidity of pyrrole increase upon introduction of nitrogen atoms in the ring?

Answer
Exchange of CH for N moieties in heterocyclic compounds stabilizes the anionic conjugate base because of the greater electronegativity of N compared with that of CH.

Problem 4.16 Pyridine reacts with electrophiles generating pyridinium salts reversibly. Substitution at the carbon centers is difficult. Clean formation of 3-nitropyridinum can be carried out under Bakke's conditions when pyridine reacts with N_2O_5 in SO_2 at 0 °C. Give a mechanism.

Answer

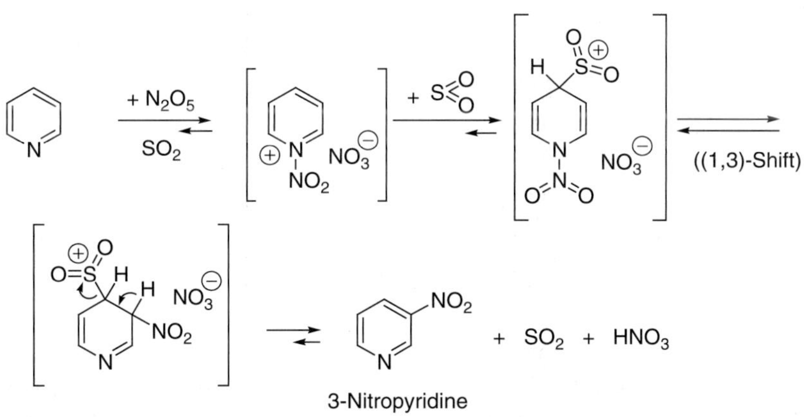

Problem 4.17 What type of stable cation in solution is formed upon treatment of 2,7-bis(*tert*-butyl)thiepin in HSO_3F/SO_2? [9]

Answer
Protonation at C(2), C(3), or C(4) of 2,7-bis(*tert*-butyl)thiepin (**A**) generates cation **B**, **C**, and **D**, respectively. These cations enjoy from stabilization by the two *tert*-butyl groups, by sulfur and π-conjugation. As for thio systems, one 2p(S:) electron pair enters in conjugation with the π-system, which realizes a homothianyl cation isoelectronic with homotropylium cation, e.g. **C** ↔ **C′** ↔ **C″**

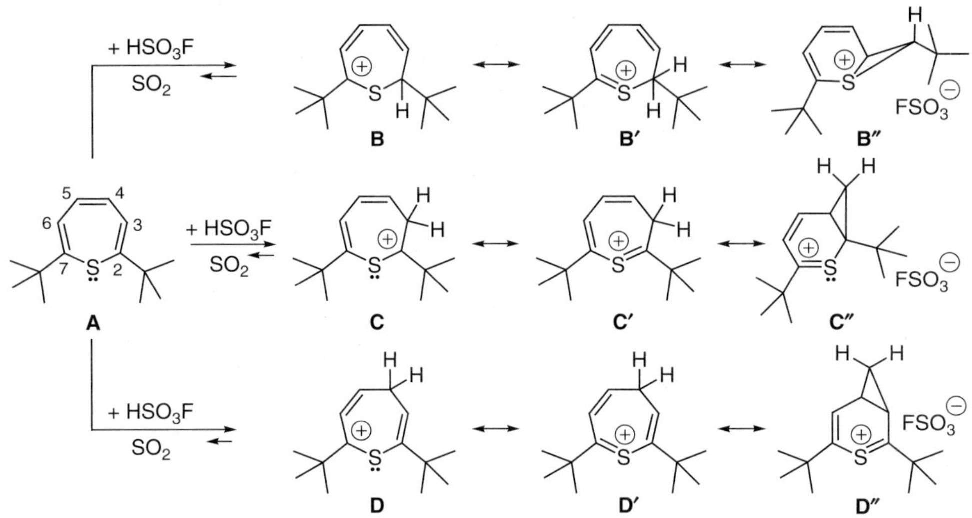

Problem 4.18 At 20 °C [2.2.2]hericene (2,3,5,6,7,8-hexamethylidenebicyclo[2.2.2]octane) adds two equivalents of SO_2 giving a bissulfolene. Under forcing conditions (high concentration, long reaction time), no trace of tris-sulfolene can be detected. Propose an explanation knowing that the cheletropic addition of SO_2 to *s-cis*-1,3-dienes gives the corresponding sulfolenes between −10 and 40 °C and that the sulfolenes undergo cheletropic elimination of SO_2 above 100 °C [10].

Answer

[2.2.2]Hericene (**A**) possesses three *s-cis*-butadiene units. In the presence of an excess of sulfur dioxide, and at 20 °C, rapid cheletropic additions give first sulfolene **B** and, then, bissulfolene **C** forms. No trace of tris-sulfolene **D** can be detected, even after prolonged staying or/and heating. The cheletropic additions **A** + SO_2 ⇆ **B** and **B** + SO_2 ⇆ **C** are reversible between 35 and 80 °C. Measurements of equilibrium constants $K(1)$ and $K(2)$ at various temperatures gave $\Delta_r H$(**A** + SO_2 ⇆ **B**) = -9.5 ± 1.2 kcal mol^{-1} and $\Delta_r H$(**B** + SO_2 ⇆ **C**) = -7.3 ± 1.2 kcal mol^{-1} in DMF. The addition **C** + SO_2 ⇆ **D** is expected to be much less exothermic due to the electronic destabilization in the barrelene derivative **D** (c. 10 kcal mol^{-1}), thus making addition **C** + SO_2 ⇆ **D** insufficiently exothermic to compensate for the entropy cost at 20–80 °C. This is why **D** is not detected.

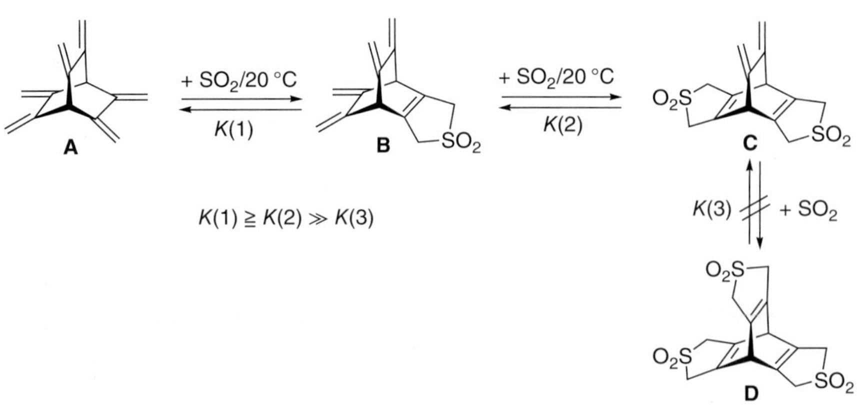

Problem 4.19 In CH_2=NH the two C—H bonds have different bond lengths. Give an explanation [11].

Answer

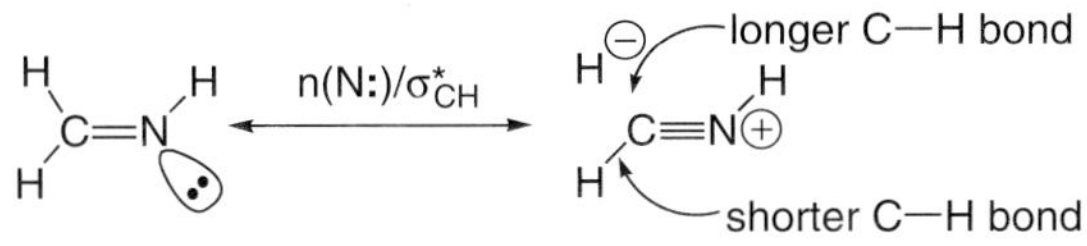

The C—H bond *anti* to the N lone pair is weakened and stretched by hyperconjugation with the N lone pair.

Problem 4.20 What are the most stable conformers of methyl fluoroformate? [12]

Answer

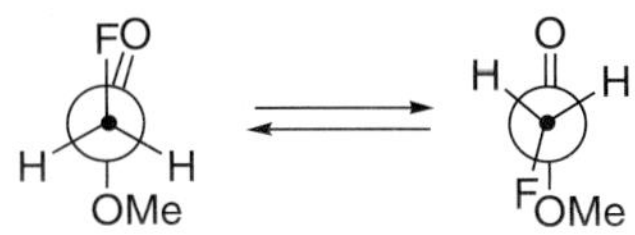

These two conformers have the same stability and are more stable than any other rotamers. The electrostatic repulsions F/C=O in the *syn* conformer are nearly the same as the electrostatic repulsion F/OMe in the *anti*-conformer. The donation $\sigma\_(FCH_2) \rightarrow \pi^*_{CO}$ is the origin of the stability of the *syn* and *anti*-conformers.

Problem 4.21 Explain the bathochromic shift introduced on the V ← N transition of cyclohexa-1,3-diene upon substitution with ethane-1,2-diyl groups at C(5) and C(6).

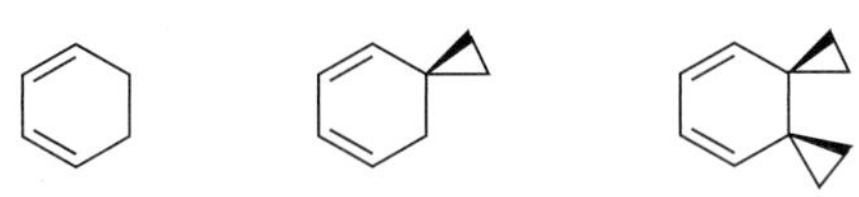

λ_{max}:	256 nm	269 nm	284 nm
ε:	4500	2200	4800

Answer
Introduction of a cyclopropane group is like adding a conjugated ethylene moiety to the diene, thus leading to a lowering of the energy of the $\pi \rightarrow \pi^*$ transition.

References

1 Bremer, M., Schleyer, P.v.R., and Fleischer, U. (1989). Nonplanarity in Huckel 2-pi aromatic systems – an Nmr–Iglo–ab initio proof of the puckered structure of cyclobutadiene dications. *J. Am. Chem. Soc.* 111: 1147–1148.
2 Firme, C.L., Antunes, O., and Esteves, P.M. (2007). Electronic nature of planar cyclobutenyl dication derivatives. *J. Phys. Chem. A* 111: 11904–11907.
3 Trabert, L. and Hopf, H. (1980). Alicyclic compounds. 3. New method of preparation and reactions of [4]radialene. *Liebigs Ann. Chem.* 1786–1800.

4 Thiele, J. (1900). Keton reactions via cyclopentadien. *Ber. Dtsch. Chem. Ges.* 33: 666–673.

5 Thiele, J. and Balhorn, H. (1906). The decendants of fulvene. 4. Condensation products of cyclopentadiene. *Liebigs Ann. Chem.* 348: 1–15.

6 Bergmann, E.D. (1968). Fulvenes and substituted fulvenes. *Chem. Rev.* 68: 41–84.

7 Bandar, J.S., Coscia, R.W., and Lambert, T.H. (2011). Demonstration of the facile reversibility of fulvene formation. *Tetrahedron* 67: 4364–4370.

8 Keegstra, M.A., Klomp, A.J.A., and Brandsma, L. (1990). Convenient synthetic procedures for 2-bromofuran and 2,5-dibromofuran. *Synth. Commun.* 20: 3371–3374.

9 Yamamoto, K., Yamazaki, S., Matsukawa, A., and Murata, I. (1984). Protonation of 2,7-di-tert-butylthiepines – the 1st observation of homothiopyrylium ions. *J. Chem. Soc., Chem. Commun.* 604–606.

10 Birbaum, J.L. and Vogel, P. (1988). Tandem cheletropic additions of sulfur-dioxide to [2.2.2]hericene – the Barrelene effect. *Helv. Chim. Acta* 71: 1474–1482.

11 Li, A.Y. (2007). Chemical origin of blue- and redshifted hydrogen bonds: Intramolecular hyperconjugation and its coupling with intermolecular hyperconjugation. *J. Chem. Phys.* 126: 154102-1–154102-11.

12 Sahnoun, R., Fujimura, Y., Kabuto, K. et al. (2006). Hyperconjugative electron-delocalization mechanism controlling the conformational preference of fluoroacetaldehyde and methyl fluoroacetate. *Bull. Chem. Soc. Jpn.* 79: 555–560.

5

Pericyclic Reactions*

Answers to Problems

Problem 5.1 Thermal isomerization of *cis*-bicyclo[3.2.0]hept-6-ene into (*Z,Z*)-cyclohepta-1,3-diene has an Arrhenius activation energy $E_a = 45.5\,\text{kcal mol}^{-1}$. In contrast, the isomerization of **A** into **B** has a much lower activation energy $E_a = 21.3\,\text{kcal mol}^{-1}$ [1]. Give an explanation.

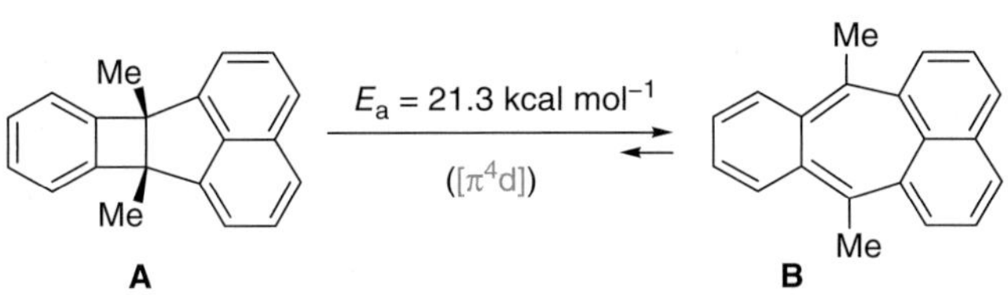

Answer

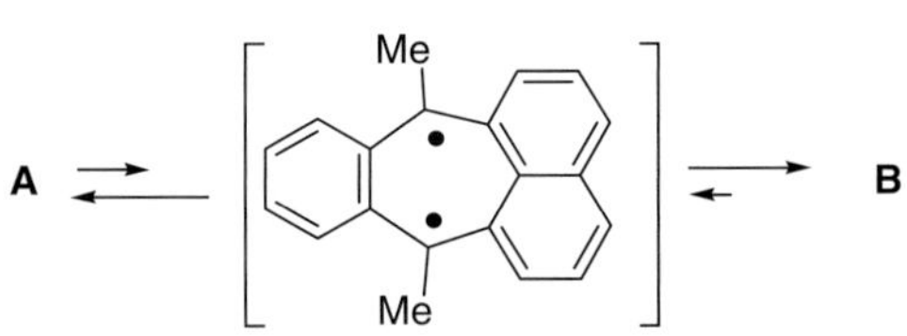

Relatively stable diradical (tertiary dibenzyl) intermediate

Problem 5.2 What are the major products formed upon thermolysis of the benzocyclobutene derivatives **A** and **B** [2]?

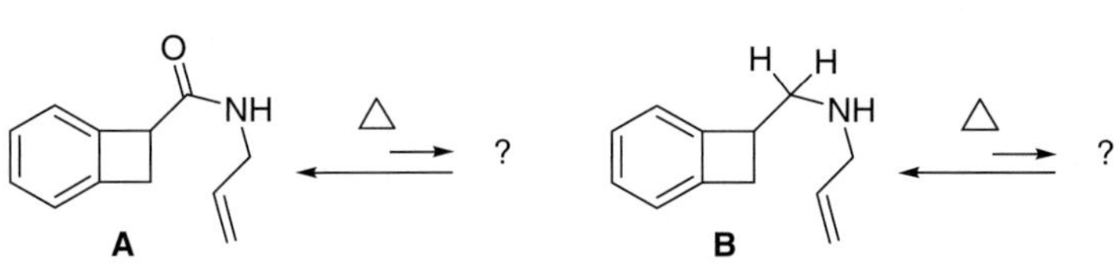

*See end of the chapter for additional material.

Organic Chemistry: Theory, Reactivity and Mechanisms in Modern Synthesis, Workbook, First Edition. Pierre Vogel and Kendall N. Houk.

Answer

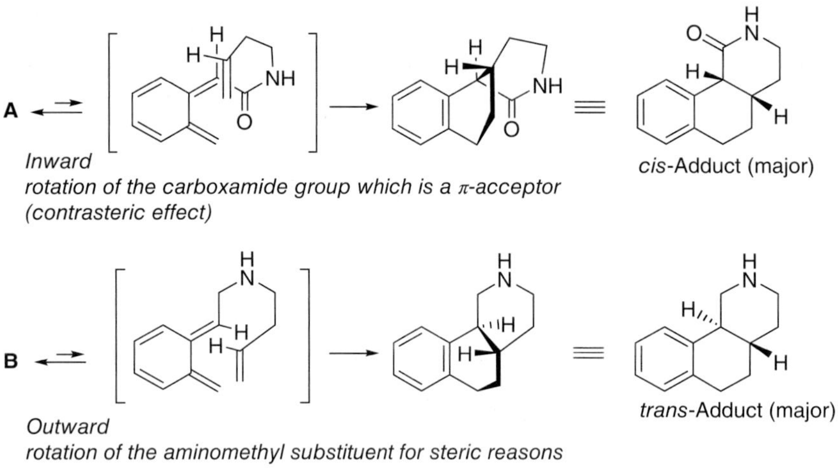

Problem 5.3 On heating to 100 °C cyclobutene **A** is isomerized into a 85 : 15 mixture of products **P** + **Q**. What are **P** and **Q** [3]?

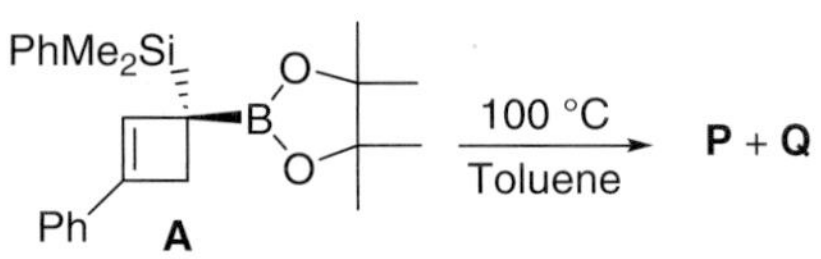

Answer

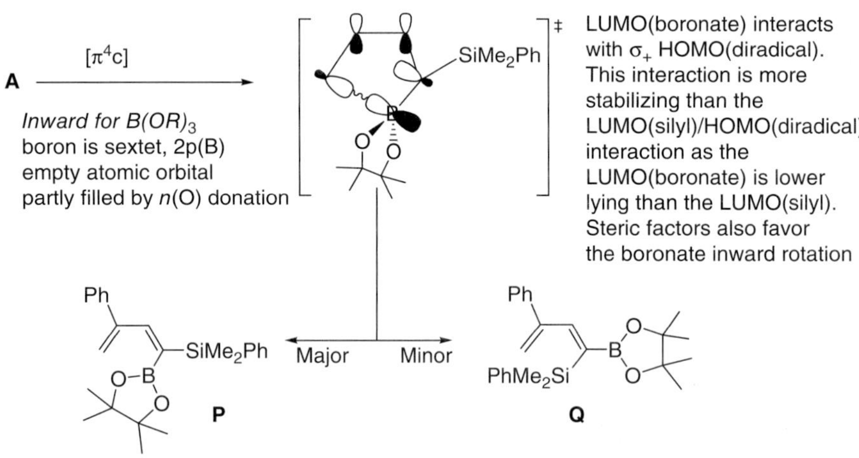

Problem 5.4 The reaction of phenylacetylenes **A** with aldehydes **B** in the presence of SbF_5 and EtOH first gives the products **P**, which upon heating in 1,2-dichloroethane provide the corresponding indanones **Q**. What are **P** and **Q**? [4]

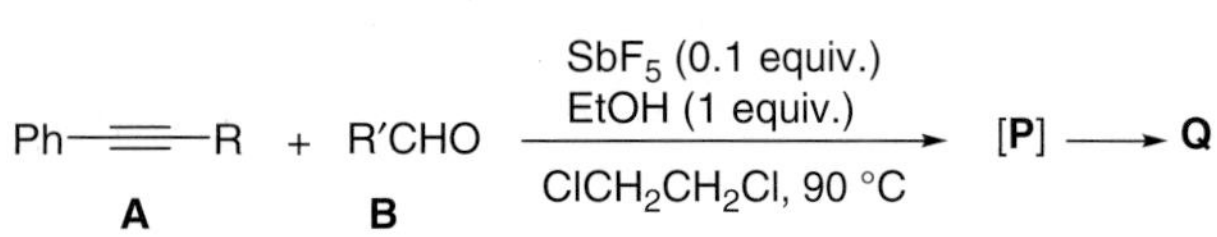

Answer

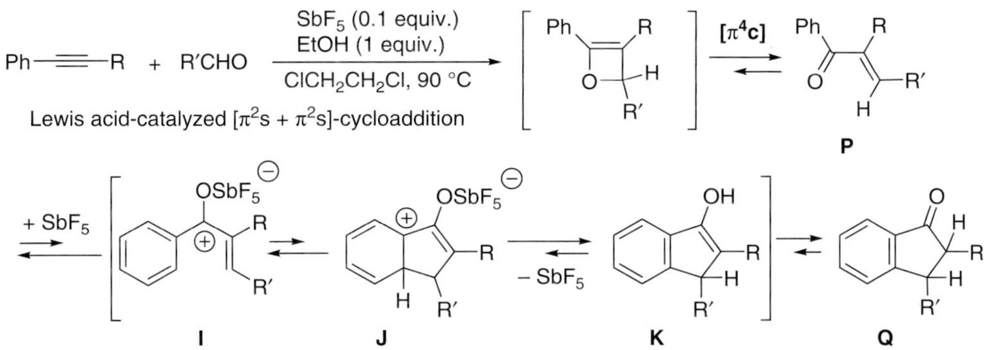

Problem 5.5 What are the products **P** + **Q** of the following reaction? [5]

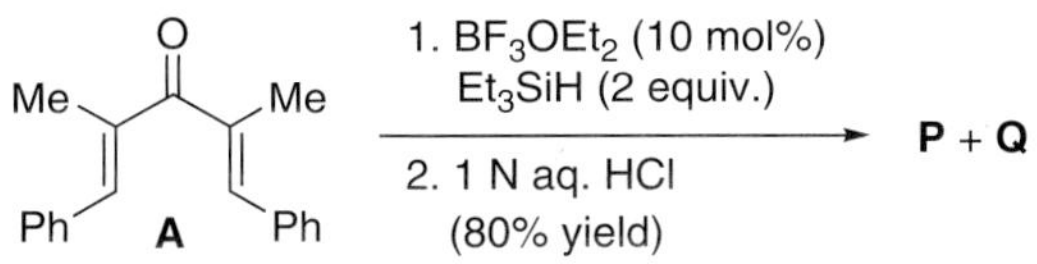

Answer

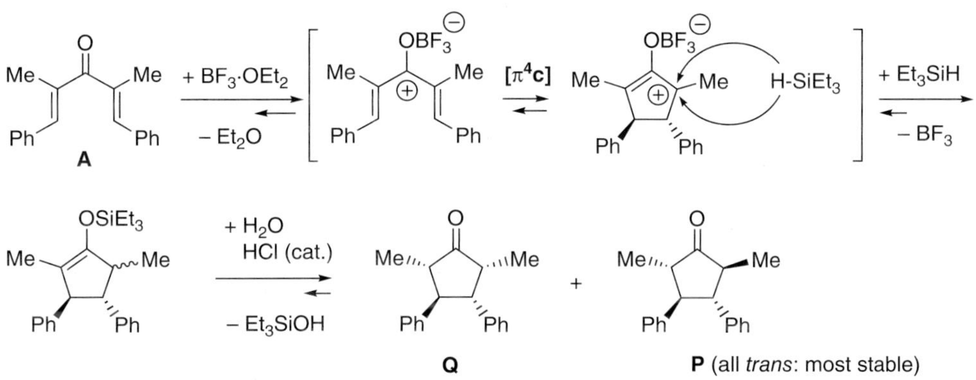

Problem 5.6 Estimate the heat of isomerization of cyclopropyl carbanion into allyl carbanion in the gas phase.

Answer

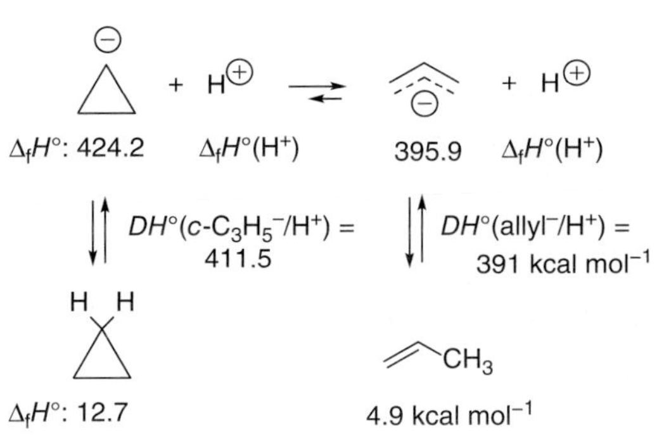

Proton affinities are defined by $DH^\circ(\text{R}^-/\text{H}^+) = \Delta_f H^\circ(\text{R}^-) + \Delta_f H^\circ(\text{H}^+) - \Delta_f H^\circ(\text{R–H})$

Thus, $\Delta_f H^\circ(c\text{-C}_3\text{H}_5^-) = DH^\circ(c\text{-C}_3\text{H}_5^-/\text{H}^+) - \Delta_f H^\circ(\text{H}^+) + \Delta_f H^\circ(c\text{-C}_3\text{H}_6)$

$\Delta_f H^\circ(\text{CH}_2{=}\text{CH–CH}_2^-) = DH^\circ(\text{CH}_2{=}\text{CH–CH}_2^-/\text{H}^+) - \Delta_f H^\circ(\text{H}^+) + \Delta_f H^\circ(\text{propene})$

$\Delta_f H^\circ(\text{CH}_2{=}\text{CH–CH}_2^-) - \Delta_f H^\circ(c\text{-C}_3\text{H}_5^-) =$

$DH^\circ(\text{CH}_2{=}\text{CH–CH}_2^-/\text{H}^+) + \Delta_f H^\circ(\text{propene}) - DH^\circ(c\text{-C}_3\text{H}_5^-/\text{H}^+) - \Delta_f H^\circ(c\text{-C}_3\text{H}_6) =$

$= 391 + 4.9 - 411.5 - 12.7 = -28.3 \text{ kcal mol}^{-1}$

Problem 5.7 Propose a mechanism for the conversion of **A** into **P** [6].

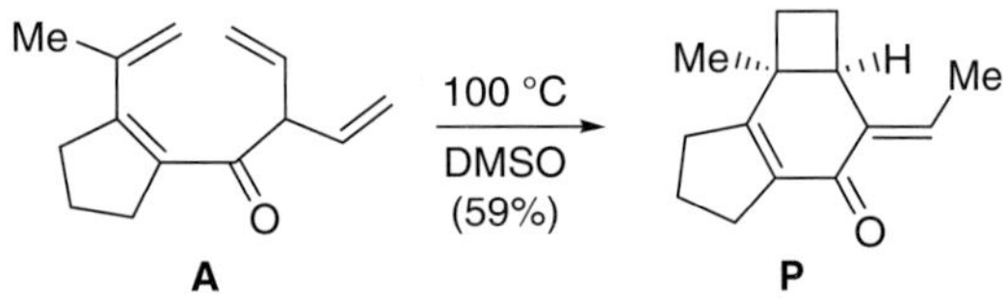

Answer

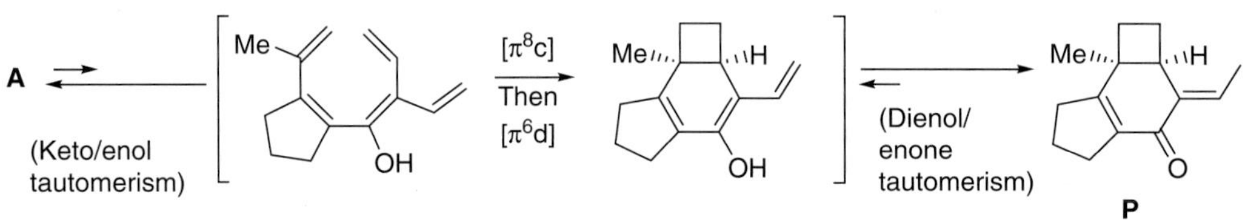

Problem 5.8 Potassium salt **B** is isomerized at −41 °C into **P**. Above 30 °C **P** is isomerized into **Q** [7]. Propose mechanisms for these isomerizations.

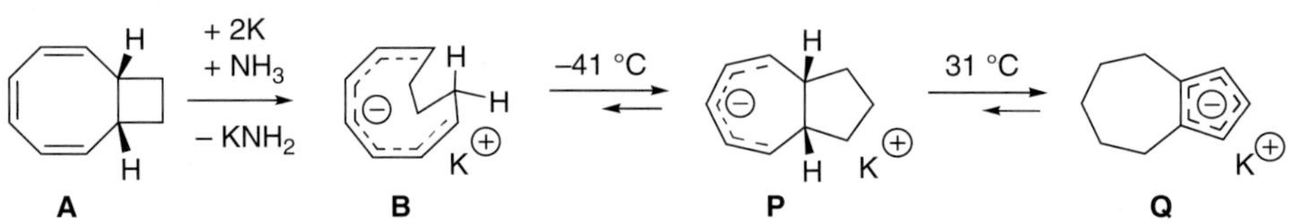

Answer

A adds formally one equivalent of KH to its strained σ(C—C) bond giving (*all-Z*)-cyclodeca-2,4,6-trien-1-yl anion/K$^+$ ion pair **B**. Direct electrocyclization of **B** into **P** would imply a "forbidden" eight electron disrotatory process. Anion/cation recombination converts **B** into the covalent intermediate **I**, which undergoes allylic rearrangement and re-ionization into (*E,Z,Z,Z,Z,Z*)-cyclodeca-2,4,6-trien-1-yl anion/K$^+$ ion pair **J**. This anion can now undergo a thermally "allowed" conrotatory electrocyclization into **P**.

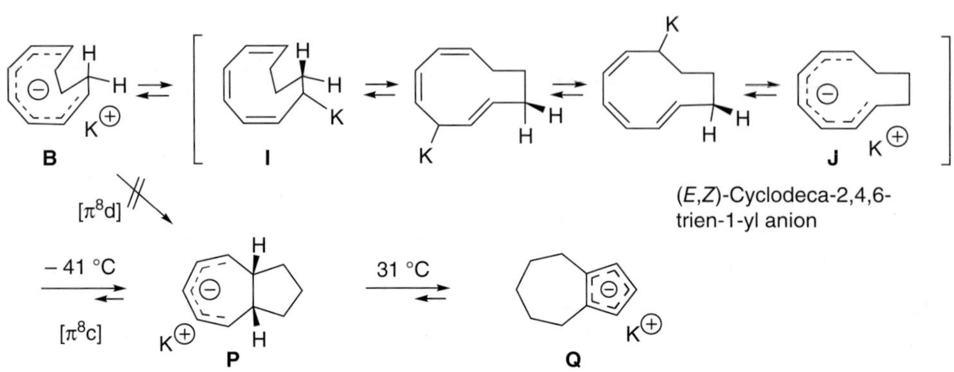

In ammonia, **P** can be protonated and deprotonated (R–K + $NH_3 \rightleftarrows KNH_2$ + R–H) into several possible isomers until **Q**, a cyclopentadienyl anion stabilized by aromaticity, is formed.

Problem 5.9 What is the major product of the reaction of **A** + **B**?

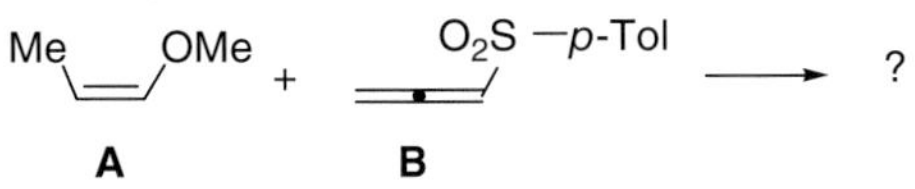

Answer

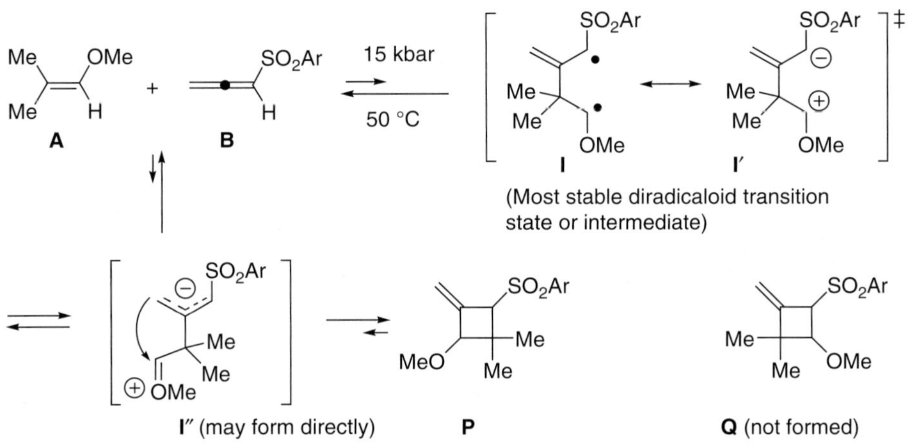

Problem 5.10 What is the product of reaction **A** + **B**? [8]

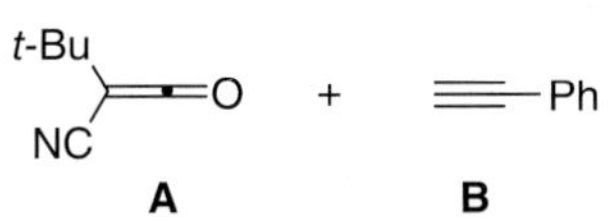

Answer

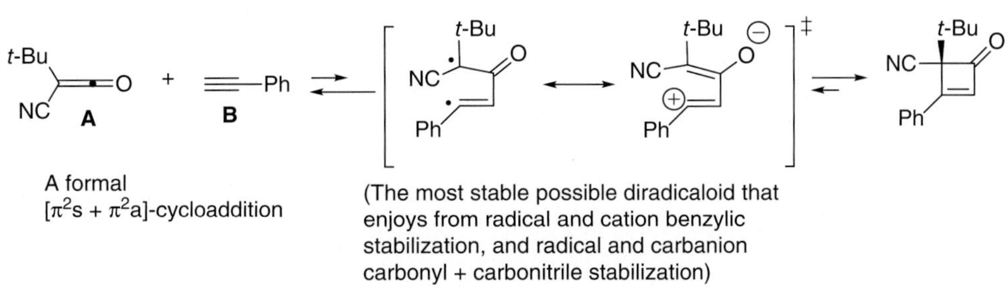

Problem 5.11 Why are *trans*- and not *cis*-disubstituted *β*-lactams formed in the Staudinger reaction used by Sheehan and Ryan in their synthesis of nocardicin D?

Answer

The imine used in this synthesis is electron-rich because of the MeS substituent. This stabilizes the iminium moiety of the zwitterionic intermediate **I** and thus permits its *Z/E* isomerization before ring closure, which leads to the major *trans*-cycloadducts.

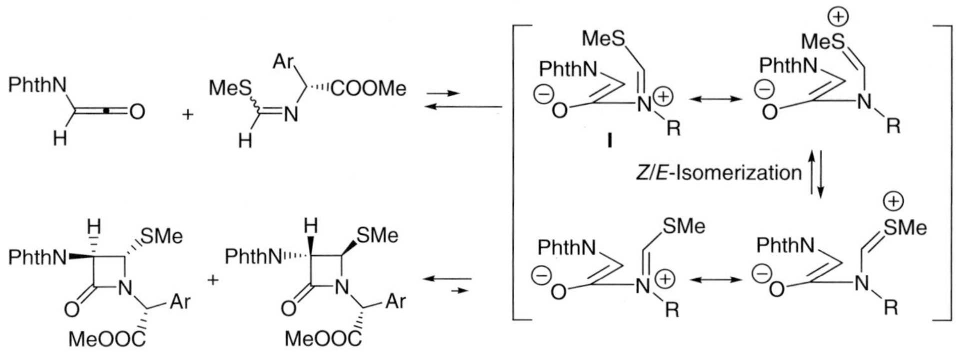

Problem 5.12 What is the product of the reaction (3 days at 60 °C) of diphenylketene with thiobenzophenone?

Answer

Ketenes react with ketones giving β-lactones. Ketenes react with thioketones giving the corresponding thio-β-lactones.

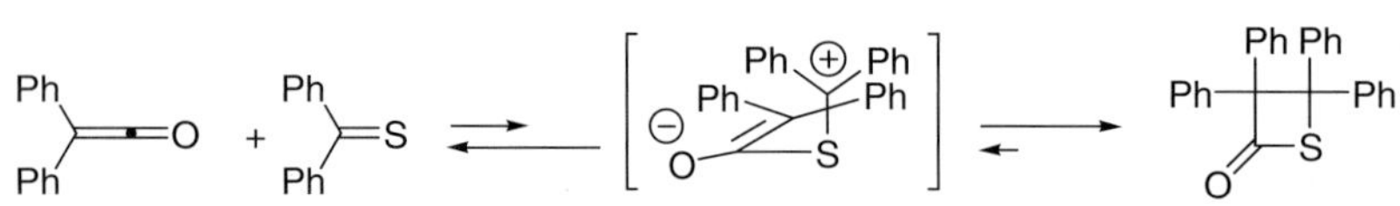

Problem 5.13 What is the major (2 + 2)-cycloadduct of reaction **A** + **B**?

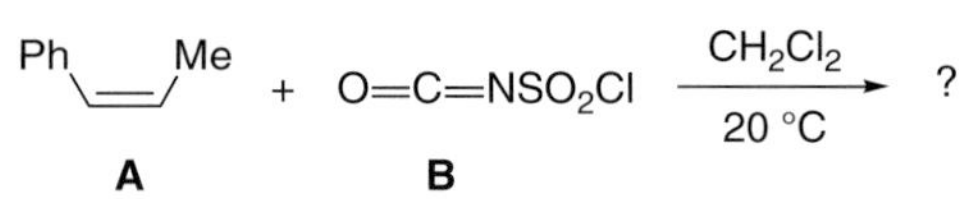

Answer

This cycloaddition is similar to the (2+2)-cycloaddition of ketenes to allenes that probably follows concerted [π^2s+π^2a] mechanisms with transition structures (TSs) resembling that of a π-complex between the allene and the carbonyl group of the isocyanate. The latter delocalizes electrons from the alkene to the heteroallene. The positive charge on the alkene prefers to reside on the carbon substituted by the phenyl group rather than on the carbon center substituted by the methyl group (for substituent effects on the stability of carbenium ions and hydride affinities, see Table 1.A.14). Thus, the nitrogen center of the heteroallene combines with the center of the alkene substituted by the phenyl group, giving the cycloadduct **P**.

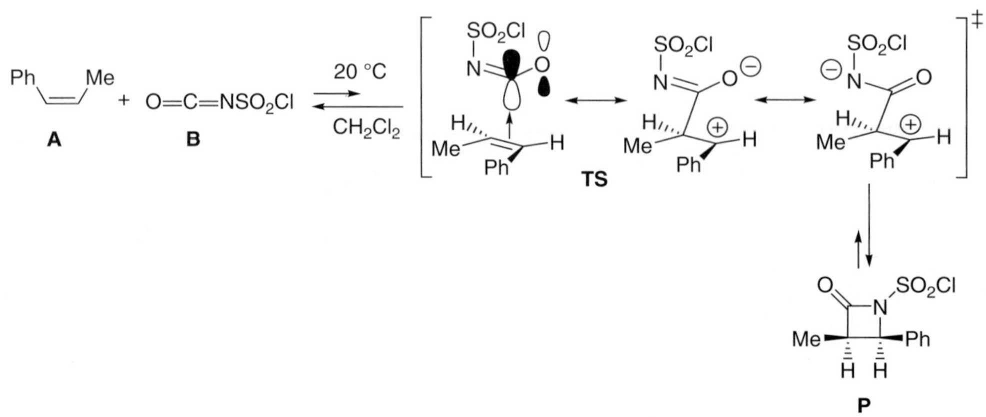

Problem 5.14 Propose a one-step conversion of aldehyde **A** into α-santalol, the principal constituent of sandalwood oil [9].

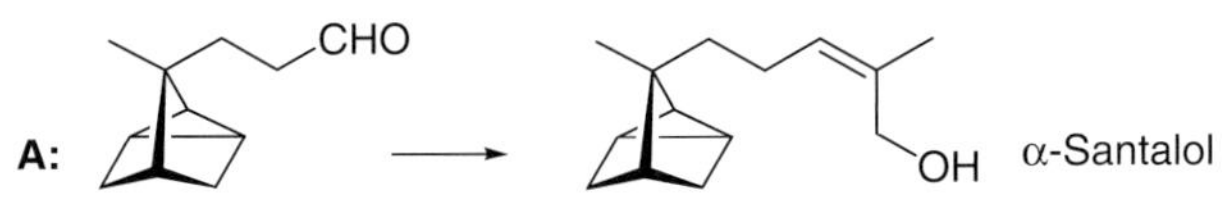

Answer

At −78 °C in tetrahydrofuran (THF) $Ph_3P{=}CHMe$ adds to aldehyde **A** and equilibrates with betaïne **B**. The latter reacts with one equivalent of *n*-BuLi in hexane at −78 °C producing ylide **Y**. Treatment of **Y** with pure CH_2O at 0 °C gives the dioxidophosphonium derivative **D**. Aqueous work-up generates α-santalol.

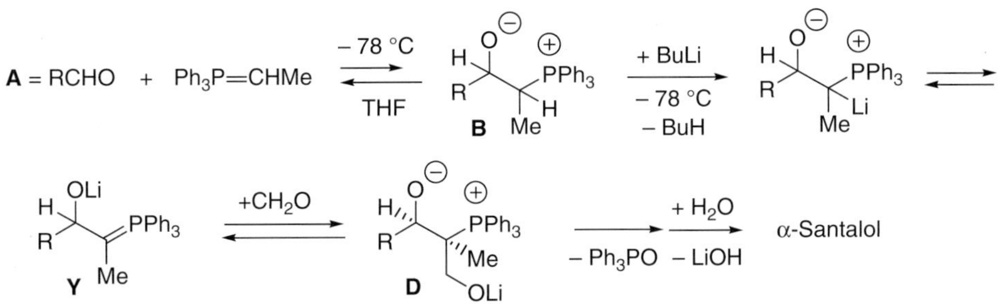

Problem 5.15 What is the product of the following reaction? [10]

Answer

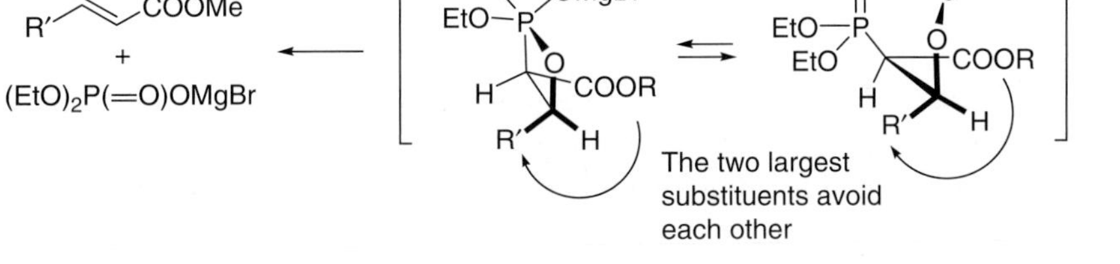

Problem 5.16 Give a possible mechanism for the following stereoselective and enantioselective reaction [11].

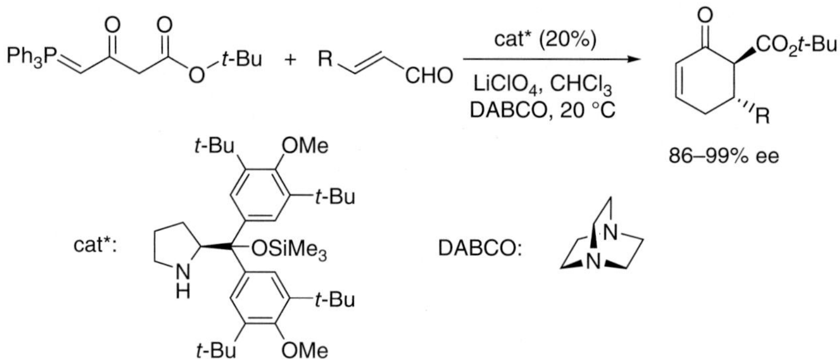

Answer

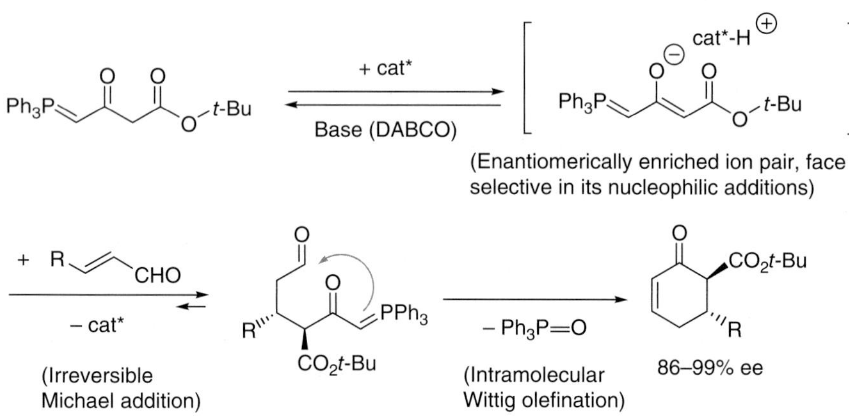

Problem 5.17 What is the main product of the reaction of [4]radialene (tetramethylidenecyclobutane) with 4-phenyl-1,2,4-triazoline-3,5-dione in large excess at 25 °C? [12]

Answer

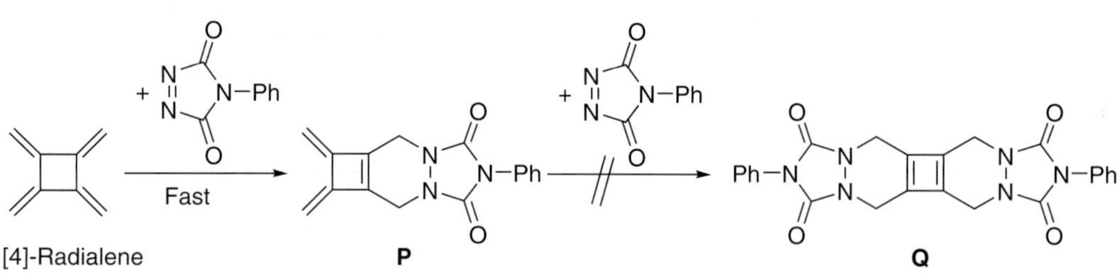

4-Phenyl-1,2,4-triazoline-3,5-dione is one of the most reactive dienophiles (Diels–Alder reaction with normal electronic demand). Even in the presence of a large excess of the dienophile only the monoadduct **P** of [4]-radialene forms; no trace of the corresponding double cycloadduct **Q** is seen at 25 °C. The addition of the second equivalent of the dienophile generates a cyclobutadiene derivative **Q**, which is unstable (Hückel anti-aromaticity).

Problem 5.18 Explain why the rate constant ratio (k_1/k_2) for the two successive Diels–Alder reactions of tetracyanoethylene (TCNE) to 2,3,5,6-tetramethylidenebicyclo[2.2.2]octane is 17 whereas that for the two successive Diels–Alder reactions of TCNE to 2,3,5,6-tetramethylidene-7-oxabicyclo[2.2.1]heptane amounts to 376 at 25 °C [13].

Answer

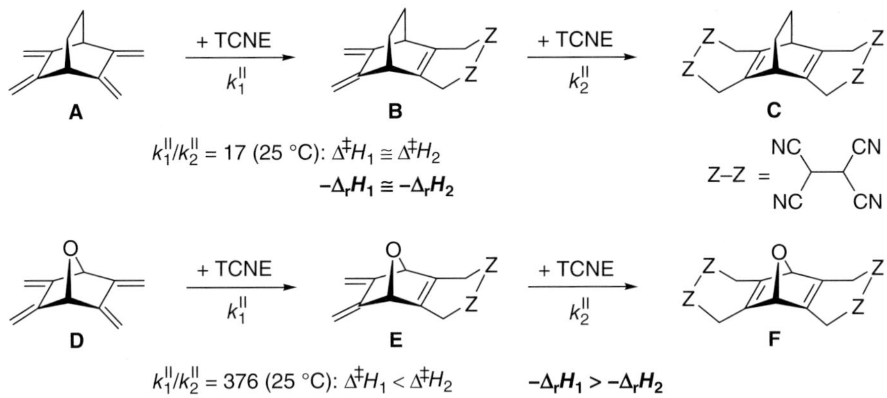

Strain increase between **F** and **E** is higher than that between **E** and **D** whereas strain remains the same between **A**, **B**, and **C**. According to the Dimroth principle, or Bell–Evans–Polanyi theory for concerted, one-step reactions ($\Delta^{\ddagger}H = \alpha\Delta_r H + \beta$) the cycloaddition rate decreases with a decrease in exothermicity.

Problem 5.19 Propose a reactivity trend for the following cycloreversions [14].

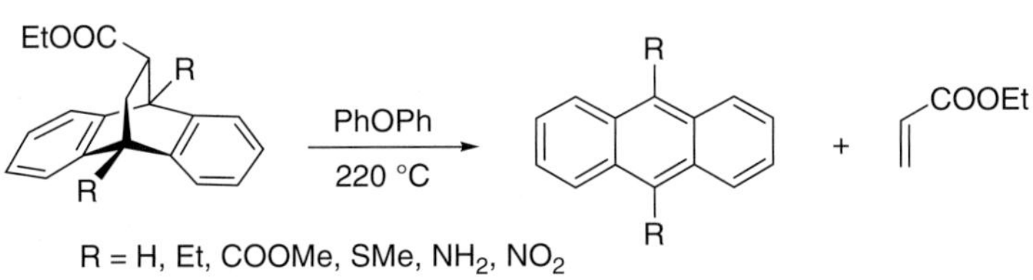

Answer

These [σ^4s+π^2s]-cycloreversions generate ethyl acrylate, an electron-poor dienophile, and 9,10-disubstituted anthracenes that are more or less electron rich depending upon the nature of the substituents R. Thus, like the Diels–Alder reactions with normal electron demand, these retro-Diels–Alder reactions see their transition structures affected by differential steric factors and by the sum EI(anthracene) – EA(dienophile). As a consequence, the more electron-rich anthracenes will form faster than the less electron-rich anthracenes. Thus the following reactivity trend can be proposed:

R:	NO_2	<	COOMe	<	H	<	Et	<	OMe	<	NH_2
Observed (10^5)k^{I}:	0.031		0.415		1.72		34.7		69		4000 s^{-1}

Problem 5.20 Propose a reactivity trend for the following Diels–Alder reactions For X = MeO, SO_2Ph_2, and $P(O)Ph_2$ [15].

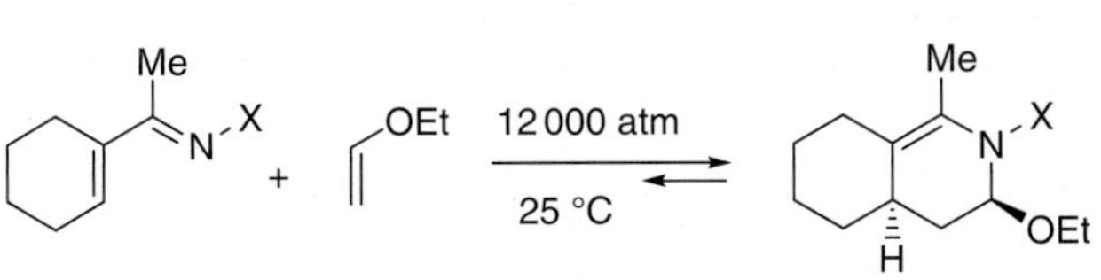

Answer

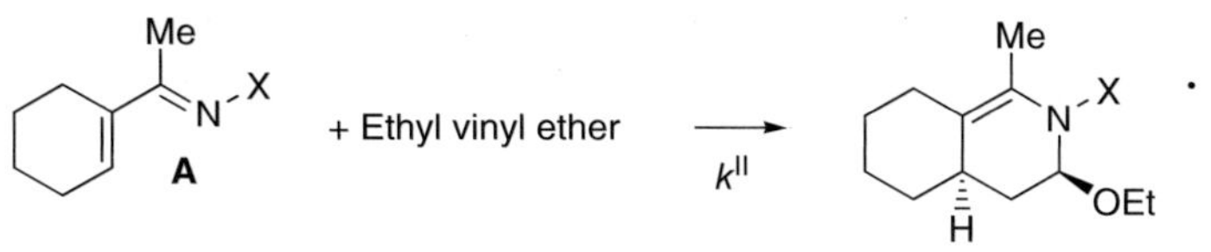

The 1-aza-1,3-dienes **A** (α,β-unsaturated imines) are electron-poor dienes like acrolein and 1-oxa-1,3-dienes. They react preferentially with electron-rich dienophiles such as ethyl vinyl ether. The more **A** is electron poor, the faster is its cycloaddition with ethyl vinyl ether. Among the three substituents X of **A**, X = MeO is the electron richest, and $PhSO_2$ the strongest electron withdrawing group. Acidities in dimethylsulfoxide (DMSO) can be used to compare the ability of $PhSO_2$ and Ph_2PO to withdraw electrons from a π-system. Table 2.A.11 gives $pK_a(PhSO_2Me) = 29.0$ and $pK_a(Ph_2(P{=}O){-}Me) = 31.3$. As $PhSO_2Me$ is a stronger acid than $Ph_2P({=}O){-}Me$, the $PhSO_2$ substituent in **A** must be more electron withdrawing than Ph_2PO. Thus, the reactivity sequence is expected to be MeO < Ph_2PO < $PhSO_2$.

Problem 5.21 Compound **A** is stable under neutral and acidic conditions but it undergoes quick cycloreversion into **P** + **Q** upon treatment with KH in THF at 25 °C. Give an explanation [16].

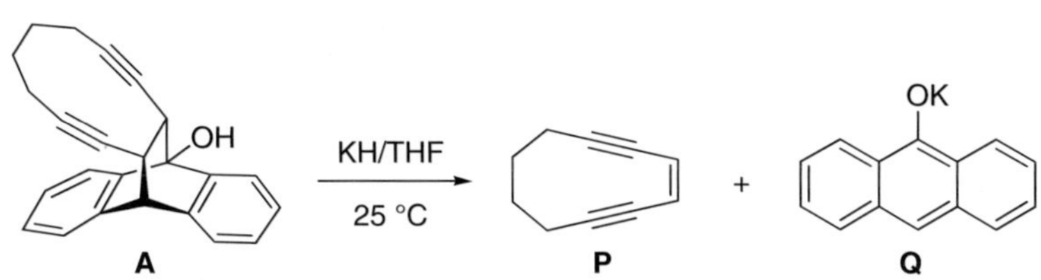

Answer

Upon deprotonation of alcohol **A**, alcoholate **B** is formed. The oxide substituent (O^-) in **B** is far more electron rich than the hydroxy substituent (OH) in **A**. In the transition state of the retro-Diels–Alder reaction, the zwitterionic limiting structure **T′** is better stabilized by the O^- than by the OH substituent. This renders the cycloreversion of **B** much quicker than that of **A** at 25 °C.

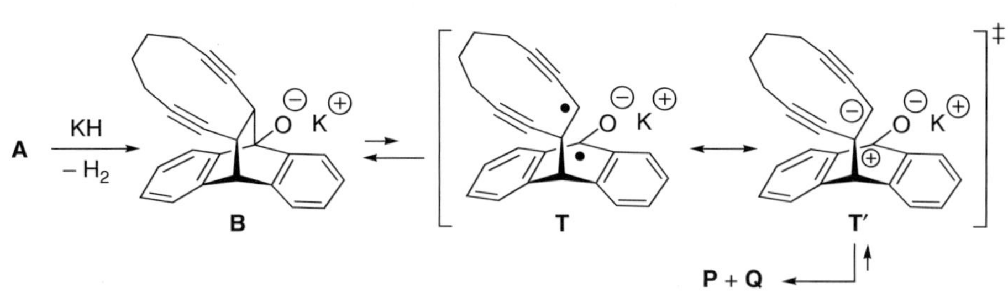

Problem 5.22 Give the reactivity trend of the Diels–Alder reactions of dimethyl 1,2,4,5-tetrazine-3,6-dicarboxylate with styrene, acrylonitrile, vinyl acetate, and ethyl ethyl ether [17],

Answer

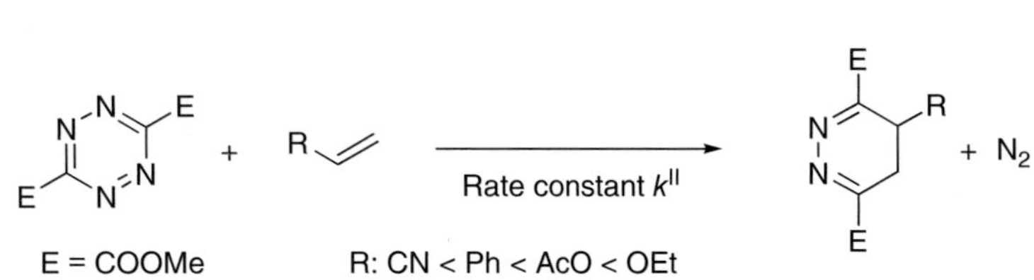

Problem 5.23 Predict the regioselectivity of the following Diels–Alder reactions.

(a) 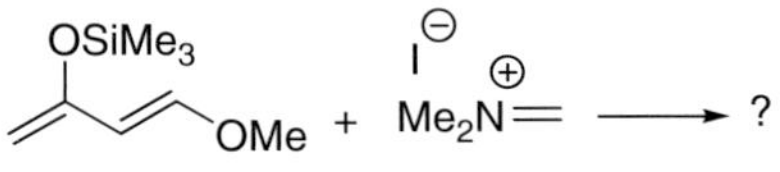

(b) 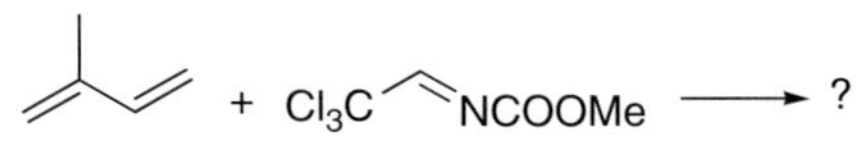

(c) 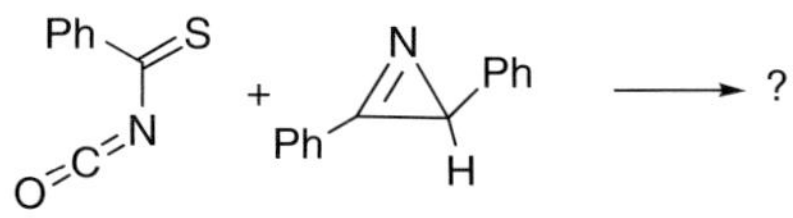

(d) 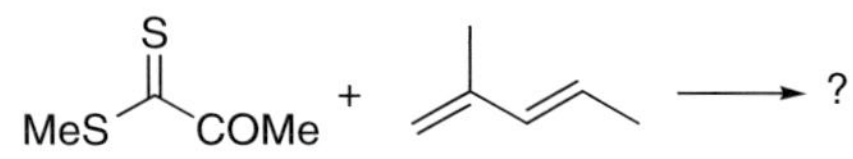

(e) 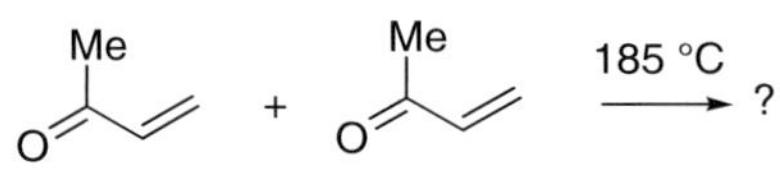

(f) 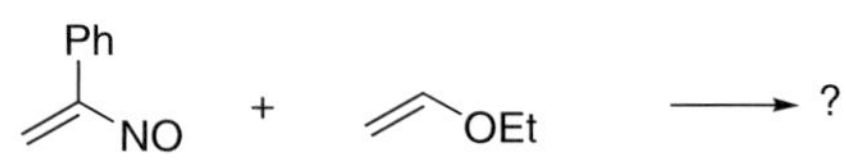

(g) 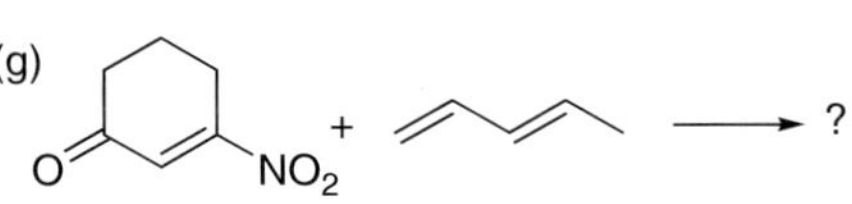

(h) 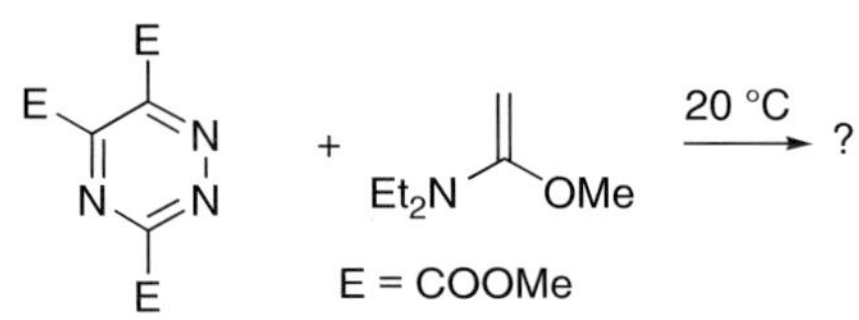

Answer

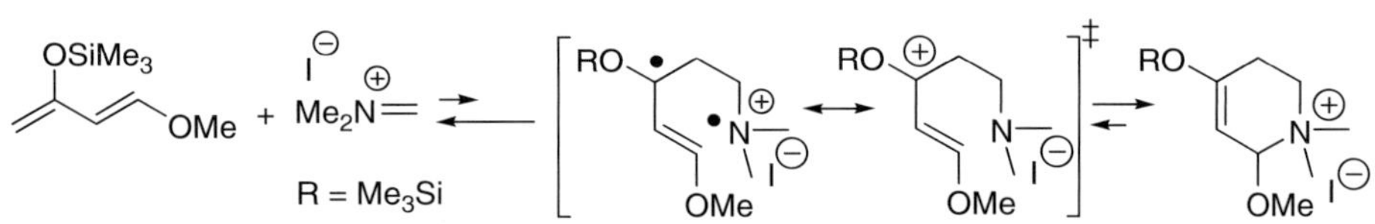

(a)

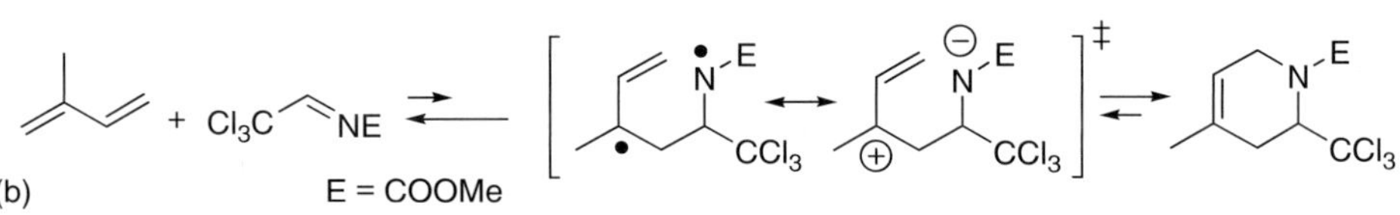

(b)

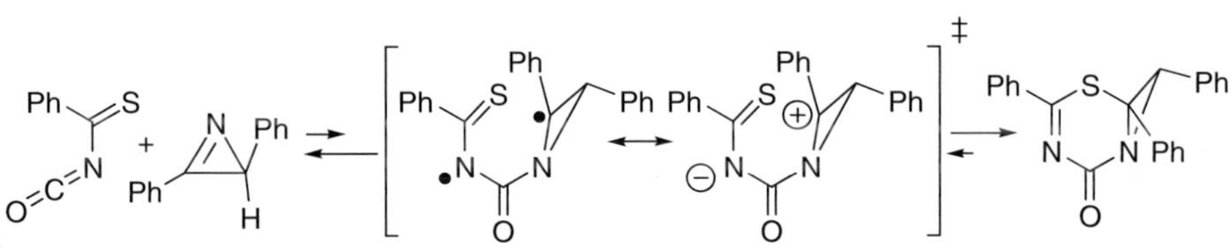

(c)

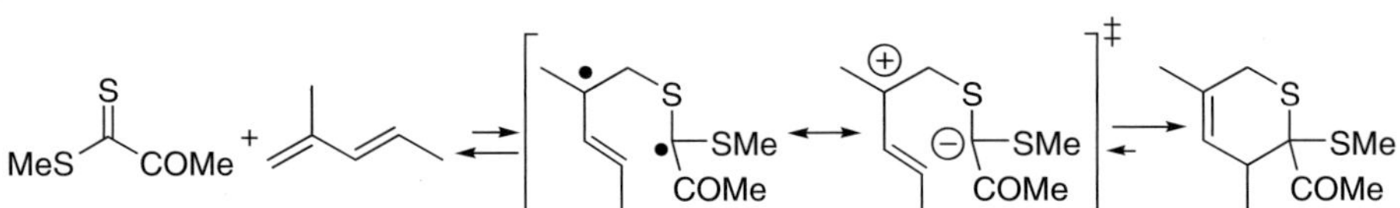

(d)

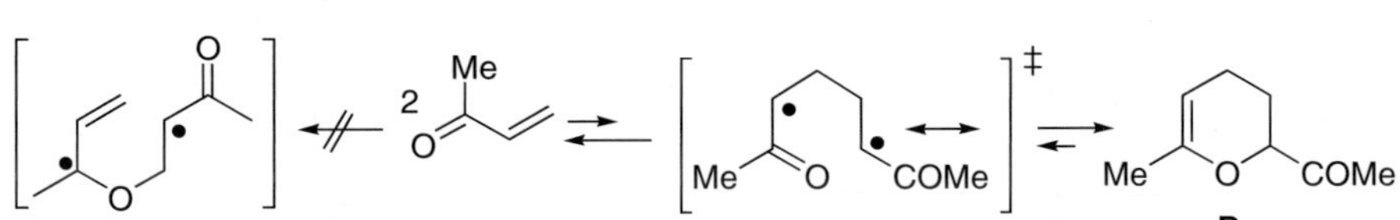

P

Does not form because it implies the conversion of a C=O double bond into a C–O single bond (about 17 kcal mol^{-1} less favorable than the conversion of a terminal C=C double bond into a C–C single bond. The stabilization of a 1-methyl-1-oxyallyl radical is not better than the stabilization of

(e) 1-acyl-1-alkylmethyl radical + 17 kcal mol^{-1}).

Involves the combination of two C=C double bonds into two C–C single bonds; the diradical is doubly stabilized by conjugation with two acetyl groups.

Jenner et al. [18] found a single adduct **P** for the cyclodimerization of methyl vinyl ketone. As the activation parameters of the reaction are slightly solvent sensitive (in heptane: $E_a = 18.3\,\text{kcal mol}^{-1}$, $\log A = 4.5$; in EtOAc: $E_a = 17.9$, $\log A = 4.0$; in CH_2Cl_2: $E_a = 17.8$, $\log A = 3.7$, in MeCN: $E_a = 17.0$, $\log A = 3.2$), the reaction may well be concerted or involve a diradical intermediate with little polarization.

(f)

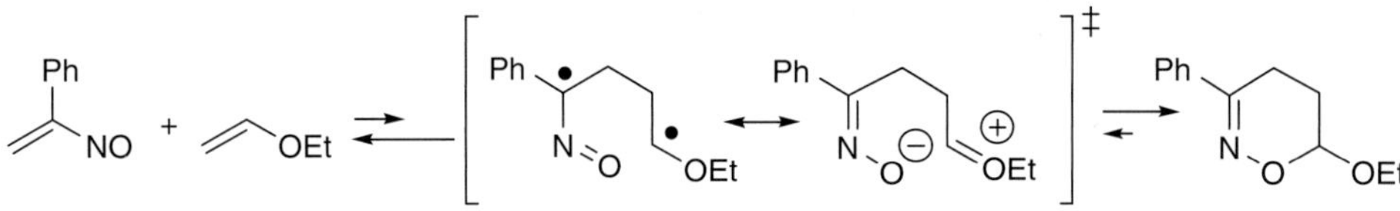

(g)

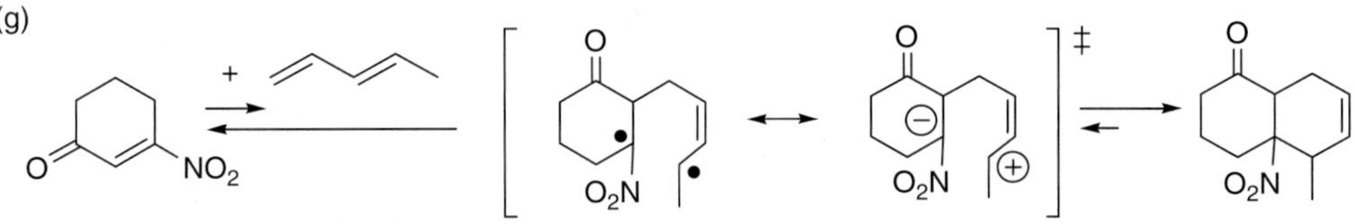

$DH°(O_2N\text{–}CH_2^-/H^+) = 356.4\ \text{kcal mol}^{-1}$

$DH°(MeCO\text{–}CH_2^-/H^+) = 369\ \text{kcal mol}^{-1}$

The nitro group stabilizes a carbanion better than an acetyl group; thus the Diels–Alder reaction is "*ortho*" regioselective with respect to the nitro substituent.

(h)

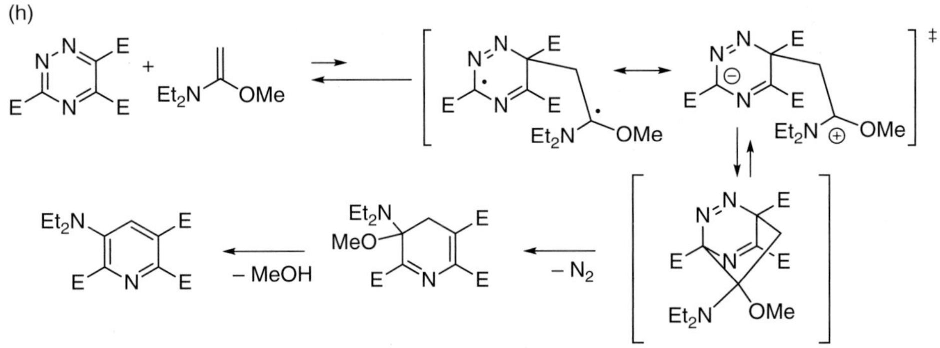

Problem 5.24 What is the major product of the following reaction? [19]

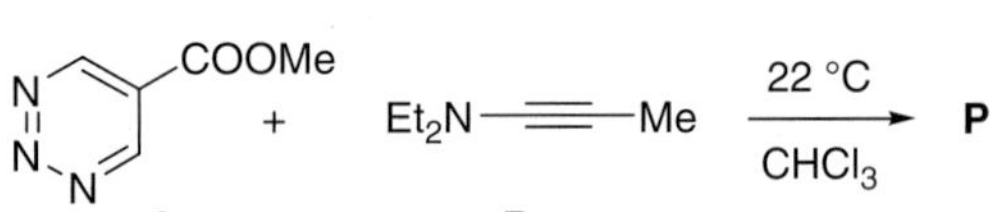

A **B**

Answer

Like 1,2,4,5-tetrazines that are electron-poor dienes reacting preferentially with electron-rich dienophiles (see Table 5.3), triazine **A** and ynemine **B** undergo a Diels–Alder reaction with inverse electron demand. There are six possible orientations for the attack of **B** onto **A**, two pairs leading to the same (4+2)-cycloadducts as shown below.

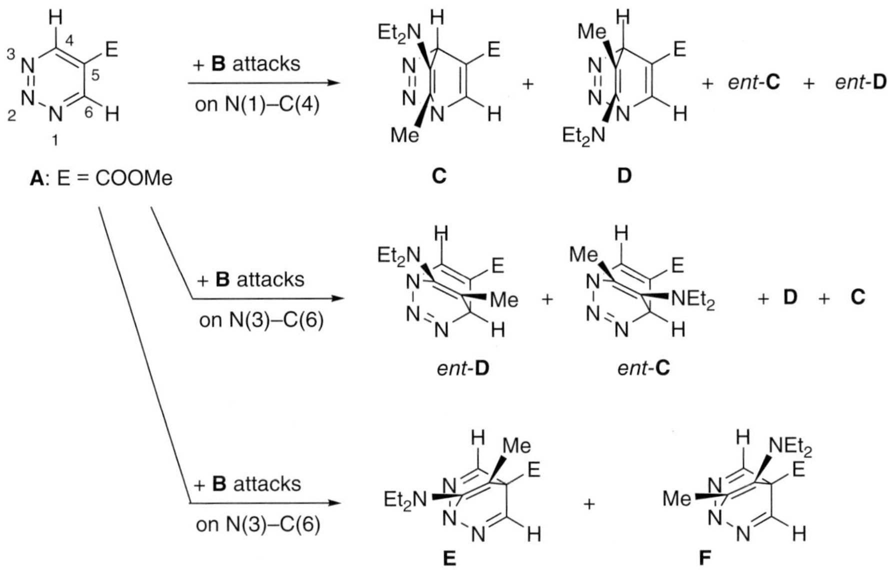

The fastest of these cycloadditions is that for which the most stable diradicaloid transition structure can be drawn. The preferred diradical limiting structures are those that contain the radical fragment Et_2N—C=C rather than Me—C=C and an ester substituted allylic carbon-centered radical. The corresponding zwitterionic limiting structures possess a cation moiety highly stabilized by the Et_2N substituent, and the anionic moiety conjugated (stabilized) with the carboxylic ester group. Accordingly, **D** is the preferred cycloadduct under kinetic control. Once formed, it undergoes fast cycloreversion providing **P** and N_2.

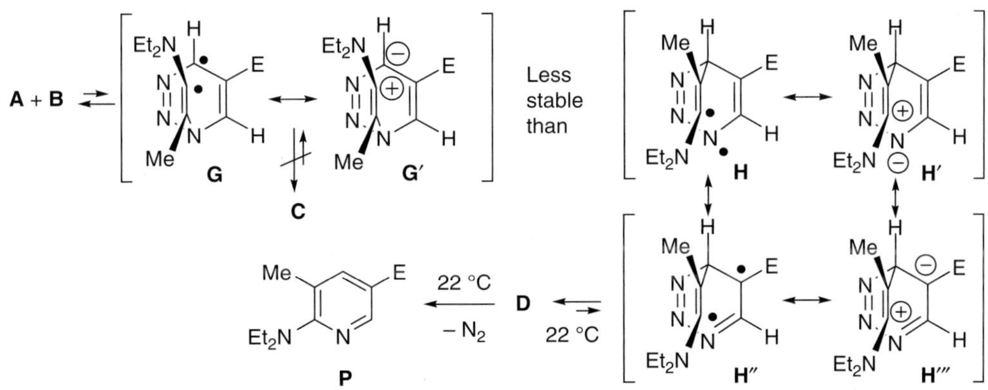

Problem 5.25 Give an explanation for the regioselectivity of the following enantioselective Diels–Alder reaction [20].

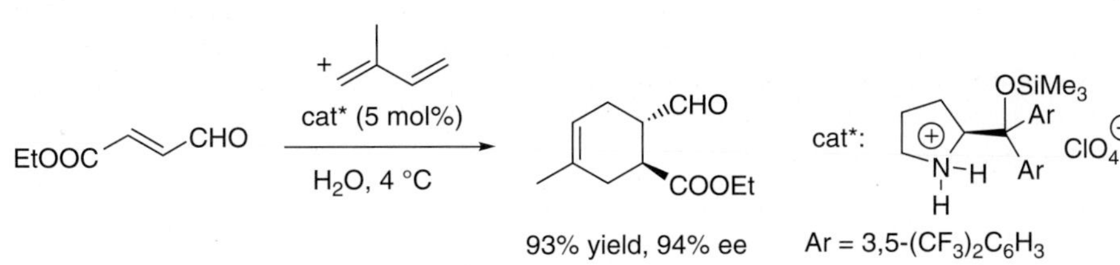

Answer

This Diels–Alder reaction is *para*-regioselective with respect to the carbaldehyde group and *meta*-regioselective with respect to the carboxylic ester moiety. As aldehydes are more acidic (α-C--H) than esters (see Table 1.A.25) the carbaldehyde group is better than the ester group in stabilizing the zwitterionic limiting structure of the transition state of this cycloaddition. The catalyst cat* is a protic acid H–A$^+$ ClO_4^- that protonates the most stable zwitterionic limiting structure (generates **T** ↔ **T′**) and consequently accelerates the reaction.

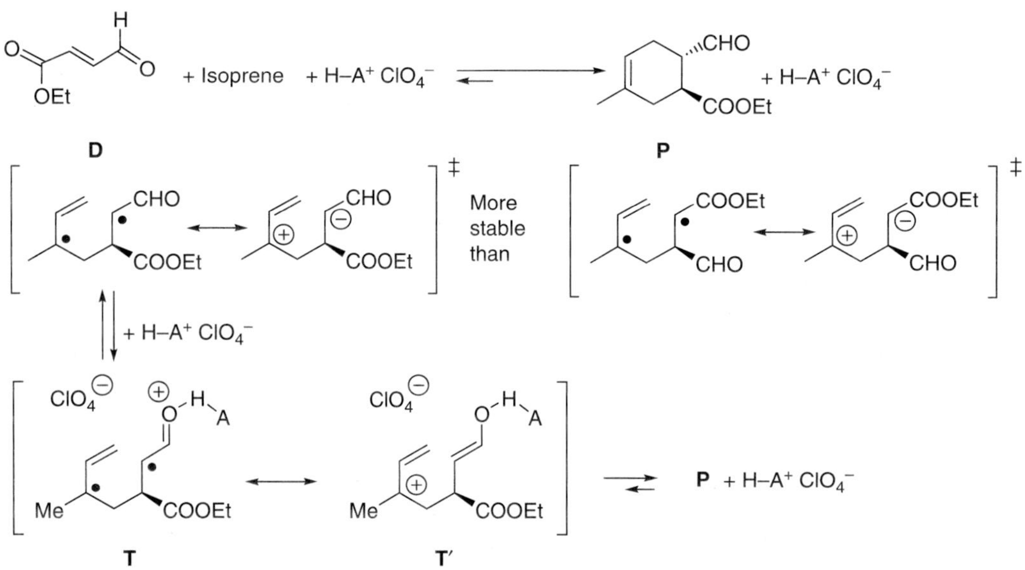

The following $TiCl_4(THF)_2$-catalyzed Diels–Alder reaction is also *para* regioselective with respect to the carbaldehyde group [21].

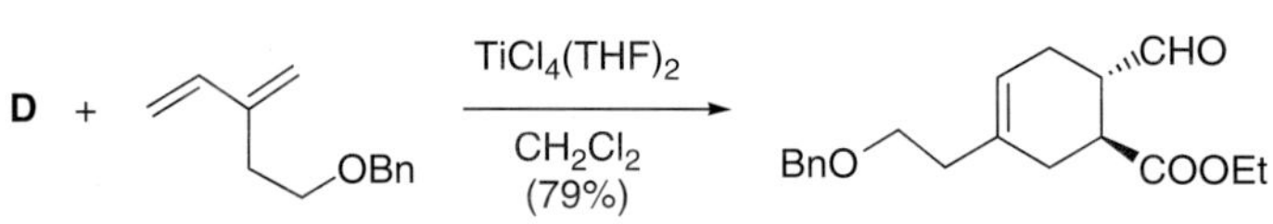

Problem 5.26 Thiophene 1-oxides add to electron-rich and electron-poor dienophiles preferentially onto the face of the diene containing the S=O moiety [22–26].
Give an explanation.

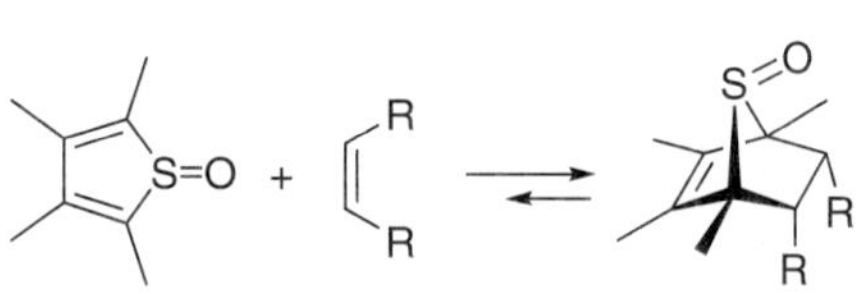

Answer

The Cieplak hyperconjugation theory suggests that the n(S:) electron pair will assist the σ(C—C) bond breaking or σ(C—C) bond making if placed antiperiplanar with this σ bond in the transition state of the cycloaddition. This is the case for the *syn* attack, not for the *anti* attack.

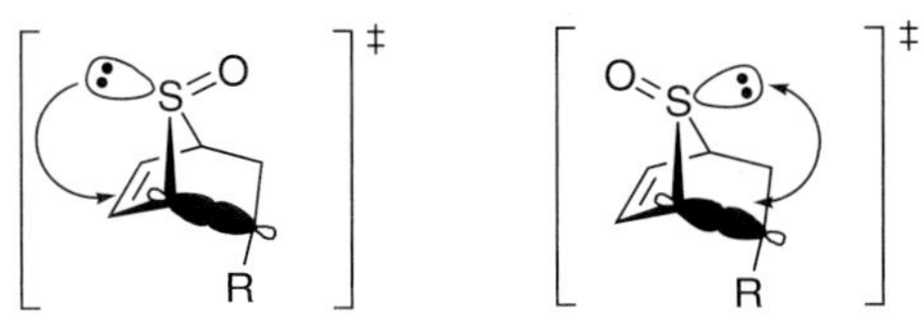

Favored Disfavored

Alternatively, the S=O function can be seen as the MeO group of 5-methoxycyclopentadiene (CpOMe). The orbital mixing rule of Fukui predicts a distortion of the highest occupied molecular orbital (HOMO) of thiophene 1-oxide similar to that described in Figure 5.30 for CpX with n(X:) lower lying in energy than π_2(diene):

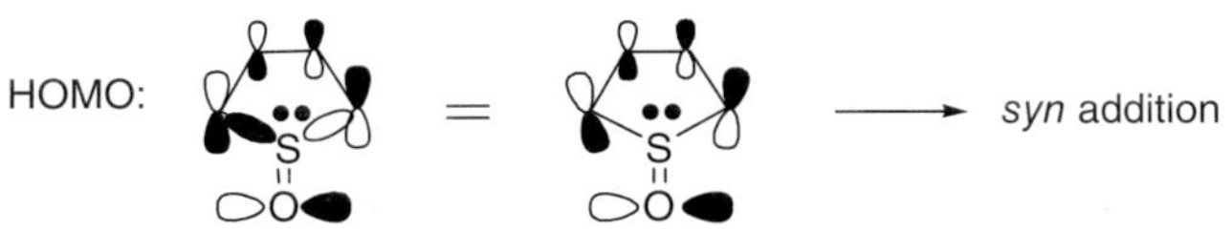

Problem 5.27 Give the major products formed in the following cycloadditions (for reactions (a)–(e) see [27]; for reaction (f), see [28]; for reaction (g), see [29]; for reaction (h), see [30])

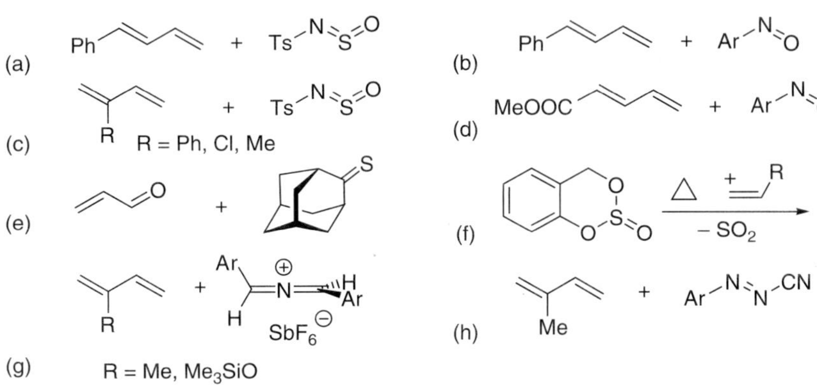

Answer

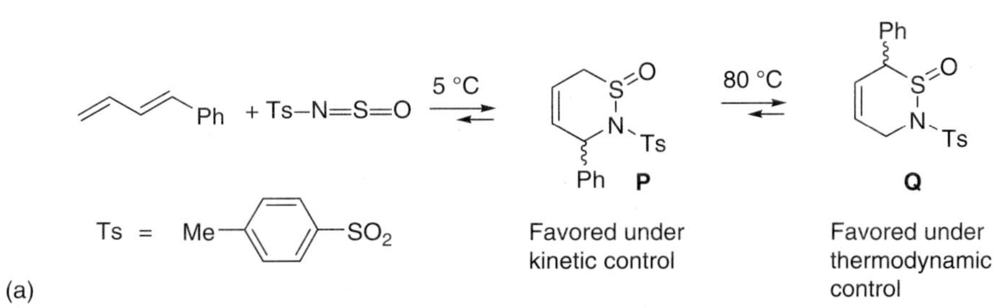

Because of the greater electronegativity difference between S/O than between S/N the S=N double bond is weaker than the S=O double bond and thus the former adds with more exothermicity than the latter. Because of steric repulsion between the Ts and Ph groups cycloadduct **P** is less stable than **Q**. Product **P** forms faster than **Q** because the diradicaloid **T** ↔ **T′** is more stable than the regioisomeric diradicaloid **U** ↔ **U′**.

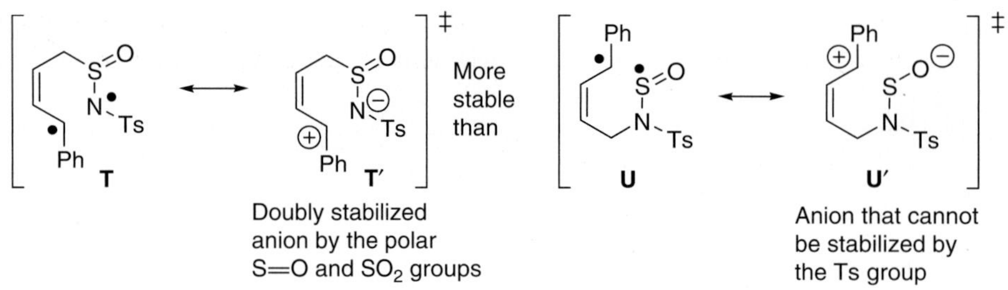

(a)

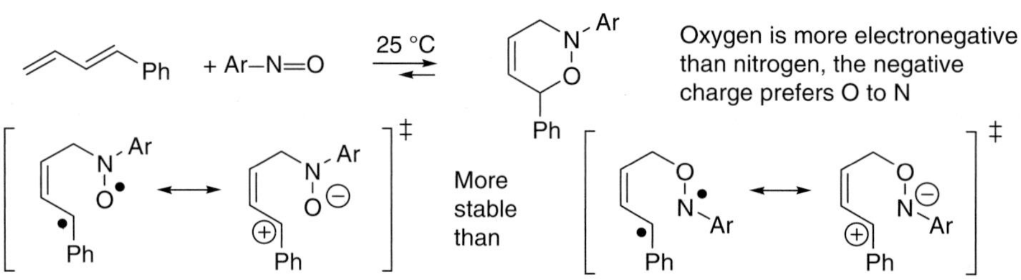

(b)

Under kinetic control, the cycloadducts with *para* Ts and R groups are favored:

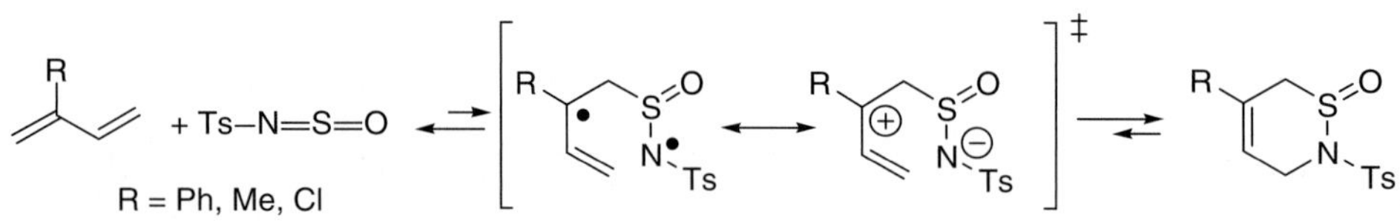

(c)

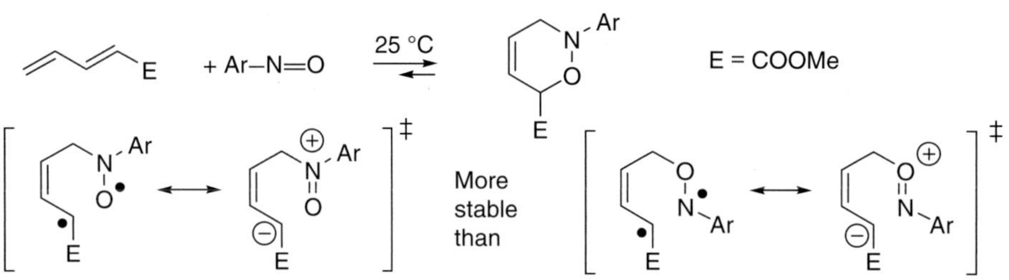

(d)

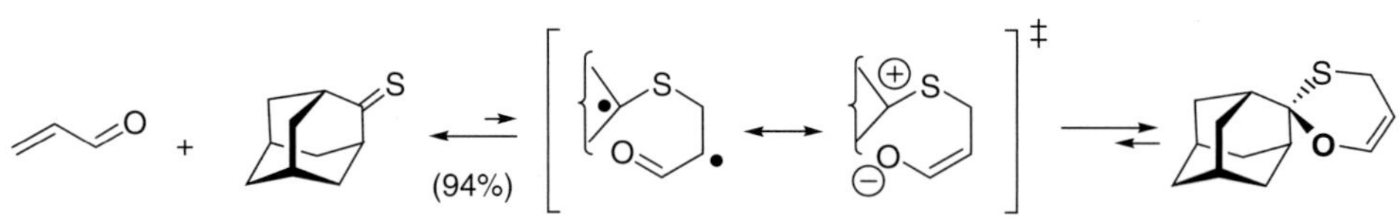

(e)

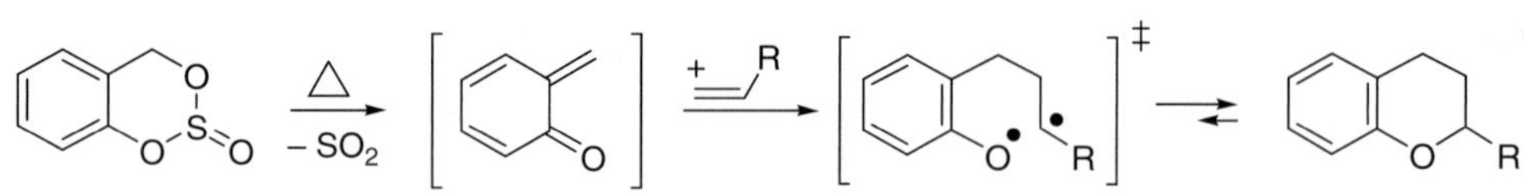

(f)

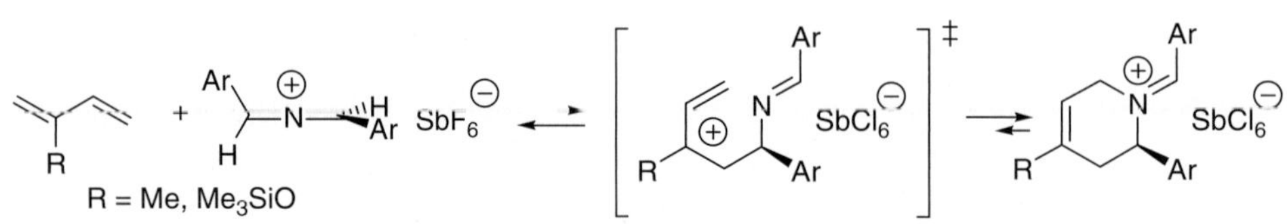

(g)

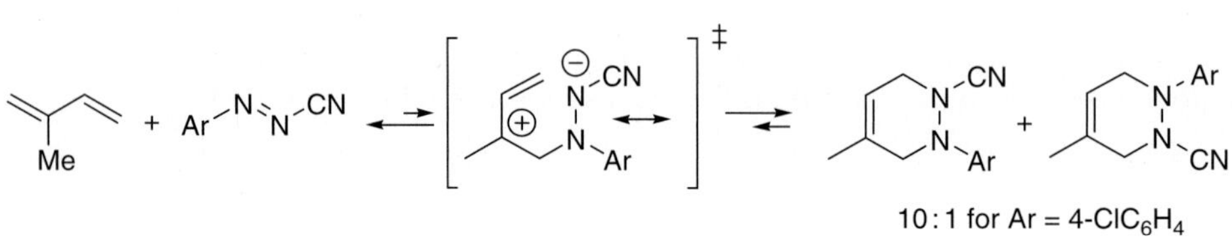

(h)

Problem 5.28 In most cases, arylnitroso and acylnitroso dienophiles lead to the same regioselectivity. However, opposite regioselectivities are observed for the following hetero-Diels–Alder reactions [31, 32]. Explain why.

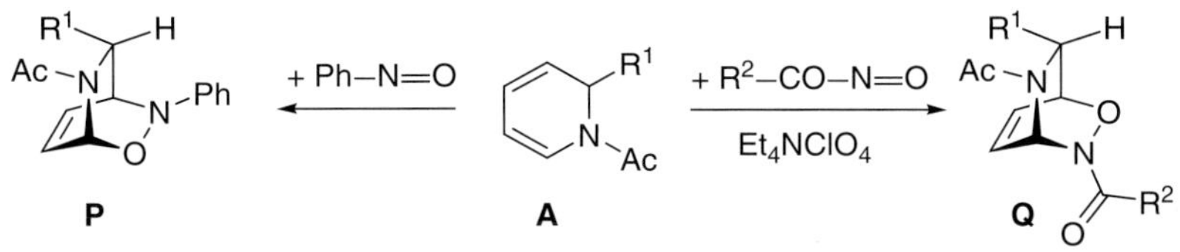

Answer

The diradicaloid model applied to reaction **A** + PhNO → **P** suggests the following limiting structures for its transition state:

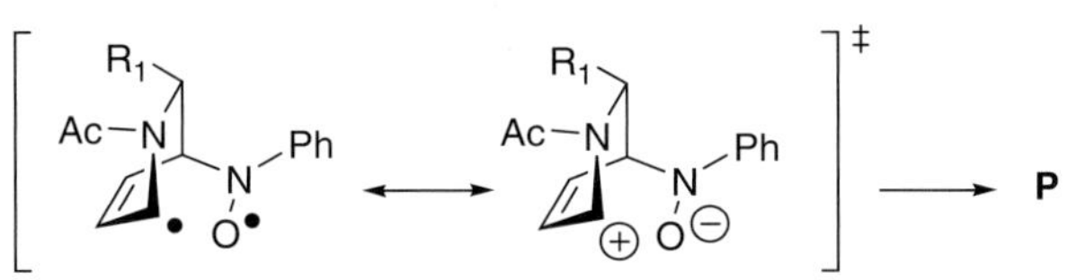

Both the diradical and zwitterionic limiting structures enjoy from the stabilization introduced by the *N*-acetylamino moiety.

In the case of reaction **A** + R^2—CO—N=O → **Q** the nitrogen center rather than the oxygen center of the nitroso group can bear the radical and the negative charge because it is substituted by the electron-withdrawing group COR^2. As a consequence the following limiting structures become important in the transition state of the reaction and favor the formation of **Q**.

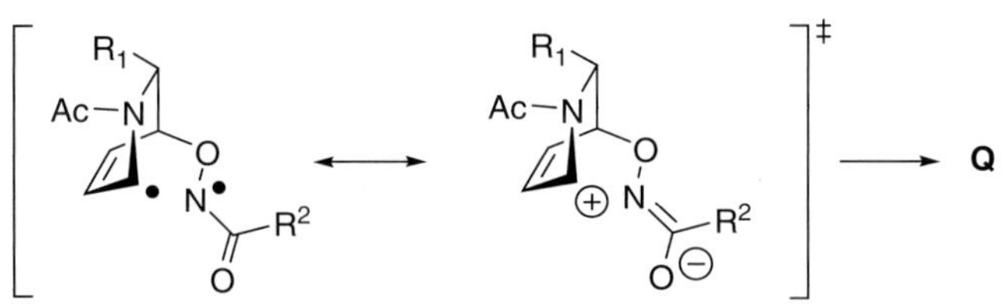

Problem 5.29 What are the products of the reaction of sydnones **B** with dimethyl acetylenedicarboxylate? [33]

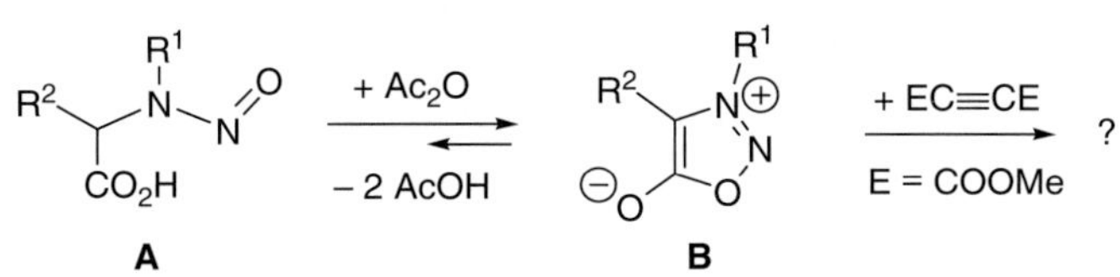

Answer

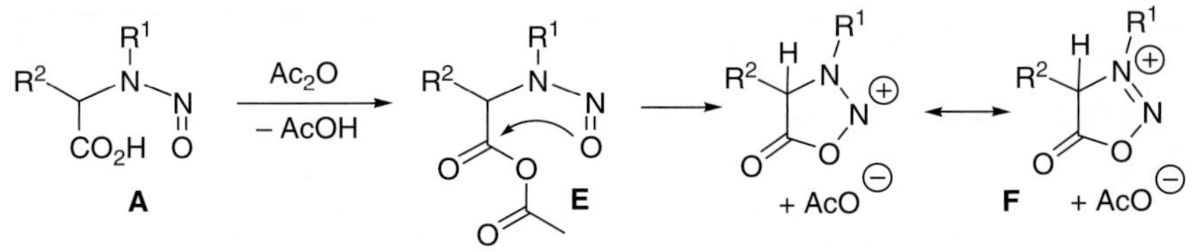

Sydnones **B** are formed upon elimination of a first equivalent of AcOH from the mixed anhydrides **E** giving the corresponding ion pairs **F**; in turn, the latter eliminate a second equivalent of AcOH providing **B**.

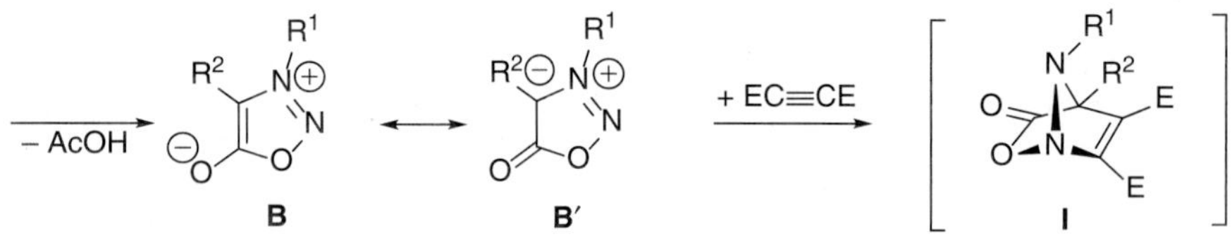

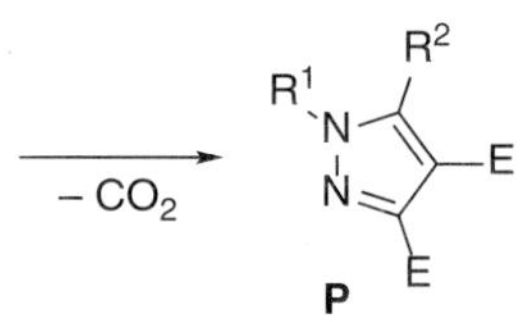

Sydnones **B** are azomethine imines (**B′**) that add to dimethyl acetylenedicarboxylate (1,3-dipolar cylo-additions) giving the bicyclic cycloadducts **I**. The latter undergo highly exothermic and exergonic $[\sigma^2 + \sigma^2 + \pi^2]$-cycloreversions as they form the stable CO_2 and the corresponding aromatic pyrazoles **P**.

Problem 5.30 What are the products of the reaction of münchnones **D** with dimethyl acetylenedicarboxylate? [34]

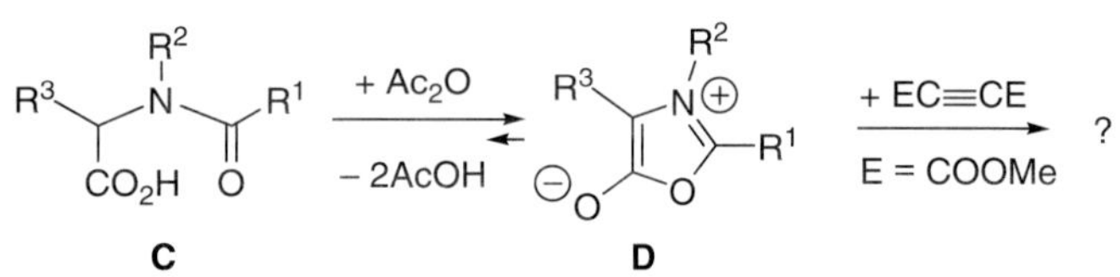

Answer

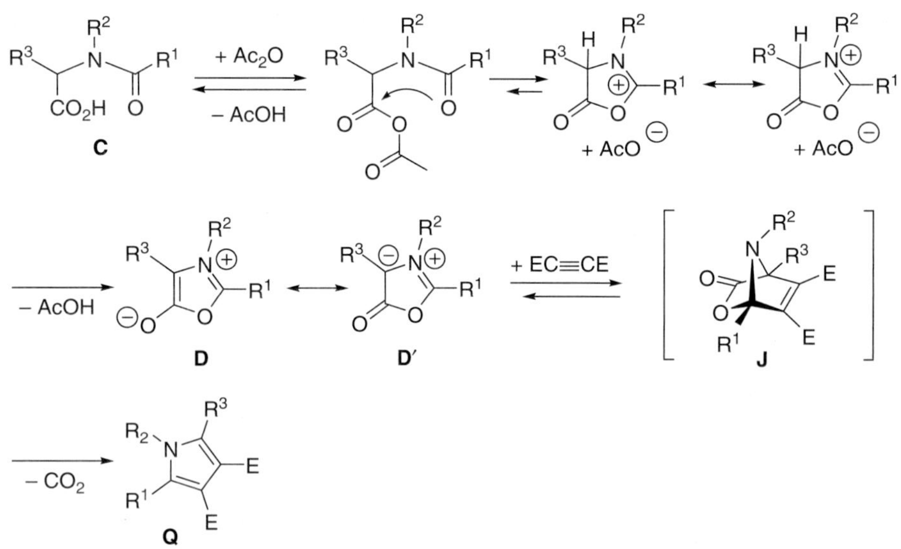

Münchnones **D** are imine ylides (**D′**) that add to dimethyl acetylenedicarboxylate (1,3-dipolar cycloadditions) giving the corresponding bicyclic cycloadducts **J**. The latter undergo highly exothermic and exergonic (4+2)-cycloreversions as they form stable CO_2 and the corresponding aromatic pyrroles **Q**.

Problem 5.31 What are the main products of the following reactions? [35]

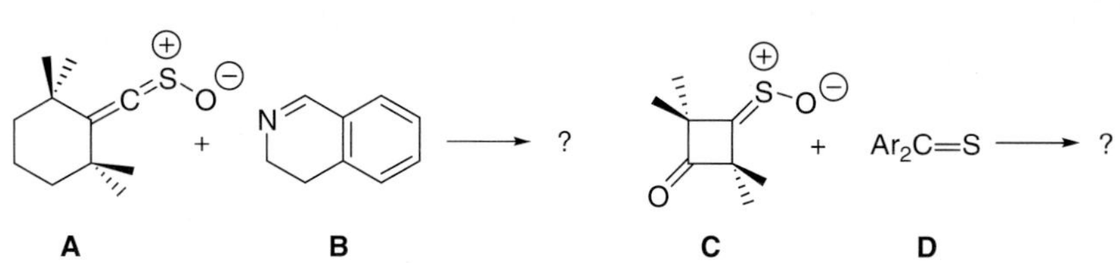

Answer

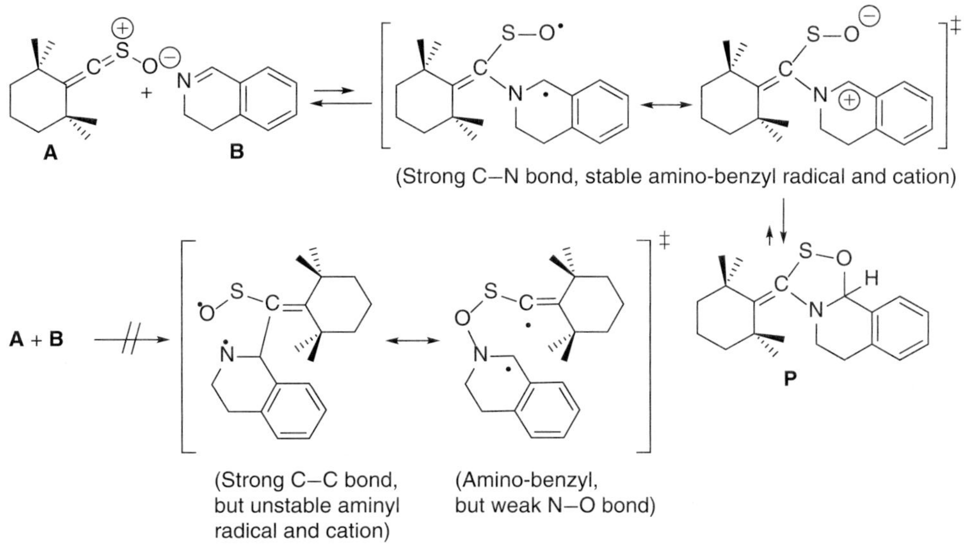

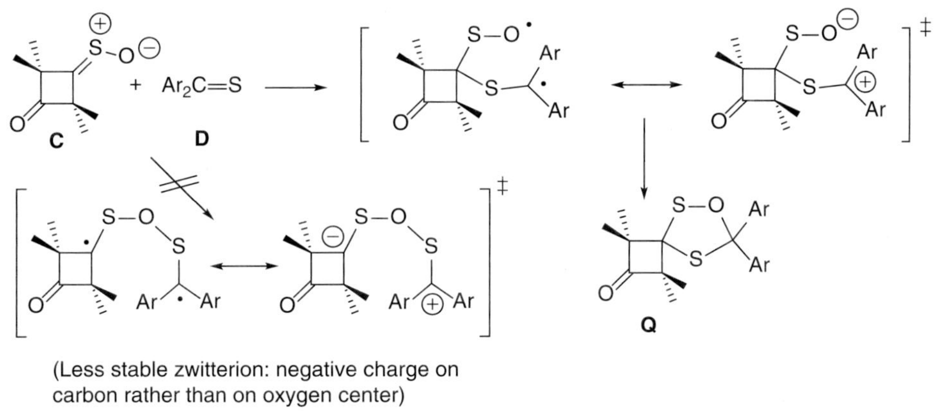

Problem 5.32 What are the major products of the following reaction? [36]

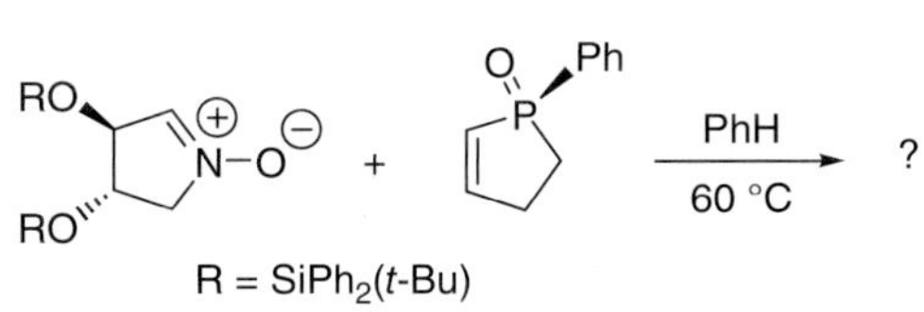

Answer

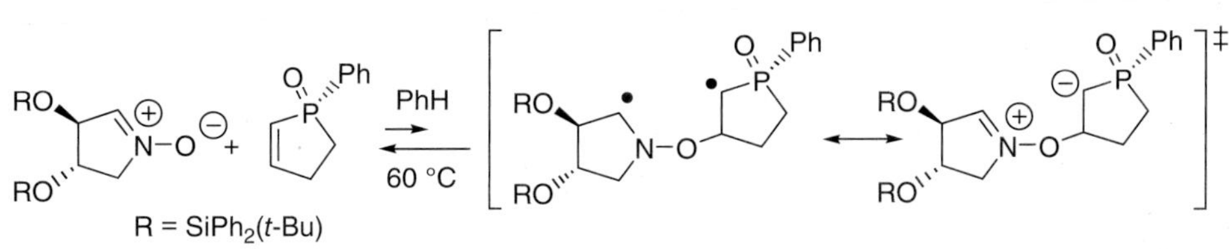

Search for the least sterically hindered face of the cycloaddents and the least sterically hindered orientation (*exo* vs. *endo*).

The nitrone adds with its face *anti* with respect to the bulky 3-silyloxy group; the face of the phosphine oxide *anti* with respect to the largest group (Ph) at the phosphorus center is preferred; the *exo* orientation is less demanding in term of steric hindrance than the endo orientation. Nevertheless, both orientations imply transition states that enjoy from an electronic stabilization, which may make both orientations to compete. The results show that the *exo* orientation is preferred by a factor of 10 : 1.

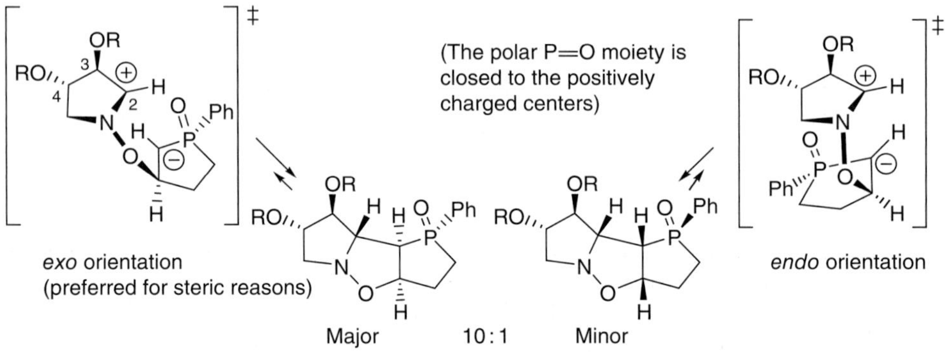

Problem 5.33 What are the products of reaction of cyclooctyne with **A**, on one hand, and with **B**, on the other hand? [37]

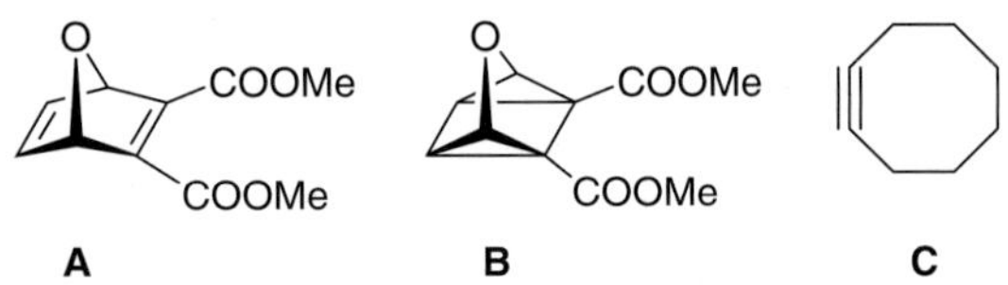

Answer

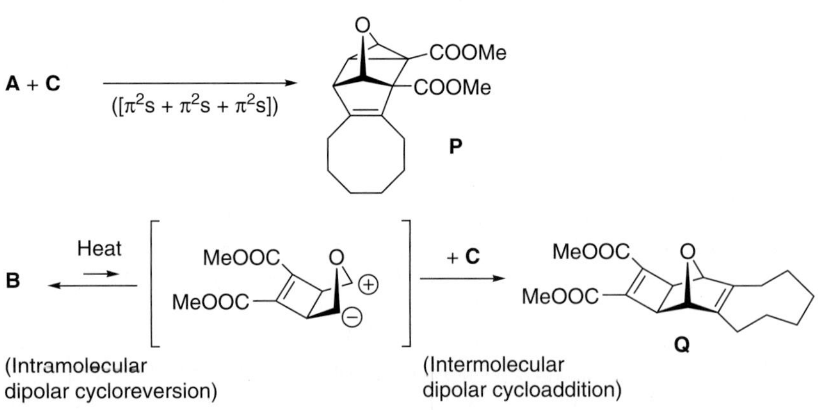

Problem 5.34 What are the possible products of the following reaction? [38]

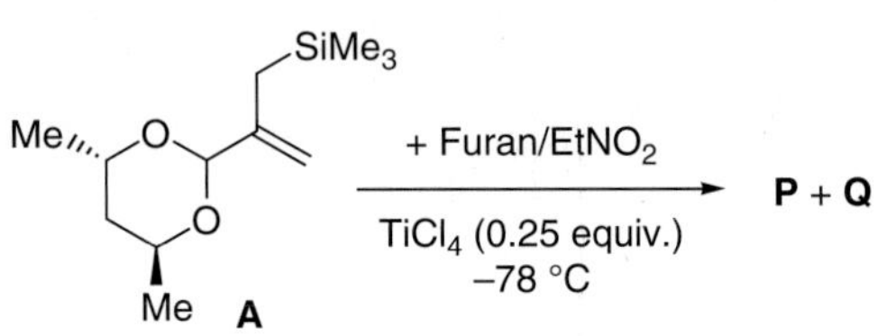

Answer

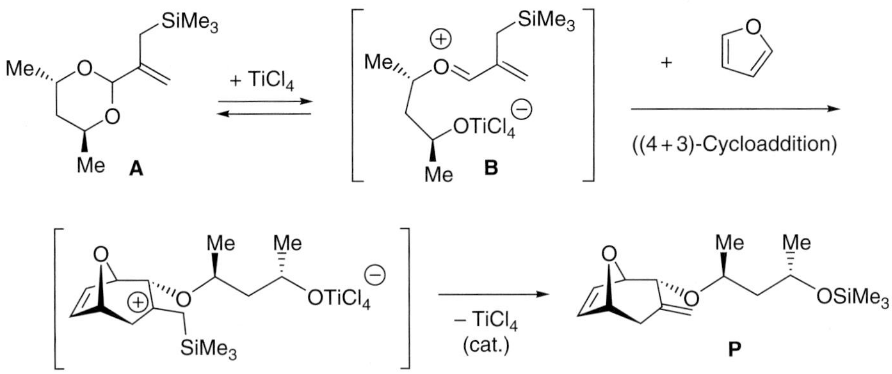

Q is a diastereomer of **P** that can form concurrently. Harmata and coworkers reported yields up to 78% for the formation of **P** + **Q**, and diastereoselectivity up to 15.8 : 1 in favor of **P**.

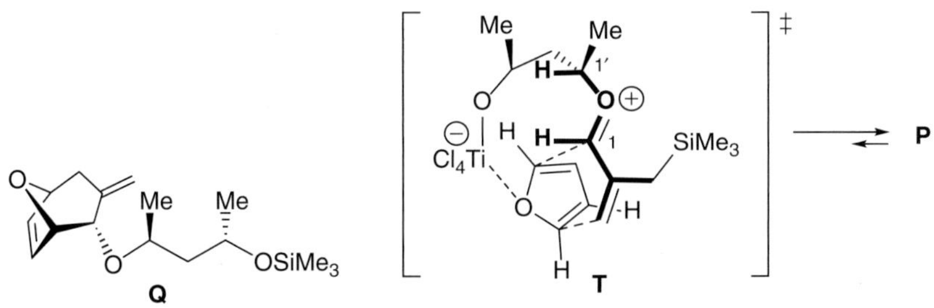

Because of the C_2-symmetry of the acetal moiety of **A**, Lewis acid induced C–O heterolysis of one or the other C—O bond leads to the same zwitterionic intermediate **B**. The high diastereoselectivity observed can be explained by invoking the transition state **T**. The 1-oxyallyl cation adopts a (*E*)-configuration in which the two C—H bonds at C(1) and C(1′) positions of the chiral auxiliary are *syn*-periplanar to minimize steric repulsions. The titanate moiety might coordinate to the oxygen atom of furan. This forces the addition of furan onto the face of the oxyallyl cation *syn* with respect to the titanate group.

Problem 5.35 (1*S*)-1-Phenylethylamine reacts with 3-chloro-3-methylbutan-2-one, Et_3N (2 equiv.), and $TiCl_4$ (0.5 equiv.) giving an imine **A**. When **A** is reacted with $AgBF_4$ in CH_2Cl_2 a salt **B** forms that reacts with furan providing an iminium salt **P**. The latter is hydrolyzed into a ketone **Q** with 60% ee. What are **A**, **B**, **P**, and **Q**? [40]

Answer

(1*S*)-1-Phenylethylamine reacts with 3-chloro-3-methylbutan-2-one, Et_3N (2 equiv.), and $TiCl_4$ (0.5 equiv.), giving imine **A**. When **A** is treated with $AgBF_4$ in CH_2Cl_2 a salt **B** is obtained that reacts with furan, providing an iminium salt **P**. The latter is hydrolyzed in aqueous acetone furnishing the bicyclic ketone **Q** with 60% ee.

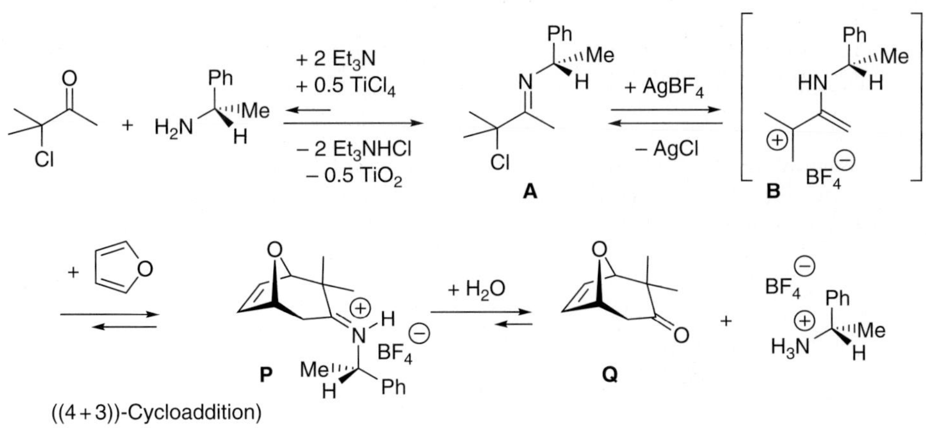

Problem 5.36 Compounds **A** eliminate one equivalent of MeOH on treatment with lithium diisopropylamide (LDA) ((i-Pr)$_2$NLi) and form the corresponding products **P**. Propose a mechanism [41].

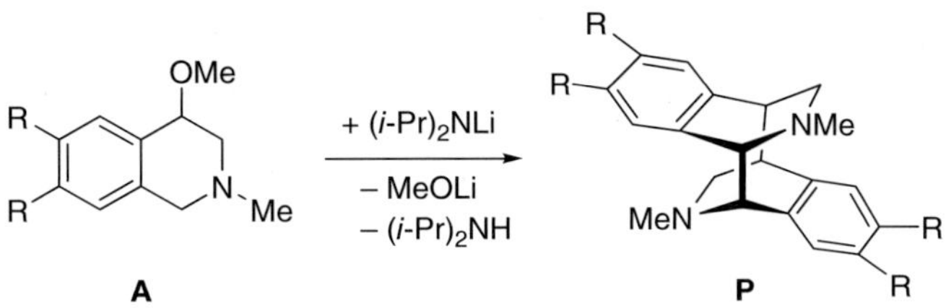

Answer

H/Li exchange of **A** occurs on the least sterically hindered benzylic center and acidified by the NMe group. This leads to 1,4-elimination of MeOLi with the formation of the orthoquinodimethane intermediate **B**. The latter dimerizes immediately into diradical **D,** which is the most stable diradical that can be obtained on combining two molecules of **B**. Each radical moiety of **D** is a benzylic radical substituted by a dialkylamino group.

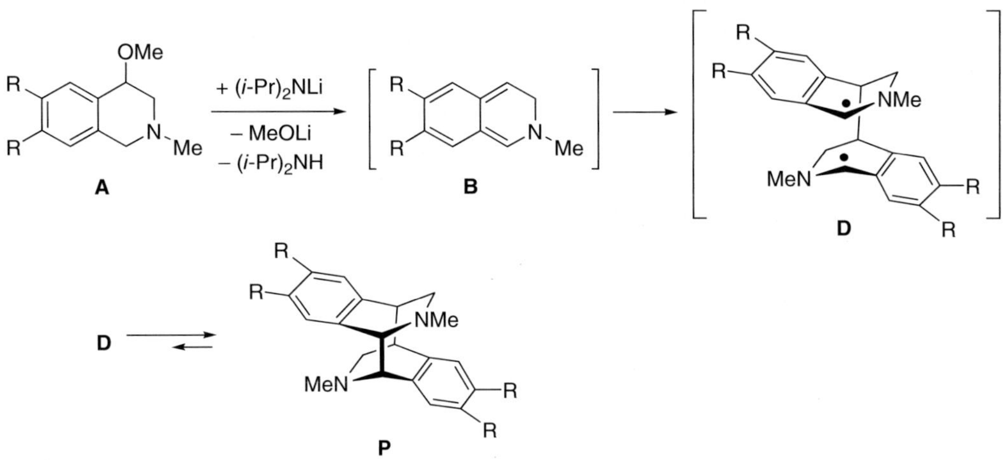

Problem 5.37 What is the major product of the thermal reaction (140 °C) of tropone with (E)-piperylene? [42, 43]

Answer

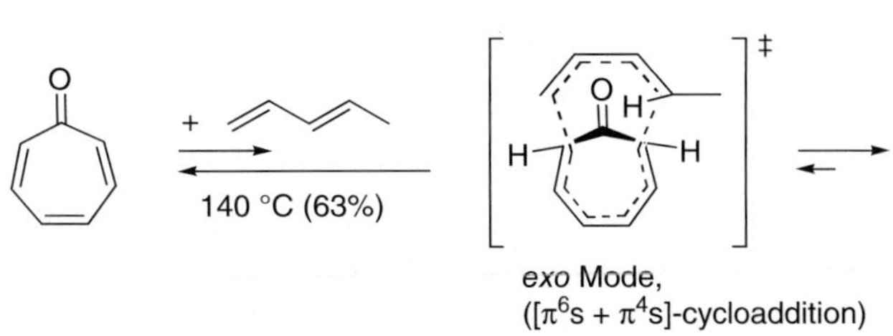

Problem 5.38 Dibromocarbene engendered by the reaction of bromoform with t-BuOK in pentane reacts with diene **D** giving a 9 : 1 mixture of cycloadducts **P** and **Q** in 80% yield. Product **P** is not isomerized into **Q** under the conditions of the reaction. What is the mechanism for the formation of **Q**? [43]

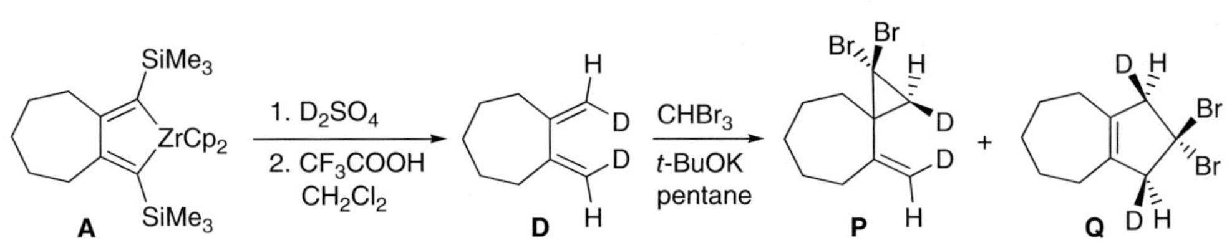

Answer

Product **P** results from a suprafacial cyclopropanation of diene **D**. **P** is isomerized via a concerted (no rotation about the C(1)—C(1′) and C(2)—C(2′) bonds) suprafacial (1,3)-sigmatropic shift into **Q′**, which is in fact **Q**. Thus, cycloadduct **Q** may not result from a suprafacial [π^4s+ω^2s]-cheletropic addition of dibromocarbene onto diene **D**.

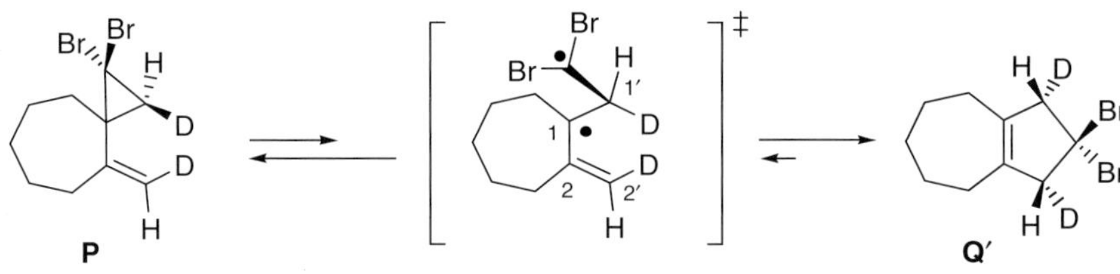

Problem 5.39 Nitric oxide (NO) is a mediator of many important biochemical pathways [44], including muscle relaxation [45], macrophage activation [46, 47], and neurotransmission [48–50]. Ingold and coworkers have used orthoquinodimethane to quench NO in a (4+1)-cheletropic addition that gives stable paramagnetic nitroxides readily detected by electron spin resonance (ESR) [51]. What is the product of the reaction of NO with the product of photoinduced electrocyclic ring opening of benzocyclobutenedione? (NO can be generated by the reaction of HNO_3 with Cu powder) [52].

Answer

Benzocylobutene-1,2-dione (**A**) undergoes photoinduced electrocyclic ring opening into diketene **B**. NO reacts with diketene **B** giving phthalimide *N*-oxide (**C**). The latter compound is an oxygen-centered radical that abstracts a hydrogen atom from the medium giving the stable *N*-hydroxyphthalimide (**D**). NO is usually contaminated by NO_2, which reacts with diketene **B** producing **E**. On heating, **E** fragmentizes into NO and phthalic anhydride. The photochemical isomerization **A** ⇄ **B** is an example of Norrish type I reaction; see Section 6.4.1. It can also be seen as an electrocyclic ring opening of a cyclobutene into a butadiene system.

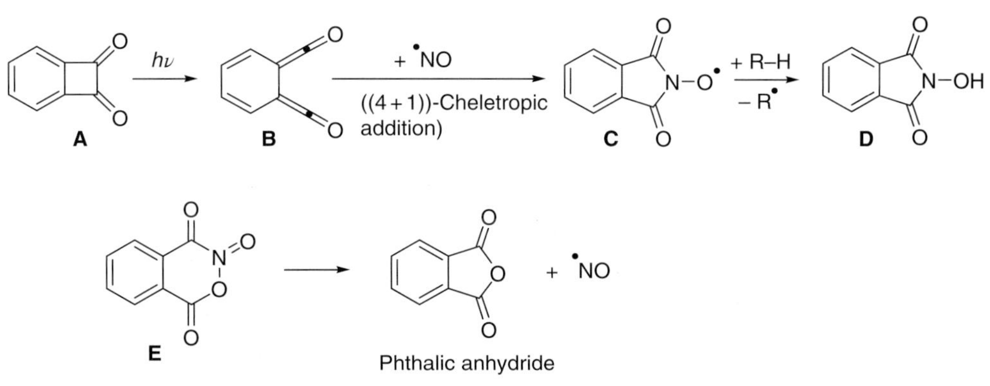

Problem 5.40 Decarbonylation of bicyclo[2.2.1]hepta-2,5-dien-7-one in PhH occurs at −60 °C and gives benzene + CO with E_a = 15.2 kcal mol^{-1} [53]. In contrast, decarbonylation of bicyclo[2.2.1]hept-2-en-7-one into 1,3-cyclohexadiene + CO is much slower and has an activation energy E_a = 34.4 kcal mol^{-1} (125 °C) [54, 55]. Both reactions are much faster than the decarbonylation of cyclopent-3-enone. Give an explanation.

Answer

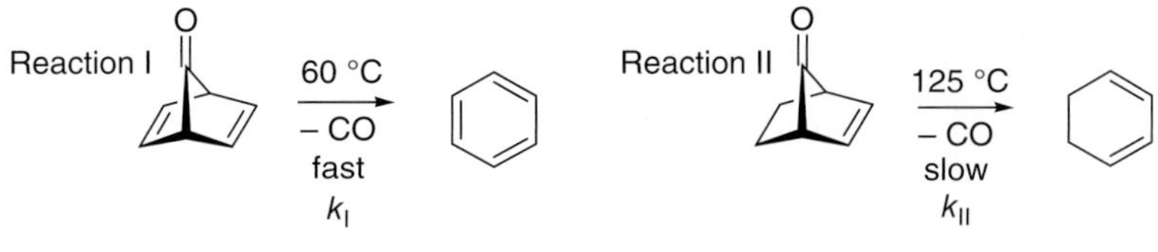

Rate constant k_I is smaller than k_{II} because reaction I is more exothermic than reaction II due to the stabilization of benzene by aromaticity, compared with cyclohexa-1,3-diene. Both reactions I and II involve starting materials that are more strained than cyclopent-3-enone, which undergoes a much slower (4+1)-cheletropic elimination into butadiene and CO. The latter reaction releases less ring strain than reaction II (and I). The intrinsic barrier of reaction II is lower than that of reaction I as predicted by the diradicaloid model. Both the diradical and zwitterionic limiting structures of the transition states are more stable for reaction I (cyclohexadienyl) than for reaction II (cyclohexenyl).

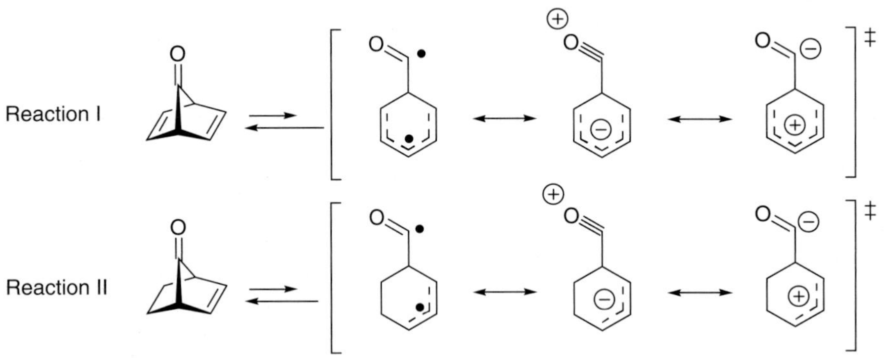

Problem 5.41 Zwitterion **A** adds to dimethyl acetylenedicarboxylate giving a unstable cycloadduct **B** that fragmentizes into **P** and phenyl isonitrile. What is **B**? Give a mechanism for the fragmentation. Propose a method to prepare **A** [56].

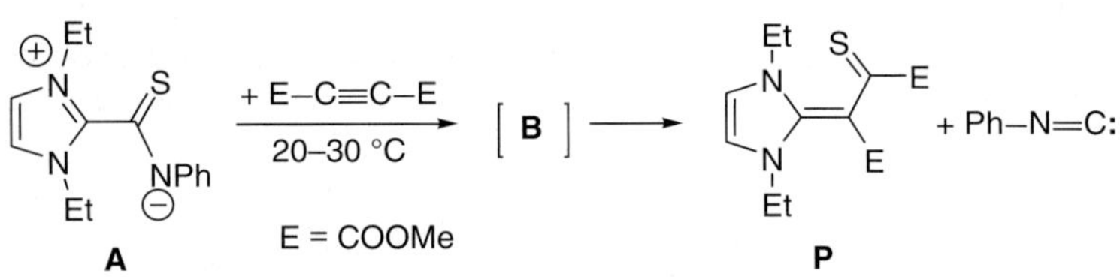

Answer

Imidazol-carbene **D** is generated by deprotonation of imidazolium salt **C** employing NaH at 20–30 °C. **D** is a carbon nucleophile that adds to the isothiocyanate (Ph—N=C=S) and forms **A**. Zwitterion **A** undergoes 1,3-dipolar cycloaddition producing the intermediate cycloadduct **B**. The latter contains the imine of a thiacyclopentenone. The imine function is isoelectronic with the carbonyl group and therefore can undergo (4+1)-cheletropic elimination, forming product **P** and phenyl isonitrile (Ph—N=C:).

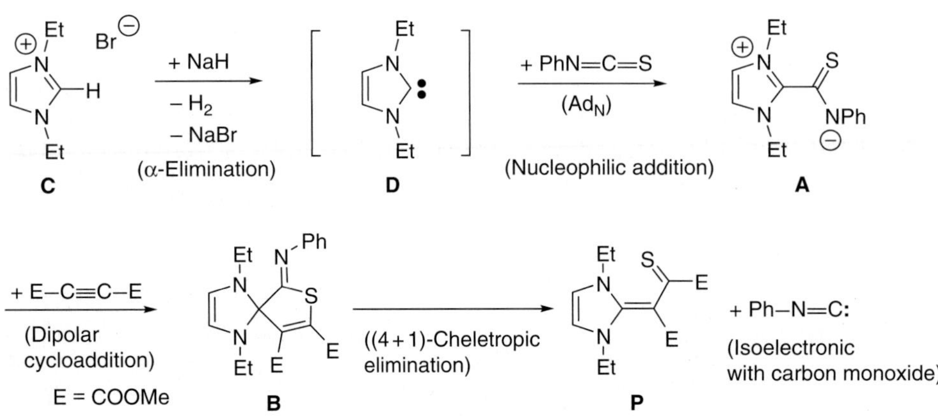

Problem 5.42 What is the major product of the following reaction? [57]

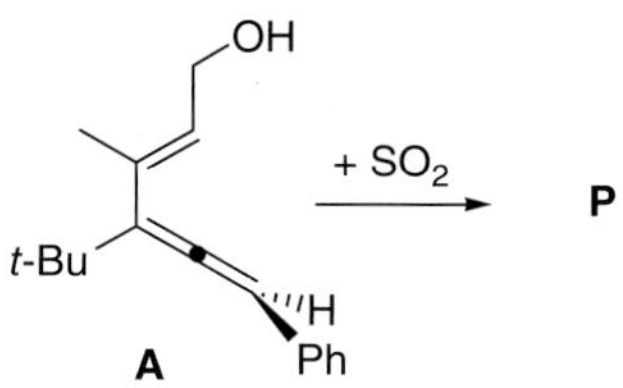

Answer

The least sterically hindered face of the vinyl allene adds to SO_2 with its conjugated diene moiety forming the corresponding sulfolene **P** in a disrotatory manner.

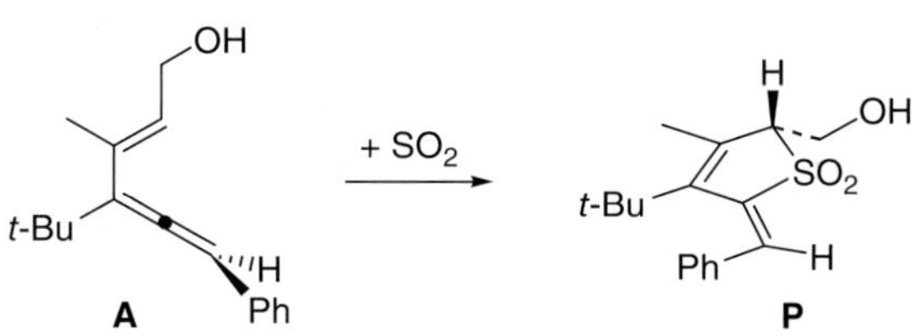

Problem 5.43 What is the major product of the following reaction? [58]

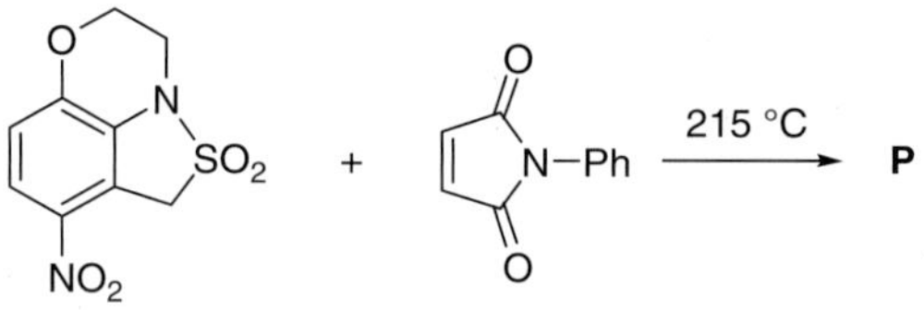

Answer

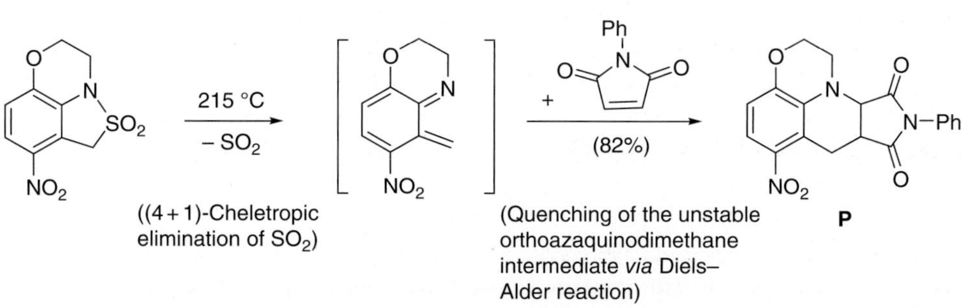

Problem 5.44 What is the main product **P** of the following reaction [59].

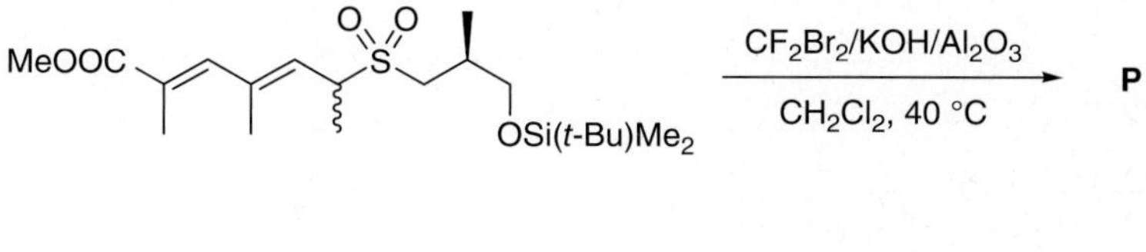

Answer

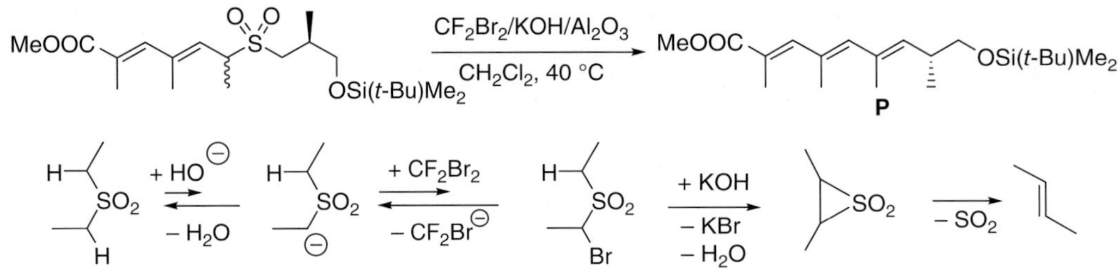

$$CF_2Br^{\ominus} + KOH \longrightarrow F_2C: + KBr + HO^{\ominus}$$

$$F_2C: + H_2O \longrightarrow CF_2(OH)H$$

$$CF_2(OH)H + 3KOH \longrightarrow HCOOK + 2KF + 2\,H_2O$$

For the chlorination with CCl_4 the following mechanism has been established:

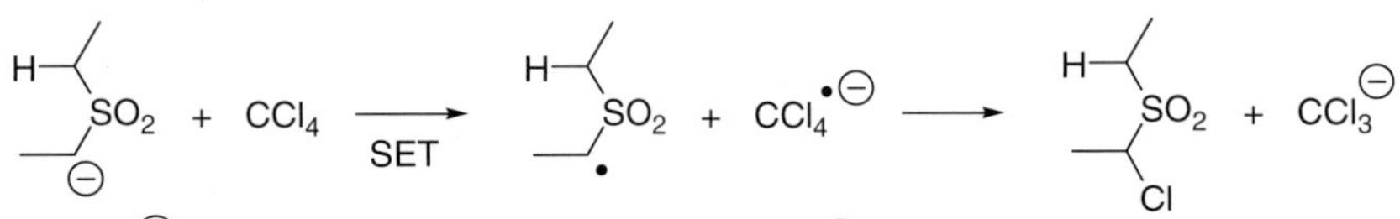

$$^{\ominus}CCl_3 + H_2O \rightleftharpoons HCCl_3 + HO^{\ominus}$$

$$^{\ominus}CCl_3 \longrightarrow Cl^{\ominus} + Cl_2C: \qquad Cl_2C: + H_2O \longrightarrow CCl_2(OH)H$$

$$CCl_2(OH)H + 3\,KOH \longrightarrow HCOOK + 2\,KCl + 2\,H_2O$$

Problem 5.45 Betaine **A** adds to dimethyl acetylenedicarboxylate giving a bicyclic adduct **B**, which, upon oxidation with *meta*-chloroperbenzoic acid (*m*CPBA), fragmentizes into cycloheptatrienone derivative **P** and nitrosobenzene. What are **B** and **P**? Explain these reactions [60].

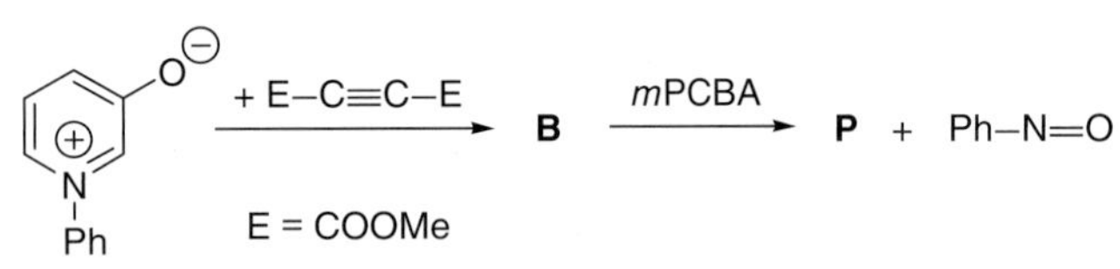

Answer

The dipolar cycloaddition of **A** to dimethyl acetylenedicarboxylate generates **B**. N-oxidation of the tertiary amine with *m*-chloroperbenzoic acid (*m*CPBA) gives *N*-amine oxide **C**. The latter can be seen as an amine, the two nonbonded electrons of which make a dative bond (N**:** donation) with the oxygen atom. Thus the *N*-amine oxide moiety is like an amine that undergoes a cheletropic elimination of a nitrene. The *N*-oxide of this nitrene is in fact a stable nitroso compound (Ph—N=O) that is liberated together with cycloheptatrienone **P** (a tropone stabilized by aromaticity: limiting structure is oxytropylium cation; see Section 4.5.13).

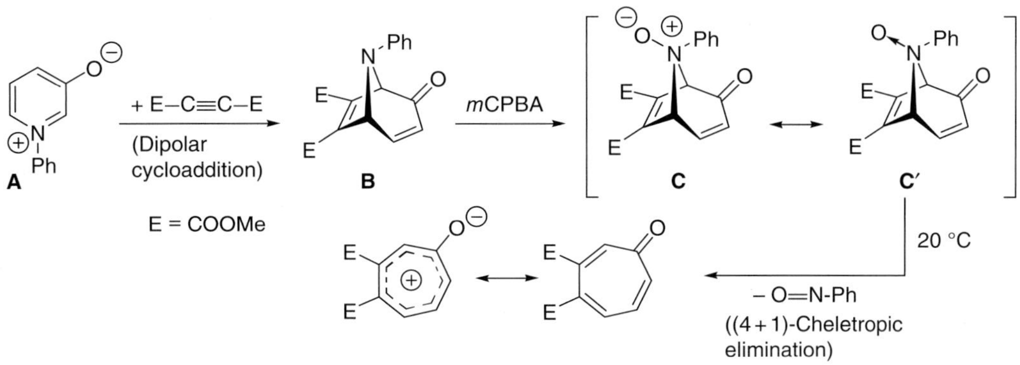

Problem 5.46 What are the products of the reaction of $[(Me_3Si)_2CH]_2Sn$: with conjugated enals, enones, and 1,2-diketones? [61]

Answer

These reactions are (4+1)-cheletropic additions. The corresponding products of (2+1)-cheletropic additions are not observed because they are unstable and are expected to be rearranged quickly into the corresponding stannocyclopentenes. Quantum mechanical calculations for the reaction of Me_2Sn: with butadiene confirm these hypotheses.

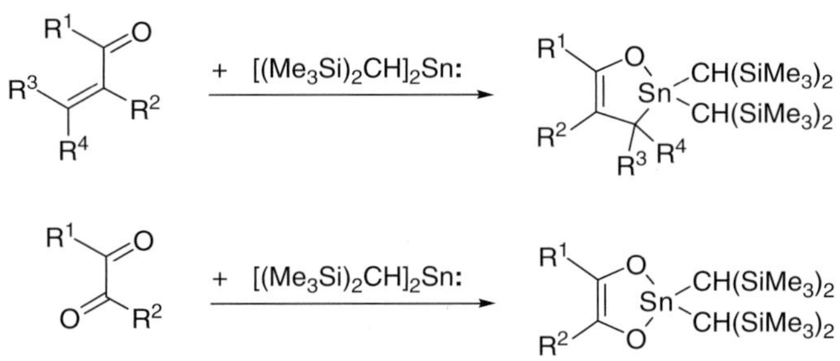

Problem 5.47 What are the products **P** and **Q** of the following reactions? [62]

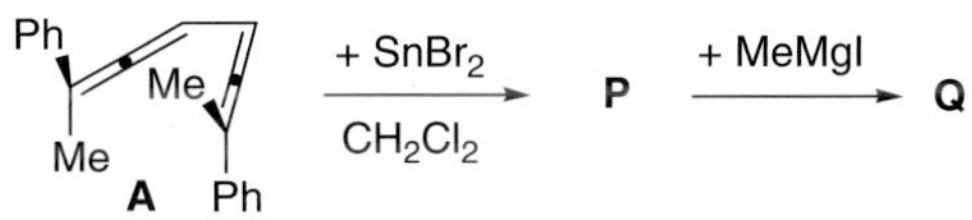

Answer

Suprafacial $[\pi^4s+\omega^2s]$-cheletropic addition of **A** to $[(Me_3Si)_2CH]_2Sn$: leads to the disrotatory rotation of the two methyl(phenyl)methylidene groups, producing **P**. Nucleophilic displacement of the tin dibromide **P** with the Grignard reagent MeMgI generates **Q** and the magnesium salts $MgI_2 + MgBr_2$.

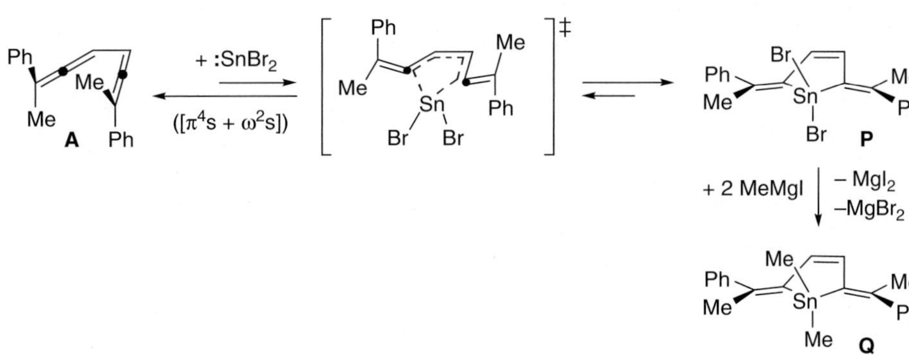

Problem 5.48 What is the product of the reaction of $[(Me_3Si)_2CH]_2Sn$: with $Fe_2(CO)_9$? With MeI? [63]

Answer

$Fe_2(CO)_9$ loses one equivalent of CO on heating, giving $Fe_2(CO)_8$. Among the various structures the complex can adopt, one is $(OC)_4Fe{=}Fe(CO)_4$ (see Section 7.7), a diferraethylene that undergoes a (2+1)-cheletropic addition with the stannylene $[(Me_3Si)_2CH]_2Sn$: giving **P**. Stannylenes are isoelectronic with carbenes and C=O. With Me—I, $[(Me_3Si)_2CH]_2Sn$: undergoes an apparent insertion into the single bond of Me-I giving **Q**. This reaction is called an oxidative insertion as the oxidation degree of the tin metal goes from +2 to +4, formally (see Section 7.7). This reaction can also be seen as a $[\sigma^2s+\omega^2s]$-cheletropic addition. Its mechanism is probably an S_N2 displacement process of Me—I by the nucleophilic stannylene that produces ion pair **W**.

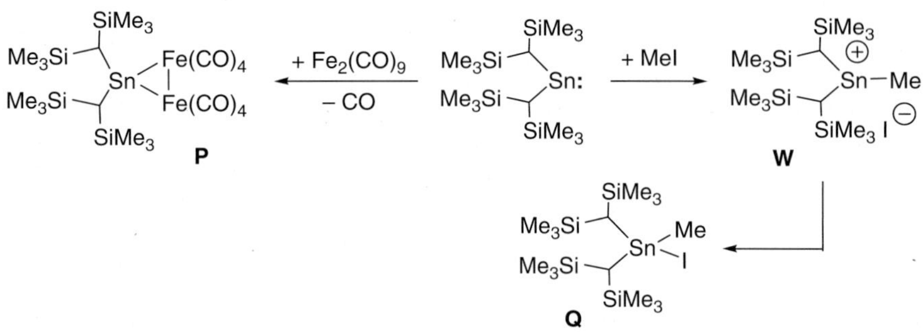

Problem 5.49 On heating the germacyclopropene **A**, a blue solution forms. The blue color fades on adding 2,3-dimethylbutadiene to this solution. Explain.

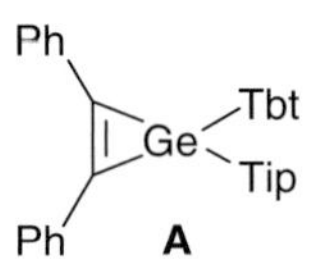

Tbt = 2,4,6-$(Me_3SiCH_2)_3C_6H_2$

Tip = 2,4,6-tri(isopropyl)phenyl

Answer

On heating, **A** gives tolane (Ph—C≡C—Ph) and germylene **B**, which is blue. The latter adds to 2,3-dimethylbutadiene in a formal (4+1)-cheletropic addition providing **P**.

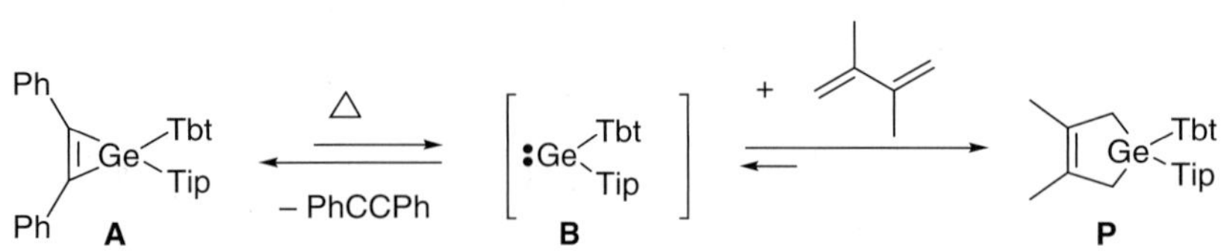

Problem 5.50 In toluene $[(OC)_5W{=}P{-}Me]$ reacts with 1,6-methano[10]annulene giving a $W(CO)_5$ complex of 3,8-methano-1-methylphosphacycloundecapentaene. Give a mechanism for this reaction [64].

Answer

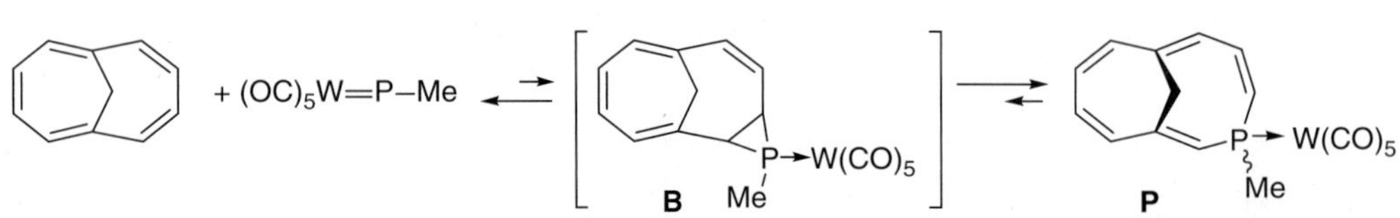

syn/anti 85 : 15

Problem 5.51 The (1,2)-migration of hydrogen in acyclic carbenium ions is very fast. Give an explanation.

Answer

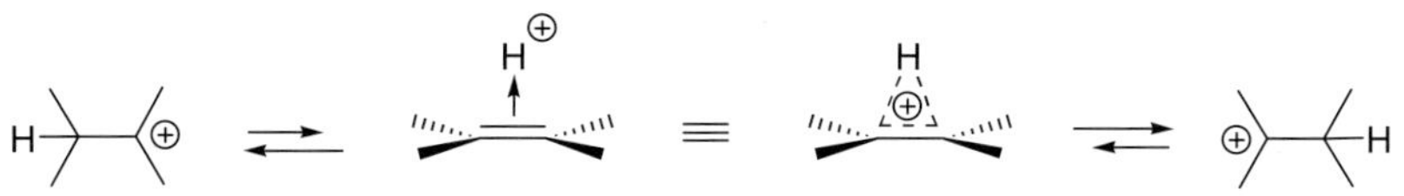

A B B′ A′

(1,2)-Hydride shift in tertiary carbenium ion **A** is expected to be an easy process because transition structure **B**, which corresponds to a π-complex of a proton with an alkene, enjoys relatively high stability arising from the (lowest lying molecular orbital) LUMO(H^+)/HOMO(alkene) interaction. As for (1,2)-shifts of alkyl groups in Wagner–Meerwein rearrangements, the suprafacial mode is much easier than the antarafacial mode. For both modes, as the LUMO(H^+) is a 1s(H) atomic orbital, it cannot lead to inversion of configuration at the migrating center. (1,2)-Shifts of hydrogen are always with retention on the migrating center.

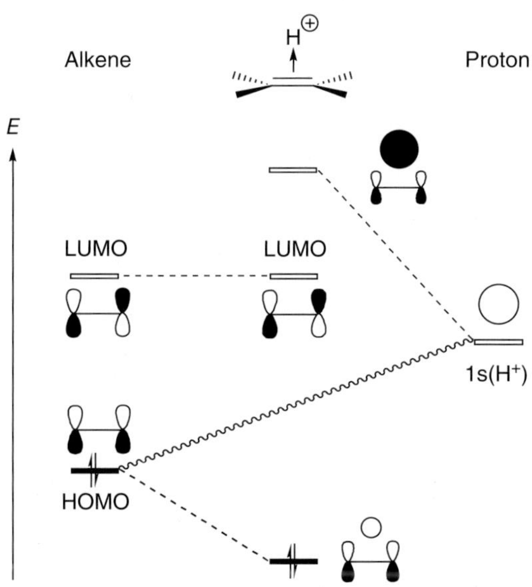

Problem 5.52 Estimate the heat of fragmentation of 2,3,3-trimethylbut-2-yl radical into methyl radical and 2,3-dimethylbut-2-ene (gas phase, 25 °C, 1 atm).

Answer

An estimate of $\Delta_r H° = +23.7 \pm 2\,\mathrm{kcal\,mol^{-1}}$ is obtained in the following manner:

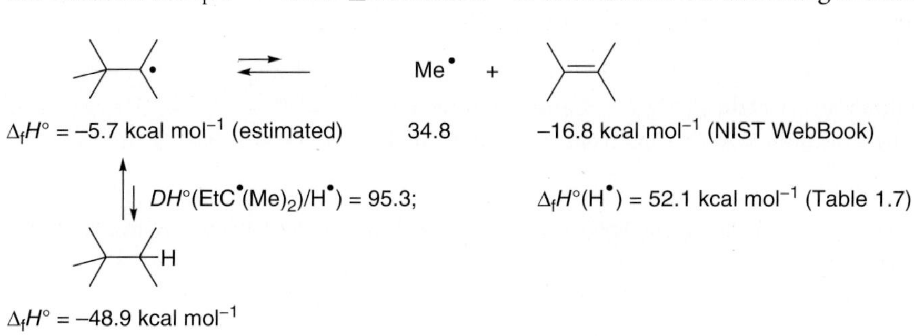

Problem 5.53 What would be the activation enthalpy of isomerization **462** → **463** at 278 °C ($k = 5 \times 10^5\ s^{-1}$) if an entropy of activation $\Delta^{\ddagger}S$ = 10 eu would be considered for the two freely rotating σ bonds of **463** (late transition state with no interaction between the radical and alkene moiety)?

Answer

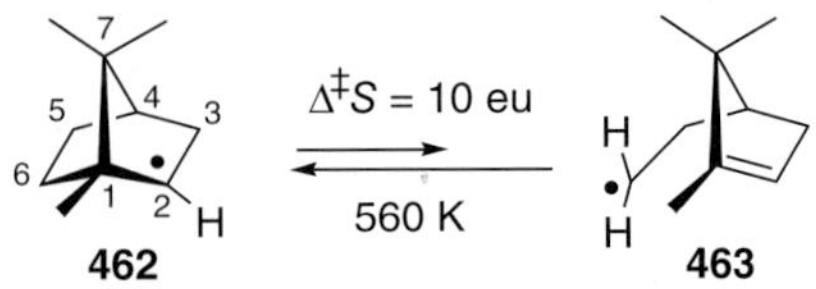

462 **463** $k = 1.5 \times 10^5\ s^{-1}$

$\Delta^{\ddagger}S$ = 10 eu corresponds to the Arrhenius pre-exponential factor $A = 10^{15}$ (see Table 3.3). Thus, $E_a = RT(\ln A - \ln k) = 1.987[560(34.54 - 13.12)] = 1112.7(21.42) = 23\,834\ \text{cal mol}^{-1}$; $E_a = 23.8\ \text{kcal mol}^{-1}$, $\Delta^{\ddagger}H = E_a - RT = 23.8 - 1.1 = 22.7\ \text{kcal mol}^{-1}$

Problem 5.54 Using $\Delta_f H°$(PhEt) = 7.1 kcal mol^{-1}, $\Delta_f H°$(1,1-dimethylcyclopropane) = −2.0 kcal mol^{-1}, $\Delta_h H°$ (cyclohexa-1,4-diene → cyclohexane) = −53.7 kcal mol^{-1} and $DH°$(cyclohexadienyl$^{\bullet}$/H$^{\bullet}$) = 76.9 kcal mol^{-1}, estimate the standard heat of isomerization 2-phenylethyl radical (**A**$^{\bullet}$) into 6,6-ethylenecyclohexa-2,4-dien-1-yl radical (**B**$^{\bullet}$).

Answer

$\Delta_r H°(\mathbf{A}^{\bullet} \rightleftarrows \mathbf{B}^{\bullet}) = \Delta_f H°(\mathbf{B}^{\bullet}) - \Delta_f H°(\mathbf{A}^{\bullet}) =$

$= \Delta_f H°(\mathbf{B}^{\bullet}) + \Delta_f H°(H^{\bullet}) - \Delta_f H°(\mathbf{B\text{–}H}) + \Delta_f H°(\mathbf{B\text{–}H}) - \Delta_f H°(\mathbf{A}^{\bullet}) - \Delta_f H°(H^{\bullet}) + \Delta_f H°(\mathbf{A\text{–}H}) - \Delta_f H°(\mathbf{A\text{–}H}) =$

$= DH°(\mathbf{B}^{\bullet}/H^{\bullet}) - DH°(\mathbf{A}^{\bullet}/H^{\bullet}) + \Delta_f H°(\mathbf{B\text{–}H}) - \Delta_f H°(\mathbf{A\text{–}H})$

$= 76.9 - 101 + \Delta_f H°(\mathbf{B\text{–}H}) - 7.1\ \text{kcal mol}^{-1}$

One has taken $DH°(\mathbf{A}^{\bullet}/H^{\bullet}) = DH°(n\text{-butyl}^{\bullet}/H^{\bullet}) = 101\ \text{kcal mol}^{-1}$

and $DH°(\mathbf{B}^{\bullet}/H^{\bullet}) = DH°(\text{cyclohexadienyl}^{\bullet}/H^{\bullet}) = 76.9\ \text{kcal mol}^{-1}$

One estimates $\Delta_f H°(\mathbf{B\text{–}H}) = \Delta_f H°(\mathbf{D}) + 53.7\ \text{kcal mol}^{-1}$ (– heat of hydrogenation of cyclohexa-1,4-diene)

One takes $\Delta_f H°(\mathbf{D}) = \Delta_f H°$(1,1-dimethylcyclopropane) – 2([C–(C)(H)$_3$]) (2 Me groups) +

+ 5([C–(C)$_2$(H)$_2$]) (5 CH$_2$ groups) = – 2.0 + 20 – 25 = – 7.0 kcal mol^{-1}

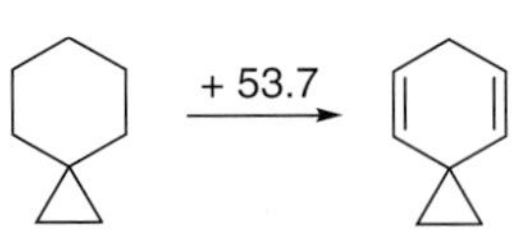

D **B–H** $\Delta_f H°(\mathbf{B\text{–}H}) = -7.0 + 53.7 = 46.7\ \text{kcal mol}^{-1}$

Finally, $\Delta_r H°(\mathbf{A}^{\bullet} \rightleftarrows \mathbf{B}^{\bullet}) = 76.9 - 101 + 46.7 - 7.1 = 15.5\ \text{kcal mol}^{-1}$

This estimate ignores the possible hyperconjugation in **B**$^{\bullet}$ between the cyclopropyl ring and the cyclohexa dienyl radical. In the case of primary alkyl radicals $DH°$(cyclopropylmethyl$^{\bullet}$/H$^{\bullet}$) = 97.4 + 1.6 kcal mol^{-1} and $DH°$(n-butyl$^{\bullet}$/H$^{\bullet}$) = 101 + 1 kcal mol^{-1} (Table 1.A.7). One assumes also that the ring strain in **D** is the same as that of 1,1-dimethylcyclopropane.

Problem 5.55 Explain the substituent effects on the activation energies measured for the following thermal (1,3)-sigmatropic rearrangements [65–68].

A 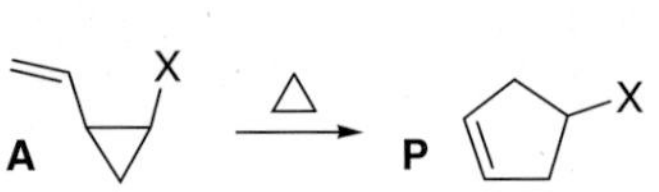

P

X = Me, E_a = 48.7; X = Ph, E_a = 44.7 kcal mol^{-1}
X = OMe, 38.7; X = NMe$_2$, 31.2

Answer

Rearrangements of **A** into **P** are (1,3)-sigmatropic shifts that imply the formation of diradical intermediates **D**. The latter are stabilized by X substitution. The greater the stabilization effect ($Me_2N > MeO > Ph > Me$), the faster is the reaction and the smaller is the Arrhenius activation energy.

A **D** **P**

Problem 5.56 Estimate the stability difference between radicals **531** and **532**.

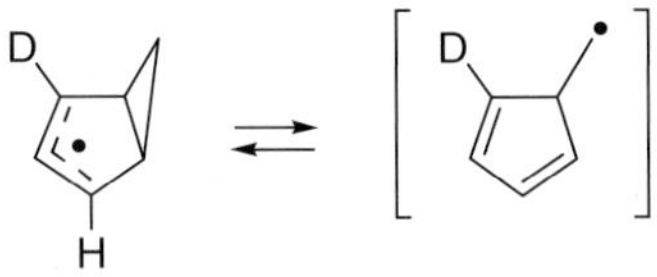

531 **532**

Answer

$\Delta_f H°(\text{CpMe}) = \Delta_f H°(\text{CpH}) + 5\,(-\,\text{CH}_2) - 2\,(+\,\text{CH}) - 10\,(+\,\text{CH}_3) = 33 - 7 = 26$ kcal mol^{-1}

$DH°(\text{CpCH}_2^{\bullet}/^{\bullet}\text{H}^{\bullet}) = DH°(\text{Et}^{\bullet}/\text{H}^{\bullet}) = 101$ kcal mol^{-1}

$\Delta_f H°(\text{CpCH}_2^{\bullet}) = 101 - \Delta_f H°(\text{H}^{\bullet}) + \Delta_f H°(\text{CpMe}) = 101 - 52 + 26 = 75$ kcal mol^{-1}

$\Delta_f H°($  $) = \Delta_f H°($ 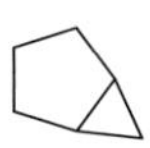 $) + 28 = 9.3 + 28 = 37.3$ kcal mol^{-1}

$DH°($  $/\text{H}^{\bullet}) =$ $DH°(\text{MeCH{=}CH{-}CH(Me)}^{\bullet}/\text{H}^{\bullet}) = 81.6$ kcal mol^{-1}

$= \Delta_f H°($ 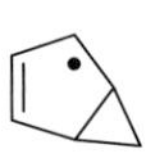 $) + \Delta_f H°(\text{H}^{\bullet}) - \Delta_f H°($  $)$

$\Delta_f H°($  $) = 81.6 - 52 + 37.3 = 67$ kcal mol^{-1}

$\Delta_f H°($ 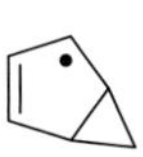 $\rightleftarrows$ 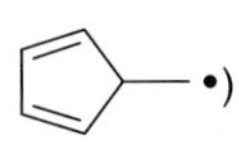 $) = 75 - 67 = 8$ kcal mol^{-1}

Problem 5.57 Explain (a) why two successive (1,3)-migrations of hydrogen around a cyclopentadiene system are much slower than a one-step, concerted (1,5)-sigmatropic rearrangement, and (b) demonstrate that the suprafacial (1,5)-sigmatropic shift of hydrogen in conjugated dienes is faster than the antarafacial mode.

Answer

(a) Suprafacial (1,3)-sigmatropic shifts of hydrogen are forced to occur with retention of the migrating center, leading to transition structures realizing Hückel orbital arrays counting four electrons. Thus the reaction is expected to be difficult (no assistance between the bond-breaking and bond-forming processes). In contrast, the suprafacial (1,5)-sigmatropic shifts are facile reactions because they have transition structures that realize Hückel orbital arrays with six electrons, thus stabilized by aromaticity (strong assistance between the bond-breaking and bond-forming processes).

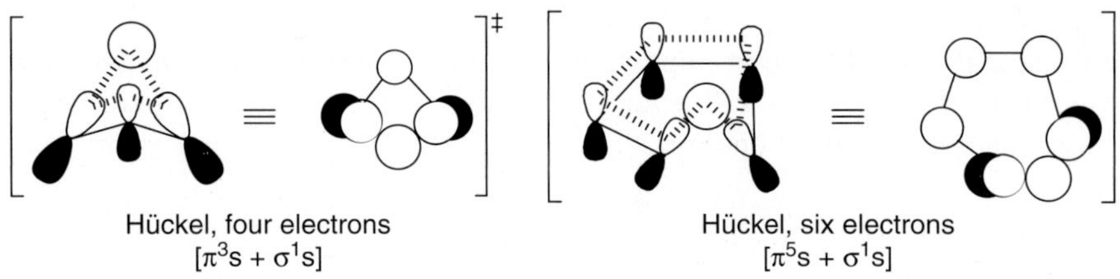

(b) Frontier molecular orbital (FMO) diagrams representing the transition structures of a suprafacial and antarafacial (1,5)-sigmatropic shift of hydrogen are given here.

For the suprafacial mode strong overlap between (singly occupied molecular orbital) SOMO(pentadienyl radical) and SOMO(H$^\cdot$) leads to a relatively strong stabilization of the transition state of the (1,5)-shift of hydrogen. For the antarafacial mode, there is no overlap between SOMO (pentadienyl radical) and SOMO(H$^\cdot$) for reasons of symmetry. Thus, this transition state is less stabilized electronically. It is further destabilized by distortion of the skeleton (a', a'' = symmetrical, antisymmetrical with respect to the mirror plane of symmetry; a, b = symmetrical, antisymmetrical with respect to the C_2 axis of rotation, respectively).

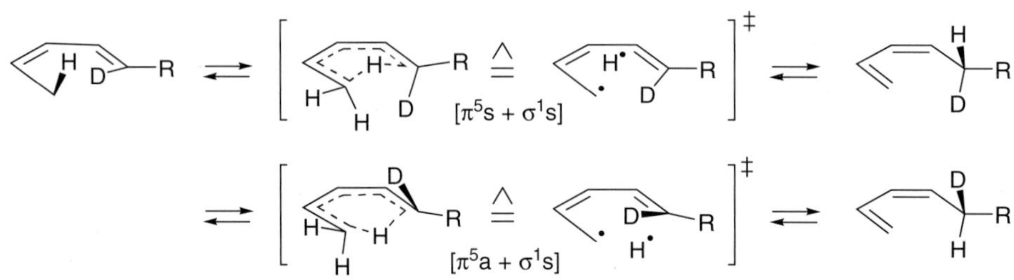

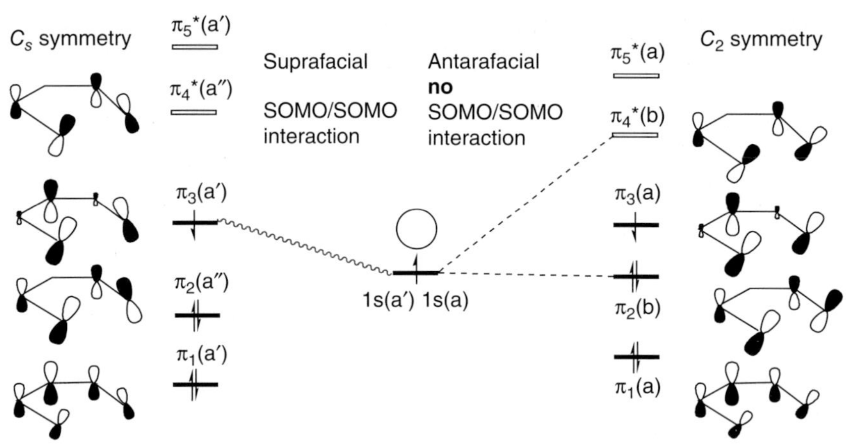

Problem 5.58 Cyclopentadiene has a pK_a = 15.0 in water (Table 1.A.24) and 18.0 in DMSO (Table 1.A.25). Is it possible that the isomerization of 5-alkylcyclopentadiene into 1- and 2-alkylcyclopentadiene can occur via deprotonation and reprotonation in CCl_4?

Answer

CCl_4 is not polar like DMSO and H_2O. Thus, the pK_a of cyclopentadiene in CCl_4 is much higher than in DMSO (18.0) and water (15.5). A pK_a value of 18.0 corresponds to $\Delta_rG(\mathrm{Cp-H} \rightleftarrows \mathrm{Cp^-} + \mathrm{H^+}) = 18(1.36) = 24.5\ \mathrm{kcal\ mol^{-1}}$. This is higher than $\Delta^{\ddagger}G = \Delta^{\ddagger}H - T\Delta^{\ddagger}S = \Delta^{\ddagger}H - 298(-0.011) = 19.3 + 3.2 = 22.5\ \mathrm{kcal\ mol^{-1}}$ measured for the (1,5)-hydrogen shift that equilibrates 5-methylcyclopentadiene with 1-methylcyclopentadiene (Scheme 5.86). Thus, it can be concluded that the mechanism of the migration of hydrogen around the cyclopentadiene systems is not a deprotonation and reprotonation in DMSO, but it could be the case in water.

Problem 5.59 Give a mechanism for isomerization **A** ⇄ **B** [69].

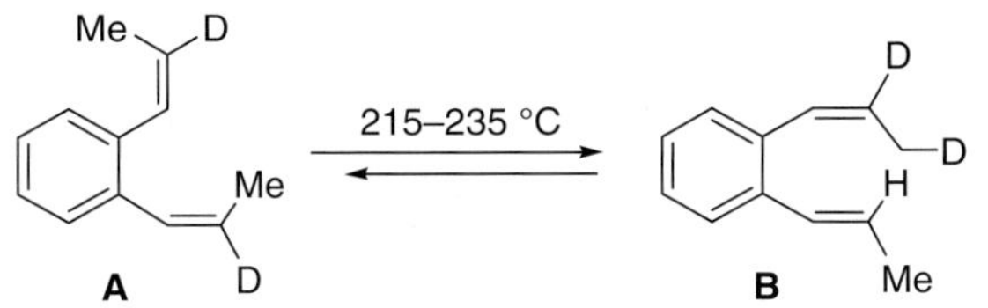

Answer

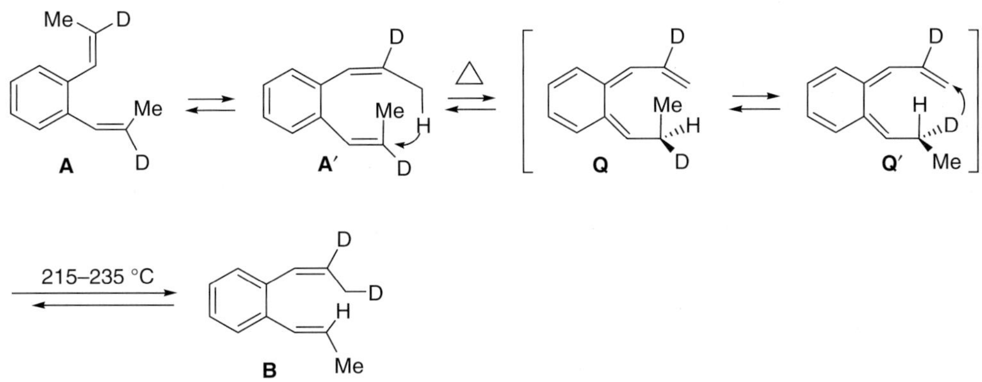

A equilibrates with rotamer **A′**, which undergoes (1,7)-hydrogen shift forming the quinodimethane derivative **Q**. This reaction is highly endothermic, which explains the high temperature to apply for the rearrangement. **Q** equilibrates with rotamer **Q′**, which undergoes a (1,7)-deuterium shift forming **B**. If hydrogen instead of deuterium migrates in **Q**, the starting material **A** forms.

Problem 5.60 Complete the following scheme of the synthesis of Yomogi alcohol.

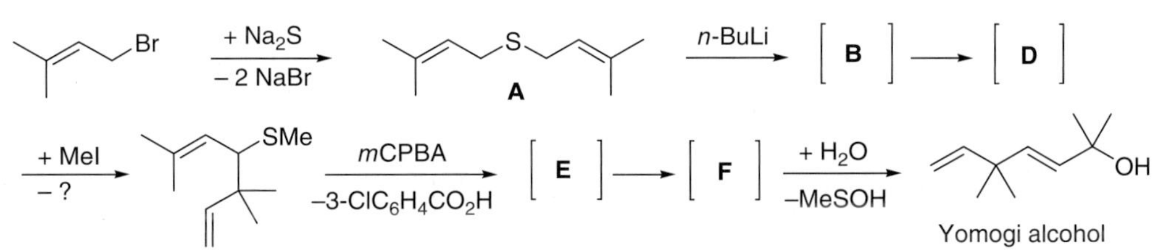

Answer

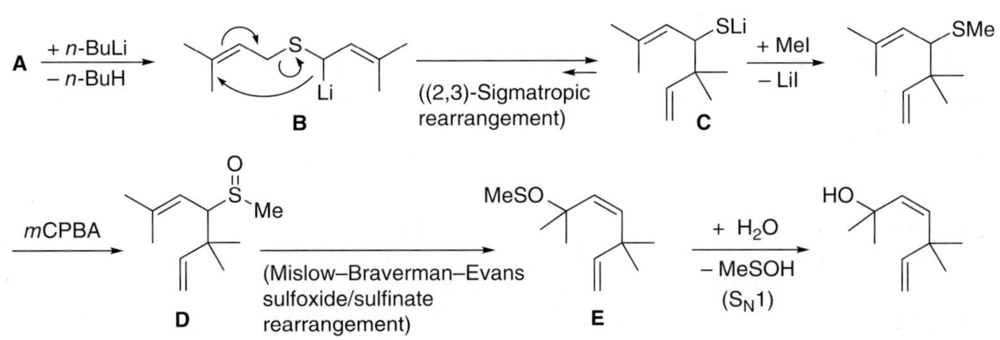

Nomenclature : MeSH: methanethiol
MeSOH: methanesulfenic acid
$MeSO_2H$: methanesulfinic acid
$MeSO_3H$: methanesulfonic acid

*m*CPBA:

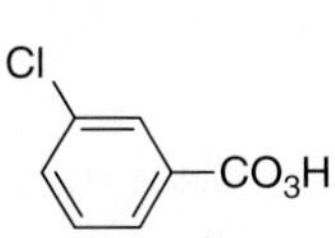

Problem 5.61 Explain the following transformations.

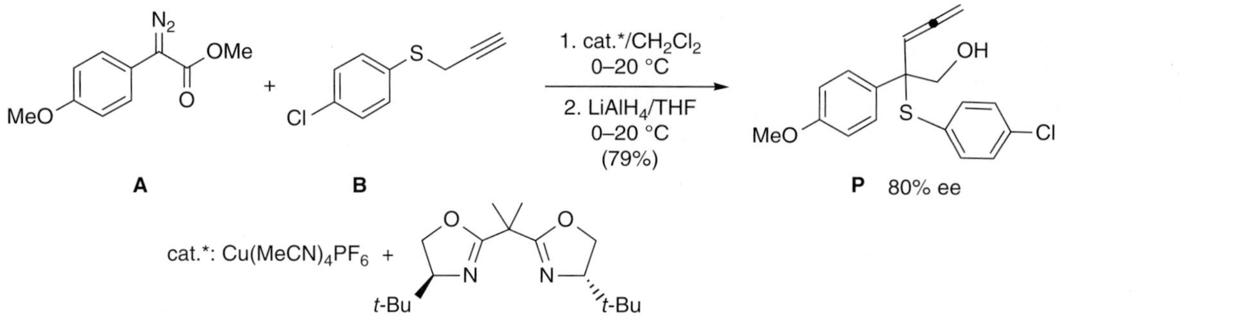

Answer

The copper-catalyzed decomposition of the diazo compound **A** generates a carbenoidcopper species with $Cu(MeCN)_4PF_6$. The MeCN ligands are exchanged with the homochiral bis(oxazolidine) ligand giving an homochiral carbenoid intermediate that reacts with the propargyl sulfide **B** to form an enantiomericallly enriched sulfonium ylide of type **C** with a trigonal S center that is chiral. Subsequent (2,3)-sigmatropic rearrangement of **C** into ester **D** and $LiAlH_4$ reduction of the latter gives alcohol **P**. As for sulfoxides ($R^1S(=O)R^2$) and sulfinylimines or sulfimides ($R^1S(=O)(=NH)R^2$) the racemization of ylide **C** is a slow process.

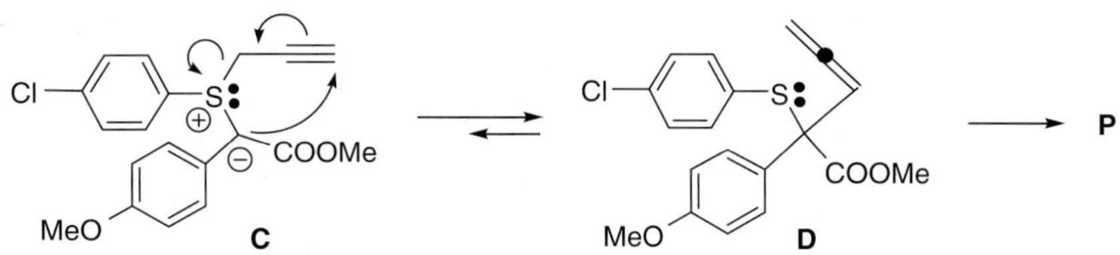

Problem 5.62 Complete the following scheme of the synthesis of **P**.

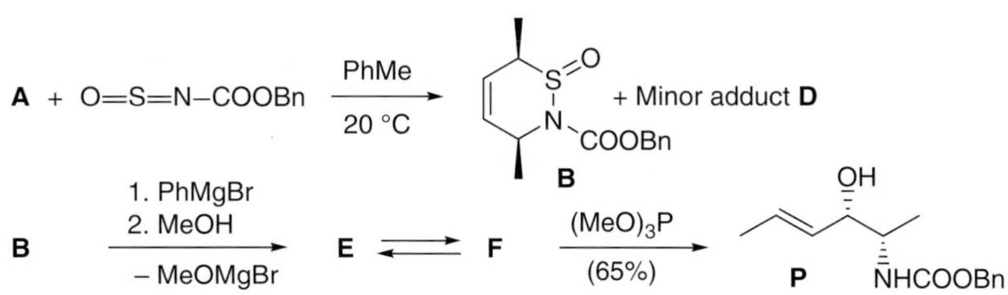

Answer

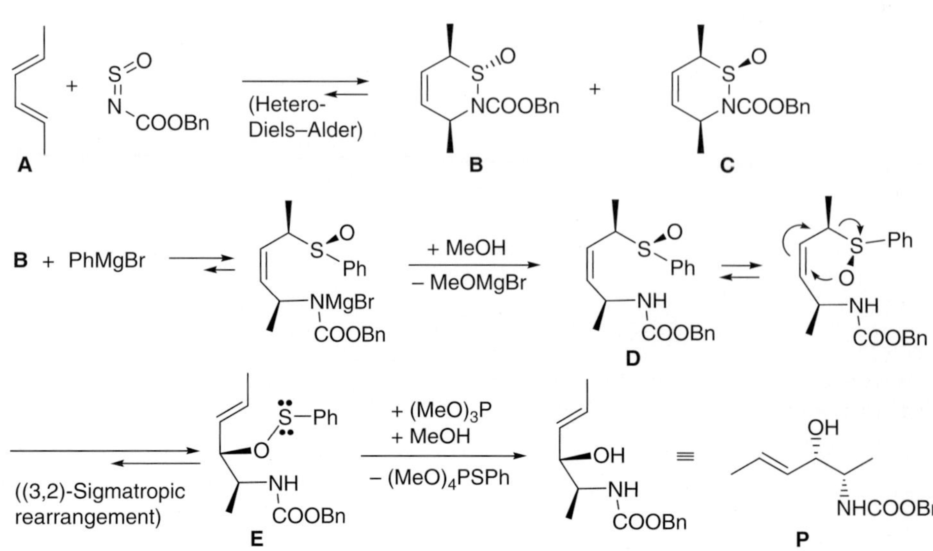

Problem 5.63 What is the main product **P** of the following reaction? [70]

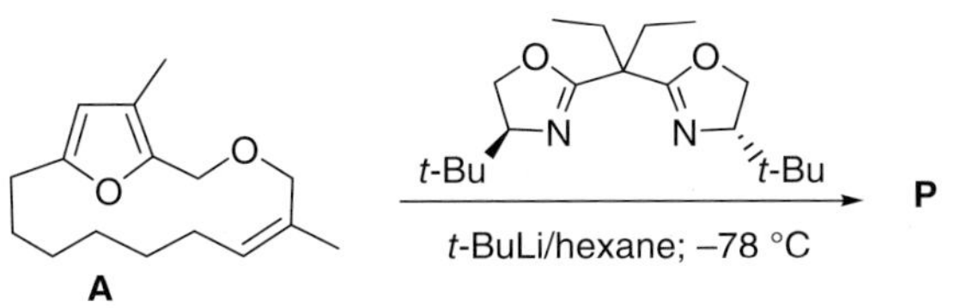

Answer

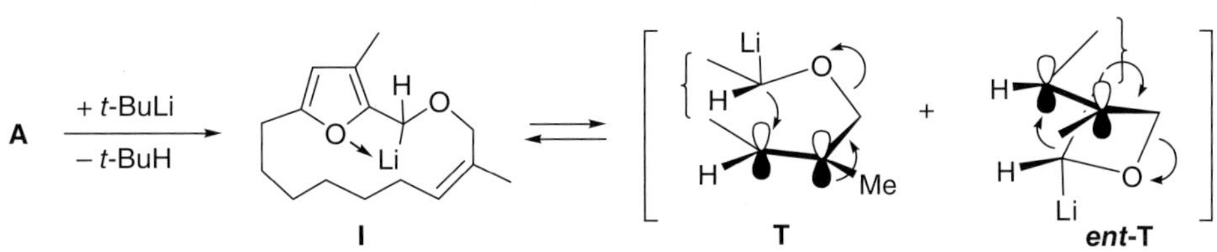

In the absence of any enantiomerically enriched chiral ligand for the organolithium species **I**, there are two possible transition structures **T** and ***ent*-T** for its (3,2)-sigmatropic rearrangement that leads to a racemic mixture of *syn*-diastereoisomers.

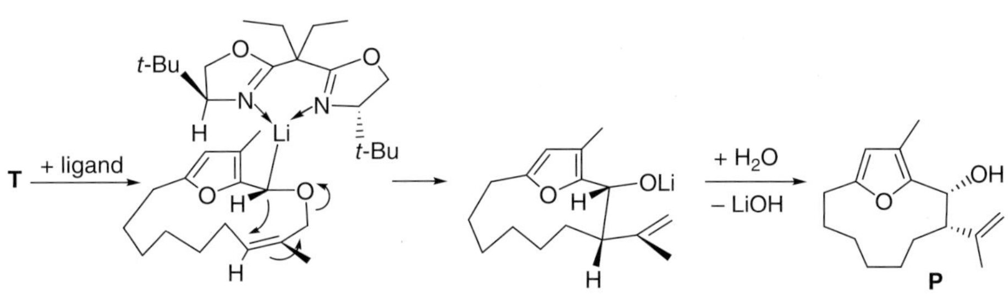

In the presence of the homochiral bis(oxazoline) ligand, the lithium of **I** coordinates to it and generates two diastereoisomeric species that do not have the same stability. The latter undergo (3,2)-sigmatropic rearrangements with the formation of an enantiomerically enriched product **P**, after aqueous work-up.

Problem 5.64 What is the heat of reaction of allyl vinyl ether ⇄ pent-4-enal?

Answer

The experimental standard heats of formation of allyl vinyl ether and of pent-4-enal are not available. The NIST WebBook of Chemistry gives $\Delta_f H°(CH_2{=}CHOCH_2CH_3) = -33.5\ \text{kcal mol}^{-1}$.

Incrementing with one CH_2 group leads to $\Delta_f H°(CH_2{=}CHOCH_2CH_2CH_3) = -33.5 - 5 = -38.5\ \text{kcal mol}^{-1}$.

Didehydrogenation of $CH_2{=}CHOCH_2CH_2CH_3$ into $CH_2{=}CHOCH_2CH{=}CH_2$ is endothermic by 30 kcal mol^{-1}; thus $\Delta_f H°(CH_2{=}CHOCH_2CH{=}CH_2) = -38.5 + 30 = -8.5\ \text{kcal mol}^{-1}$.

One finds $\Delta_f H°(CH_3CH_2CH_2CHO) = -50.6\ \text{kcal mol}^{-1}$.

Incrementing by one CH_2 group and didehydrogenation to generate a terminal olefin gives $\Delta_f H°(CH_2{=}CHCH_2CH_2CHO) = -50.6 - 5 + 30 = -25.6\ \text{kcal mol}^{-1}$.

Finally, the Claisen rearrangement $CH_2{=}CHOCH_2CH{=}CH_2 \rightleftarrows CH_2{=}CHCH_2CH_2CHO$ is estimated to be exothermic by $-25.6 - (-8.5) = -17.1\ \text{kcal mol}^{-1}$

Problem 5.65 What is the intermediate **B** and the product **P** of the following reaction sequence? [71]

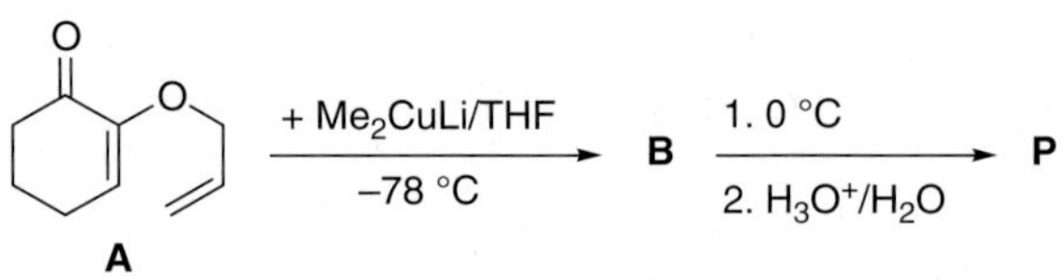

Answer

Conjugate addition (1,4-addition) of the methylcuprate to the cyclohexenone moiety of **A** generates **B**.

At 0 °C, **B** undergoes an easy anionic oxy-Claisen rearrangement into alcoholate **C**, which is converted into product **P** upon acidic aqueous work-up.

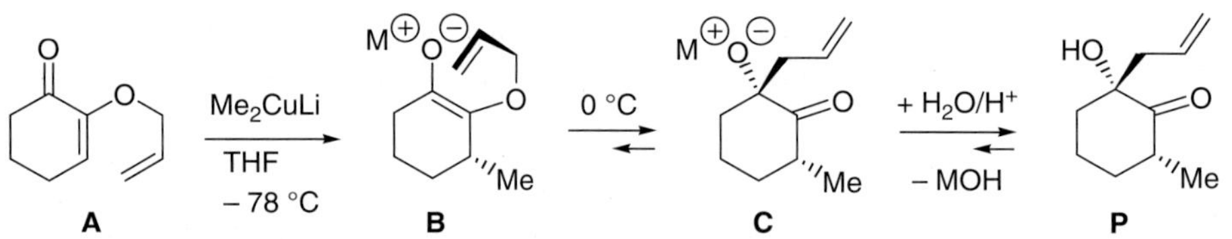

Problem 5.66 What is the main product **P** of the following reaction sequence? [72]

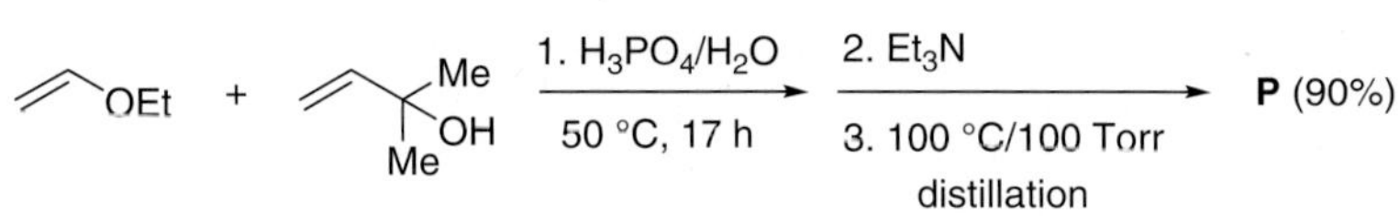

Answer

Acid-catalyzed addition of an alcohol onto an alkenyl ether generates an acetal. The regioselectivity of this electrophilic addition is controlled by the relative stability of the two possible regioisomeric carbenium ion intermediates that form upon addition of a proton to the alkenyl ether. For obvious reasons the oxycarbenium ion is the most stable isomeric carbocation that can form. It is then quenched by the alcohol. Under acidic conditions, and upon heating, the acetal eliminates one equivalent of ethanol, giving an allyl vinyl ether that undergoes a Claisen rearrangement.

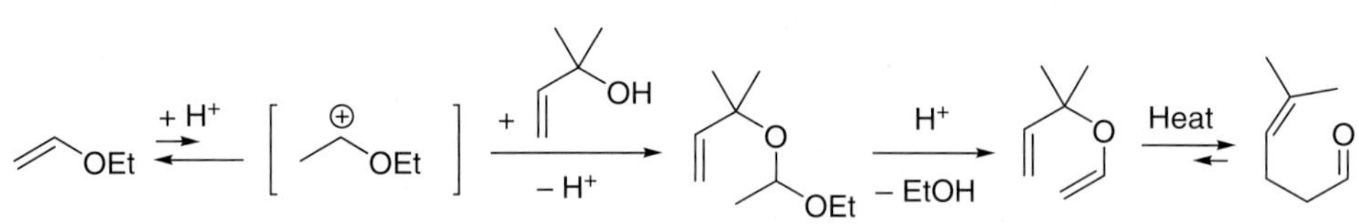

Problem 5.67 What is the main product **P** of the following reaction? [73]

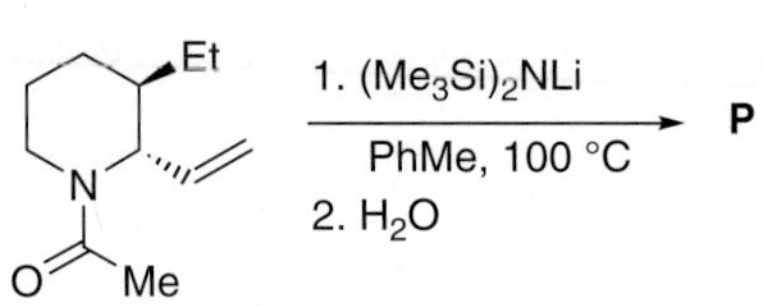

Answer

A 3-aza-2-oxyhexa-1,5-diene **B** is generated upon enolization of acetamide **A**. On heating to 100 °C the lithium enolate **B** undergoes a (3,3)-sigmatropic rearrangement into a lithium (1*Z*,5*E*)-azadodec-1,5-diene-2-alcoholate

D. The driving force of this rearrangement is the conversion of 1-aminoenolate **B** into a 1-oxyimine (imidate **D**). The difference in stability arises in part from the difference in n(O$^-$)/π^*(C=C) and n(O$^-$)/π^*(C=N) interactions, the latter being more stabilizing than the former because the π^*(C=N) LUMO is lower lying in energy than π^*(C=C) (effect of the greater N electronegativity compared with that of C). **D** produces product **P** upon aqueous work-up.

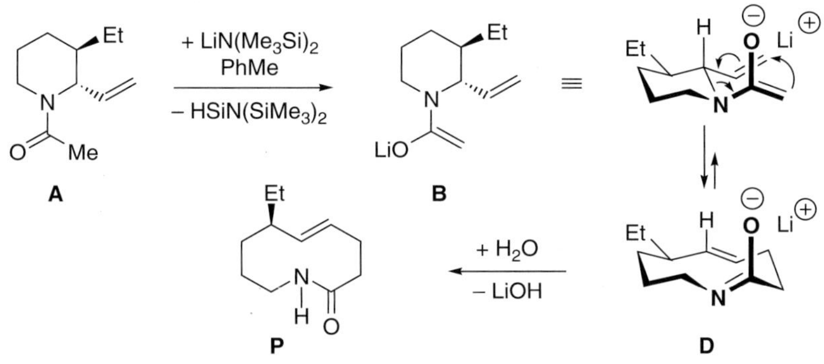

Problem 5.68 What is the intermediate **B** and the major product **P** of the following reaction? [74]

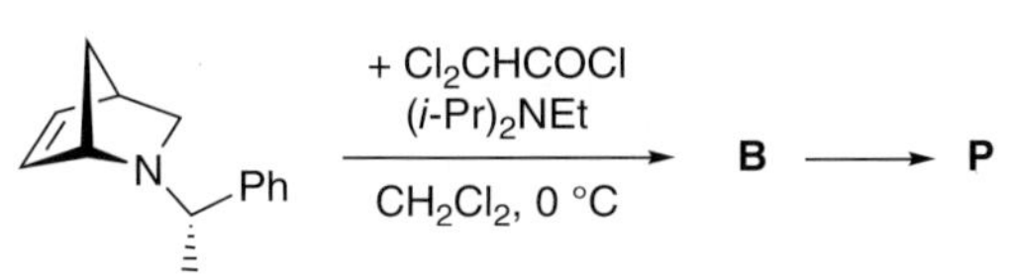

Answer

This is an example of ketene-aza-Claisen rearrangement, which is analogous to the ketene-Claisen rearrangement of Malherbe–Bellus (Reaction (5.69), Figure 5.52).

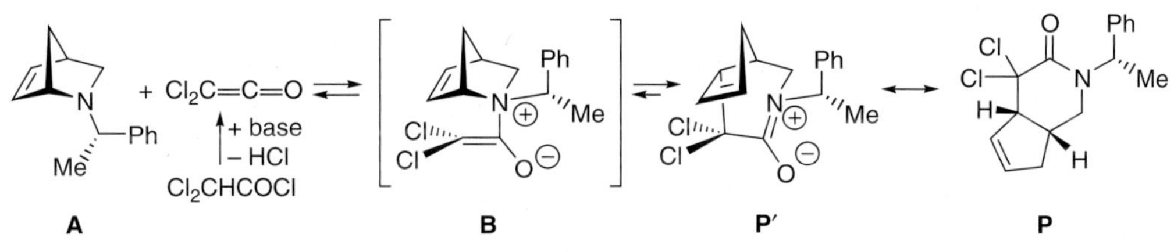

Problem 5.69 What is the major product **P** of the following reactions? [75]

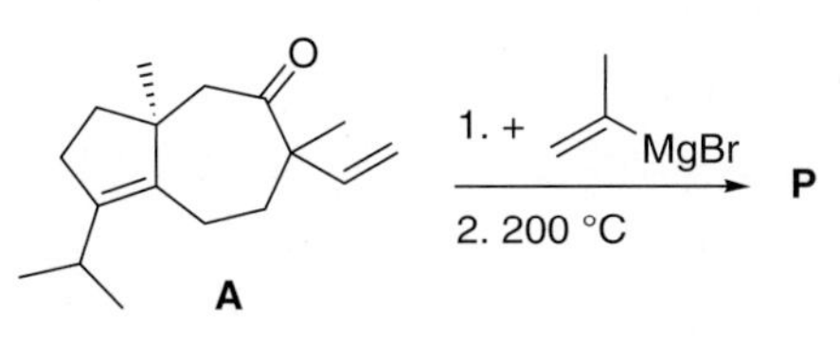

Answer

Addition of the Grignard reagent to ketone **A** gives the allylic alcoholate **B**. On heating to 200 °C **B** undergoes an oxy-Cope rearrangement to form with the more stable enolate **D**. After aqueous work-up product **P** is obtained.

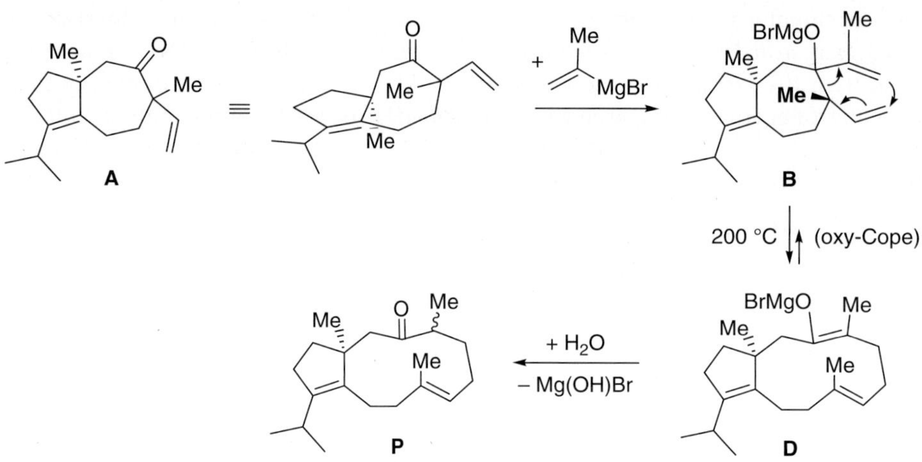

Problem 5.70 What is the product **P** of the following reactions? [76]

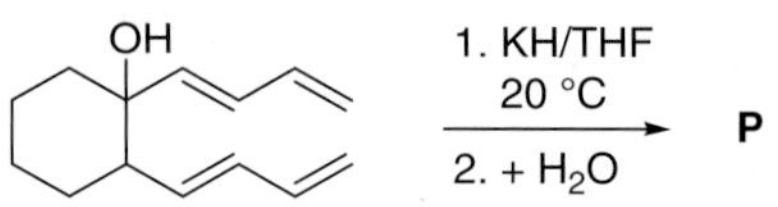

Answer

Treatment of **A** with KH generates potassium alcoholate **B** and H_2. This compound contains a 3-oxy-1,6-divinyl-1,5-diene moiety. Both vinyl substituents in **B** are expected to accelerate the oxy-Cope rearrangement (diradicaloid model) of **B,** which equilibrates easily with enolate **D**, a compound more stable than **B** because it contains an enolate moiety (conjugation between the alcoholate and alkene moieties) in place of a non-conjugated alkene + alcoholate moieties. Enolate **D** is a 4,5-divinyl-1-oxy-1,5-diene, which is also expected (diradicaloid model) to undergo a facile oxy-Cope rearrangement, and produces the vinyl enolate **E**, which is more stable than **D** (conjugation enolate/alkene).

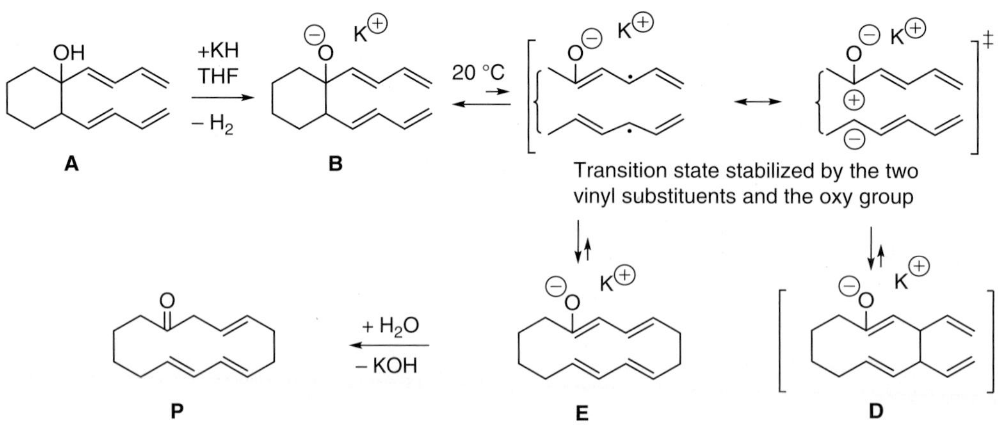

Problem 5.71 What are the major products **P** and **Q** of the following reactions? [77]

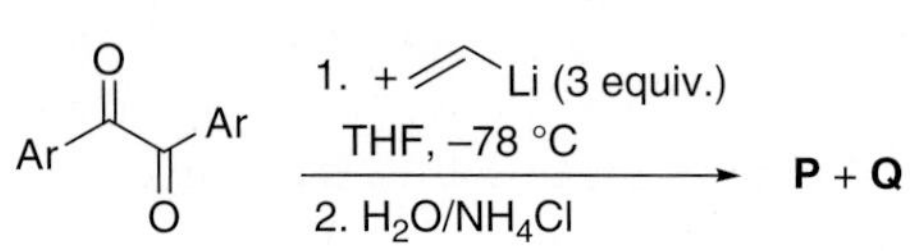

Answer

Diaryl α-diketones add vinyllithium to generate 3,4-dioxy-1,5-dienes **B** that undergo fast dioxy-Cope rearrangements into the corresponding bis(enolates) **D**, as expected from the diradicaloid model. The bis-allyl diradical limiting structures are highly stabilized by the two aryl and two oxy substituents. As a consequence, **D** is formed below room temperature. After aqueous work-up diketones **P** are obtained. As the medium is alkaline, they can equilibrate with aldols **Q** through intramolecular aldol reactions (see Section 5.7.6).

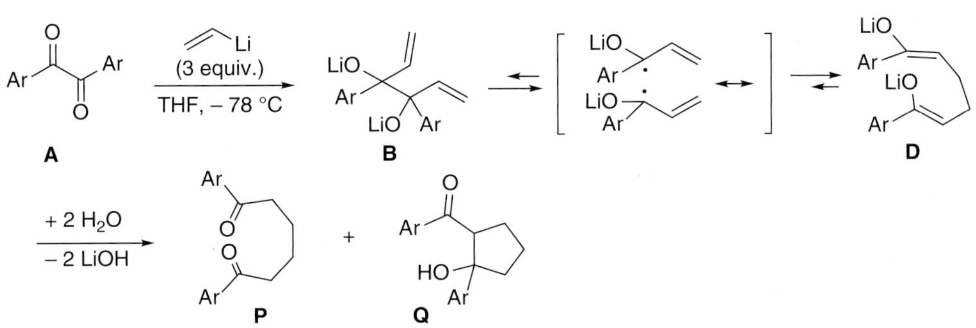

Problem 5.72 Explain the following reaction [78].

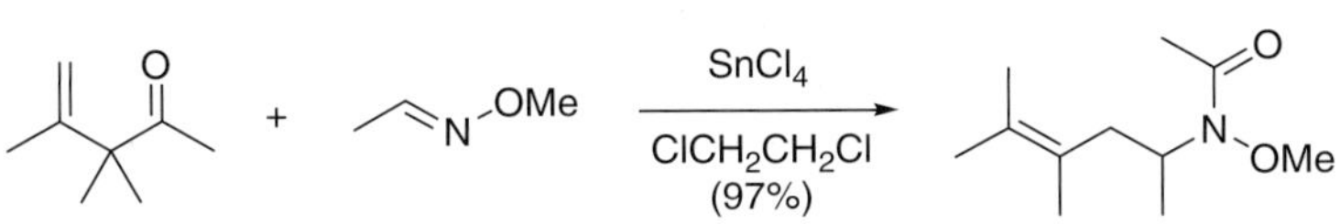

Answer

The Lewis acid activates the addition of the *O*-methyl oxime to ketone **A**, generating zwitterion **D**. The latter undergoes a facile 2-azonia-Cope rearrangement furnishing **E**, which equilibrates with **P**.

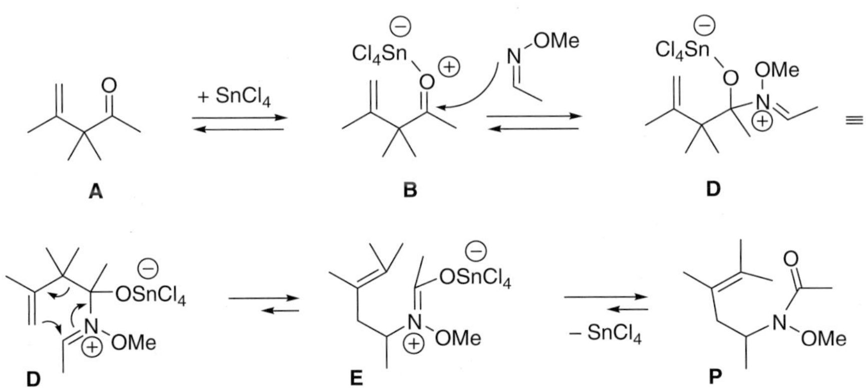

Problem 5.73 Explain the following reactions [79].

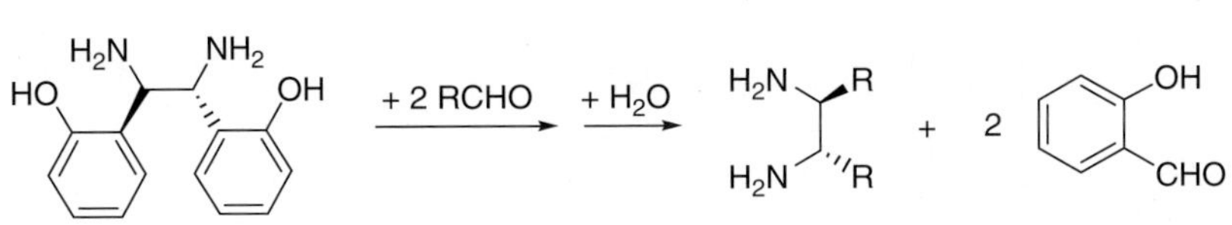

Answer

Enantiomerically enriched C_2-diamines are very useful in asymmetric synthesis. They are used as chiral bases, as chiral ligands, or in the preparation of chiral ligands of various metallic complexes that catalyze reactions enantioselectively. Chiral diamine **A** reacts with 2 equiv. of aliphatic aldehyde RCHO to give diimine **D** and two equivalents of water. Diimine **D** is a 2,5-diaza-1,5-diene that can undergo a 2,5-diaza-Cope rearrangement into **E**. The reaction is relatively fast as **D** equilibrates with **D'** in which one of the phenol groups protonates intramolecularly one of the two imine moieties, making the reaction analogous to a 2-azonia-Cope rearrangement. Diimine **E** is more stable than diimine **D** because of π-conjugation between the aryl and imine groups, conjugation that is not present in **D**. As for other Cope rearrangements, the transition state adopts a chair-like conformation in which steric repulsion between the Ar and R substituents are minimized. This is the case if they reside in equatorial positions. These features render the conversion of **A** into **E** highly stereoselective, with nearly complete transfer of chirality. Hydrolysis of **E** provides the aliphatic C_2-diamines **P**.

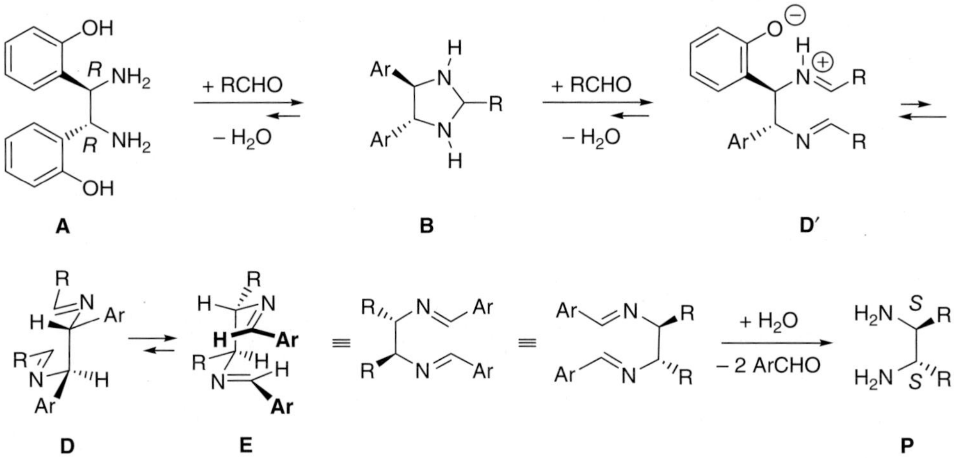

Problem 5.74 What is product **P** in the following reaction? [80]

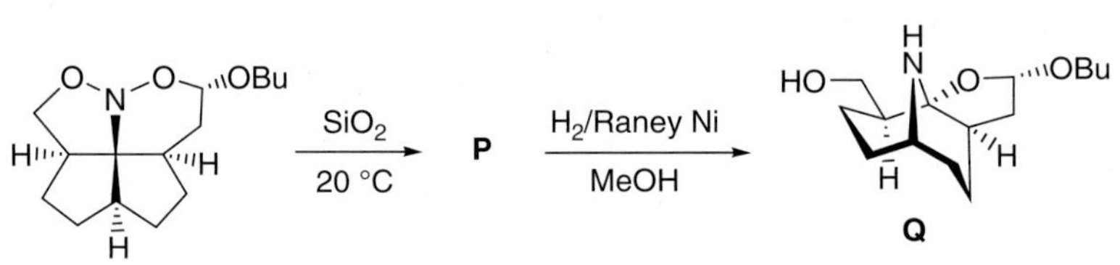

Answer

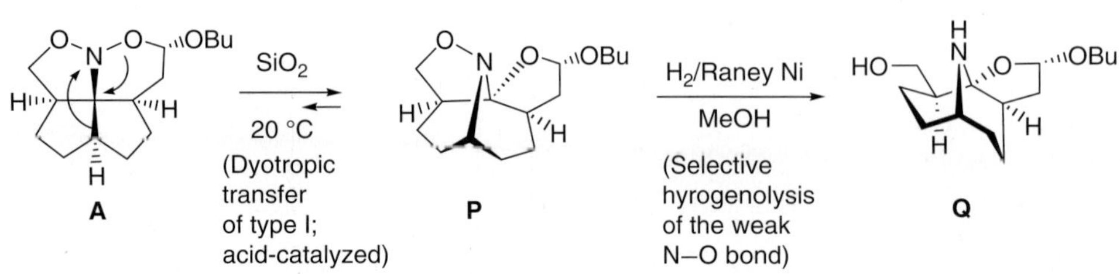

Problem 5.75 What is the isomer of diphenyl(trimethylsilyl)methyl acetate obtained on heating to 190 °C? [81]

Answer

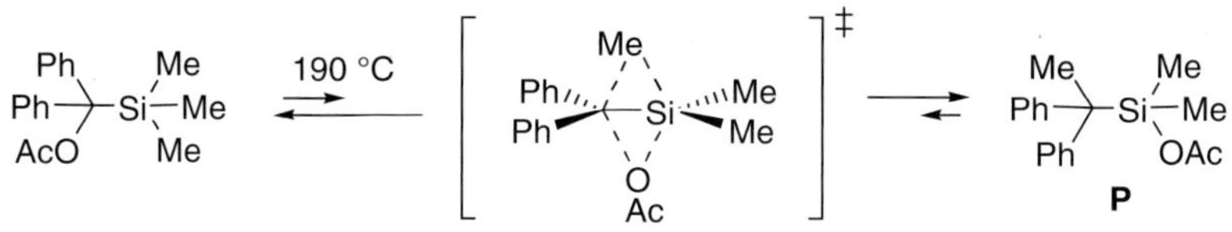

The driving force of this dyotropic rearrangement of type I is the formation of the Si—O bond at the expense of a Si—C bond that overcompensates the bond strength difference between the C—C and C—O bonds.

Problem 5.76 What is the product of the following reaction?

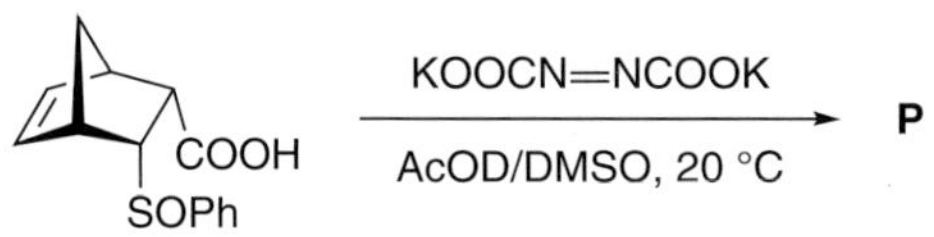

Answer

Potassium azodicarboxylate reacts with deuterated acetic acid producing deuterated azodicarboxylic diacid. The latter decomposes at room temperature into carbon dioxide and deuterated diazene.

Deuterated diazene hydrogenates the norbornene double bond in an exclusive *cis* manner preferentially on its least sterically crowded side, the *exo* face.

KOOC–N=N–COOK + 2 AcOD ⇌ 2 AcOK + DOOC–N=N–COOD

DOOC–N=N–COOD ⟶ 2 CO_2 + DN=ND

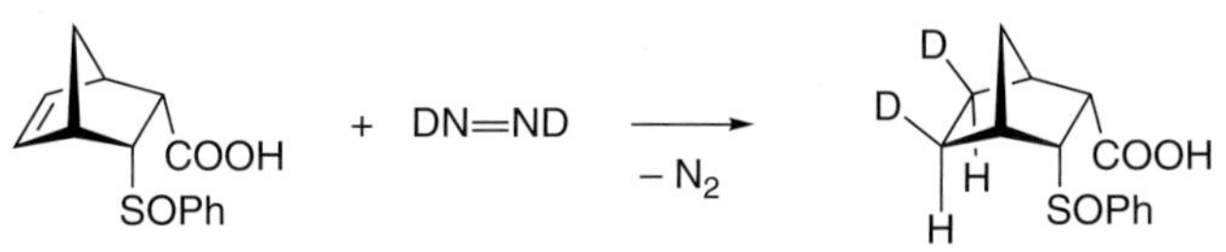

(Concerted dyotropic rearrangement of type II: [$\pi^2s + \sigma^2s + \sigma^2s$])

Problem 5.77 What are the intermediate compounds **B**, **D** and the final product **P** of the following reaction sequence?

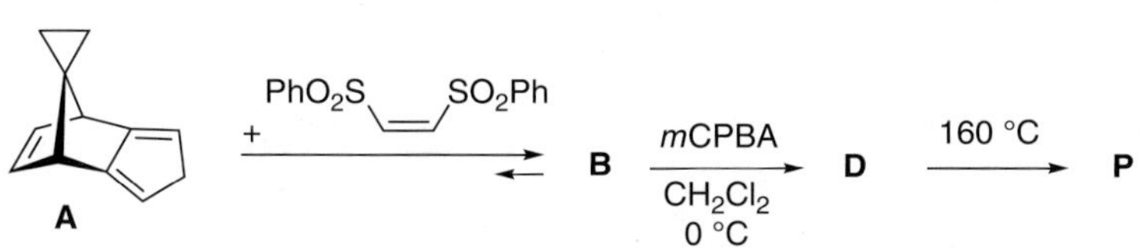

Answer

Cyclopentadiene **A** adds to dienophiles preferentially onto its *endo* face (Section 5.3.15) because the cycloadducts resulting from the *endo* face attacks generate *syn*-sesquinorbornene systems that are more stable than the corresponding *anti*-sesquinorbornenes. This thermodynamic property is encountered also in the transition state of the Diels–Alder reactions, as predicted by the Dimroth principle and the Bell–Evans–Polanyi theory on concerted reactions (the transition states of these cycloadditions resemble the cycloadducts). The cycloadditions of **A** are irreversible below 100 °C. Nevertheless, there is a kinetic preference for the *endo* face additions of **A**. Thus, cycloadduct **B** forms selectively under kinetic control. The reaction does not follow the endo Alder rule for steric reasons. The double C=C bond of **B** that is annulated with both norbornene rings is polarized toward the *exo* face of both norbornene moieties (one of the "norbornene effects," this one being a consequence of hyperconjugative interactions). This makes it more reactive than the other C=C double bond of **B**. Thus, epoxidation of **B** with *m*-chloroperbenzoic acid (*m*CPBA) is both chemo- and *exo*-face selective giving epoxide **D**. In **D**, the *endo* $(PhSO_2)C—H$ bonds are almost in contact with the C=C double bond of the norbornene unit. A dyotropic rearrangement of type II becomes possible and occurs on heating to 160 °C giving the final product **P**. The latter is more stable than **D** because of π-conjugation between the sulfonyl and alkene moieties in **P**, which is not present in **D**.

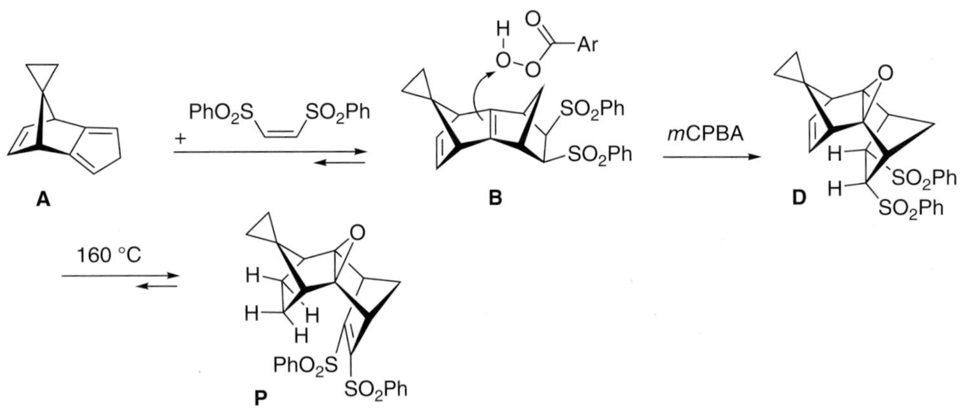

Problem 5.78 On heating methylidenecyclobutane with an excess of maleic anhydride to 195 °C for 30 hours a 1 : 2 adduct is obtained. Explain [82].

Answer

Maleic anhydride (**A**) and methylidenecyclobutane (**B**) react in an Alder ene-reaction giving a first product **D** that is a cyclobutene derivative. Its alkene moiety can take part, as that of **B**, in a H-ene-reaction with a second equivalent of **A**. Three possible products **P**, **Q**, and **W** can be formed. **P** arises from H-transfer from an exocyclic CH_2 group that is freely rotating, and thus can adapt easily to the geometry requirements of the cyclic transition state of the ene-reaction. Product **Q** arises from the H-transfer from an endocyclic CH_2 group and is much more constrained than the exocyclic CH_2 moiety, and thus cannot adapt as easily to the geometry requirements of the ene-reaction. Furthermore, the alkene moiety of **P** is exocyclic (alkylidenecyclobutane) whereas that of **Q** is endocyclic (cyclobutene). One thus predicts that **Q** is less stable than **P**. At 195 °C, the reaction temperature, it is not excluded that **P** and **Q** are equilibrated by retro-ene-reactions.

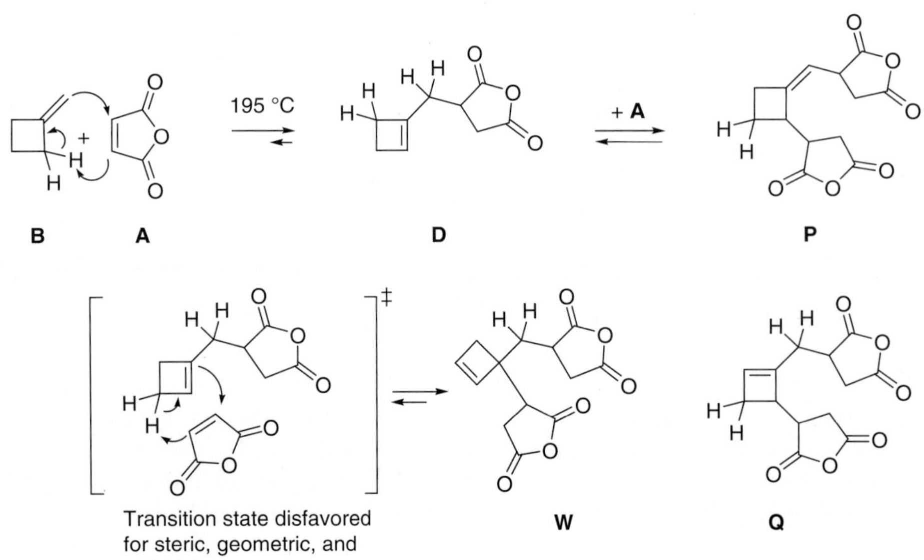

Transition state disfavored for steric, geometric, and enthalpic reasons

The regioselectivity of the second Alder ene-reaction **D** + **A** depends on steric and geometric factors. **P** is favored over **W**, which is destabilized by *gem*-disubstitution by relatively large substituents. Furthermore, **W** is a cyclobutene system: **P** is not, and thus, **W** is more strained. These steric, geometrical, and enthalpic factors that can be readily foreseen for the difference in stability between **P** and **W** also intervene in the transition states of these one-step, concerted pericyclic reactions. If **Q** and **W** should not be equilibrating with **D**, under kinetic control their rates of formation would be lower than that of the formation of **P** because of the Dimroth principle and the Bell–Evans–Polanyi theory of concerted, one-step reactions.

Problem 5.79 Explain the following reaction [83].

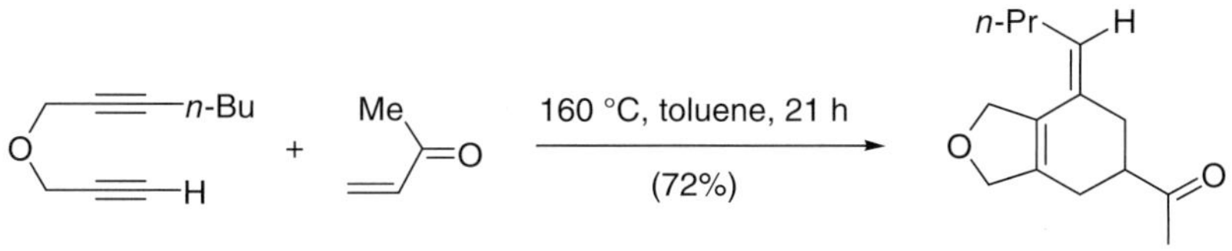

Answer

The bis(acetylene) **A** undergoes an intramolecular H-ene-reaction on heating to 160 °C giving the triene **B**. **B** reacts with methyl vinyl ketone in an intermolecular Diels–Alder reaction producing **P**. The high regioselectivity of this cycloaddition is easily understood by the diradicaloid model of the Diels–Alder reaction transition states. A classical rule predicts that *ortho*-cycloadducts are preferred kinetically over their *meta*-regioisomers. Accordingly, **P** should not be the preferred cycloadduct. For the *ortho* orientation **D1** is the most stable diradical that one can write. Its charge transfer configuration represented by zwitterion **Z1** enjoys stabilization from allylic conjugation for its cationic part, and from enolate conjugation for its negatively charged part. In this case, only one of the two double bonds of the allenyl unit contributes to the allylic conjugation since the two π systems are orthogonal in the allene moiety. For the *meta* orientation both the diradical **D2** and zwitterion **Z2** limiting structures are more stable than **D1** and **Z1**, respectively, because both double bonds conjugate with the radical and the cation (pentadienyl system instead of an allyl system).

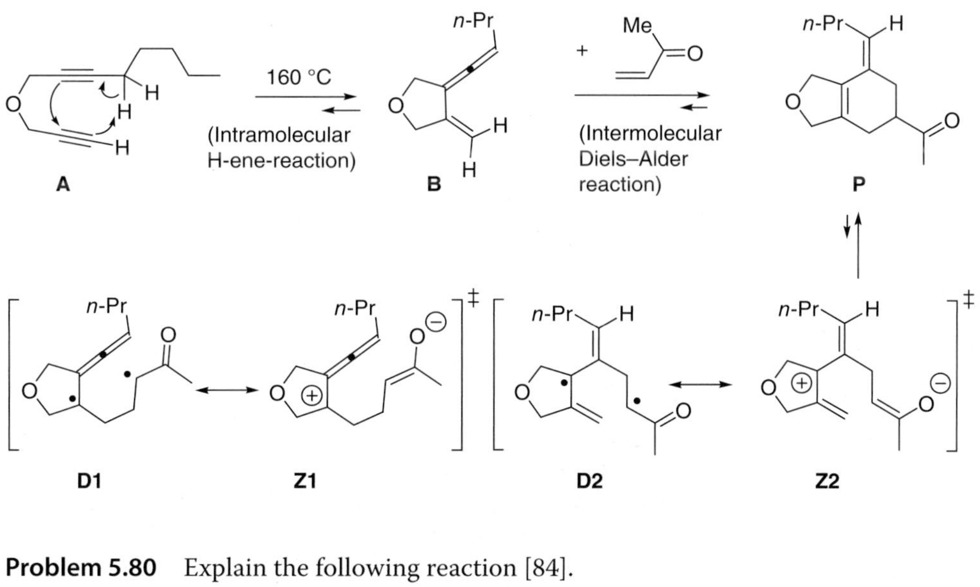

Problem 5.80 Explain the following reaction [84].

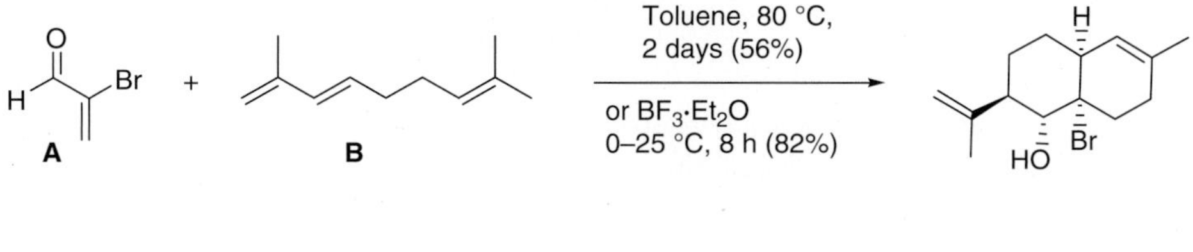

Answer

The Diels–Alder reaction of **A** + **B** is *para* and *ortho* regioselective, as predicted by the diradicaloid model. The same model predicts also that the preferred stereoselectivity is that following the Alder endo rule (stabilizing electrostatic interaction between the negatively charged enolate moiety and the positively charged alkyl substituted allyl cation moiety of the zwitterionic limiting structure **T1** of the transition state). This gives cycloadduct **D**. Under the conditions of its formation an intramolecular carbonyl ene-reaction is possible, generating product **P**. The stereoselectivity of this reaction is controlled by steric and geometrical factors.

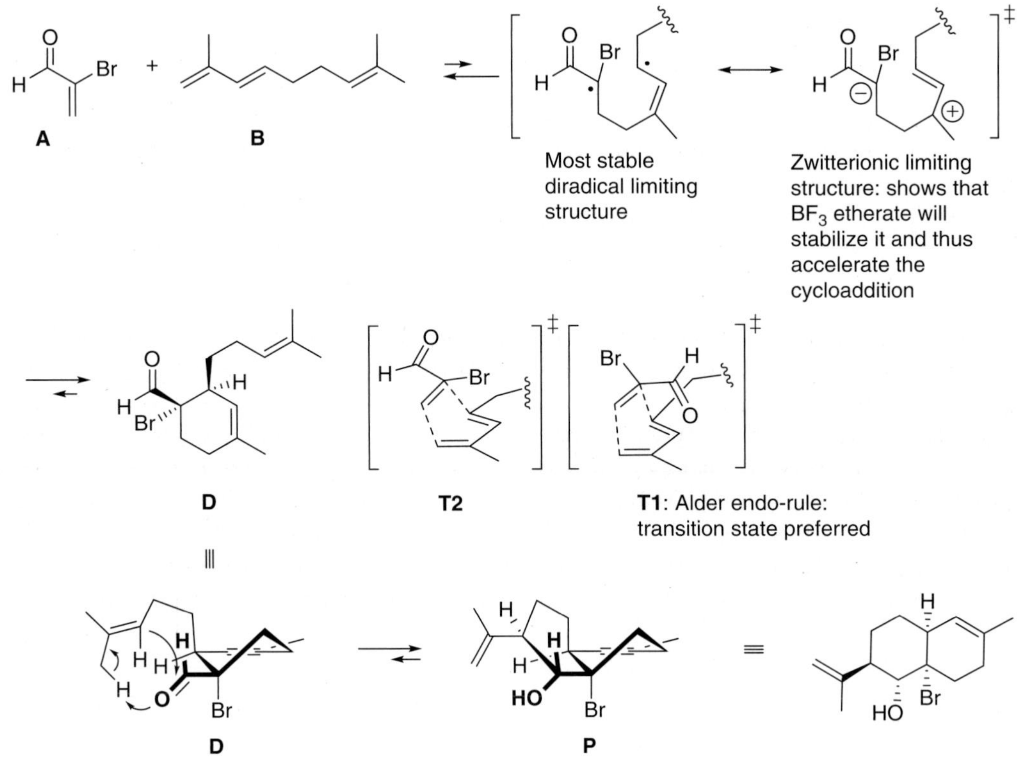

Problem 5.81 Can one use the carbonyl ene-reaction to capture CO_2 with an alkene?

Answer

The NIST WebBook of thermochemical data gives the following gaseous standard heats of formation (25 °C, 1 atm.):

	CO_2	propene	butyric acid	butyrolactone
$\Delta_f H°$:	−94.05	4.9	−113.7	−87.0 kcal mol^{-1}

Considering a standard heat of hydrogenation of terminal alkenes that amounts to −30 kcal mol^{-1}, one estimates $\Delta_f H°(CH_2 = CHCH_2COOH) = -113.7 + 30 = -83.7$ kcal mol^{-1}.

Thus, the oxa ene-reaction of propene with CO_2 is endothermic by $\Delta_r H° = -83.7 - 4.9 + 94.05 = 5.2$ kcal mol^{-1}. Being a condensation, its entropy is negative; thus, this reaction is also endergonic.

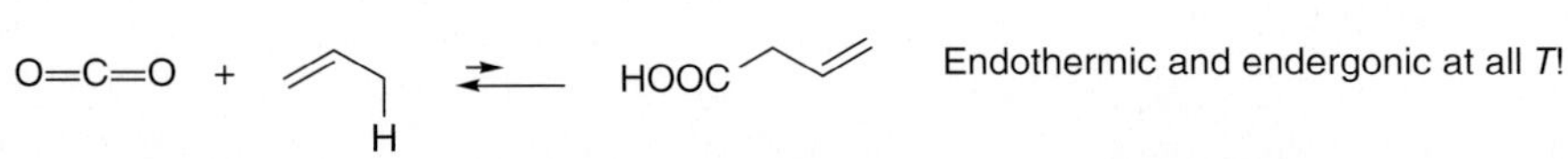

If one takes $\Delta_r S°$(propene + CO_2 = but-3-enoic acid) = −30 eu, one calculates at 25 °C a free energy change of reaction $\Delta_r G° = \Delta_r H° - T\Delta_r S° = 5.4 + 298(0.030) = 14.4$ kcal mol^{-1} (gas, 1 atm). If a catalyst would catalyze the oxa ene-reaction of CO_2 and propene at 25 °C and under 1 atm, the equilibrium constant K for the formation of but-3-enoic acid would be smaller than 10^{-10} M^{-1}. In the condensed state, under pressure, a 1 : 1 mixture of CO_2 and propene reaches concentration of c. 11 M. Thus the amount of but-3-enoic acid that could be formed under these conditions would not be higher than 10^{-8} M. Thus, for the oxa ene-reaction of CO_2 with an alkene to be a possible means to capture CO_2 one needs to stabilize but-3-enoic acid selectively without stabilizing propene or/and CO_2. Formation of a salt with an amine such as NH_3 (does not react with CO_2 in the absence of water) might help realize a solution, and precipitation of the ammonium carboxylate could be exploited to displace the equilibrium. Isomerization of but-3-enoic acid into butyrolactone is exothermic by $\Delta_r H^\circ = -87.0 - (-83.7) = -3.3$ kcal mol^{-1}. This cyclization has a negative entropy (two single bond rotations are being blocked) of c. −10 eu, which leads to $\Delta_r G^\circ$(but-3-enoic acid ⇄ butyrolactone) = −3.3 − 298(−0.01) = −0.3 kcal mol^{-1}.

Obviously, this reaction alone cannot displace the disfavored oxa ene-reaction equilibrium at 25 °C and under 1 atm. Possible isomerization of but-3-enoic acid into (*E*)-but-2-enoic acid does not help very much as it brings not more than 2 kcal mol^{-1} of exothermicity. Any condensation reaction that would combine but-3-enoic acid or (*E*)-but-2-enoic acid with themselves or with propene would have to pay for the entropies of condensation <−33 eu; thus, they would contribute to c. 10 kcal mol^{-1} to the free energy of the process. To compensate for the 14.4 kcal mol^{-1} of the initial carbonyl ene-reaction the heat of the condensation reaction must be c. −14.4 − 10 = −24.4 kcal mol^{-1}. This is not provided by the formation of an ester by carboxylic acid addition to an alkene. The NIST WebBook gives $\Delta_r H^\circ$(MeCOOH + CH_2 = CHMe ⇄ MeCOO-*i*-Pr) = −8.4 kcal mol^{-1}. If the ester or polyester can precipitate from the reaction mixture, this reaction might displace the carbonyl ene-reaction equilibrium!

The Alder ene-reaction of two alkene units is more exothermic than the formation of an ester by combination of carboxylic acid with an alkene. With $\Delta_f H^\circ$(but-1-ene) = −0.15 kcal mol^{-1} and $\Delta_f H^\circ$(EtCH(Me)-*n*-Bu) = −50.8 kcal mol^{-1} one estimates $\Delta_f H^\circ$(EtCH(Me)—CH—CH=CH—Me, the product of ene-reaction of two but-1-ene units) = −50.8 + 28 = −22.8 kcal mol^{-1} and $\Delta_f H^\circ$(2 but-1-ene ⇄ EtCH(Me)—CH=CH—Me) = −22.8 + 2(−0.15) = −22.5 kcal mol^{-1}. This is not enough to make the overall process $2CO_2$ + 2 propene = HOOC—CH=CH—CH(Me)—CH_2—COOH) exergonic at 25 °C and under 1 atm, but is not far off. Again, if the product of condensation or a polymer would precipitate, the process might be used to capture CO_2. It remains to find a suitable catalyst for the carbonyl ene-reaction of CO_2. Several catalysts are available for the C—C formation between two alkenes, and for their polymerization.

Problem 5.82 Explain the following transformations [85, 86].

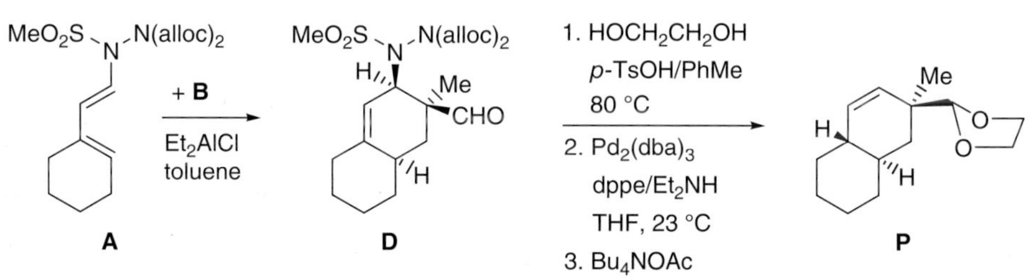

alloc = CH_2=$CHCH_2OCO$; dba = (*E*,*E*)-PhCH=CHCOCH=CHPh
dppe = 1,2-bis(diphenylphosphino)ethane, *p*-TsOH = *p*-toluenesulfonic acid

Answer

In the presence of the strong Lewis acid catalyst Et_2AlCl **A** and **B** (methacrolein) undergo a highly *ortho*-regioselective, and Alder stereoselective Diels–Alder reaction (23 °C, toluene) giving cycloadduct **D**. Under acidic conditions and azeotropic distillation (elimination of water co-product of the reaction), **D** reacts with ethyleneglycol forming acetal **E**. Transition metal complexes such as that resulting from the reaction of $Pd_2(dba)_3$ and diphosphine dppe form π-complexes with the allylic groups of the alloc moieties. The latter undergo cleavage of their allyl C-carbonate bonds with liberalization of the free NH_2 group as shown with **F** (formation of allylpalladium complexes that are displaced by the nucleophile Et_2NH). In the presence of a base such as Bu_4NOAc compound **F** undergoes a β-elimination of $MeSO_2H$ providing diazene **G**. The latter intermediate undergoes a facile retro ene-reaction with elimination of N_2 and formation of product **P**. The stereoselectivity of the latter reaction is given by the configuration of the Diels–Alder cycloadduct **D**.

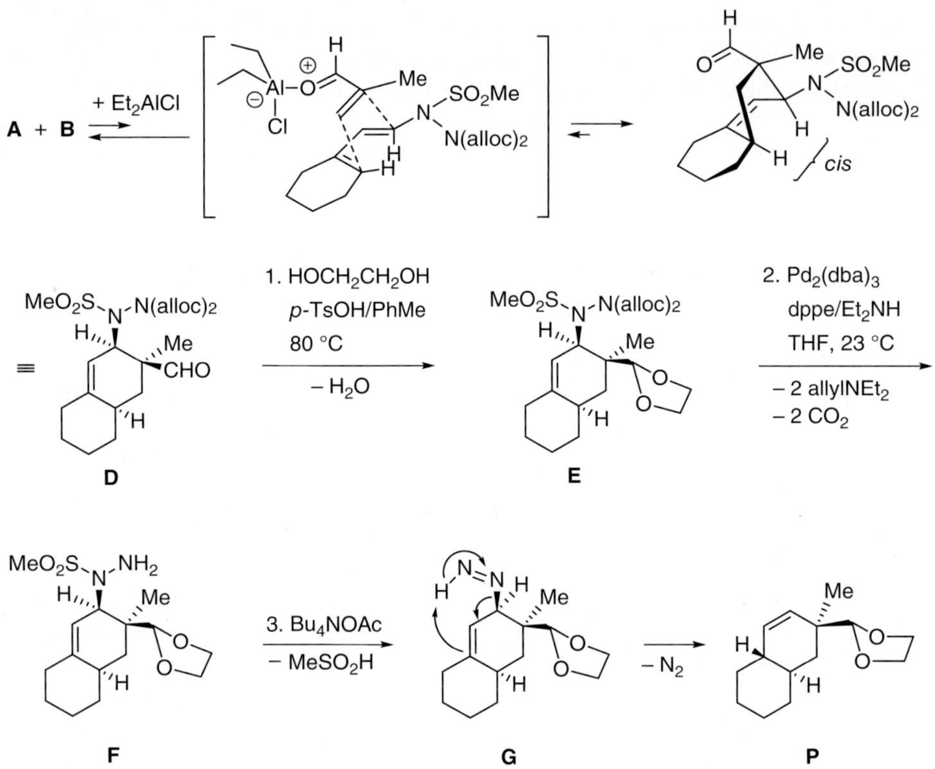

Problem 5.83 What are the products of the reaction of allylsilanes with sulfur dioxide? [87]

Answer

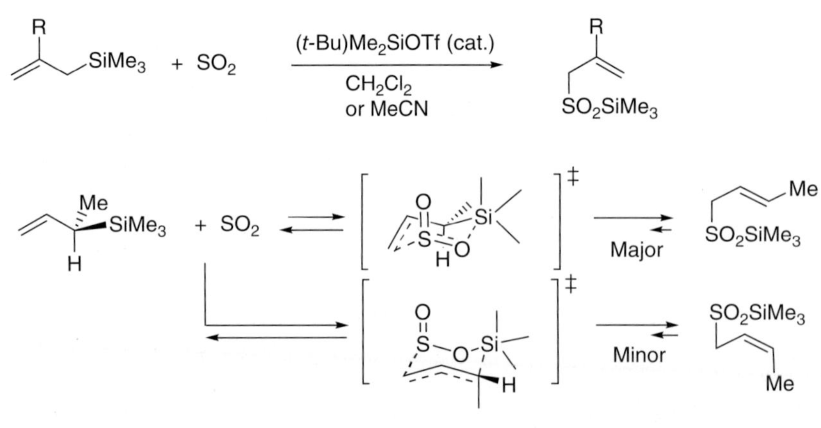

Problem 5.84 What is the major product **P** of the following reaction? [88]

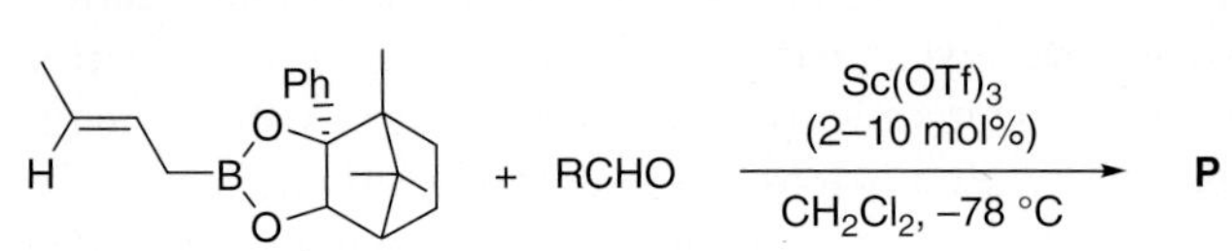

Answer

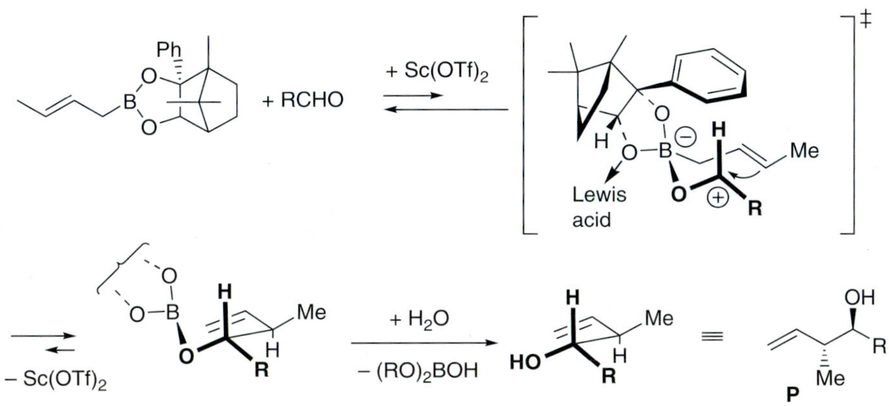

The probable role of the Lewis acid $Sc(OTf)_2$ is to enhance the electrophilicity of boron of the borinic ester and reinforce its coordination with the aldehyde. This also lowers the activation energy of the bora ene-reaction.

Problem 5.85 Explain the following reactions. Are the stereoselectivities expected? [89]

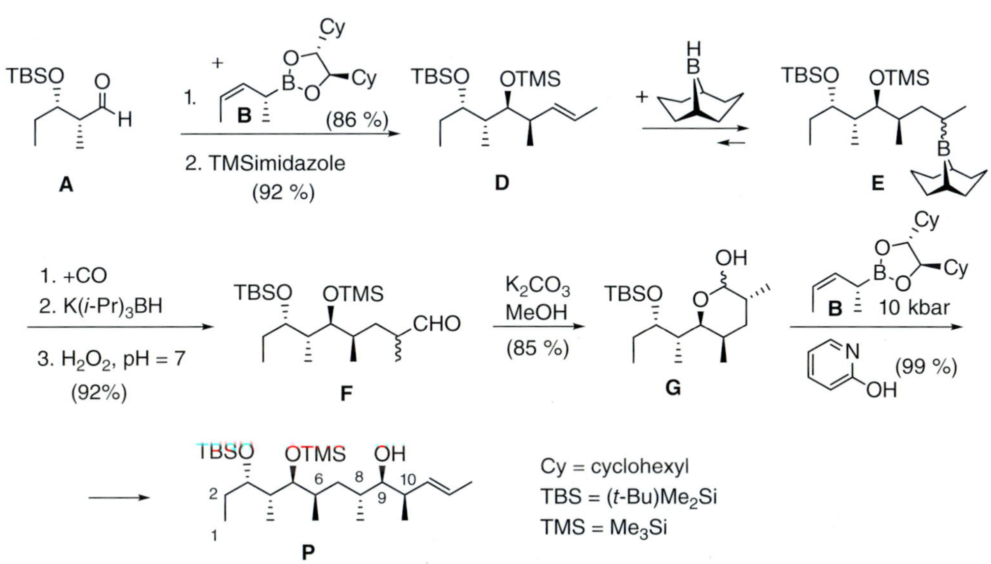

Answer

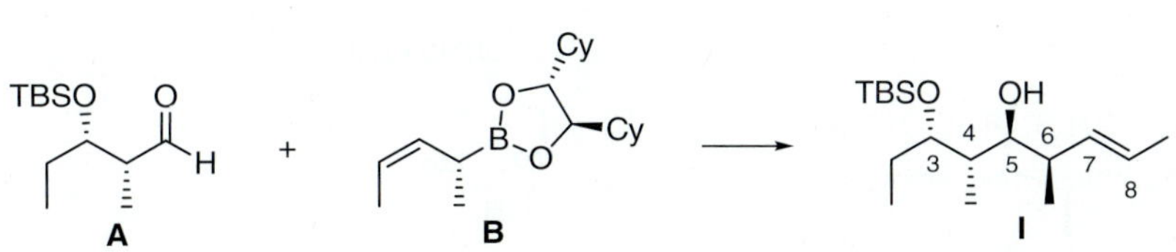

This carbonyl bora ene-reaction adopts transition state **T1,** which minimizes steric repulsions including 1,5-*syn* interactions. The alternative transition state **T2** is not as favorable.

A = RCHO

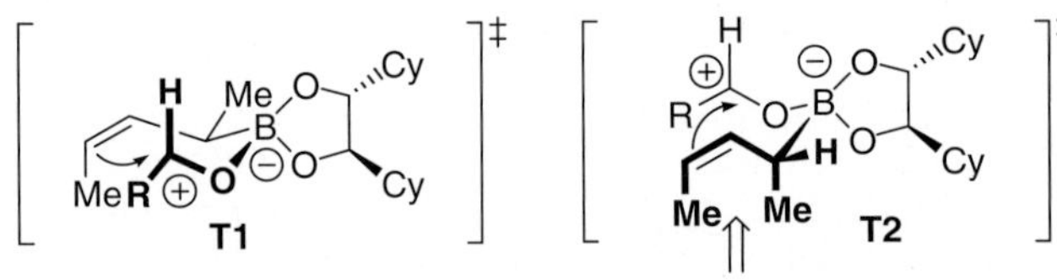

T1 ignores the facial selectivity of the carbonyl group of aldehyde **A**: the anti-Cram (3,4-*anti*) forms because of **T1**. The Cram product (3,4-*syn*) would be formed if **T2** should be preferred. Silylation of homoallylic alcohol **I** gives **D**. Its hydroboration is regioselective for steric reasons. The voluminous 9-BBN recognizes the bulk of the Me—C(6) group and thus the boron adds to C(6) yielding **E**.

Steric factor renders hydroboration **D** into **E** regioselective, but not stereoselective. The latter borane reacts with CO giving aldehyde **F** after suitable work-up. Selective methanolysis of the least sterically hindered silyl ether, Me_3SiO, gives an alcohol that adds intramolecularly to the aldehyde forming hemiacetal **G**. Under basic conditions, α-epimerization occurs and finally the most stable stereoisomer obtained is that which places a maximum number of its substituents in equatorial positions as shown below:

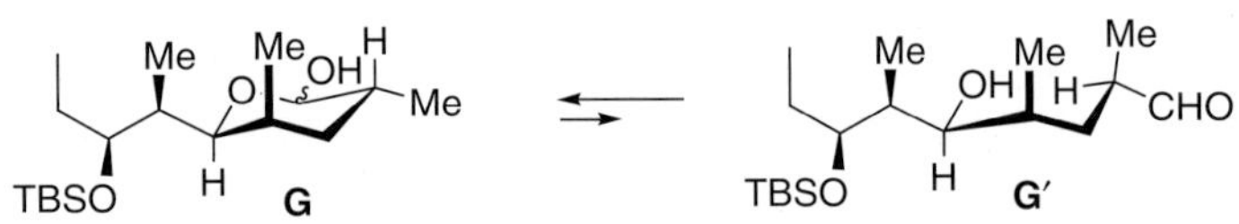

Crotylation **F** + **B** implies the ring-opened aldehyde **G′** that equilibrates with **G**. The absolute configuration of **B** controls the diastereoselectivity of this carbonyl bora ene-reaction, as in the case of reaction **A** + **B**, thus producing the 8,9-*anti*-9,10-*syn* product **P**, preferentially.

Problem 5.86 Using propionamide *ent*-**696**, prepare the polypropionate fragment **P** [90].

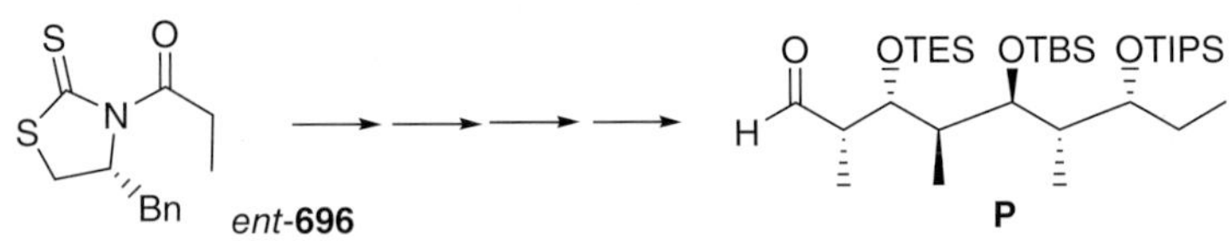

TES = Et_3Si
TBS = $(t\text{-Bu})Me_2Si$
TIPS = $(i\text{-Pr})_3Si$

Answer

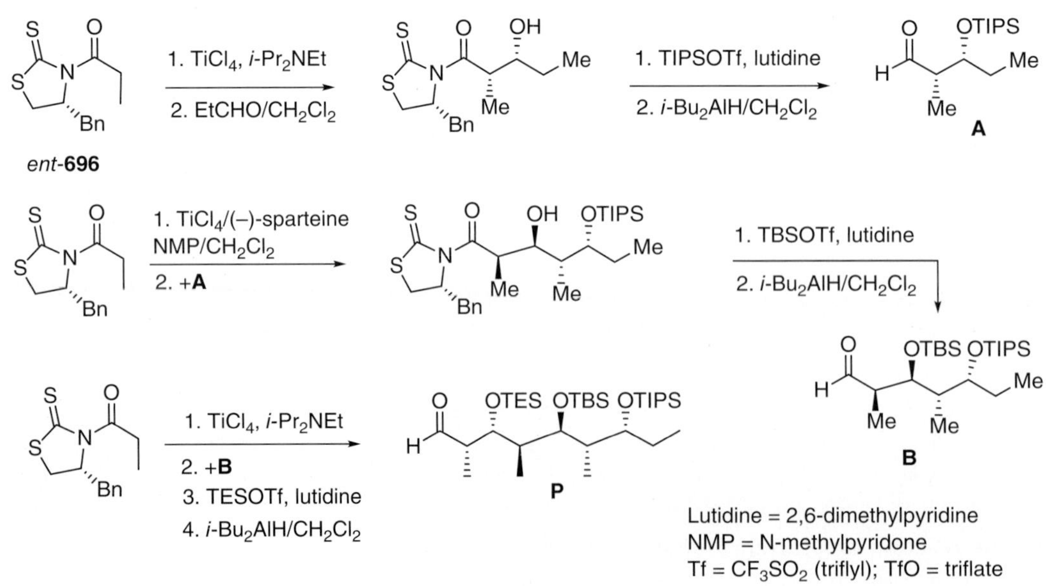

Lutidine = 2,6-dimethylpyridine
NMP = N-methylpyridone
Tf = CF_3SO_2 (triflyl); TfO = triflate

Problem 5.87 What is the major product **P** of the following reaction? [91]

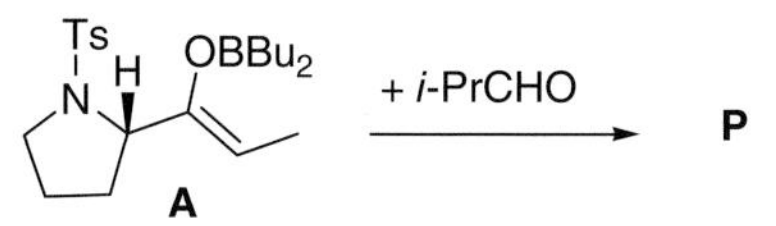

Answer

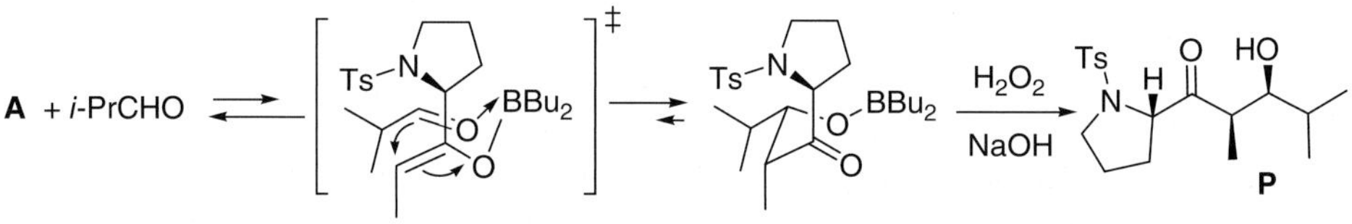

Zimmerman–Traxler chair-like transition state with minimization of steric repulsions

Problem 5.88 What are the major products of the following cross-aldol reactions all starting with **A** and finishing with a work-up using $H_2O/H_2O_2/KOH$? [92]

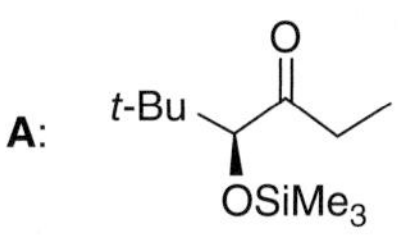

(a) 1. (*i*-Pr)$_2$NLi/THF, –78 °C; 2. TMEDA, then + PhCHO.
(b) Bu$_2$BOTf/(*i*-Pr)$_2$NEt, then PhCHO.
(c) MTMP/THF, –5 °C, then PhCHO.
(d) MTMP/THF/HMPA/dioxane, then (*i*-PrO)$_3$TiCl and PhCHO.

TMEDA = $Me_2NCH_2CH_2NMe_2$; HMPA = $(Me_2N)_3P{=}O$; Tf = CF_3SO_2; MTMP = 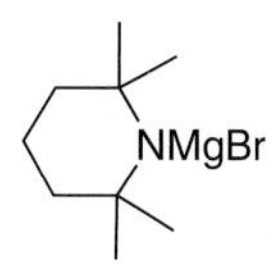

Answer

(a)

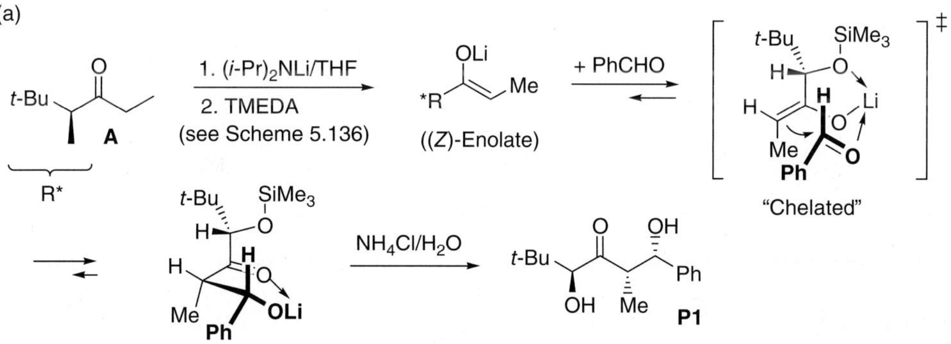

(b)

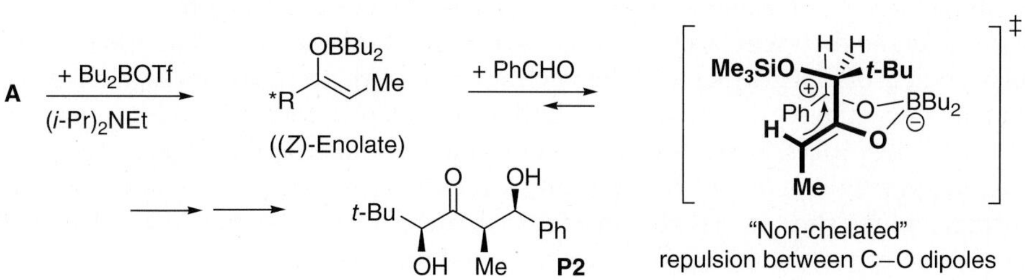

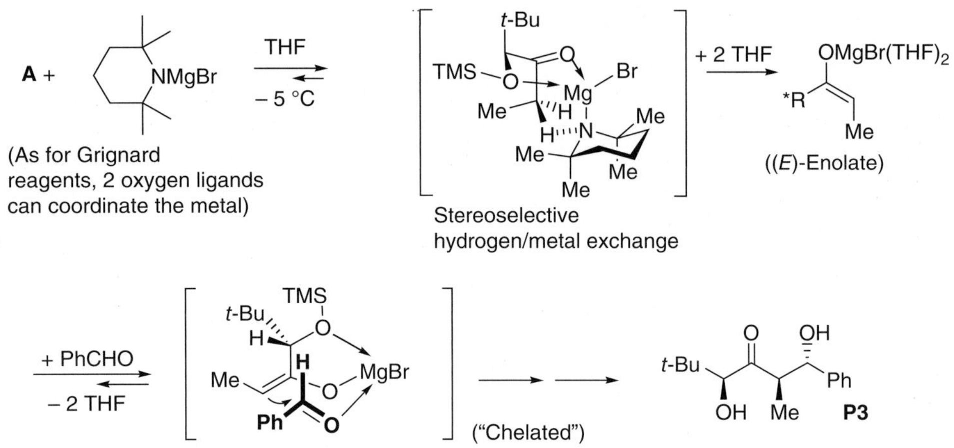

(c)

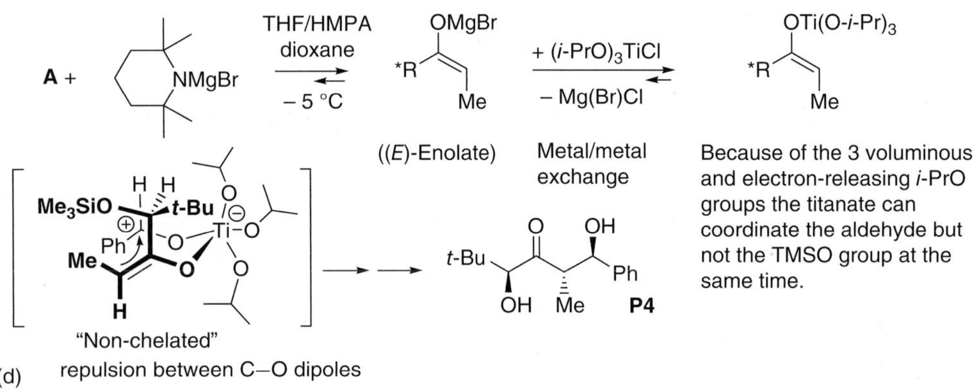

(d)

Additional Material

The Wigner–Witmer rule announced in 1928 paved the way to Longuet–Higgins theories of correlation of electronic configurations and to the Woodward–Hoffmann rules of pericyclic reactions.

In classical mechanics, one body, or a group of bodies in motion follow pathways that conserve the total motion momentum. To make these objects deviate from their "natural" trajectories, energy has to be given to them. This conservation principle applied to electrons and nuclei of molecules becomes the principle of conservation of the total angular momentum. Considering the Born–Oppenheimer approximation that assumes that the motion of electrons can be separated from the motion of nuclei, the total angular momentum of a chemical system can be taken as the sum of the momentum of the electrons and nuclei. If one can neglect the dynamic effects of the nuclei and electron spin interactions, one can state that thermal reactions will follow pathways conserving the angular momentum of the electrons. This is the Wigner–Witmer rule announced in 1928 [39, 93].

The wave function describing the state of reactants being converted into products thus concerns the electronic orbitals of reactants and products and their special characteristics, especially orbital shape, and orbital symmetry derived from the molecular point group of geometry of the nuclear framework. In other words, if an element of symmetry (e.g. a mirror plane of symmetry σ, a C_2-axis, or both elements of symmetry as in C_{2v} point group symmetry) is maintained during the process that converts reactants into products, and vice versa, the transition state of the reaction is associated with one or several transition structures (the point of highest energy on the potential energy hypersurface; lifetime shorter than the time for a vibration $<10^{-13}$ seconds) that might or might not enjoy stabilization due to electron exchange between reactants and products.

References

1 Steiner, R.P. and Michl, J.P. (1978). Pi-biradicaloid hydrocarbons – pleiadene family. 3. Facile symmetry-forbidden thermal conversion of a polycyclic butadiene moiety to a cyclobutene. *J. Am. Chem. Soc.* 100 (20): 6413–6415.

2 Jefford, C.W., Bernardinelli, G., Wang, Y. et al. (1992). Torquoselectivity in the electrocyclic conversion of benzocyclobutenes to ortho-xylylenes. *J. Am. Chem. Soc.* 114 (4): 1157–1165.

3 Murakami, M., Usui, I., Hasegawa, M., and Matsuda, T. (2005). Contrasteric stereochemical dictation of the cyclobutene ring-opening reaction by a vacant boron p orbital. *J. Am. Chem. Soc.* 127 (5): 1366–1367.

4 Saito, A., Umakoshi, M., Yagyu, N., and Hanzawa, Y. (2008). Novel one-pot approach to synthesis of indanones through Sb(V)-catalyzed reaction of phenylalkynes with aldehydes. *Org. Lett.* 10 (9): 1783–1785.

5 Giese, S. and West, F.G. (2000). Ionic hydrogenation of oxyallyl intermediates: the reductive Nazarov cyclization. *Tetrahedron* 56 (52): 10221–10228.

6 Lawrence, A.L., Wegner, H.A., Jacobsen, M.F. et al. (2006). Investigations towards the synthesis of (−)-coprinolone, via a thermal 8 pi-6 pi electrocyclization cascade of 1,5,7-trien-4-ones. *Tetrahedron Lett.* 47 (49): 8717–8720.

7 Staley, S.W. and Heyn, A.S. (1975). Electrocyclic reaction of cyclodecatrienyl anion. *J. Am. Chem. Soc.* 97 (13): 3852–3854.

8 Alhusaini, A.H., Muqtar, M., and Ali, S.A. (1991). Study of ketene–alkene cycloadditions and cycloreversions of cyclobutanones. *Tetrahedron* 47 (36): 7719–7726.

9 Corey, E.J. and Yamamoto, H. (1970). Modification of Wittig reaction to permit stereospecific synthesis of certain trisubstituted olefins. Stereospecific synthesis of α-santalol. *J. Am. Chem. Soc.* 92 (1): 226–228.

10 Claridge, T.D., Davies, S.G., Lee, J.A. et al. (2008). Highly (*E*)-selective Wadsworth–Emmons reactions promoted by methylmagnesium bromide. *Org. Lett.* 10 (23): 5437–5440.

11 Liu, Y.K., Ma, C., Jians, K. et al. (2009). Asymmetric tandem Michael addition-Wittig reaction to cyclohexenone annulation. *Org. Lett.* 11: 2848–2851.

12 Trabert, L. and Hopf, H. (1980). Alicyclic compounds. 3. New method of preparation and reactions of [4]radialene. *Liebigs Ann. Chem.* (11): 1786–1800.

13 Vogel, P. (2008). Combinatorial synthesis of linearly condensed polycyclic compounds, including anthracyclinones, through tandem Diels–Alder additions. In: *Anthracycline Chem. Biol. I*, vol. 282 (ed. K. Krohn), 187–214. Heidelberg, Berlin, Germany: Springer Verlag.

14 Chung, Y.S., Duerr, B.F., Nanjappan, P., and Czarnik, A.W. (1988). Diene-substituent effects on the rate of the retro-Diels–Alder reaction – cycloreversion reactivity varying over a range of 5 powers of 10. *J. Organomet. Chem.* 53 (6): 1334–1338.

15 Boger, D.L., Corbett, W.L., Curran, T.T., and Kasper, A.M. (1991). Inverse electron demand Diels–Alder reactions of *N*-sulfonyl α,β-unsaturated imines – a general-approach to implementation of the 4-pi participation of 1-aza-1,3-butadienes in Diels–Alder reactions. *J. Am. Chem. Soc.* 113 (5): 1713–1729.

16 Bunnage, M.E. and Nicolaou, K.C. (1997). The oxide anion accelerated retro-Diels–Alder reaction. *Chem. Eur. J.* 3 (2): 187–192.

17 Thalhammer, F., Wallfahrer, U., and Sauer, J. (1990). Reactivity of simple open-chain and cyclic dienophiles in inverse-type Diels–Alder reactions. *Tetrahedron Lett.* 31 (47): 6851–6854.

18 Jenner, G., Rimmelin, J., Libs, S., and Antoni, F. (1976). Pericyclic reactions under pressure. 4. Thermal cyclodimerization of acrolein and methylvinyl ketone. *Tetrahedron* 32: 1107–1111.

19 Anderson, E.D. and Boger, D.L. (2011). Inverse electron demand Diels–Alder reactions of 1,2,3-triazines: pronounced substituent effects on reactivity and cycloaddition scope. *J. Am. Chem. Soc.* 133 (31): 12285–12292.

20 Hayashi, Y., Samanta, S., Gotoh, H., and Ishikawa, H. (2008). Asymmetric Diels–Alder reactions of α,β-unsaturated aldehydes catalyzed by a diarylprolinol silyl ether salt in the presence of water. *Angew. Chem. Int. Ed.* 47 (35): 6634–6637.

21 Sun, T.-W., Ren, W.-W., Xiao, Q. et al. (2012). Diastereoselective total synthesis of (+/−)-schindilactone A, part 1: Construction of the ABC and FGH ring systems and initial attempts to construct the CDEF ring system. *Chem. Asian J.* 7: 2321–2333.

22 Naperstkow, A.M., Macaulay, J.B., Newlands, M.J., and Fallis, A.G. (1989). Pi-facial diastereoselectivity in Diels–Alder reactions of 2,5-dimethylthiophene oxide. *Tetrahedron Lett.* 30 (38): 5077–5080.

23 Treiber, A., Dansette, P.M., ElAmri, H. et al. (1997). Chemical and biological oxidation of thiophene: preparation and complete characterization of thiophene S-oxide dimers and evidence for thiophene S-oxide as an intermediate in thiophene metabolism in vivo and in vitro. *J. Am. Chem. Soc.* 119 (7): 1565–1571.

24 Li, Y.Q., Thiemann, T., Sawada, T. et al. (1997). Lewis acid catalysis in the oxidative cycloaddition of thiophenes. *J. Organomet. Chem.* 62 (23): 7926–7936.

25 Furukawa, N., Zhang, S.Z., Horn, E. et al. (1998). X-ray crystallographic analysis of 2,5-bis(diphenylmethylsilyl)thiophene monooxide and the Diels–Alder reaction of thiophene monooxide with dienophiles. *Heterocycles* 47 (2): 793–809.

26 Takayama, J., Sugihara, Y., Takayanagi, T., and Nakayama, J. (2005). syn-pi-Face- and endo-selective, inverse electron-demand Diels–Alder reactions of 3,4-di-tert-butylthiophene 1-oxide with electron-rich dienophiles. *Tetrahedron Lett.* 46 (24): 4165–4169.

27 Boger, D.L. and Weinreb, S.N. (1987). *Hetero-Diels–Alder Methodology in Organic Synthesis.* New York, NY: Academic Press.

28 Deguin, B. and Vogel, P. (1992). Hetero-Diels–Alder addition of sulfur-dioxide to 1,3-dienes- suprafaciality, regioselectivity, and stereoselectivity. *J. Am. Chem. Soc.* 114 (23): 9210–9211.

29 Geisler, A. and Wurthwein, E.U. (1994). Diels–Alder-reaction and ene-reaction of 2-azaallenium salts. *Tetrahedron Lett.* 35 (1): 77–80.

30 Tietze, L.F. and Kettschau, G. (1997). Hetero Diels–Alder reactions in organic chemistry. *Top. Curr. Chem.* 189: 1–120.

31 Dubey, S.K. and Knaus, E.E. (1985). Some regiochemical and stereochemical aspects of the Diels–Alder reaction of nitrosocarbonyl compounds with *N*-substituted 1,2-dihydropyridines. *J. Organomet. Chem.* 50 (12): 2080–2086.

32 Bodnar, B.S. and Miller, M.J. (2011). The nitrosocarbonyl hetero-Diels–Alder reaction as a useful tool for organic syntheses. *Angew. Chem. Int. Ed.* 50 (25): 5629–5646.

33 Gotthard, H. and Huisgen, R. (1968). 1,3-Dipolar cyclo additions .35. Delta2-pyrazolines from sydnones and lakenes. *Chem. Ber. Recl.* 101 (2): 552–563.

34 Brunn, E., Funke, E., Gotthard, H., and Huisgen, R. (1971). 1.3-Dipolar cycloadditions. 61. Cycloadditions of *N*-substituted oxazolium-5-olates to nitriles, nitro, nitroso and azo compounds. *Chem. Ber. Recl.* 104 (5): 1562–1572.

35 Huisgen, R., Mloston, G., and Polborn, K. (1995). 1,3-Dipolar activity in cycloaddtions of an aliphatic sulfine. *J. Organomet. Chem.* 61: 6570–6574.

36 Nguyen, T.B., Martel, A., Gaulon-Nourry, C. et al. (2012). 1,3-Dipolar cycloadditions of nitrones to hetero-substituted alkenes Part 2: sila-, thia-, phospha- and halo-substituted alkenes. *Org. Prep. Proced. Int.* 44 (1): 1–81.

37 Matsumoto, K., Taketsuna, H., Ikemi, Y. et al. (1998). Cycloaddition reactions of oxaquadricyclane, oxanorbornadiene, and related compounds with cyclooctyne. *Heterocycles* 49: 79–83.

38 Harmata, M., Jones, D.E., Kahraman, M. et al. (1999). Chiral cations in 4+3 cycloadditions. *Tetrahedron Lett.* 40 (10): 1831–1834.

39 Wigner, E. and Witmer, E. (1928). About the structure of the diatomic molecule spectra according to quantum mechanics. *Z. Angew. Phys.* 51 (11–12): 859–886.

40 Kende, A.S. and Huang, H. (1997). Asymmetric [4+3] cycloadditions from chiral α-chloro imines. *Tetrahedron Lett.* 38 (19): 3353–3356.

41 Simig, G. and Schlosser, M. (1994). The spontaneous cyclodimerization of 2,3-dihydroisoquinolines after base-promoted elimination of methanol from 4-methoxy-1,2,3,4-tetrahydroisoquinolines. *Tetrahedron Lett.* 35 (19): 3081–3082.

42 Rigby, J.H. and Fleming, M. (2002). Construction of the ingenane core using an Fe(III) or Ti(IV) Lewis acid-catalyzed intramolecular [6+4] cycloaddition. *Tetrahedron Lett.* 43 (48): 8643–8646.

43 Kraakman, P.A., Dewolf, W.H., and Bickelhaupt, F. (1989). The stereochemistry of the 1,4-addition of dibromocarbene to 1,2-dimethylenecycloheptane. *J. Am. Chem. Soc.* 111 (22): 8534–8535.

44 Moncada, S., Palmer, R.M.J., and Higgs, E.A. (1991). Nitric-oxide – physiology pathophysiology, and pharmacology. *Pharmacol. Rev.* 43 (2): 109–142.

45 Palmer, R.M.J., Ashton, D.S., and Moncada, S. (1988). Vascular endothelial-cells synthesize nitric-oxide from L-arginine. *Nature* 333 (6174): 664–666.

46 Marletta, M.A., Yoon, P.S., Iyengar, R. et al. (1988). Macrophage oxidation of L-arginine to nitrite and nitrate - nitric-oxide is an intermediate. *Biochemistry* 27 (24): 8706–8711.
47 Hibbs, J.B., Taintor, R.R., Vavrin, Z., and Rachlin, E.M. (1988). Nitric-oxide - a cyto-toxic activated macrophage effector molecule. *Biochem. Biophys. Res. Commun.* 157 (1): 87–94.
48 Snyder, S.H. Nitric-oxide – 1st in a new class of neurotransmitters. *Science* 1992 (5069): 257, 494–496.
49 Culotta, E. and Koshland, D.E. (1992). No news is good-news. *Science* 258 (5090): 1862–1865.
50 Schuman, E.M. and Madison, D.V. (1991). A requirement for the intercellular messenger nitric-oxide in long-term potentiation. *Science* 254 (5037): 1503–1506.
51 Korth, H.G., Sustmann, R., Lommes, P. et al. (1994). Nitric-oxide cheletropic traps (Nocts) with improved thermal-stability and water solubility. *J. Am. Chem. Soc.* 116 (7): 2767–2777.
52 Adam, W., Bottle, S.E., Grice, I.D. et al. (1996). The cheletropic trapping of nitric oxide by the bis-ketene 1,2-dicarbonylcyclohexa-3,5-diene and the diene 3,4-diphenyl-2,5-dimethyl-2,4-hexadiene. *Tetrahedron Lett.* 37 (12): 2113–2116.
53 Birney, D.M. and Berson, J.A. (1985). Norborna-2,5-dien-7-one – a covalent benzene carbon-monoxide adduct – a new point on the cycloreversion structure reactivity correlation curve. *J. Am. Chem. Soc.* 107 (15): 4553–4554.
54 Rubio, M., Hernandez, A.G., Daudey, J.P. et al. (1980). Thermal decarbonylation of some strained ketones. *J. Organomet. Chem.* 45 (1): 150–154.
55 Buxton, J.P. and Simpson, C.J.S.M. (1986). Thermal decarbonylations of unsaturated cyclic-ketones – kinetics and dynamics. *Chem. Phys.* 105 (3): 307–316.
56 Zhu, Q., Liu, M.-F., Wang, B., and Cheng, Y. (2007). An unprecedented tandem 1,3-dipolar cycloaddition-cheletropic elimination: a facial approach to novel push-pull olefins. *Org. Biomol. Chem.* 5 (8): 1282–1286.
57 Souto, J.A., Lopez, C.S., Faza, O.N. et al. (2005). 2-Alkylidenesulfol-3-enes by (regio-and) stereoselective cheletropic addition of SO_2 to (di)vinylallenes. *Org. Lett.* 7 (8): 1565–1568.
58 Wojciechowski, K. (1993). Reactions of aza-ortho-xylylenes generated from 2,1-benzisothiazoline 2,2-dioxides. *Tetrahedron* 49 (33): 7277–7286.
59 Bouchez, L.C. and Vogel, P. (2005). Synthesis of the C(1)–C(11) polyene fragment of apoptolidin with a new sulfur dioxide-based organic chemistry. *Chem. Eur. J.* 11 (16): 4609–4620.
60 Roberts, V.A. and Garst, M.E. (1985). Synthesis of 4-substituted cycloheptatrienones by oxidative cheletropic elimination of nitrosobenzene from 6-substituted 8-phenyl-8-azabicyclo[3.2.1]octa-3,6-dien-1-ones. *J. Organomet. Chem.* 50 (6): 893–894.
61 Hillner, K. and Neumann, W.P. (1986). Additions of a stable stannylene R2Sn to vinyl carbonyl-compounds and 1,2-diketones. *Tetrahedron Lett.* 27 (44): 5347–5350.
62 Marx, R., Neumann, W.P., and Hillner, K. (1984). [2+4] Cheletropic cyclo-additions of stannylenes alkyl-2Sn, amino-2Sn, and halogen-2Sn. *Tetrahedron Lett.* 25 (6): 625–628.
63 Tokitoh, N. and Okazaki, R. (2000). Recent topics in the chemistry of heavier congeners of carbenes. *Coord. Chem. Rev.* 210: 251–277.
64 Bulo, R.E., Trion, L., Ehlers, A.W. et al. (2004). C–C bond insertion of a complexed phosphinidene into 1,6-methano[10]annulene. *Chem. Eur. J.* 10 (21): 5332–5337.
65 Ellis, R.J. and Frey, H.M. (1964). Thermal unimolecular isomerizations of cis- and trans-1-methyl-2-vinylcyclopropanes. *J. Chem. Soc.* S: 5578–5583.
66 Roth, W.R. and Konig, J. (1965). Wasserstoff-Verschiebungen .3. Thermische Isomerisierung von cis-hexadien-(1.4). *Liebigs Ann. Chem.* 688 (OCT): 28–39.
67 Simpson, J.M. and Richey, H.G. (1973). Effects of methoxyl and phenyl substituents on thermal rearrangements of vinylcyclopropane. *Tetrahedron Lett.* (27): 2545–2548.
68 McGaffin, G., Grimm, B., Heinecke, U. et al. (2001). On the substituent effects of the thermal ethenyleyclopropane-to-cyclopentene rearrangement: gas-phase kinetics of ethoxy-, methylthio- and trimethylsilyl-substituted ethenylcyclopropanes. *Eur. J. Org. Chem.* (18): 3559–3573.
69 Heimgartner, H., Schmid, H., and Hansen, H.J. (1972). Thermal behavior of 1,2-dipropenylbenzenes. *Helv. Chim. Acta* 55 (5): 1385–1403.
70 Tsubuki, M., Takahashi, K., and Honda, T. (2003). Stereoselective construction of beta-isopropenyl alcohol moiety at the C(2) and (3) of kallolide A and pinnatin A using a [2,3] wittig rearrangement of cyclic furfuryl ethers. *J. Organomet. Chem.* 68 (26): 10183–10186.

71 Koreeda, M. and Luengo, J.I. (1985). Anionic oxy-claisen rearrangement of enolates of alpha-allyloxy ketones – a remarkable rate-accelerating effect exhibited by the nature of the counterion. *J. Am. Chem. Soc.* 107 (19): 5572–5573.

72 Marbet, R. and Saucy, G. (1967). Über die Reaktion von tertiaren Vinylcarbinolen with Vinyläthern – eine neue Methode zur Herstellung von gamma,delta-ungesättigen Aldehyden. *Helv. Chim. Acta* 50 (7): 2095–2100.

73 Suh, Y.G., Kim, S.A., Jung, J.K. et al. (1999). Asymmetric total synthesis of fluvirucinine A(1). *Angew. Chem. Int. Ed.* 38 (23): 3545–3547.

74 Cid, M.M. and Pombovillar, E. (1993). Enantioselective synthesis of 3-azabicyclo[4.3.0]nonane alkaloids. *Helv. Chim. Acta* 76 (4): 1591–1607.

75 Mehta, G., Karra, S.R., and Krishnamurthy, N. (1994). An oxy-cope rearrangement route for the enantioselective construction of 5,11-fused framework of dolabellane diterpenes. *Tetrahedron Lett.* 35 (17): 2761–2762.

76 Wender, P.A., Sieburth, S.M., Petraitis, J.J., and Singh, S.K. (1981). Macroexpansion methodology – medium ring synthesis based on an 8 unit ring expansion process. *Tetrahedron* 37 (23): 3967–3975.

77 Clausen, C., Wartchow, R., and Butenschon, H. (2001). Reactions of 1,2-diketones with vinyllithium: addition reactions and dianionic oxy cope rearrangements of cyclic and acyclic substrates. *Eur. J. Org. Chem.* (1): 93–113.

78 Zhou, L., Li, Z., Zou, Y. et al. (2012). Tandem nucleophilic addition/oxy-2-azonia-cope rearrangement for the formation of homoallylic amides and lactams: total synthesis and structural verification of motuporamine G. *J. Am. Chem. Soc.* 134 (49): 20009–20012.

79 So, S.M., Mui, L., Kim, H., and Chin, J. (2012). Understanding the interplay of weak forces in [3,3]-sigmatropic rearrangement for stereospecific synthesis of diamines. *Acc. Chem. Res.* 45 (8): 1345–1355.

80 Denmark, S.E. and Montgomery, J.I. (2005). Synthesis of cis,cis,cis,cis-[5.5.5.4]-1-azafenestrane with discovery of an unexpected dyotropic rearrangement. *Angew. Chem. Int. Ed.* 44 (24): 3732–3736.

81 Reetz, M.T. and Greif, N. (1977). Dyotropic rearrangements. 14. Thermal rearrangement of silymethyl acetates. *Angew. Chem. Int. Ed. Engl.* 16 (10): 712–713.

82 Alder, K. and Dortmann, H.A. Uber substituierende addition und dien-synthese beim methylen-cyclobutan. 9. Uber substituierende additionen. *Chem. Ber. Recl.* 1952 (6): 85, 556–565.

83 Robinson, J.M., Sakai, T., Okano, K. et al. (2010). Über substituirende Addition und Dien-synthese beim Methylencyclobutan. 9. Über substituierende Additionen. *J. Am. Chem. Soc.* 132 (32): 11039–11041.

84 Kraus, G.A. and Kim, J. (2004). Tandem Diels–Alder/ene reactions. *Org. Lett.* 6 (18): 3115–3117.

85 Sammis, G.M., Flamme, E.M., Xie, H. et al. (2005). Design, synthesis, and reactivity of 1-hydrazinodienes for use in organic synthesis. *J. Am. Chem. Soc.* 127 (24): 8612–8613.

86 Jabbari, A., Sorensen, E.J., and Houk, K.N. (2006). Transition states of the retro-ene reactions of allylic diazenes. *Org. Lett.* 8 (14): 3105–3107.

87 Bouchez, L. and Vogel, P. (2002). One-pot, three-component synthesis of open-chain, polyfunctional sulfones. *Synthesis (Stuttgart)* (2): 225–231.

88 Lachance, H., Lu, X.S., Gravel, M., and Hall, D.G. (2003). Scandium-catalyzed allylboration of aldehydes as a practical method for highly diastereo- and enantioselective construction of homoallylic alcohols. *J. Am. Chem. Soc.* 125 (34): 10160–10161.

89 Hoffmann, R.W. and Rolle, U. (1994). Synthesis of A C-15 C-27 segment of venturicidine. *Tetrahedron Lett.* 35 (27): 4751–4754.

90 Crimmins, M.T. and Slade, D.J. (2006). Formal synthesis of 6-deoxyerythronolide B. *Org. Lett.* 8 (10): 2191–2194.

91 Evans, D.A. and Taber, T.R. (1980). Enantioselective aldol condensations via boron enolates – a steric model for asymmetric induction. *Tetrahedron Lett.* 21 (49): 4675–4678.

92 Heathcock, C.H. (1990). Understanding and controlling diastereofacial selectivity in carbon–carbon forming reactions. *Aldrichim. Acta* 23: 99–111.

93 Silver, D.M. (1974). Hierarchy of symmetry conservation rules governing chemical-reaction systems. *J. Am. Chem. Soc.* 96 (19): 5959–5967.

94 Rigby, J.H., Ateeq, H.S., Charles, N.R. et al. (1993). Metal-promoted higher-order cycloaddition reactions – stereochemical, regiochemical, and mechanistic aspects of the [6-pi-plus4-pi] reaction. *J. Am. Chem. Soc.* 115 (4): 1382–1396.

6

Organic Photochemistry

Answers to Problems

Problem 6.1 Direct photoexcitation of (*E*,*E*)-1-deuteropenta-1,3-diene results in the selective isomerization of the CH=C(D)H double bond, giving (*Z*,*E*)-1-deuteropenta-1,3-diene. Is this observation compatible with the formation of diradical intermediates? [1]

Answer

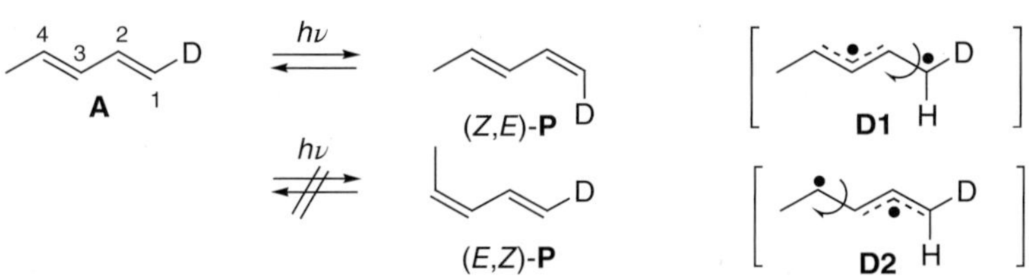

Diradicals **D1** and **D2** are envisioned as intermediates. They would permit rotation about the π(C(1)—C(2)) and π(C(3)—C(4)) double bonds giving (*Z*,*E*)-**P** and (*E*,*Z*)-**P**, respectively. **D1** is a 1,3-dialkylallyl radical + primary alkyl radical. **D2** is a 1-alkylallyl radical + secondary alkyl radical. Table 7.1 gives $DH°(n\text{-Pr}^{\cdot}/\text{H}^{\cdot}) = 101\,\text{kcal mol}^{-1}$ and $DH°(i\text{-Pr}^{\cdot}/\text{H}^{\cdot}) = 99.4\,\text{kcal mol}^{-1}$; the relative stability between a primary and a secondary alkyl radical is c. 2 kcal mol^{-1} in favor of the secondary radical. Table 7.1 gives $DH°(\text{CH}_2{=}\text{CH—C(Me)H}^{\cdot}/\text{H}^{\cdot}) = 83.6\,\text{kcal mol}^{-1}$ and $DH°(\text{MeCH}{=}\text{CH—C(Me)H}^{\cdot}/\text{H}^{\cdot}) = 81.6\,\text{kcal mol}^{-1}$. Thus, the stability difference between a 1-methylallyl radical and a 1,3-dimethylallyl radical also amounts to c. 2 kcal mol^{-1}. This suggests that diradicals **D1** and **D2** should have the same enthalpy and should be formed to similar extents. Accordingly, one would expect both product (*Z*,*E*)-**P** and (*E*,*Z*)-**P** to be formed upon irradiation of **A**, which is not the case! The results show that **D1** is favored, perhaps because its zwitterionic limiting structure **Z1** is more stable than **Z2**, the zwitterionic limiting structure of **D2**. Upon rotation, **Z1** reaches transition state **T1**, which is a 1,3-dialkylallyl cation, whereas **Z2**, upon rotation, reaches transition state **T2**, which is a 1-alkylallyl cation less stable than a 1,3-dialkyallyl cation.

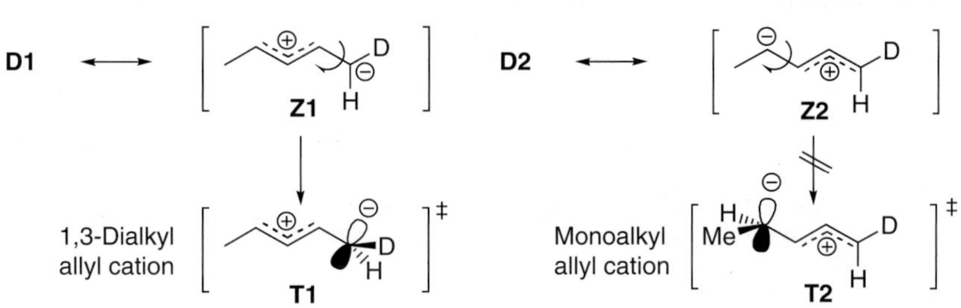

Organic Chemistry: Theory, Reactivity and Mechanisms in Modern Synthesis, Workbook, First Edition. Pierre Vogel and Kendall N. Houk.

Problem 6.2 What are the products formed upon irradiation of 1-methylcyclohexene in toluene containing methanol? [2]

Answer

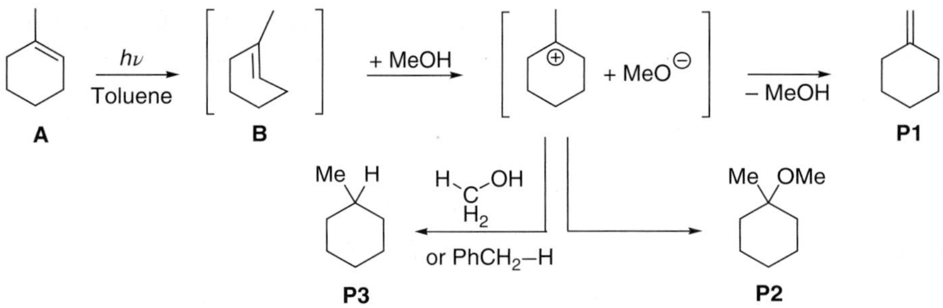

Toluene (solvent) absorbs the UV light and generates a triplet state that transfers its energy to **A**. This induces rotation about the C=C double bond with the formation of unstable **B**, a highly reactive alkene that is protonated readily by methanol giving 1-methylcyclohexyl cation and methanolate anion as intermediates. They react to produce methylidenecyclohexane (**P1**) through elimination, or form 1-methoxy-1-methylcyclohexane (**P2**) through nucleophilic quenching. Intermediate 1-methylcyclohexyl cation can be reduced competitively through hydride transfer from either methanol or the solvent (toluene). These hydride transfers are exothermic as the hydroxymethyl cation and the benzyl cation that form are much more stable than 1-methylcyclohexyl cation.

Problem 6.3 On direct photoirradiation of cyclooct-2-enone (**A**) that shows a λ_{max} at 321 nm one obtains an isomer **B** with a λ_{max} = 283 nm. What is isomer **B**? [3]

Answer

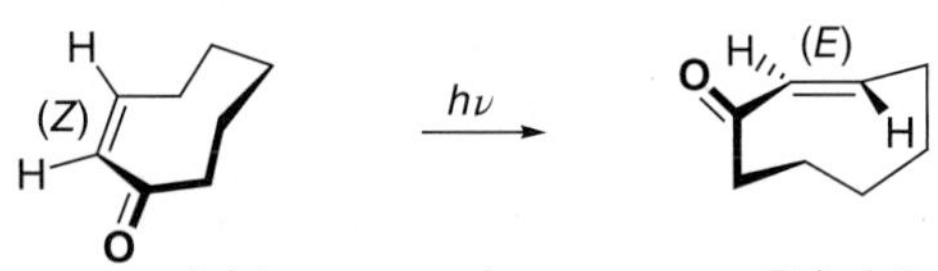

A (planar enone) **B** (twisted enone)

(*Z*)-Cyclooct-2-enone (**A**) can adopt a planar enone moiety with optimal conjugation between the carbonyl and alkene functions. Upon irradiation of **A,** (*E*)-cyclooct-2-enone (**B**) forms. In the latter, the enone chromophore cannot adopt a planar conformation for geometry reasons. As a consequence, the LUMO/HOMO energy gap is smaller in **A** than in **B**, and thus λ_{max}(**A**) > λ_{max}(**B**).

Problem 6.4 Does the transition state of the thermal isomerization of bicyclo[1.1.0]butane enjoy some aromatic stabilization?

Answer

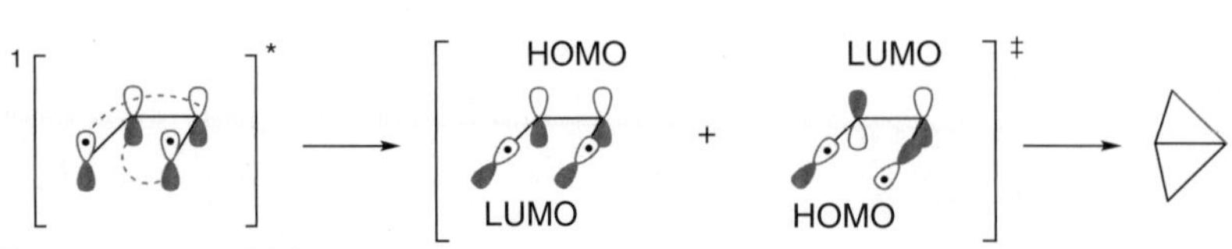

(*Z*)-but-2-ene-1,4-Diyl diradical

The photoinduced *s-cis*-butadiene → bicyclo[1.1.0]butane isomerization can be seen as an intramolecular antarafacial/antarafacial cycloaddition denoted [π^2a+π^2a]. It is allowed according to the Woodward–Hoffmann rules. This reaction can be seen also as the conrotatory ring closure of a (*Z*)-but-2-ene-1,4-diyl diradical (the excited state of butadiene) for which LUMO(alkene)/HOMO(diradical) and LUMO(diradical)/HOMO(alkene) stabilizing interactions are possible, thus rendering the reaction relatively facile (low activation energy for the double cyclization).

Problem 6.5 What is the major product of the direct irradiation of benzobarrelene? What is the main product of irradiation of benzobarrelene in the presence of acetophenone? [4, 5]

Answer

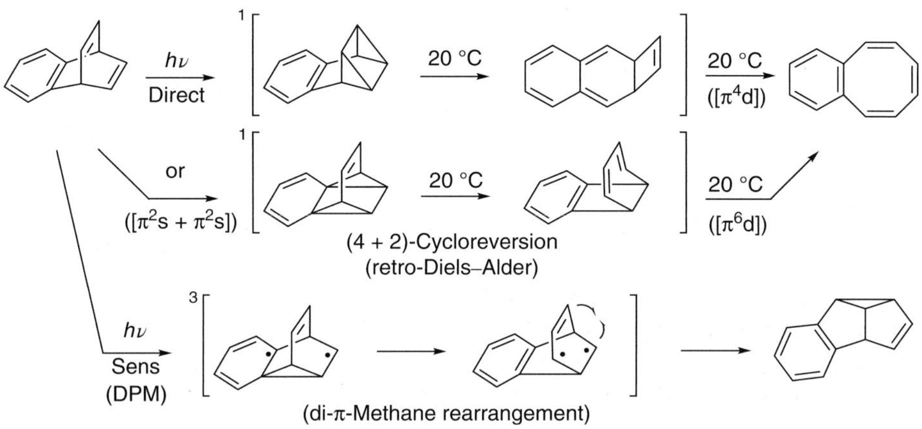

The direct irradiation of benzobarrelene gives benzocyclooctatetraene via a $[\pi^2s+\pi^2s]$ cycloaddition of its two homoconjugated alkene units, followed by an exothermic, thermally allowed, retro-Diels–Alder cycloreversion and a subsequent disrotatory electrocyclic ring opening. The latter pericyclic reaction is thermally "forbidden" but highly exothermic because of the conversion of an orthoquinodimethane system into a divinylbenzene derivative. In the same time, a cyclobutene unit is converted into a butadiene derivative, another exothermic reaction because of the cyclobutene ring strain relief. An alternative route is the $[\pi^2s+\pi^2s]$-cycloaddition of the benzo and one alkene unit, followed by a highly exothermic (4+2) cycloreversion and a thermally allowed $[\pi^6d]$-electrocyclic ring opening.

The sensitized excitation of benzobarrelene gives a triplet diradical, which rearranges into another more stable triplet diradical that finally generates benzosemibullvalene (typical di-π-methane rearrangement).

Problem 6.6 What is the major product of the sensitized irradiation of benzobarrelene-5-carbonitrile? [6]

Answer

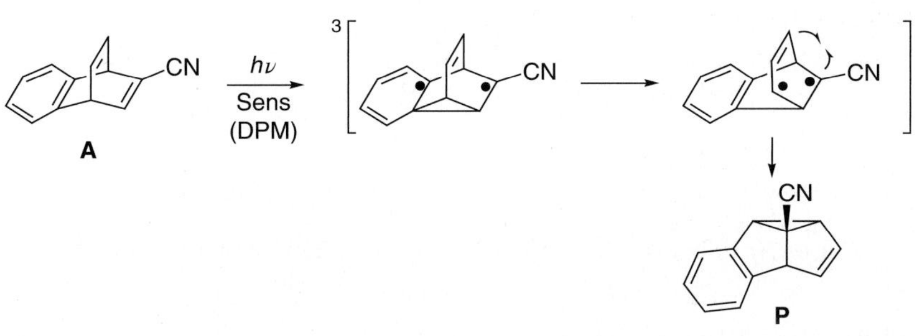

The di-π-methane rearrangement of benzobarrelene **A** is regioselective. The most stable diradicals are expected to be formed on the photoreaction hypersurface and the latter enjoys the radical stabilizing effect of the cyano substituent. This leads to the major product **P**.

Problem 6.7 What is the major product of the photoexcitation of 1,4-methano-1,4-dihydronaphthalene-5-carbonitrile? [7–9]

Answer

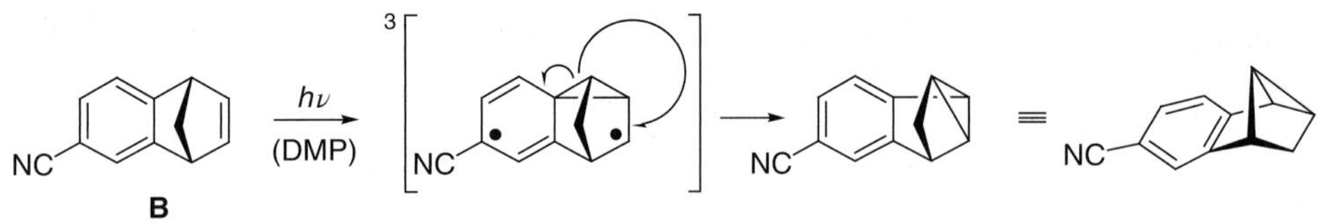

The di-π-methane rearrangement of **B** involves the formation of diradical intermediates. The most stable one that forms preferentially enjoys radical stabilization by the remote cyano substituent because of conjugation.

Problem 6.8 Propose a mechanism for the photoisomerization of 1,3-diphenylpropene into *cis*- and *trans*-1,2-diphenylcyclopropane and 1-phenylindane [10].

Answer

In their original paper Griffin and co-workers reported an unusual (1,2)]-shift of phenyl group that must be invoked to explain the formation of *cis*- and *trans*-1,2-diphenylcyclopropane upon irradiation of (*E*)-1,3-diphenylpropene. This photoinduced (1,2)-phenyl shift is in fact the first example of di-π-methane (DPM) rearrangement ever reported.

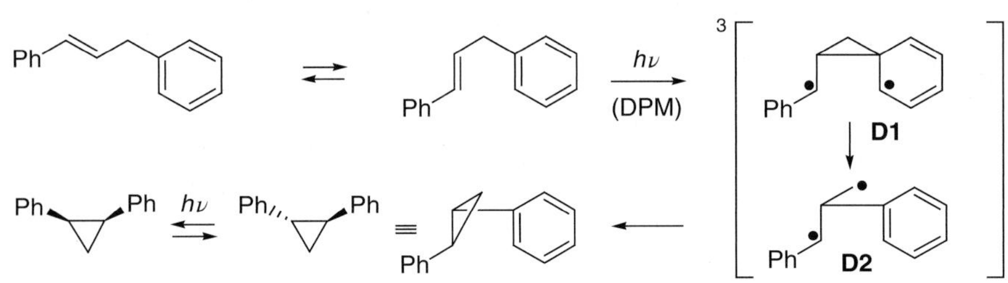

cis- and *trans*-1,2-Diphenylcyclopropane are equilibrated under photochemical excitation. The formation of 1-phenylindane as coproduct can be explained in the following way:

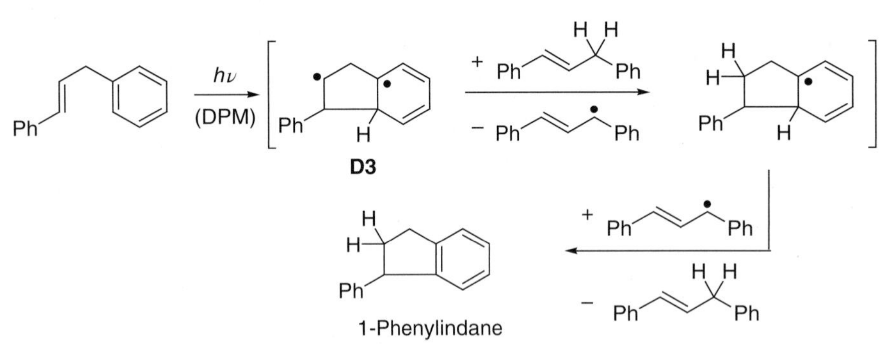

Alternatively, 1-phenylindane could arise from diradical **D3** through an intramolecular (1,3)-shift of a hydrogen atom, a highly exothermic reaction.

Problem 6.9 What are products **B** and **P** of the following photoreaction? [11]

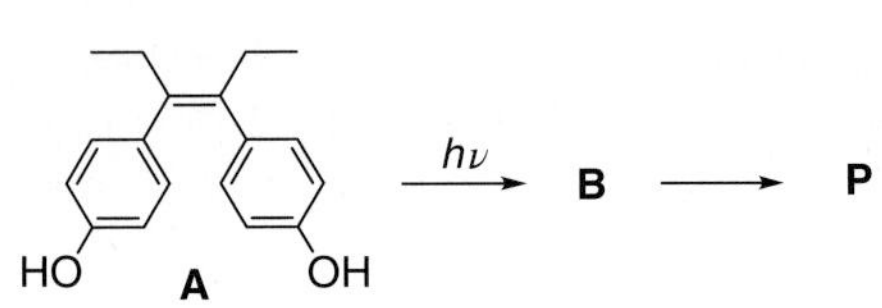

Answer

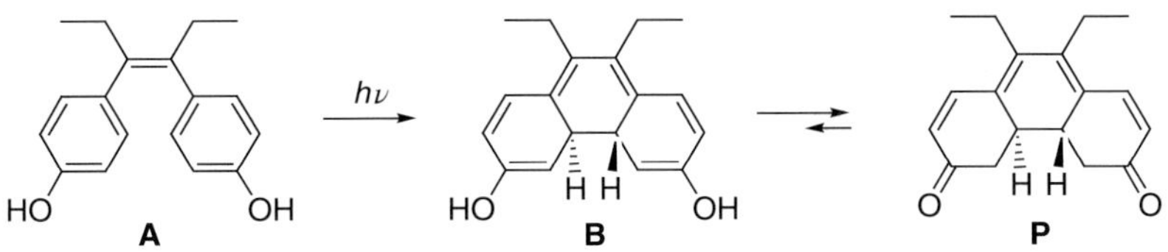

Problem 6.10 What is the product formed upon irradiation of diphenyl ether in the presence of I_2 [12].

Answer

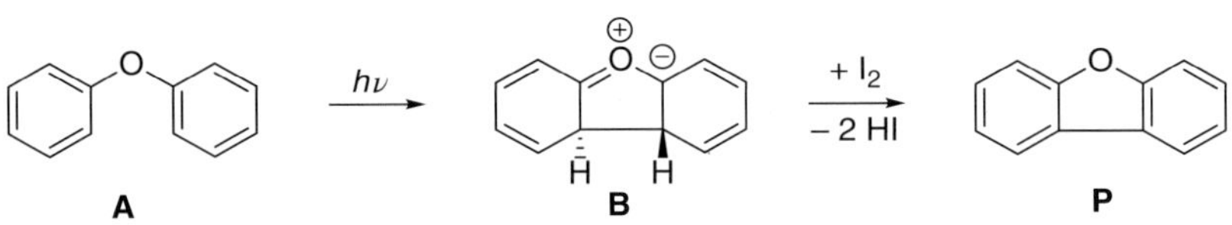

Diphenyl ether (**A**) is isoelectronic with (*Z*)-stilbene. It undergoes a photoinduced [π^6d] electrocyclic ring closure into carbonyl ylide **B**. Oxidation of the latter with iodine provides **P**.

Problem 6.11 What are the major products of the irradiation ($\lambda_{irr} > 300$ nm) of 2,2,4,4-tetramethylcyclobuta-1,3-dione (i) in degassed benzene and (ii) in MeOH at 20 °C [13].

Answer

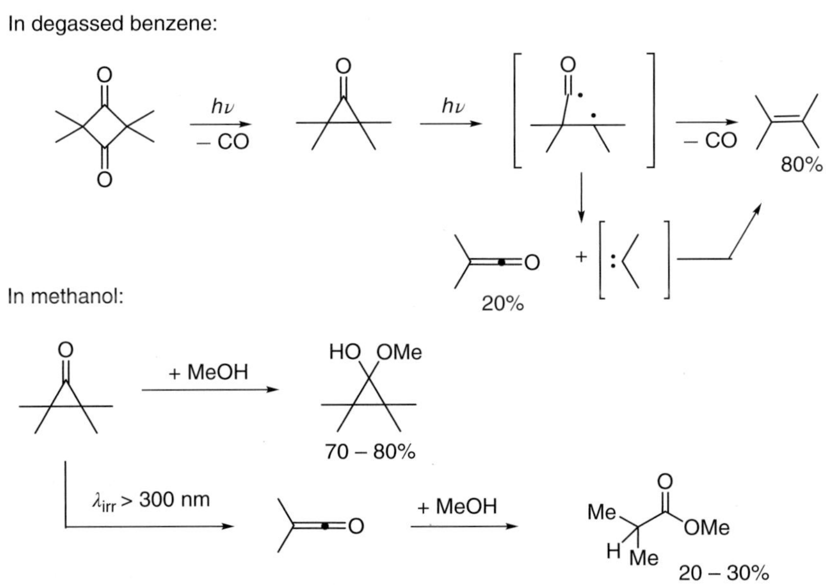

Problem 6.12 Photolysis of *trans*-1-benzoyl-1-methyl-4-t-butylcyclohexane (**A**) gives benzaldehyde and its α-epimer **B**. Photolysis of **B** gives an isomeric cyclobutanol **P**. What is **P**? Explain these reactions [14].

Answer

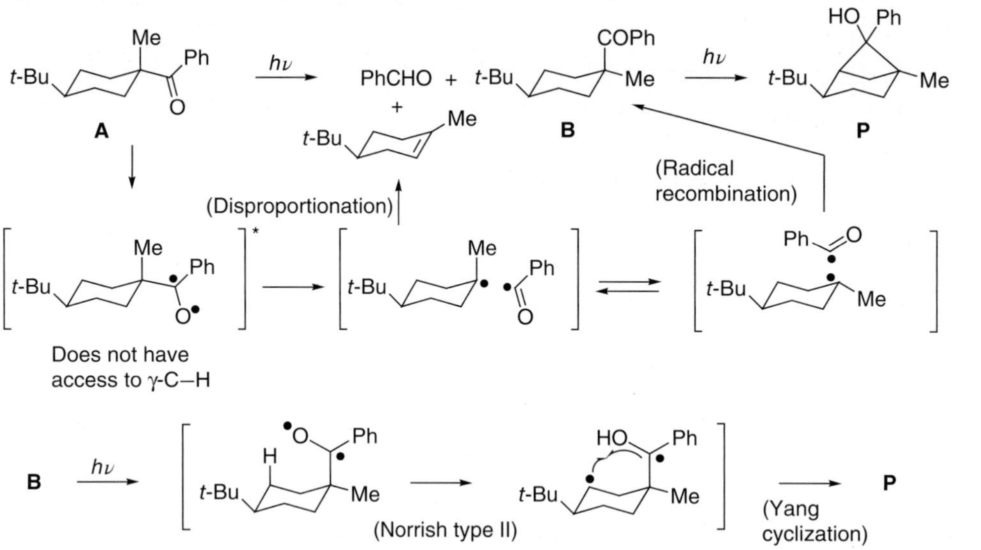

Problem 6.13 What products are formed upon irradiation (λ_{irr} = 275–380 nm) of gaseous heptanal (mixed with air, 1 atm, 23 °C)? [15]

Answer

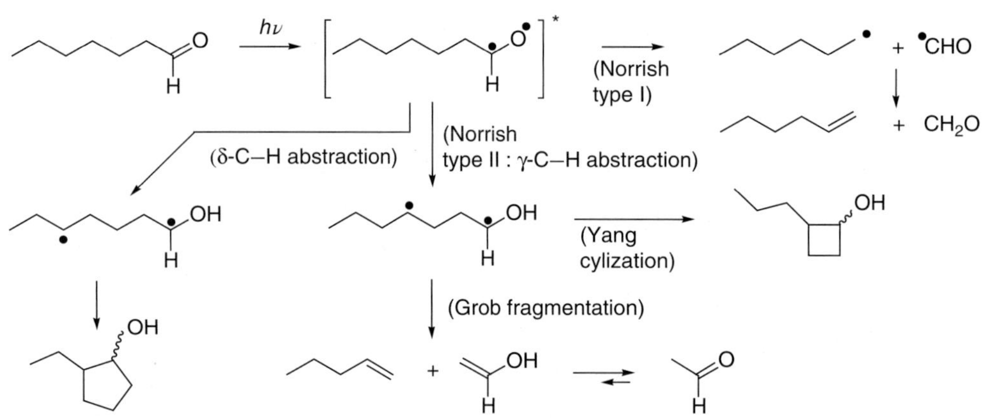

In the presence of oxygen the following reactions will occur:

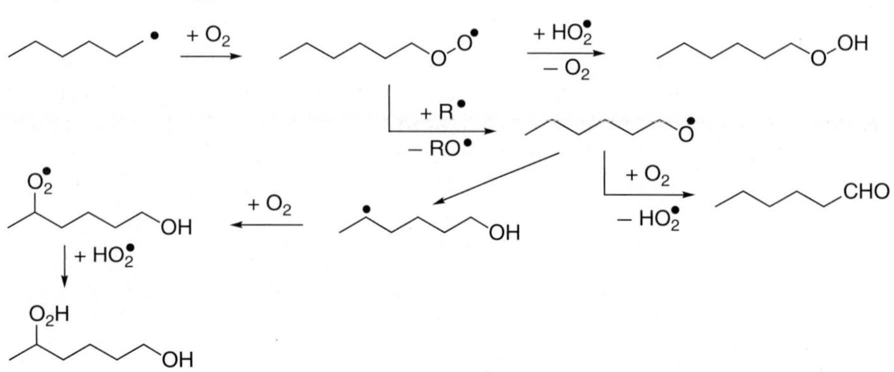

Problem 6.14 Irradiation (in MeCN, Pyrex vessel, medium pressure Hg lamp, $\lambda_{irr} > 290\,nm$) of 2,5-dimethylbenzoyloxirane results in 2-hydroxymethyl-6-methylindan-1-one. Give a mechanism for this photochemical isomerization [16].

Answer

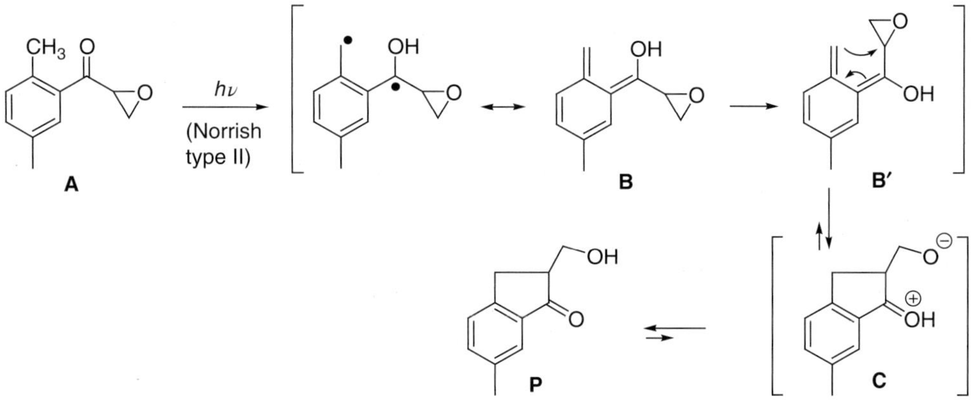

Intermediate **B** resulting from a Norrish type II reaction is a highly reactive, nucleophilic dienol that displaces the oxirane moiety giving intermediate **C** that immediately undergoes prototropic rearrangement into product **P**.

Problem 6.15 Photolysis of 4-methyl-4-(3-methylbutyl)cyclopent-2-en-1-one leads to three products involving a diradical intermediate resulting from an intramolecular hydrogen transfer. What are these three products? [17]

Answer

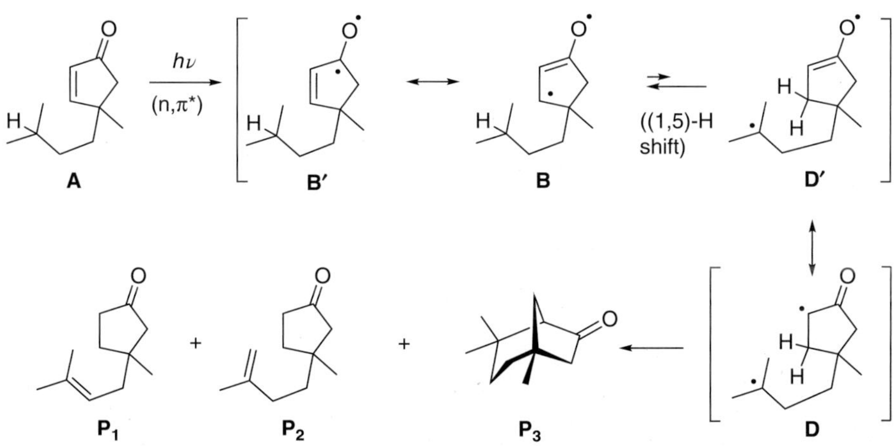

The first excited state of cyclopentenone **A** can be represented as a delocalized diradical **B,** which bears an important part of the spin at the β-position. If a nearby alkyl group can approach this center and provide a hydrogen atom through a 1,5-H transfer (six-membered transition state) this transfer might occur. This is particularly easy in the case of **B** as it generates an oxo-alkyl/tertiary alkyl diradical intermediate **D**. Intramolecular hydrogen transfer and diradical combination lead to the mixture of products **P1** + **P2** + **P3**.

Problem 6.16 Explain the following photoisomerization [18].

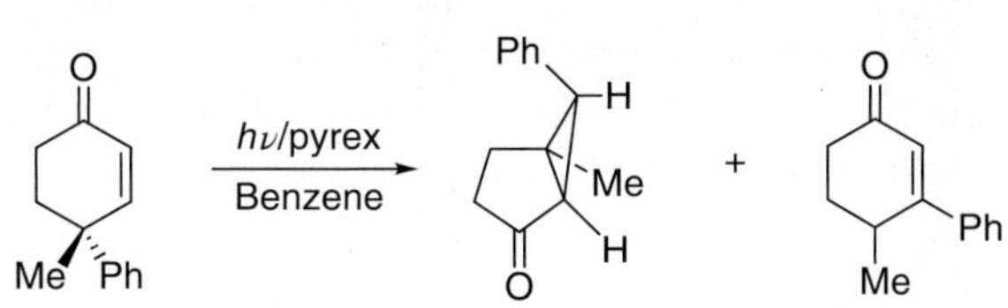

(*R*)-(+)-**A** (−)-**B**

Answer

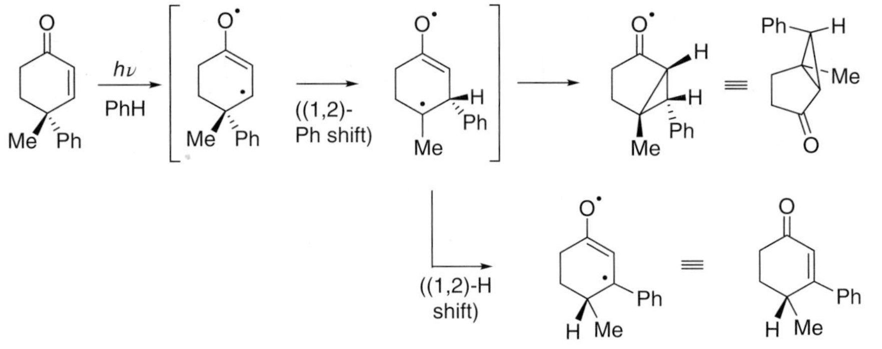

Problem 6.17 What are the products of direct and sensitized photolysis of bicyclo[2.2.2]oct-5-en-2-one? [19].

Answer

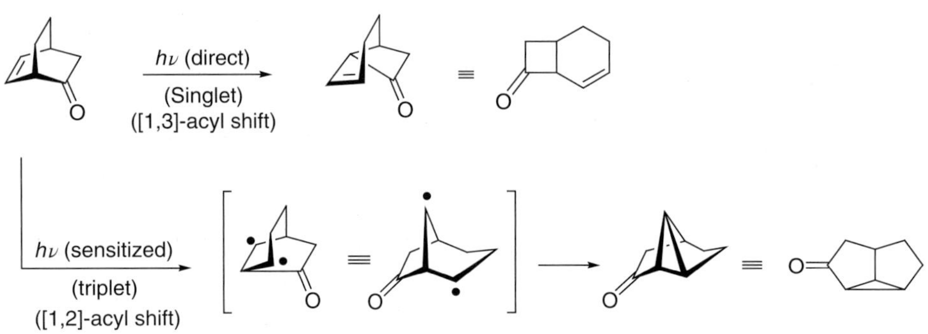

Problem 6.18 What is the main product of photolysis of 5,6-dimethylidenebicyclo[2.2.1]heptan-2-one? [20].

Answer

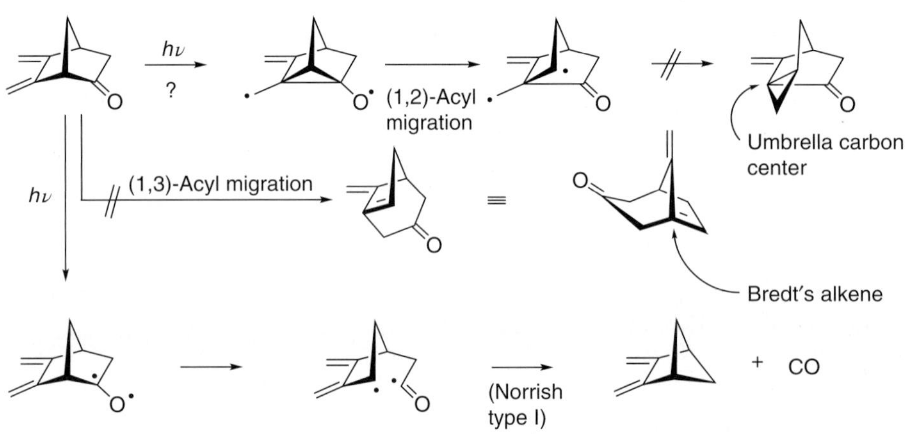

(1,2)- and (1,3)-acyl transfer are prohibited because they would lead to highly strained isomers. Thus, the β,γ-unsaturated ketone follows the Norrish I fragmentation, producing 2,3-dimethylidenebicyclo[2.1.1]hexane. The Norrish II reaction is not possible for geometric reasons: there is no C—H bond that can approach the carbonyl group.

Problem 6.19 Give a possible mechanism for the photoisomerization of 1,3,4-trideuteriobenzene into its isomeric trideuteriobenzenes [21].

Answer

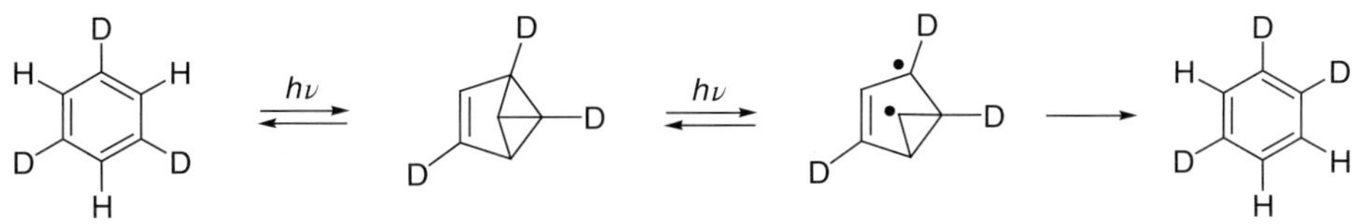

Problem 6.20 Photohydration of 1,4-dimethylpyrimidium chloride in KOH/H_2O gives three products. What are they? [22]

Answer

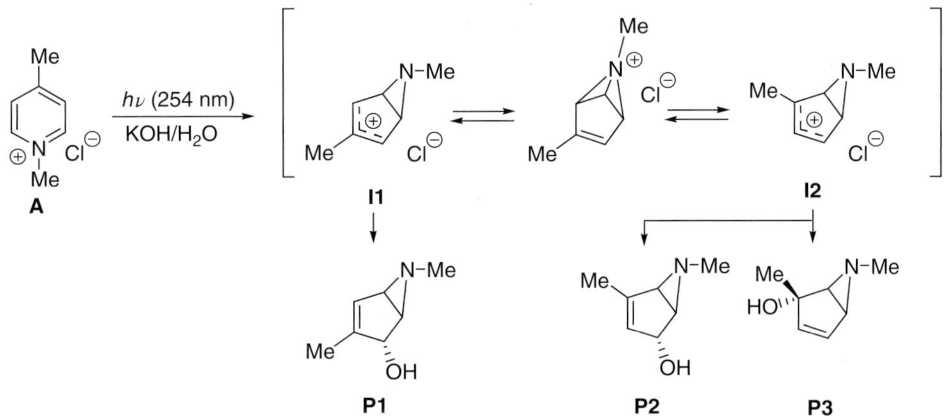

Problem 6.21 What products are formed in the alkaline photohydration of 3-alkoxypyridinim salts in water? [23]

Answer

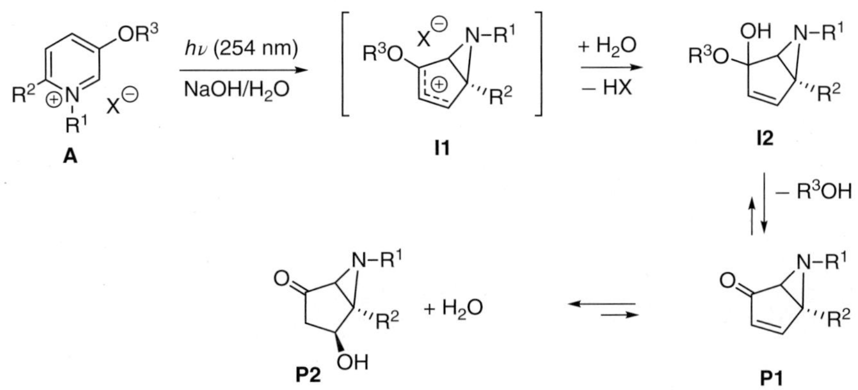

Problem 6.22 What is the product of photoisomerization of 3-oxido-1-phenylpyridinium betaine? [24]

Answer

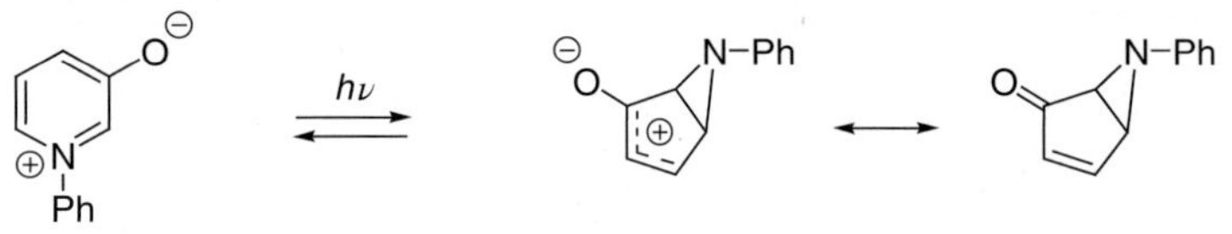

Problem 6.23 Propose a mechanism for the photochemical isomerization of 5-methylpyrrole-2-carbonitrile into its regioisomers [25].

Answer

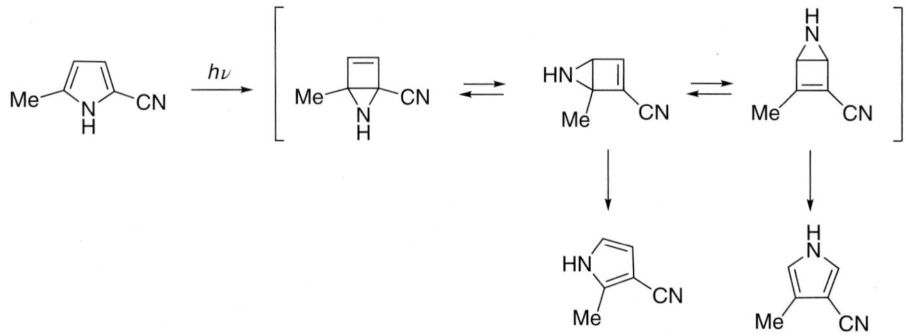

Problem 6.24 On irradiating thiophenes with primary amines the corresponding pyrroles are obtained. Give an explanation [26].

Answer

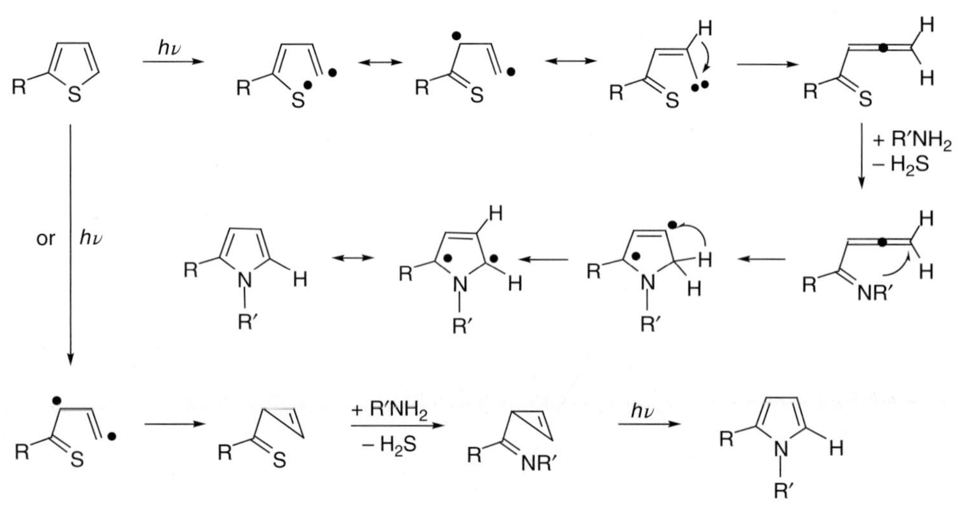

Problem 6.25 Propose a mechanism for the formation of $CF_3C{\equiv}CCF_3$ in the photochemical fragmentation (6.46).

Answer

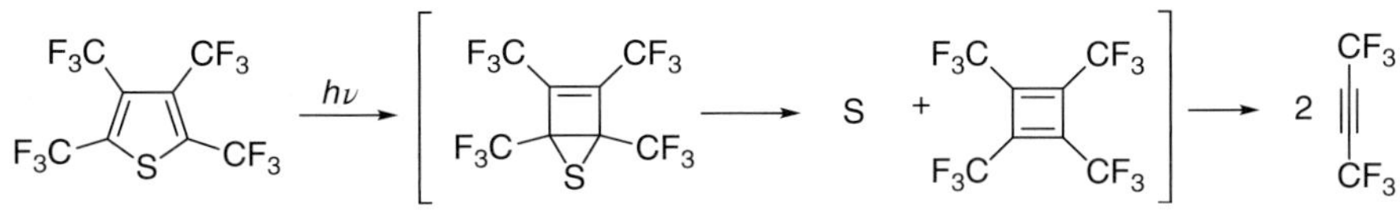

Problem 6.26 Explain the following one-pot, two-step preparation of quinolines **Q** [27].

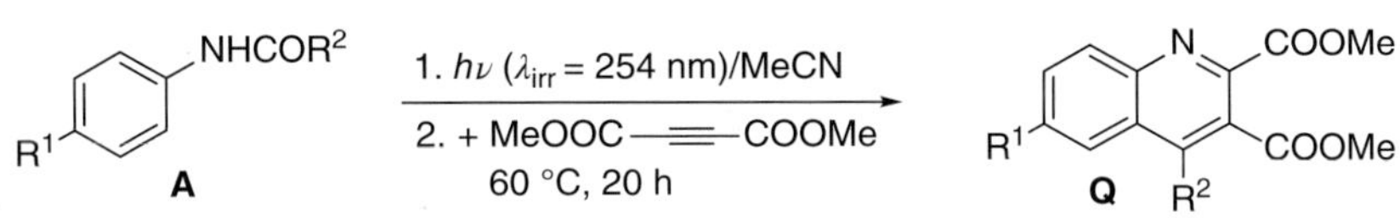

Answer
Photo-aza-Fries rearrangements convert acetanilides **A** into anilines **B**:

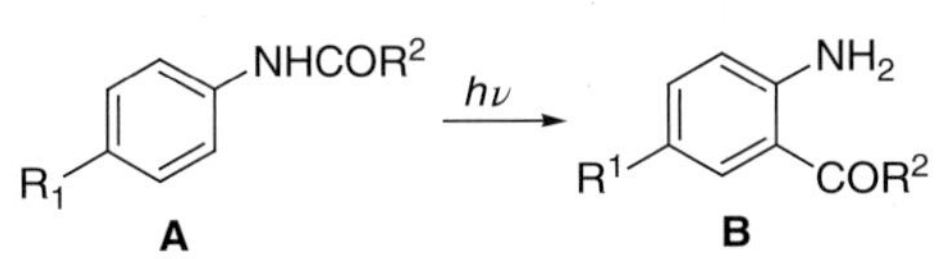

Anilines **B** are nucleophiles that undergo conjugate additions (1,4-additions) with α,β-unsaturated carboxylic esters such as dimethyl acetylenedicarboxylate generating intermediates **C**, which, in turn, undergo intramolecular nucleophilic additions to the carbonyl group giving alcohols **D** that eliminate 1 equiv. of water providing quinolines **Q**. The latter are highly exergonic because of aromatization.

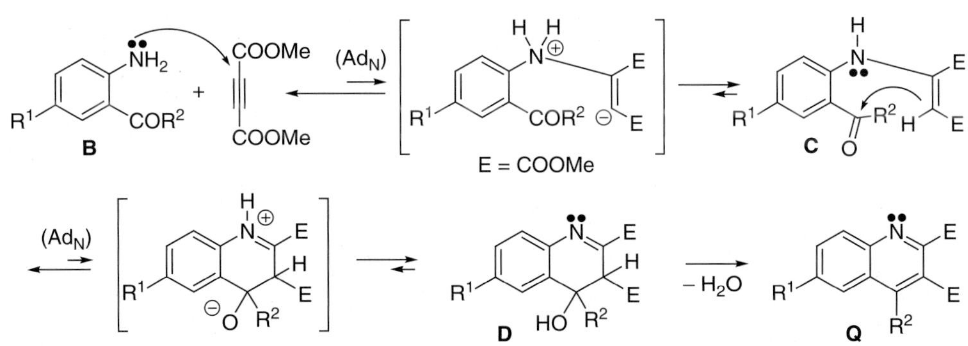

Problem 6.27 What is the main product of the photolysis of benzyl 2,4,6-trimethylphenyl ether? [28]

Answer

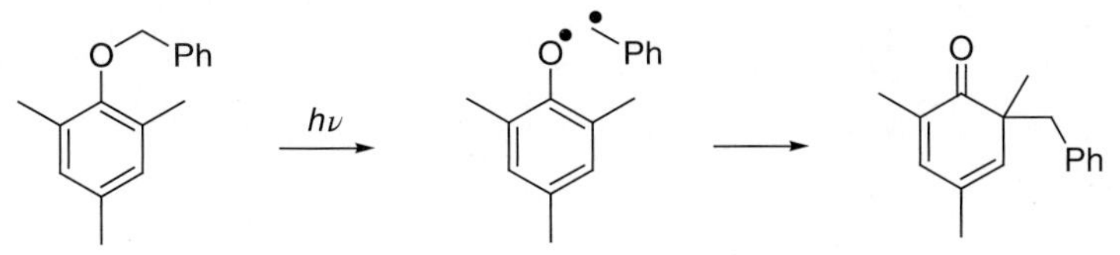

Problem 6.28 What are the products of irradiation at 350 nm of 9-diazo-1,8-diazafluorene in MeOH? [29]

Answer

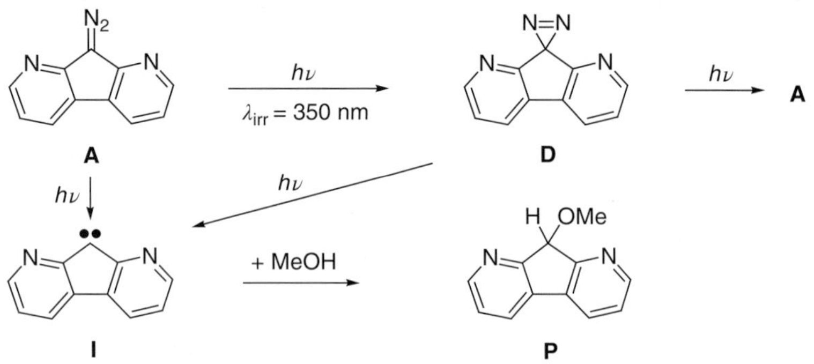

A photostationary state **A** ⇄ **D** is attained upon irradiation of **A**. Continuous irradiation generates carbene intermediate **I** that cannot undergo any facile (1,2)-shift. It thus reacts with the medium (MeOH) producing **P**.

Problem 6.29 What is the main product formed upon irradiation of 2,3-diazabicyclo[2.2.1]hept-2-ene?

Answer

Problem 6.30 Photolysis of **A** in the presence of fumarodinitrile gives a mixture of products including a 2*H*-azirine resulting from a rare case of carbene cheletropic addition to a carbonitrile. What is this product and what are the other products of the reaction?

A: 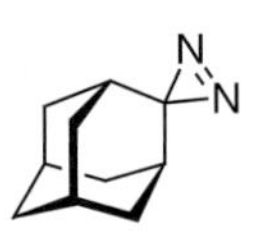

Answer

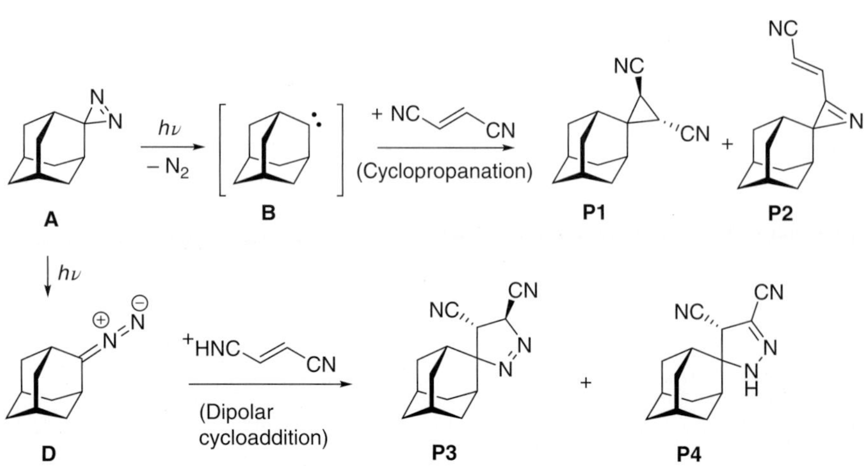

Problem 6.31 What are the products formed upon photolysis of 1,3-dichloro-1,3-diphenylpropane in cyclohexane [30].

Answer

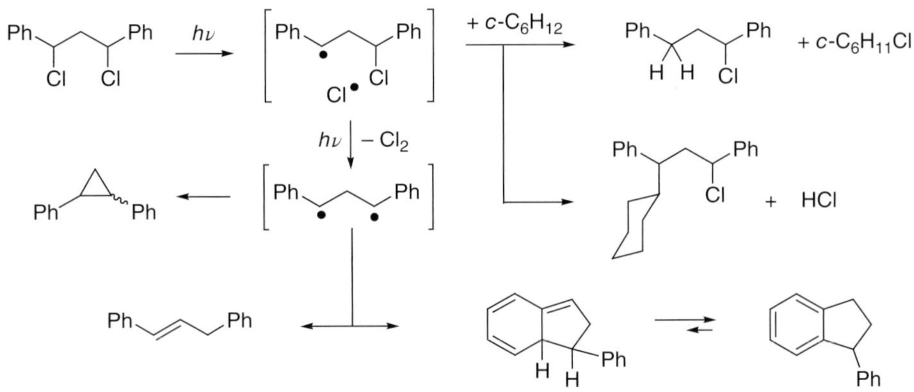

Problem 6.32 The photolysis (quartz, high pressure Hg burner) of iodobenzene in anisole gives a 71 : 15 : 14 mixture of *ortho-*, *meta-*, and *para*-phenylanisole. Similarly, photolysis of 2-iodothiazole in anisole gives a 63 : 14 : 23 mixture of *ortho-*, *meta-*, and *para*-(2-tiazolyl)anisole. Similar product distributions are obtained in the photolysis of iodobenzene and 1-iodothiazole in benzonitrile. Propose a mechanism for these photoreactions [31].

Answer

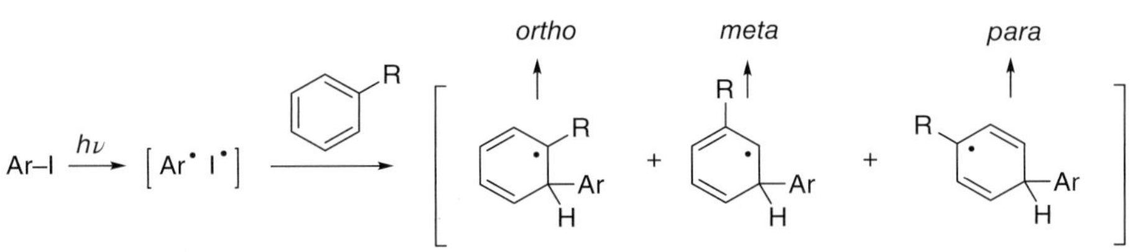

Since the same *ortho/meta/para* regioselectivity is observed for all these photolyses, one cannot adopt the hypothesis of aromatic electrophilic substitutions. If it were the mechanism one would expect less *meta* than *ortho* + *para* products in the case of anisole (electron-rich arene, activated arene) and more *meta* product than *ortho* + *para* products in the case of benzonitrile (electron-poor arene, deactivated arene). Thus, phenyl and 2-thiophenyl radical are the species that react with anisole and benzonitrile following a mechanism that implies the formation of substituted cyclohexadienyl radical intermediates.

Problem 6.33 What are the major products formed upon irradiation of 1,2-diiodobenzene in cyclohexane, in benzene, and in benzene in the presence of tetraphenylcyclopentadienone (tetracyclone)? [32]

Answer

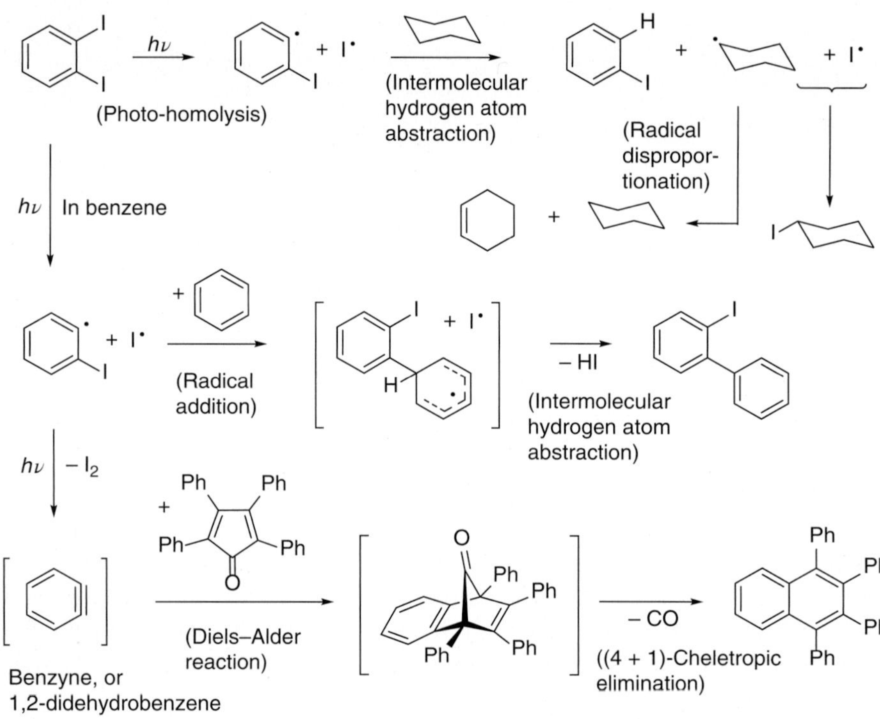

Problem 6.34 Propose a sequence of reactions that convert 1,3-diaminobenzene into (*E*,*E*)-3-azahexa-3,5-dien-1-yne-6-carbonitrile (NC—CH=CH—CH=N—C_2H) [33].

Answer

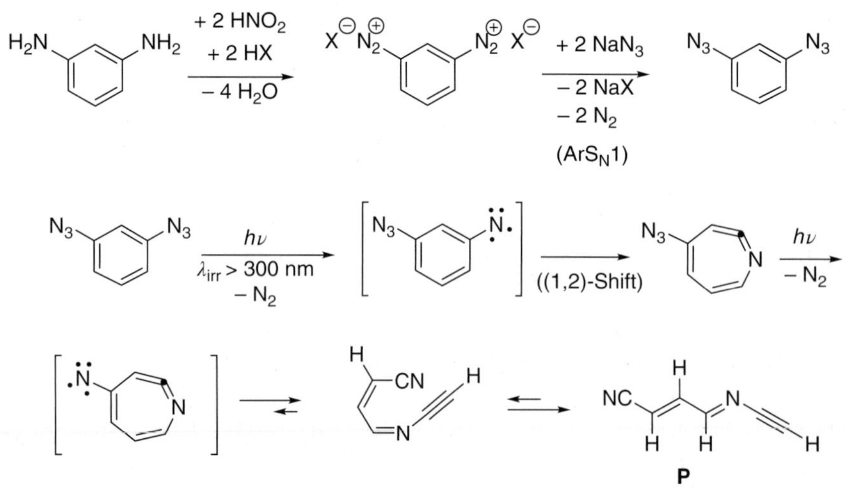

P

Problem 6.35 What products result from the irradiation (Hg burner, Pyrex vessel) of 1,1-diazido-2,3,4,5,6-penta-*O*-benzyl-D-glucose in CH_2Cl_2? [34] and of 2,3,4,6-tetra-*O*-acetyl-D-glucopyranosylidene diazide in the presence of acrylonitrile in quartz tubes? [35]

Answer

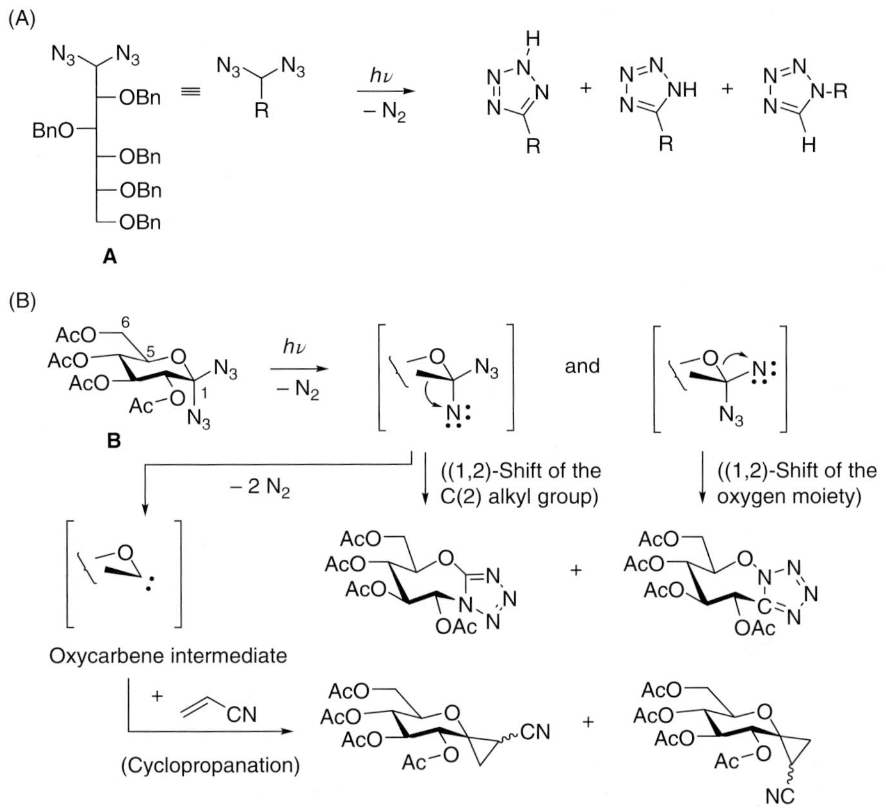

Problem 6.36 Propose a mechanism for the photolysis of 1-methyl-4-phenyl-1*H*-tetrazole-5(4*H*)-thione that produces methylphenylcarbodiimide [36].

Answer

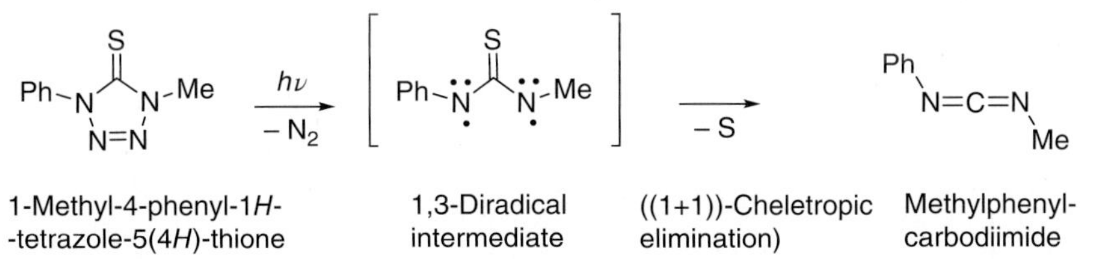

Problem 6.37 What are the products formed upon irradiation (254 nm) of 2,4-diphenyl-2*H*-tetrazole? [37]

Answer

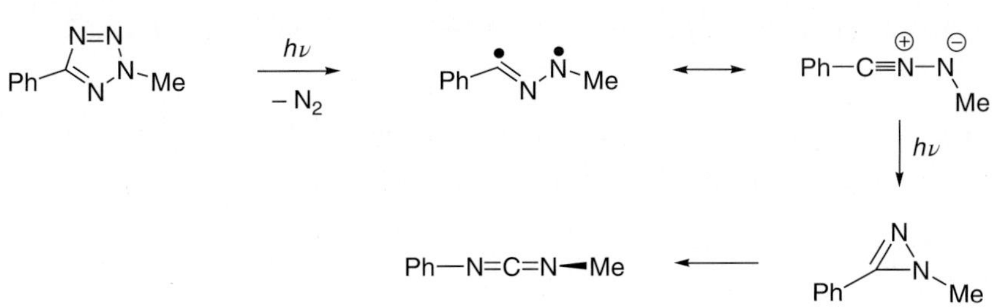

Problem 6.38 What are the major products of photochemical intramolecular cycloadditions of cycloocta-1,4-diene [38] and of cyclodeca-2,6-dienone? [39]

Answer

(a)

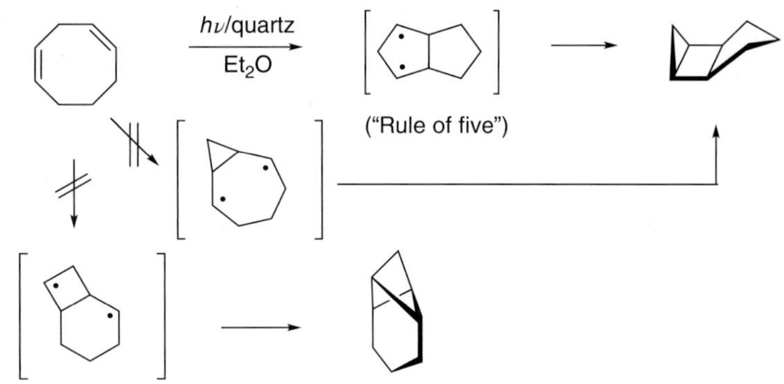

(b)

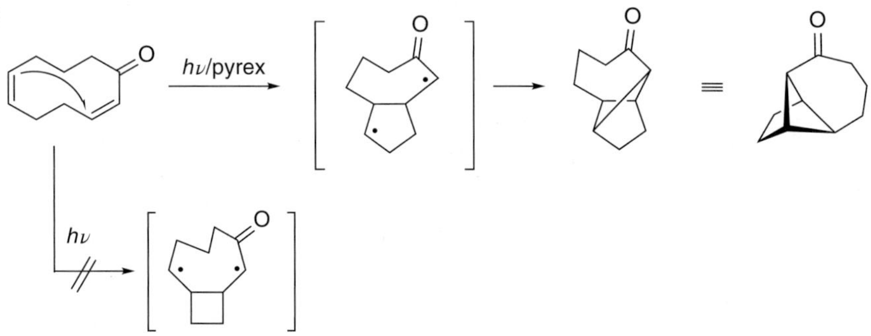

Problem 6.39 Propose a two-step synthesis of **P** [40].

P:

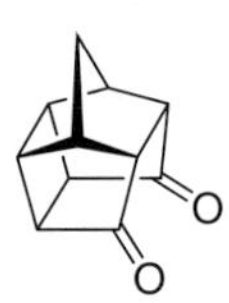

Answer

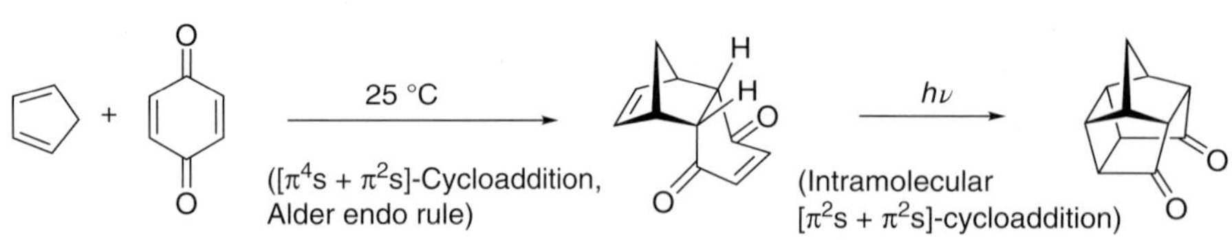

Problem 6.40 What is the product of UV irradiation of 4-(*t*-butyl)-2-(5-methylhex-4-en-1-yl)cyclohex-2-en-1-one? [41]

Answer

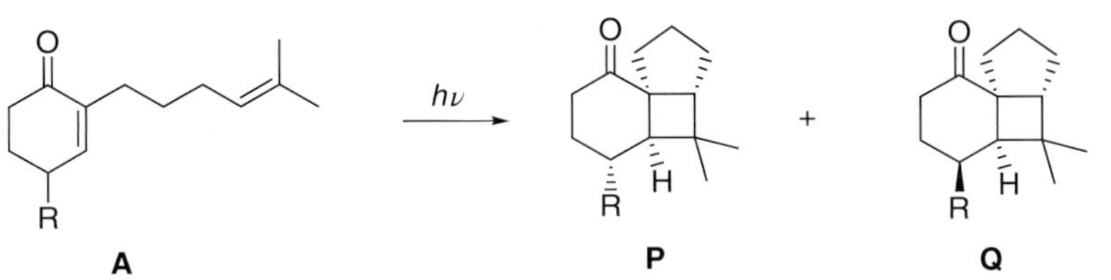

If R = *t*-Bu, cycloadduct **P** is the unique product obtained quantitatively.

The bulk of the 4-*t*-Bu substituent makes the enone cycloaddition highly face selective.

If R = Me, a 3 : 1 mixture of cycloadducts **P** + **Q** is obtained. For steric reasons, the face of the enone *anti* with respect to the 4-substituent remains favored.

Problem 6.41 What is the major product **P** of irradiation of **A**? [42]

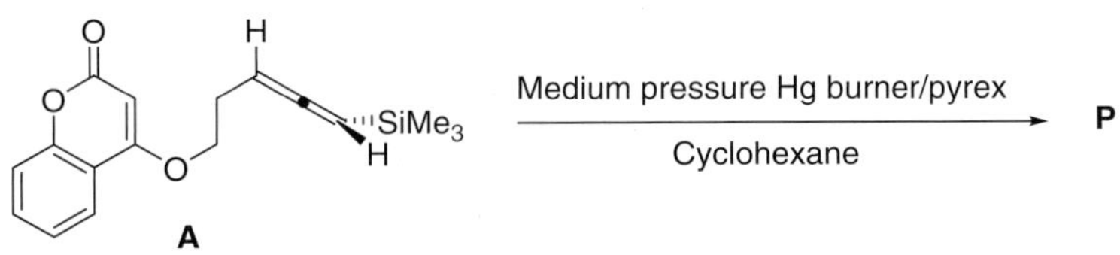

Answer

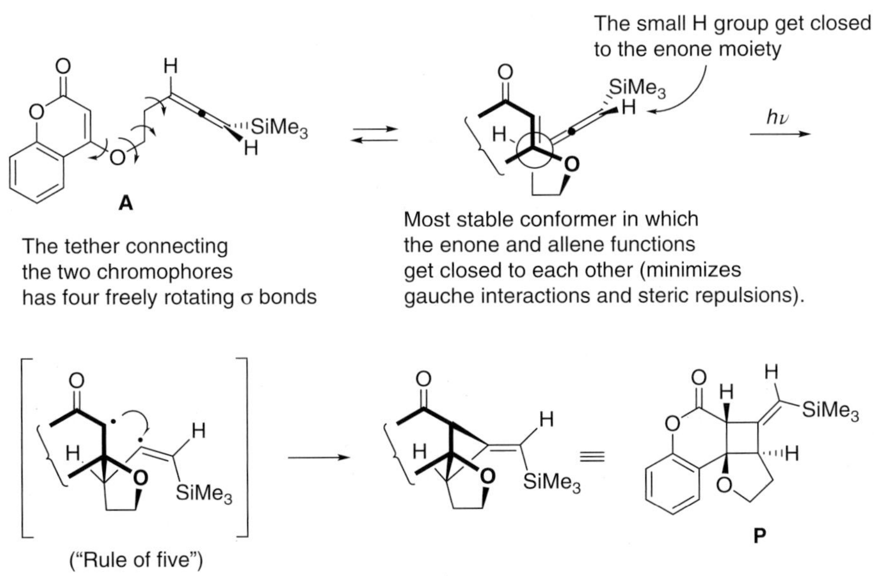

Problem 6.42 What are the intermediate products formed in the sequence of reactions that converts **A** into **P**? [43]

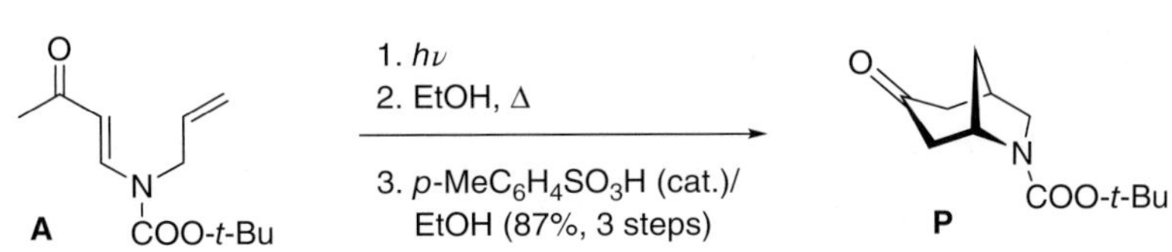

Answer

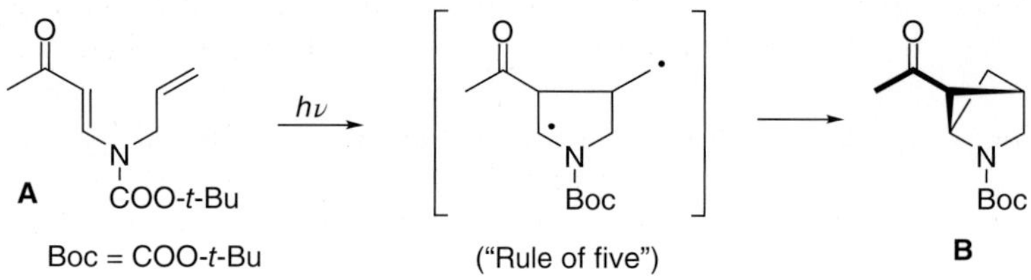

Boc = COO-t-Bu

Alternatively:

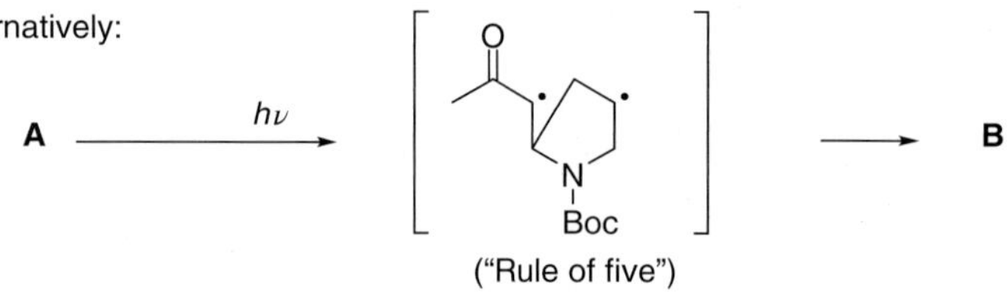

Both alkanediyl-1,4-diradicals follow the "rule of five." The second one is secondary–secondary and is stabilized by the carbonyl group. The first one is also relatively stable although it is primary–secondary, but stabilized by the BocN group.

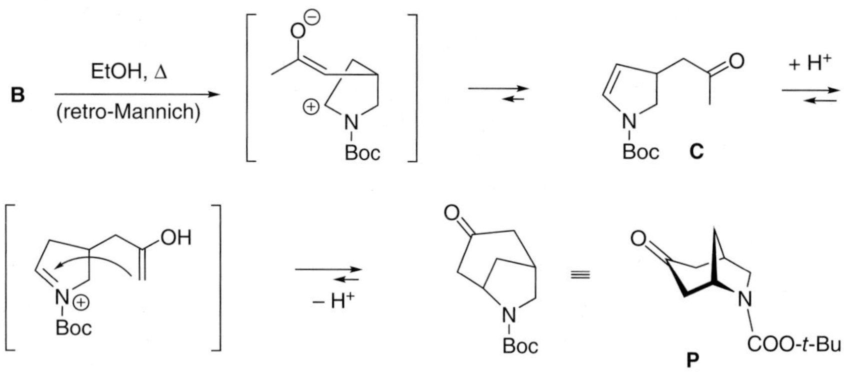

Problem 6.43 What are the main products of triplet sensitized irradiation of (A) dimethyl octa-2,6-dienedioate and (B) dimethyl nona-2,7-dienedioate? [44]

Answer

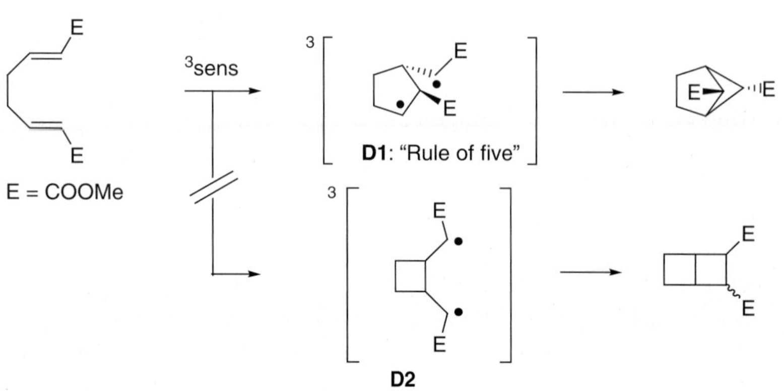

The "rule of five" here is consistent with the large difference in stability between diradicals **D1** and **D2** that manifests itself on the rates of their formation (Dimroth principle or Bell–Evans–Polanyi theory on concerted reactions: the least exothermic, or most endothermic cyclization is the slowest). Both diradicals **D1** and **D2** are secondary–secondary alkanediyl diradicals.

D1 is stabilized by c. 8 kcal mol^{-1} because of the conjugation of one of the alkyl radicals with the ester group and **D2** enjoys twice this stabilization effect (c. 16 kcal mol^{-1}). However, ring strain in **D1** (6 kcal mol^{-1}) is much less than ring strain in **D2** (26.5 kcal mol^{-1}). Thus, **D1** is estimated to be 20.5 to 8 kcal mol^{-1} more stable than **D2**.

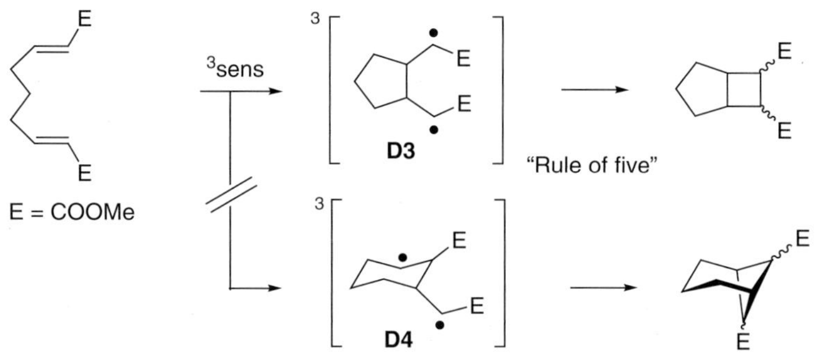

The "rule of five" states that diradical **D3** forms faster than **D4** and cyclizes efficiently into the bicyclo[3.2.0]heptane products. Diradical **D3** is more strained (c. 6 kcal mol^{-1}) than diradical **D4** but enjoys from twice the stabilization energy due to radical conjugation with the ester group. Thus, **D3** and **D4** are expected to have similar enthalpies of formation. Nevertheless, for entropic reasons, the five-membered ring diradical **D3** is expected to be formed faster than the six-membered diradical **D4**.

Problem 6.44 What are the major cycloadducts of the photochemical (2+2)-cycloadditions of (A) cyclohex-2-enone to acrylonitrile, [45, 46] (B) to isobutylene, and (C) to 1,1-dimethoxyethylene? [47]

Answer

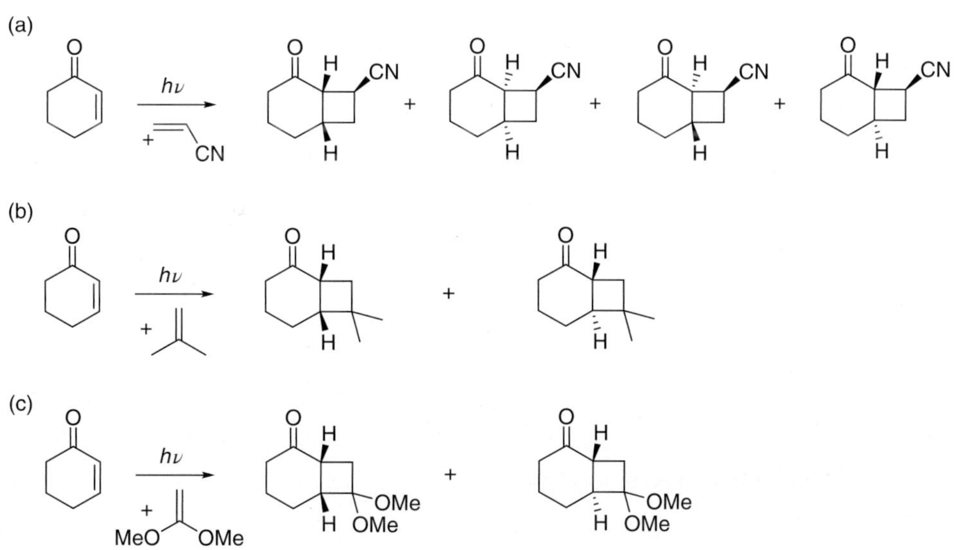

Cyclohex-2-enone undergoes (*Z*)/(*E*)-photoisomerization. Thus the (2+2)-cycloaddition give mixtures of *cis*- and *trans*-bicyclo[4.2.0]oct-3-en-2-one derivatives. Alkenes with electron-withdrawing substituents lead to head-to-head cycloadducts, whereas alkenes with electron-releasing substituents favor the head-to-tail cycloadducts.

Problem 6.45 Devise a synthetic scheme that converts **A** into **P** [48].

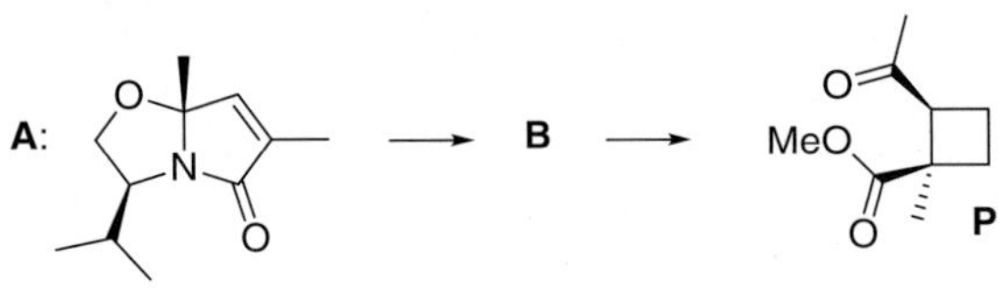

Answer

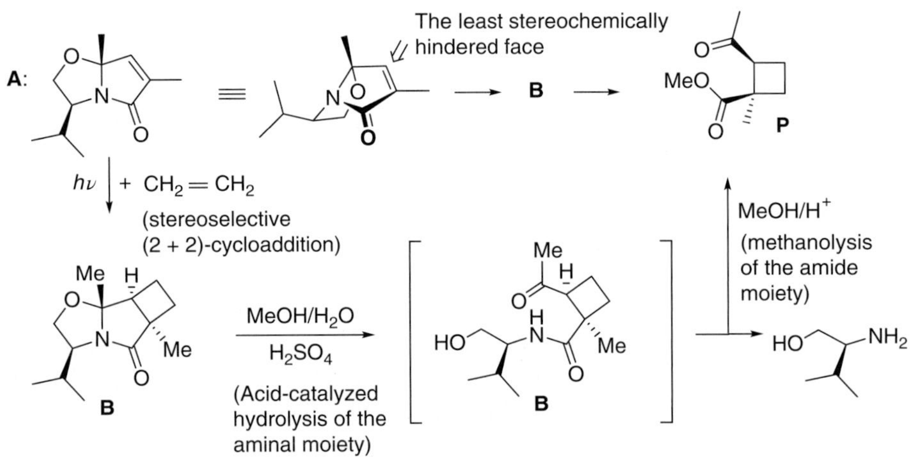

Problem 6.46 What is the major product **P** of UV irradiation of **A** + **B**? [49]

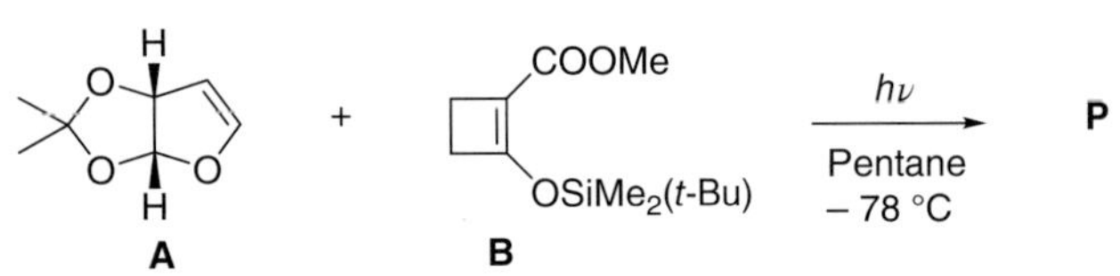

Answer

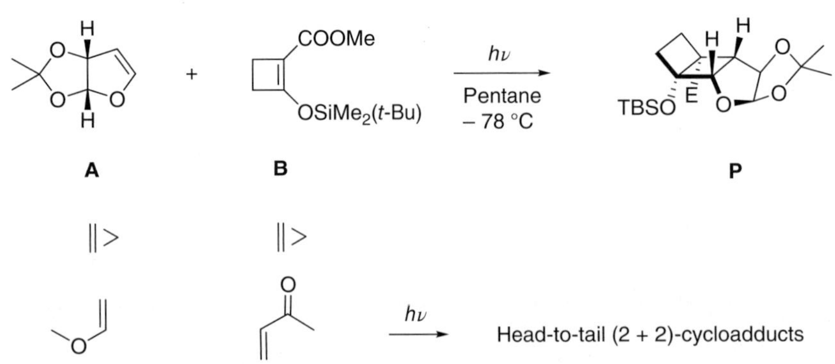

The least sterically hindered face of **A** adds to the excited state of acrylate **B**

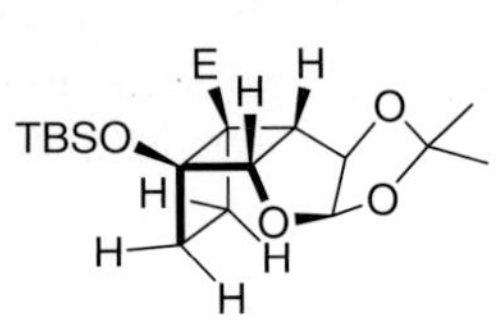

Less favored: the rigidity of the cyclobutane ring forces its C–H bonds to bump into the dihydrofuran ring. The COOMe and TBSO substituents can rotate to escape steric repulsion between them and the dihydrofuran ring

E = COOMe; TBS = (*t*-Bu)Me_2Si

Problem 6.47 What is the major product **P** of the photolysis of **A** with acetylene? [50]

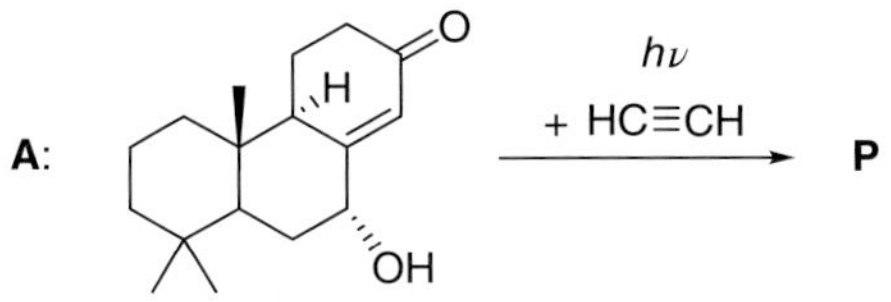

Answer

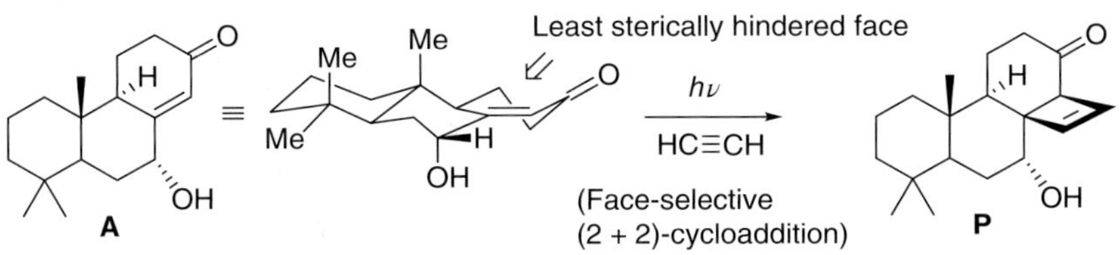

Problem 6.48 What are the major photocycloadducts of the irradiation of cyclopentene with the methyl carbamate of isoindoline? [51]

Answer

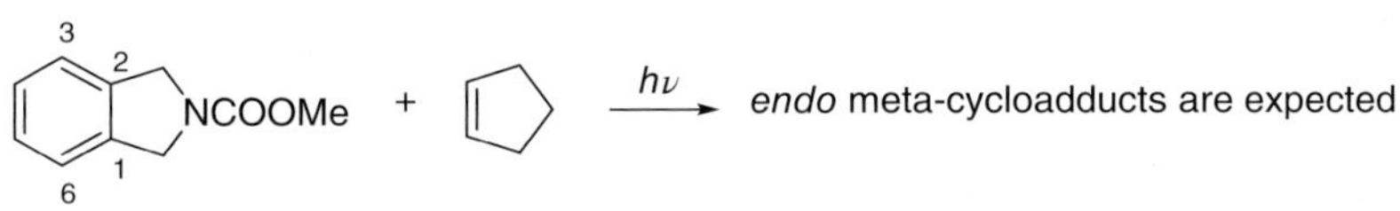

The 1,2-dialkyl substituents of benzene should favor (3 + 2)-cycloaddition across C(1) and C(3), or C(2) and C(6):

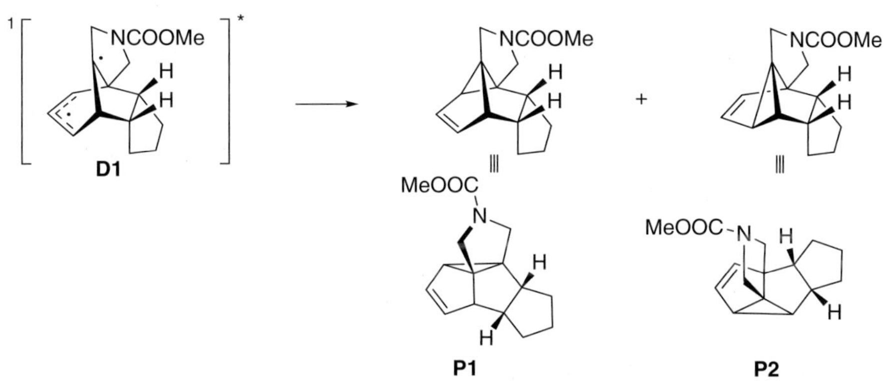

Diradical **D1** enjoys stabilization from the alkyl substituents. Compare with the photo-cycloaddition of toluene with alkenes. All other regioisomeric diradicals **D2–D4** are less stable. Substitution of the allyl radical moiety by an alkyl group is less stabilizing since the secondary allyl radical is intrinsically more stable than a secondary alkyl radical (the spin is delocalized onto two carbon centers in the allyl radical, not in the alkyl radical). This theory is verified also with the photochemical (3+2)-cycloaddition of 1,3-dimethylbenzene with cyclopentene (Scheme 6.61).

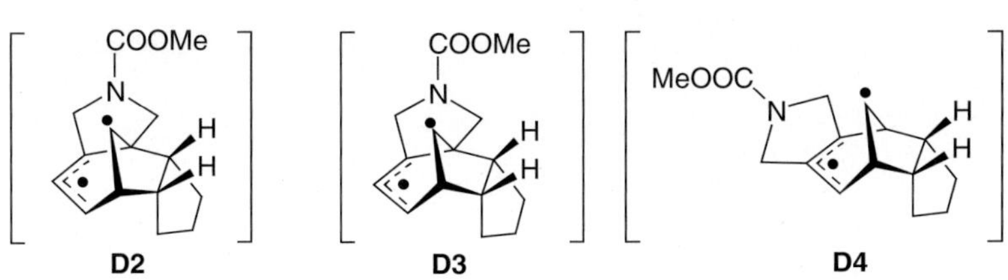

Problem 6.49 What is the major photocycloadduct of acrylonitrile with 2,6-dimethoxybenzonitrile? [52]

Answer

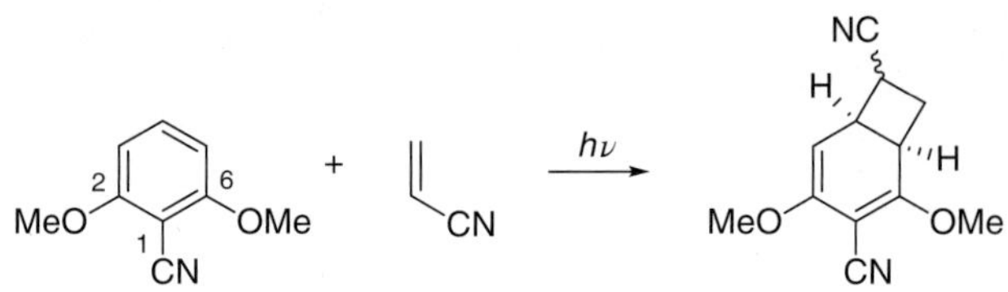

A P (70%)

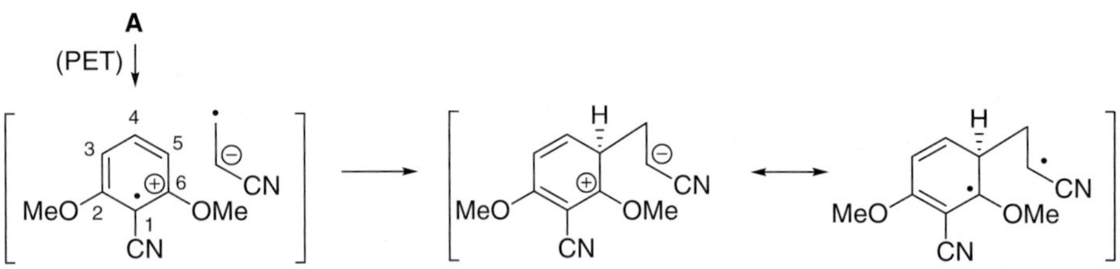

I1 Z1 D1

Arene cycloaddent **A** is electron rich (2,6-dimethoxy) and alkene cyloaddent is electron poor (acrylonitrile). Thus, photoinduced electron transfer from the arene to acrylonitrile is favored and generates the radical-anion/radical-cation pair **I1**. Coulombic attraction in **I1** maintains the cyano group of acrylonitrile in the vicinity of one of the two methoxy groups of the arene. The strongest C—C bond forms first. It is the one that involves C(5), not C(1) of the arene (unsubstituted single C—C bonds are stronger than substituted ones. In general, the strongest C—C bonds are those giving unstable radicals upon homolysis (Section 1.8). This generates zwitterion **Z1**, a relatively stable species since its cationic part is a 1,3-dimethoxycyclohexadienyl cation and its anionic part a carbanion substituted by a cyano group (stabilization of the anion by charge/dipole interaction and π-conjugation, polarizability effect of the cyano group).

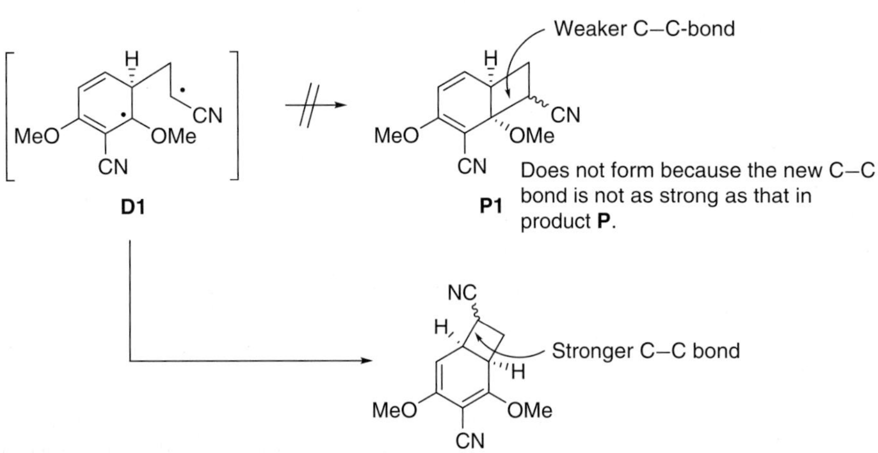

P (70%)

Problem 6.50 What products are formed upon irradiation of benzanthracene in the presence of cyclohexa-1,3-diene? [53–55]

Answer

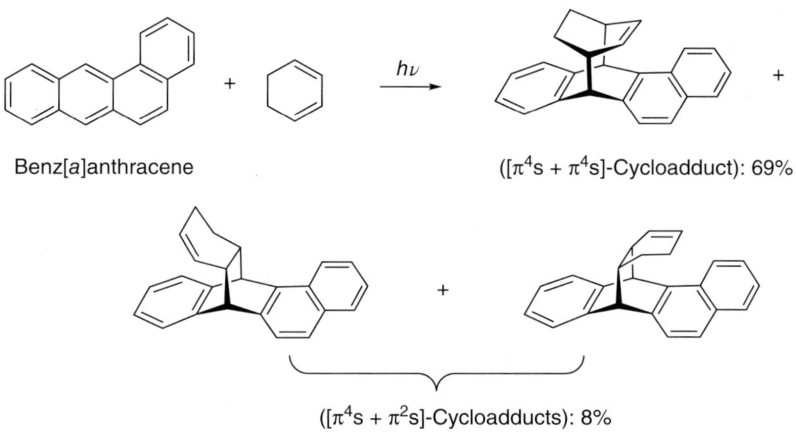

Problem 6.51 What are the products of Paternó–Büchi reaction of but-2-yne with benzaldehyde at −45 °C? [56]

Answer

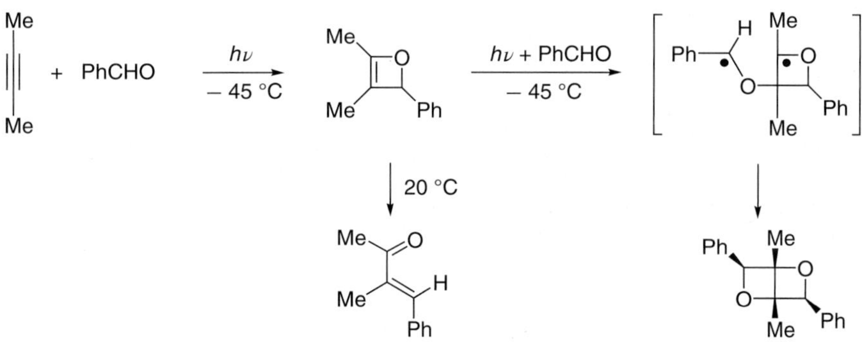

Problem 6.52 What is the major product of the photochemical reaction of furan with benzaldehyde? [57–60]

Answer

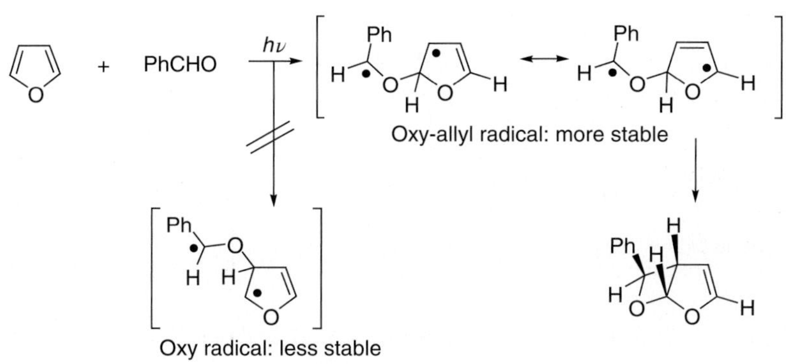

Problem 6.53 What is the major product of irradiation (λ_{irr} = 300 nm) of 3,4,5-trimethylisoxazole with benzaldehyde in MeCN at −10 °C? [61]

Answer

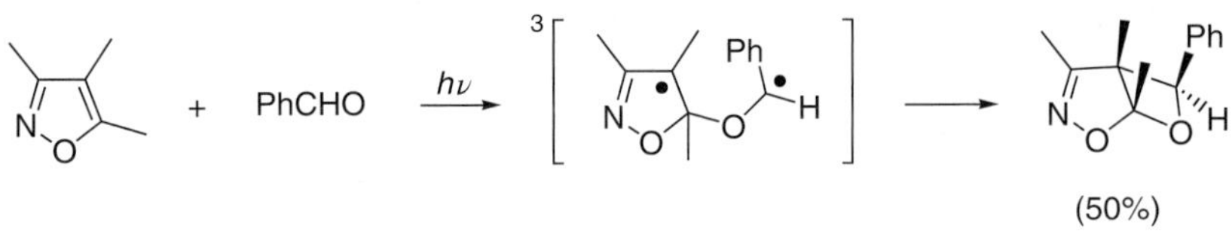

Problem 6.54 What is the product of direct irradiation (λ_{irr} = 254 nm, quartz vessel, MeCN, −20 °C) or sensitized excitation (high pressure Hg burner, solidex filter, acetone, 60 °C) of compound **A**? [62]

Answer

A:

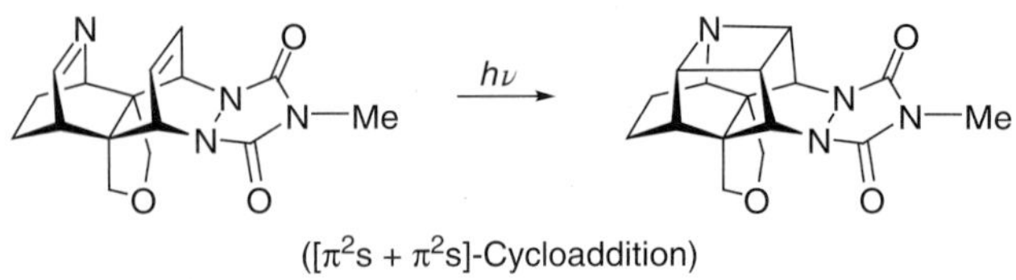

Problem 6.55 Why is the lifetime of $^1O_2(^1\Sigma^+{}_g)$ much shorter than that of $^1O_2(^1\Delta_g)$?

Answer

The electronic transition $^1O_2(^1\Sigma^+{}_g) \rightarrow {}^1O_2(^1\Delta_g)$ is spin allowed, whereas transition $^1O_2(^1\Delta_g) \rightarrow {}^3O_2(^3\Sigma^-{}_g)$ is not.

Problem 6.56 What is the product of photo-oxygenation of 2,3,5,6-tetramethylidene-7-oxabicyclo[2.2.1]heptane? [63]

Answer

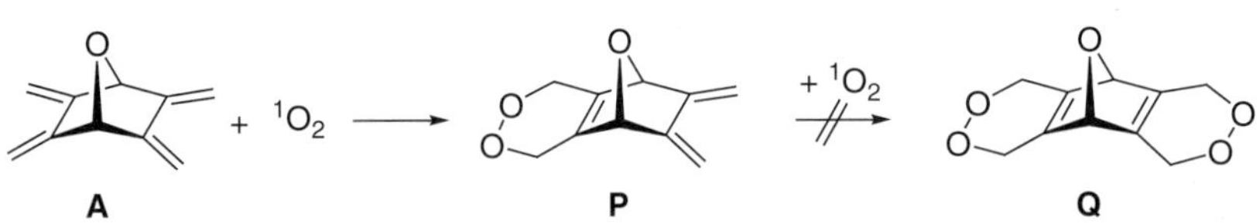

Endoperoxide **P** forms but not the double endoperoxide **Q** after prolonged exposure to $^1O_2$. As for 5,6-dimethylidene-7-oxabicyclo[2.2.1]hept-2-ene (**278**, Table 6.5), physical and chemical deactivation of singlet oxygen is very efficient with triene **P**. Homoconjugation between the endocyclic double bond and the exocyclic *s-cis*-butadiene moieties of **P** increases its IE slightly with respect to IE(**A**).

Problem 6.57 What is the major product of the following reaction sequence? [64]

A:

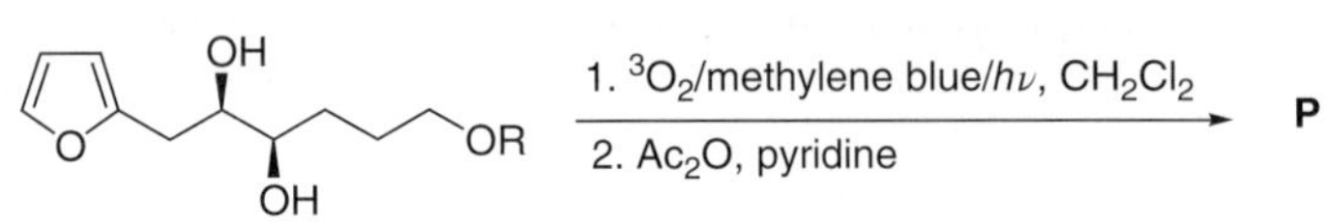

Answer

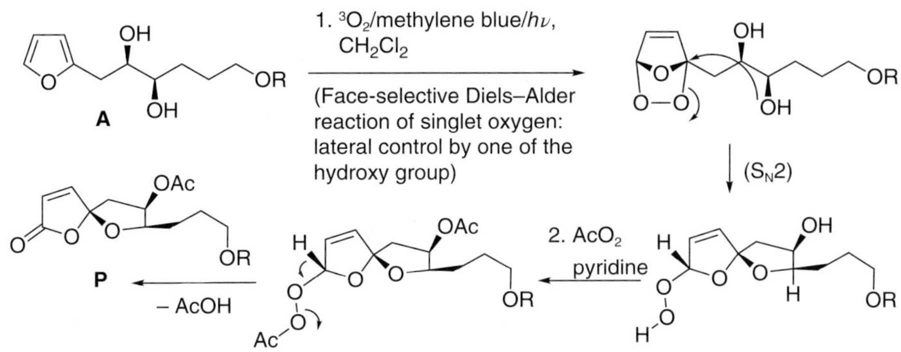

Problem 6.58 Convert **A** into **P** [65].

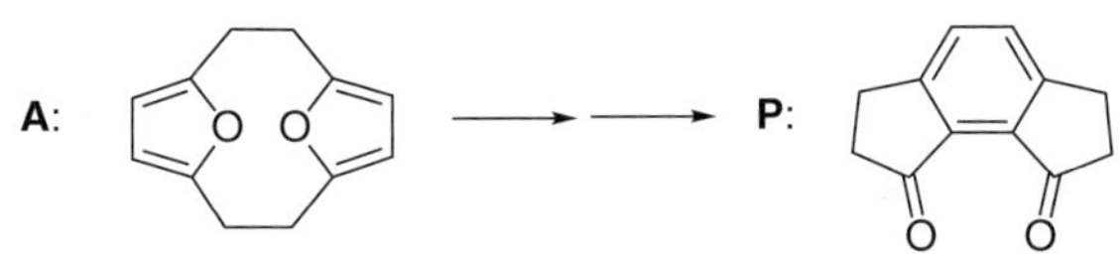

Answer

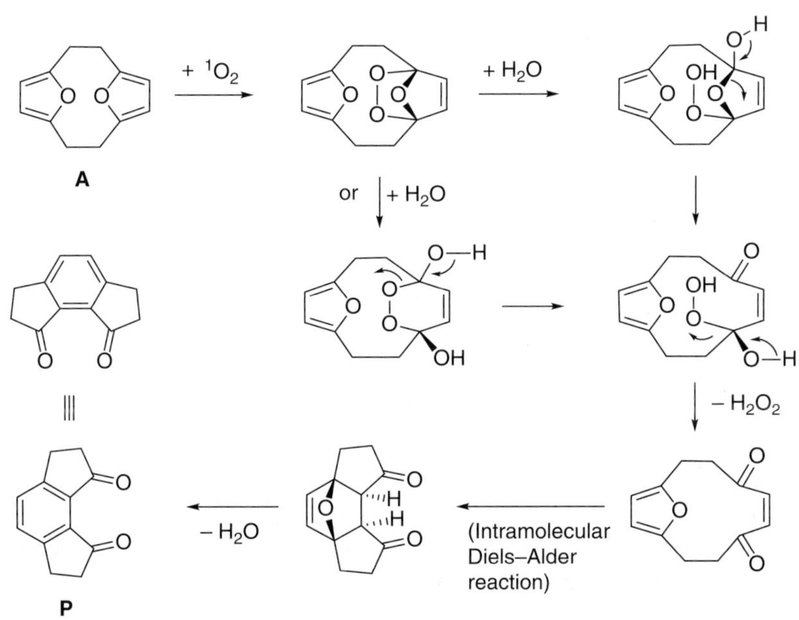

Problem 6.59 What are the products of photo-oxygenation of adamantylideneadamantane (**A**) in the presence of PhSOMe? [66]

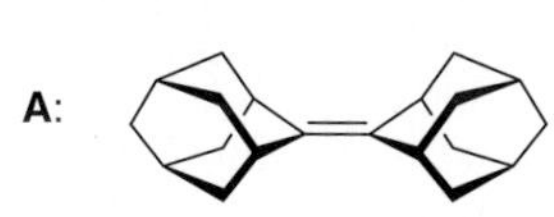

Answer

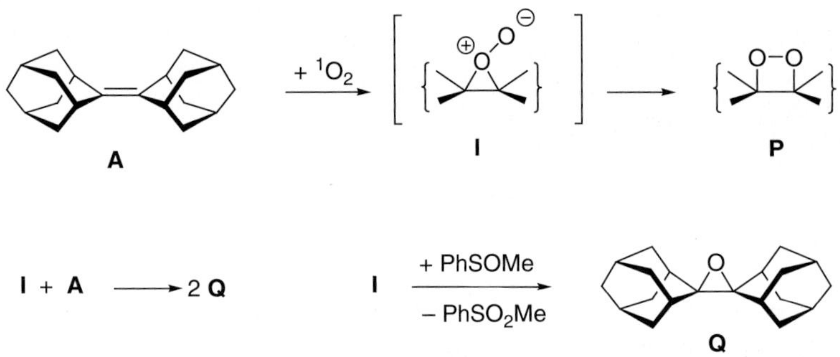

Photo-oxygenation of alkene **A** generates dioxetane **P** (major) and epoxide **Q** (minor). In the presence of sulfoxide PhSOMe more epoxide **Q** forms, which indicates that the latter results from the perepoxide intermediate **I**. The photo-oxygenation of **A** is inhibited by 1,4-diaza-bicyclo[2.2.2]octane (DABCO), an efficient quencher of singlet oxygen, but not by 2,6-bis(*t*-butyl)cresol, an efficient inhibitor (quencher) of radicals. Thus, reaction **I** + **A** → 2**Q** can be seen as the epoxidation of alkenes by hydroperoxides (ROOH) and peracids (RCO_3H).

For geometrical reasons **A** cannot undergo dioxa-ene reaction with $^1O_2$.

Problem 6.60 Photo-oxygenation of 1,2,3,4,5,6,7,8-octahydronaphthalene (**A**) gives a product **B** that is esterified with PhCOCl/pyridine into **C**. Product **C** undergoes a Criegee's rearrangement into **Q**. The latter reacts in alkaline water giving a diketone **P**. What are **B**, **C**, **Q**, and **P**? [67]

Answer

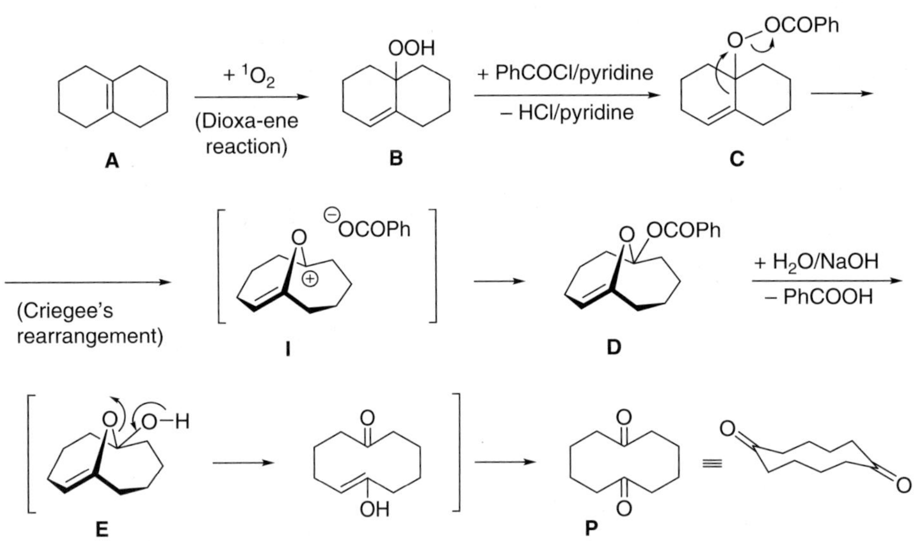

Dioxa-ene reaction **A** + $^1O_2$ gives allylic hydroperoxide **B**, which is esterified into perbenzoate **C**. As for the Baeyer–Villiger oxidation of ketone into esters by peracids, the O—O bond in perester **C** is relatively weak and polarized. This permits a (1,2)-sigmatropic shift **C** → **I** and quenching of ion pair **I** into hemiacetal ester **D**. Upon base-catalyzed hydrolysis **D** liberates hemiacetal **E** that equilibrates with diketone **P**.

Problem 6.61 What are the major products of the following photo-oxygenations? [68]

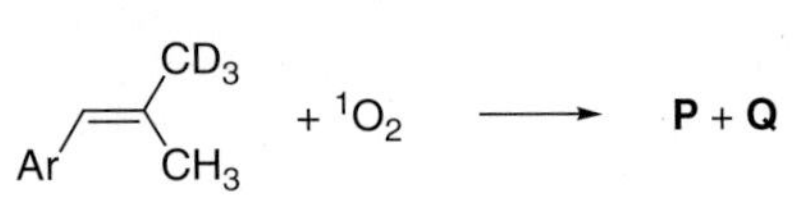

(a) Ar = 4-$MeOC_6H_4$
(b) Ar = 4-$CF_3C_6H_4$

Answer

(a)

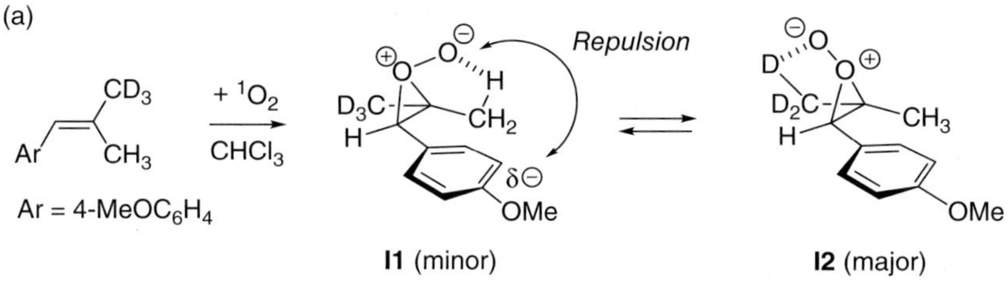

Electrostatic repulsion between the terminal oxygen center of the perepoxide (does not have to be symmetrically bridged) and the methoxy group of $MeOC_6H_4$ renders **I1** less stable than **I2**.

(b)

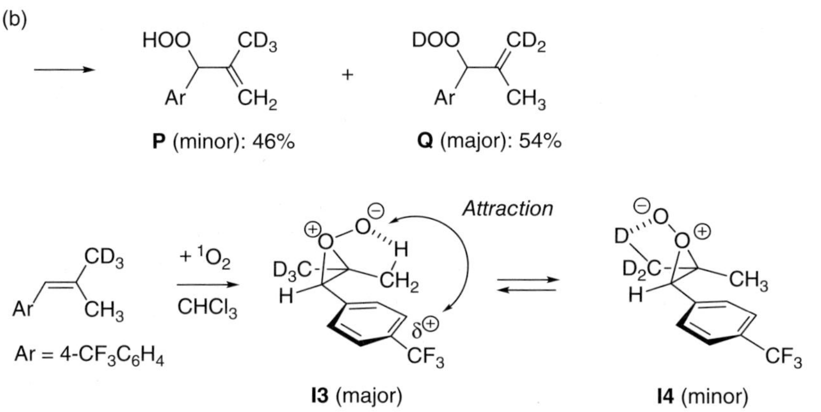

Electrostatic attraction between the terminal oxygen center of the perepoxide and the CF_3 group of $CF_3C_6H_4$ renders **I3** more stable than **I4**.

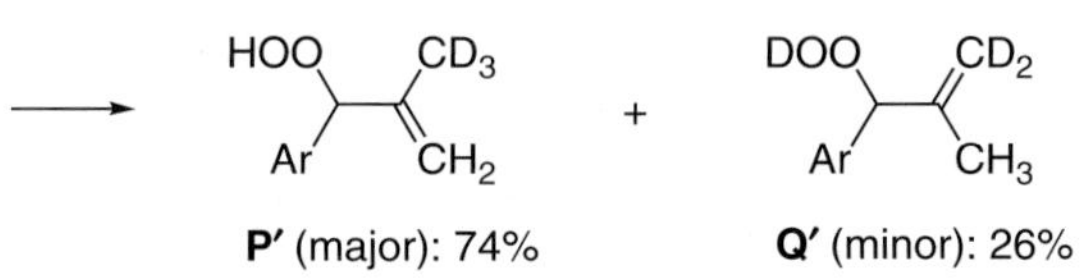

Problem 6.62 What is the major product of photo-oxygenation of the enol trimethylsilyl ether derived from methyl isobutyrate? [69]

Answer

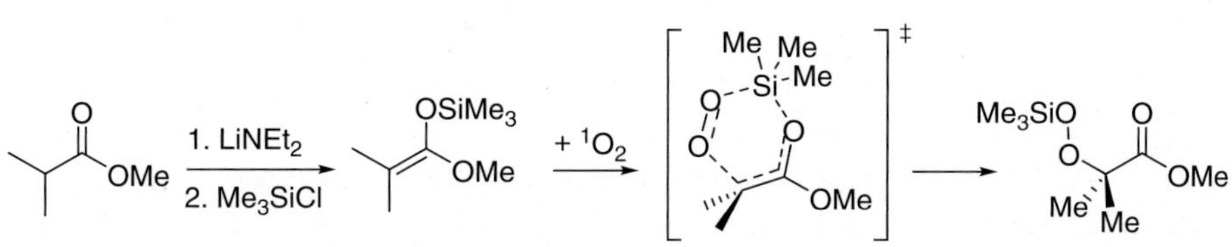

(Siloxy-dioxa-ene reaction) **T**

The alternative alkoxy-dioxa-ene reaction that would involve the migration of the Me group instead of the Me_3Si group does not compete because of the stronger affinity of oxygen for silicon than for carbon. The transition state **T** depicted above might be preceded by non-symmetrical perepoxide intermediate **I1**, or by zwitterionic intermediates **I2** or **I3**.

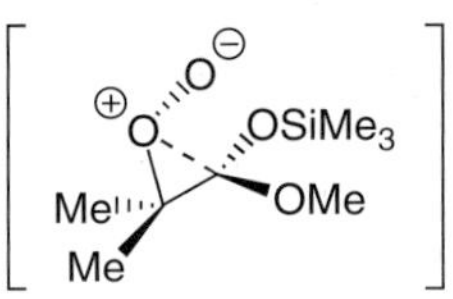

I1

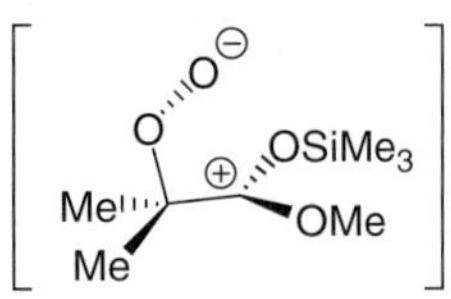

I2

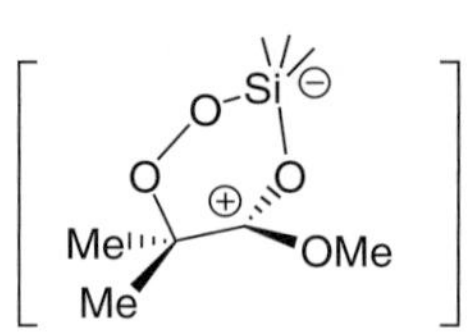

I3

Problem 6.63 Explain the following reaction sequence [70].

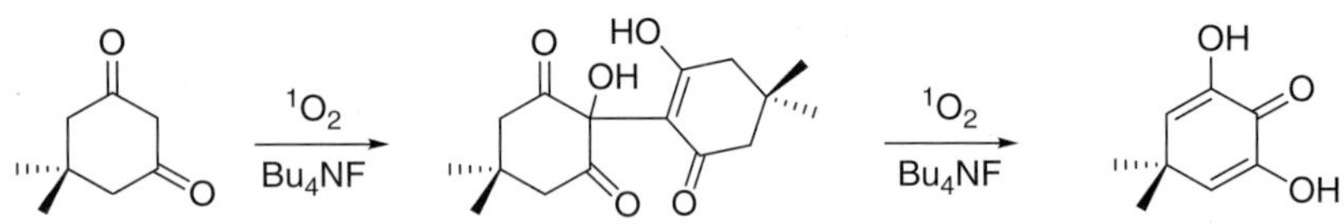

Answer

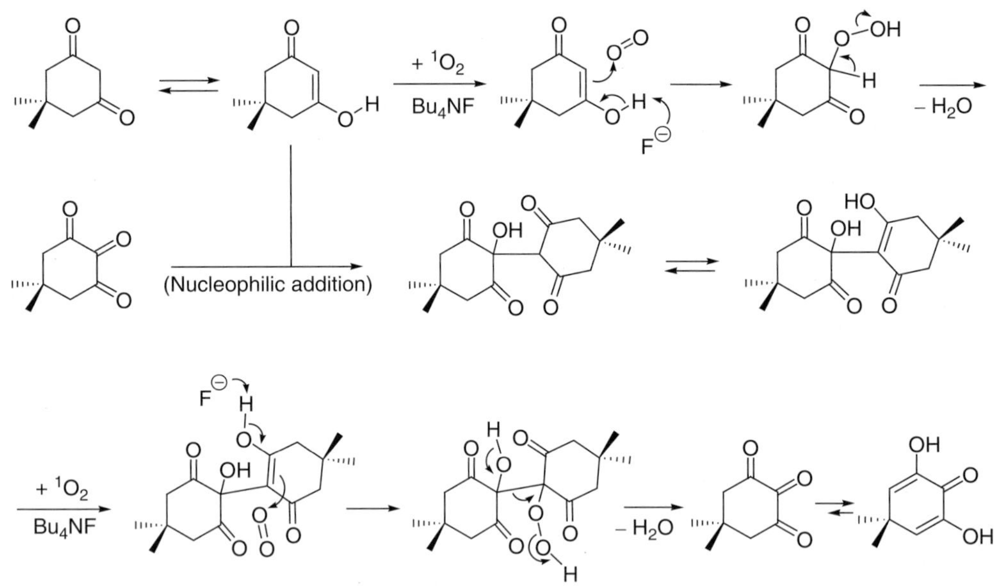

TBAF(Bu_4NF) catalyzes the keto/enol equilibria and renders the enol that reacts with singlet oxygen more nucleophilic because of the F^- hydrogen bridging. Singlet oxygen is a strong electrophile:
$^1O_2 \leftrightarrow {}^+O\text{–}O^-$.

Problem 6.64 What are the main products of photo-oxygenation of **A**? [71]

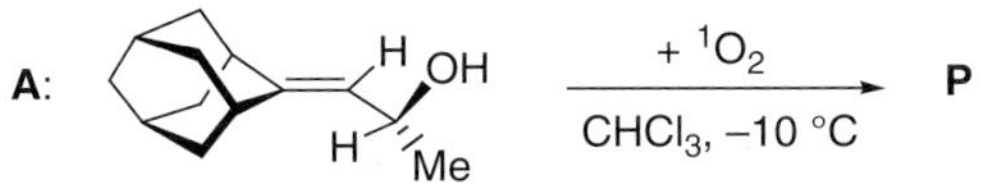

Answer

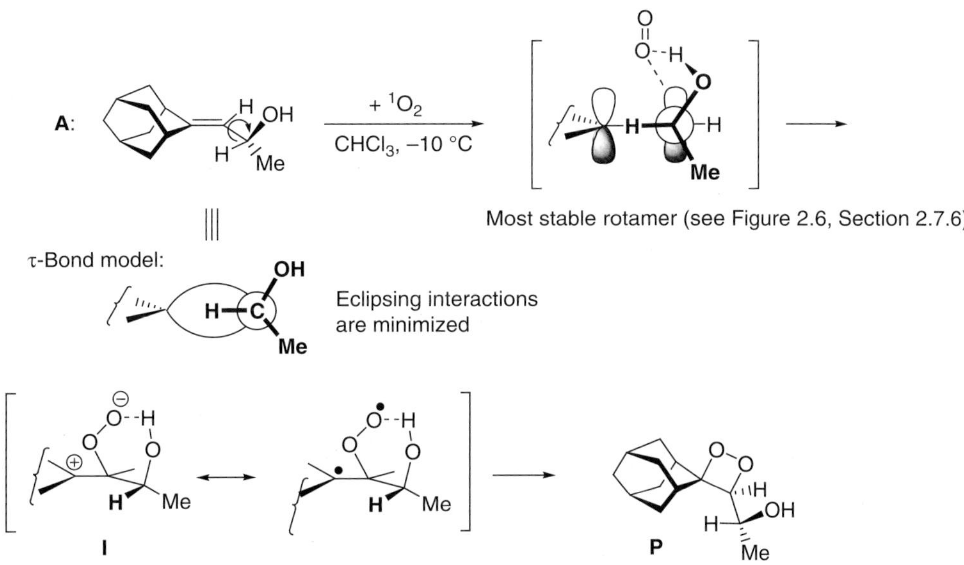

Alkene **A** is not particularly electron rich. It adds to singlet oxygen forming probably intermediate **I** that is a polarized 1,4-alkyl-peroxyl diradical (diradical with zwitterion character) with the terminal oxygen atom bearing a relatively important partial negative charge. The latter can chelate the alcoholic moiety (make an intramolecular hydrogen bridge) favoring one face of the alkene for its reaction with $^1O_2$. Cyclization of intermediate **I** produces 1,2-dioxetane **P**.

Problem 6.65 *Trans*-resveratrol (**A**) is an antioxidant agent found in several plants and wine, with several beneficial health effects such as cardiovascular modulating effect, antiaging and antitumor effects. The photo-oxygenation of **A** in CD_3CN using methylene blue as sensitizer gives two aldehydes (**P1** and **P2**) and moracin M after heating the reaction mixture to 80 °C. Explain these transformations [72].

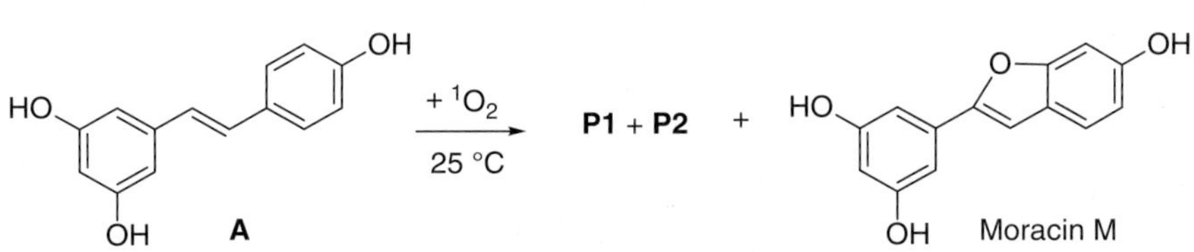

Answer

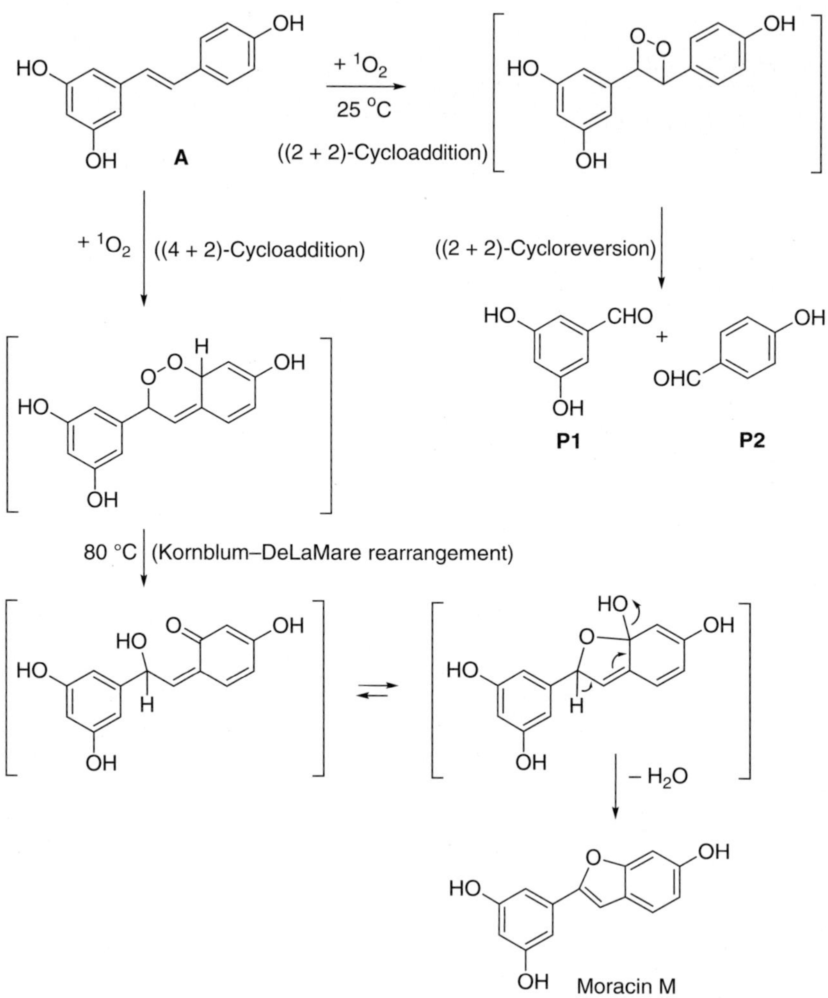

A cannot undergo any dioxa-ene reaction. Although the Diels–Alder reaction destroys the aromaticity of one of the phenol rings, it competes with the (2+2)-cycloaddition.

Problem 6.66 Propose a mechanism for the visible light photocatalyzed conversion of **A** into **P**, a formal intramolecular (3+2)-cycloaddition [73].

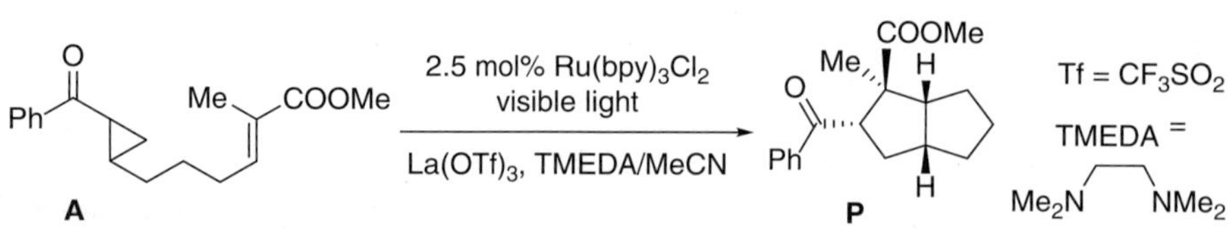

Answer

This is a photocatalyzed (3+2)-cycloaddition. After excitation of the photocatalyst, the Ru(II) complex takes an electron from the tertiary amine generating the reduced form of the catalysts, a Ru(I) species and an aminimium radical-cation. The Ru(I) complex reduces the ketone coordinated to the Lewis acid giving the corresponding ketyl radical-anion **I1**. The latter undergoes a quick cyclopropyl/homoallyl rearrangement equilibrating with the radical-anion intermediate **I2**. The radical moiety of the latter adds intramolecularly to the methacrylate part forming radical-anion **I3**. A second ring-closing reaction (intramolecular addition of a radical to an alkene) converts **I3** into ketyl radical-anion **I4**. The overall process is highly exothermic since two new C—C bonds are formed compared with **I2**. Ketyl radical-anion **I4** then reduces the aminium radical-cation formed during the photoreduction of the photocatalyst.

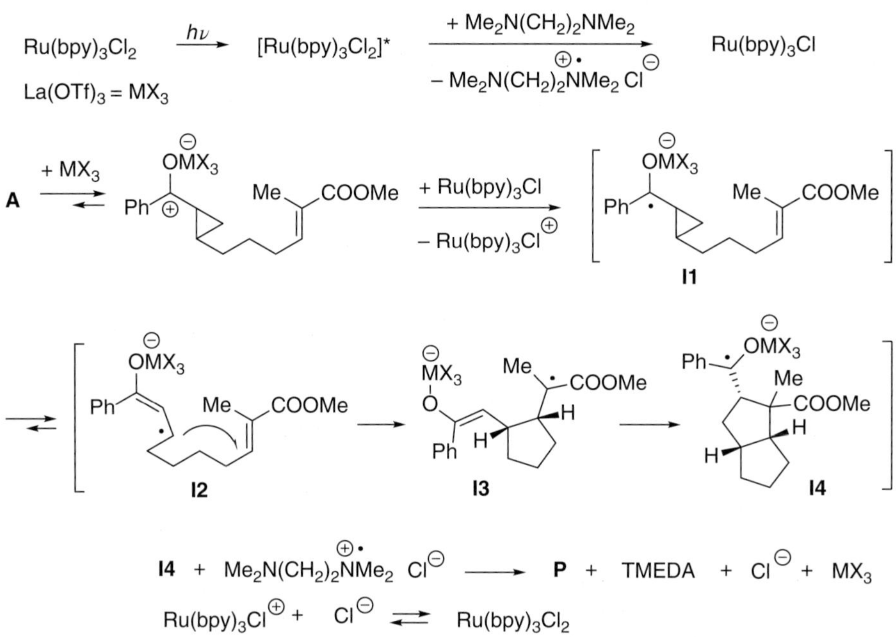

Problem 6.67 What are the possible products of irradiation ($\lambda_{irr} = 254$ nm) of 7-oxabicyclo[2.2.1]heptan-2-one in MeCN containing 5 equiv. of Et_3N? [74]

Answer

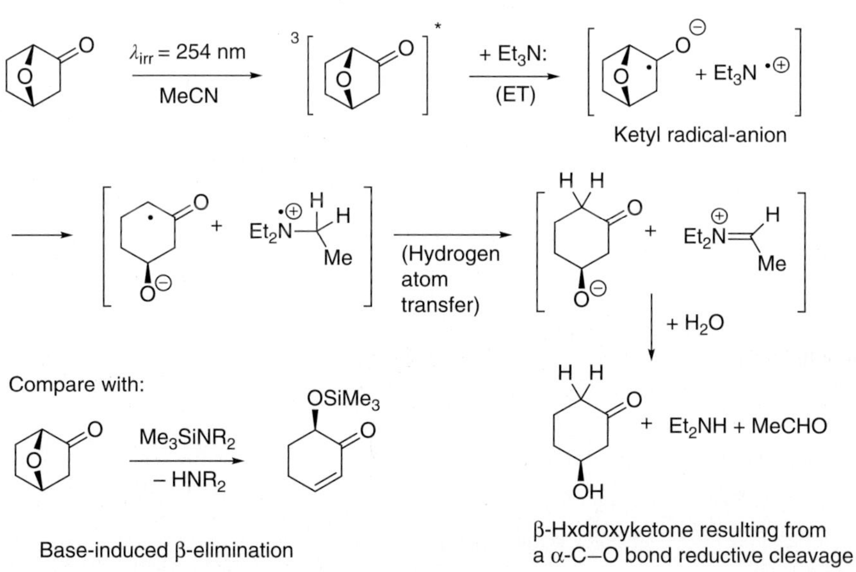

Tertiary amine such as Et_3N: gives an electron to the ketone in its triplet excited state giving a pair of ketyl radical-anion and aminimium radical-cation. The latter is a good hydrogen atom donor as a stable iminium ion ($Et_2N^{(+)}$=CHMe) is produced after the transfer. Thus, hydrogen atom transfer from the amininium ion to the ketyl radical-anion generates an alcoholate that can be seen as the result of the reduction of the carbonyl compound by hydride anion displacing the a-C-O ethereal group. After aqueous work-up the alcoholate is converted into the corresponding alcohol, and the iminim ion is hydrolyzed into a secondary amine (Et_2NH) and an aldehyde (MeCHO).

If the hydrogen atom transfer occurs before the a-C—O bond cleavage of the ketyl radical-anion, an alcoholate forms that is protonated by the water giving an alcohol, the product of reduction of the ketone. Alternatively, the initial ketyl radical-anion can take a proton from the amininium ion generating a pair of relatively stable hydroalkyl radical and aminoalkyl radical. The latter can undergo all kinds of radical reactions, included their combination (heterocoupling) into aminoalcohols and, after diffusion out of the solvent cage, products of homocoupling such as the corresponding pinacol. This is illustrated below with the photoreaction of acetone with Et_3N.

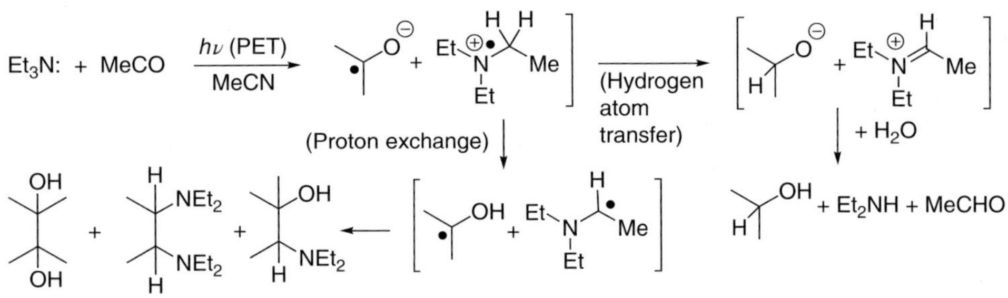

Problem 6.68 Explain the following transformations [75]:

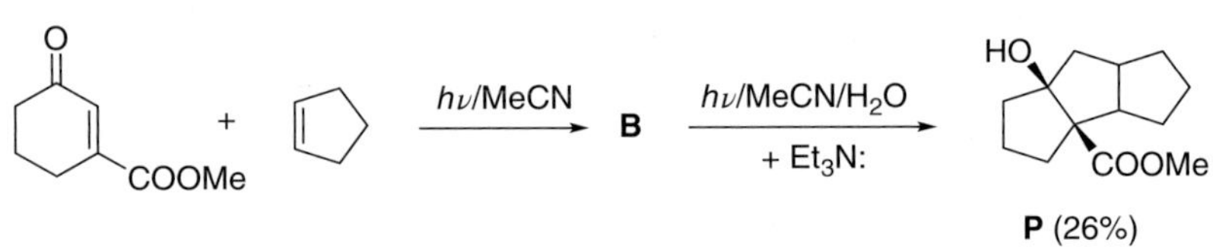

Answer

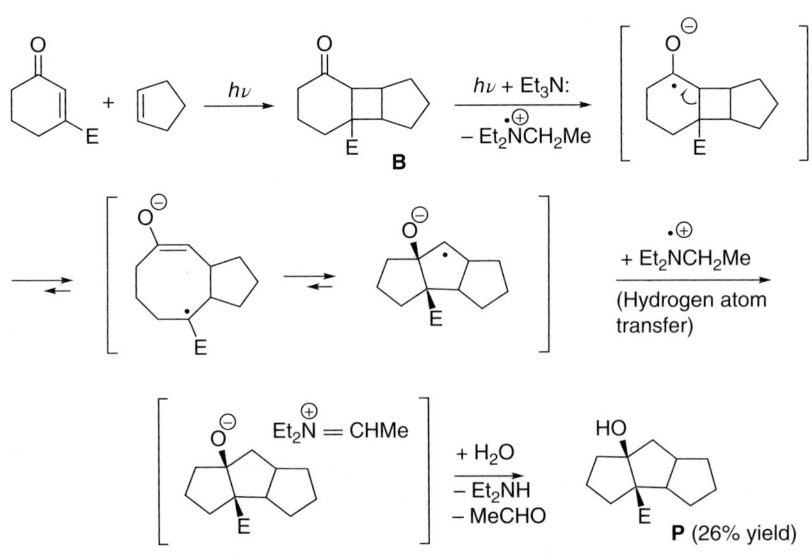

Problem 6.69 What are the main products **P** (94%) + **Q** of the following photoreaction (Ts = tosyl = *p*-toluenesulfonyl = 4-$MeC_6H_4SO_2$)? [76]

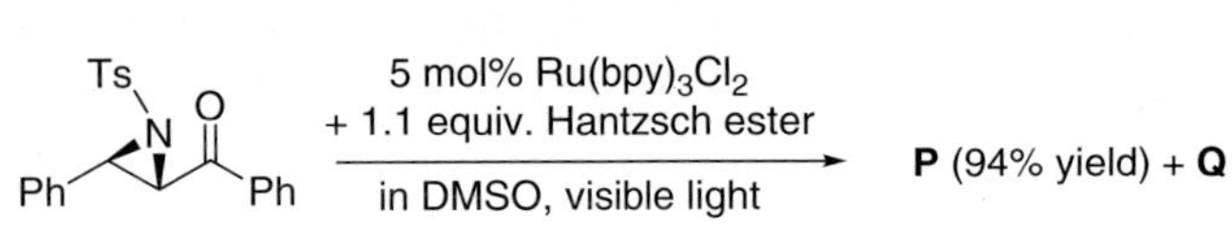

Answer

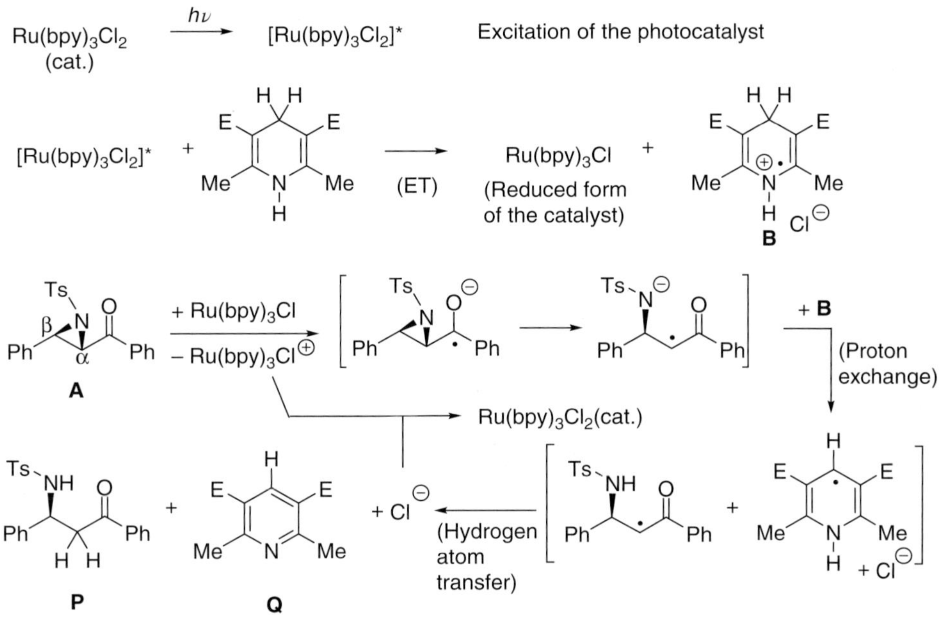

Formally, the Hantzsch ester provides H_2 that hydrogenolyzes the α-C—N bond, not the benzylic β-C—N bond of **A**!

Problem 6.70 What is the major product **P** (69%) of the following photoreaction (E = COOMe)? [77]

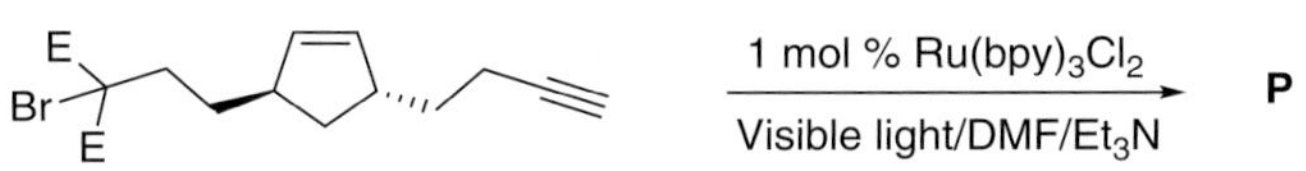

Answer

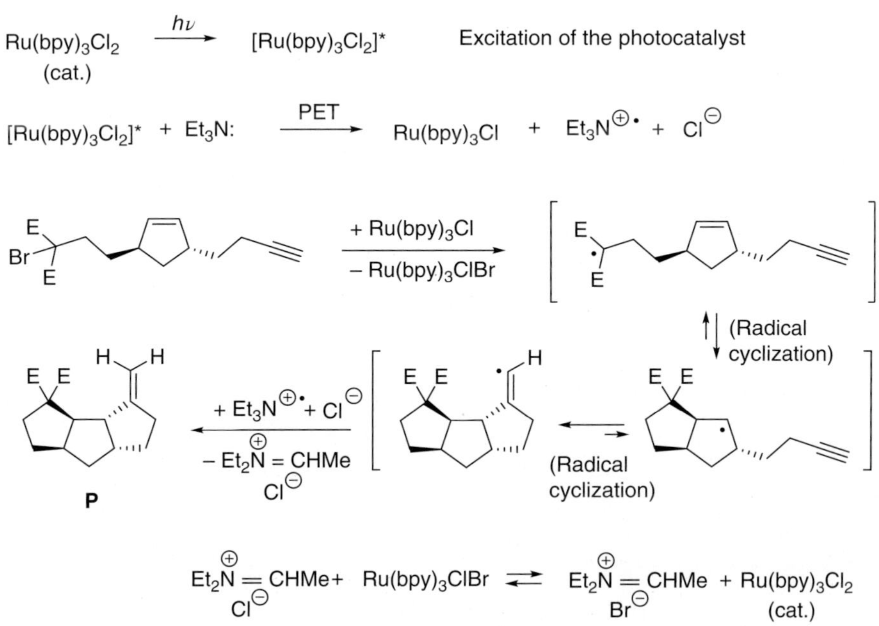

Problem 6.71 Propose a mechanism for the photocatalyzed reductive cleavage of carbamate **B** (protected form of amines) using ascorbic acid as stoichiometric reductant [78].

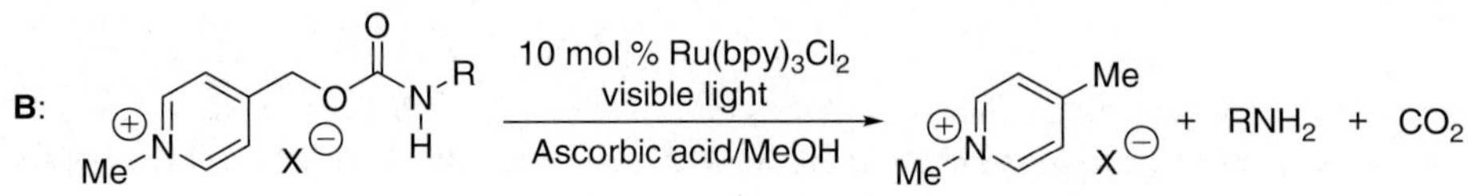

Answer

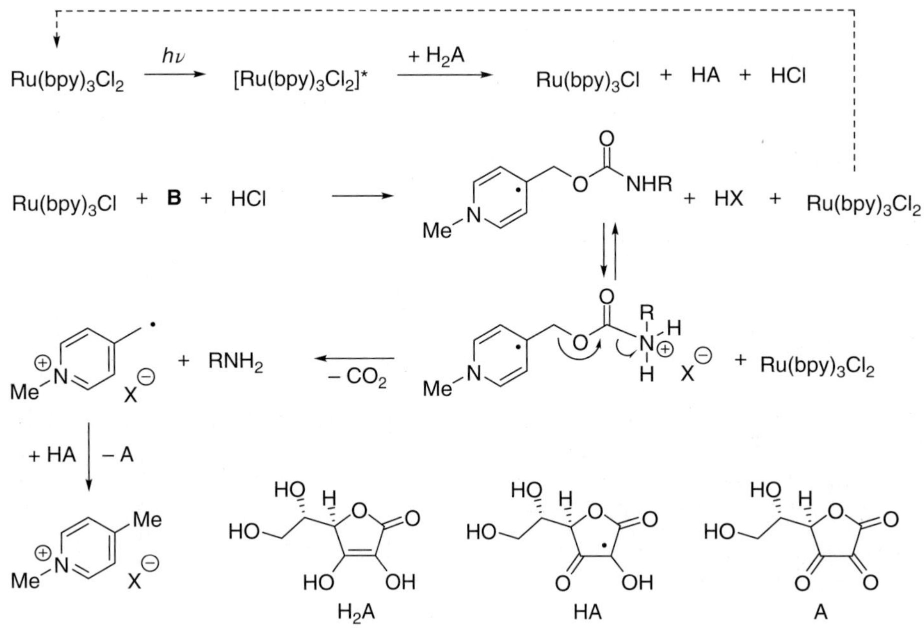

Problem 6.72 Explain the following reaction (TPT, 2,4,6-triphenylpyrilium tetrafluoroborate) [79].

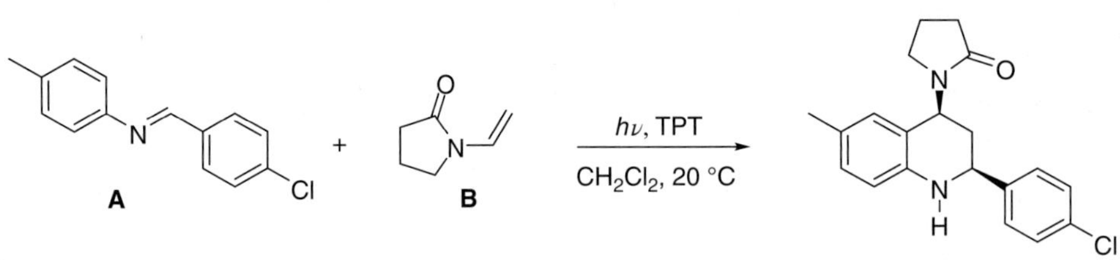

Answer

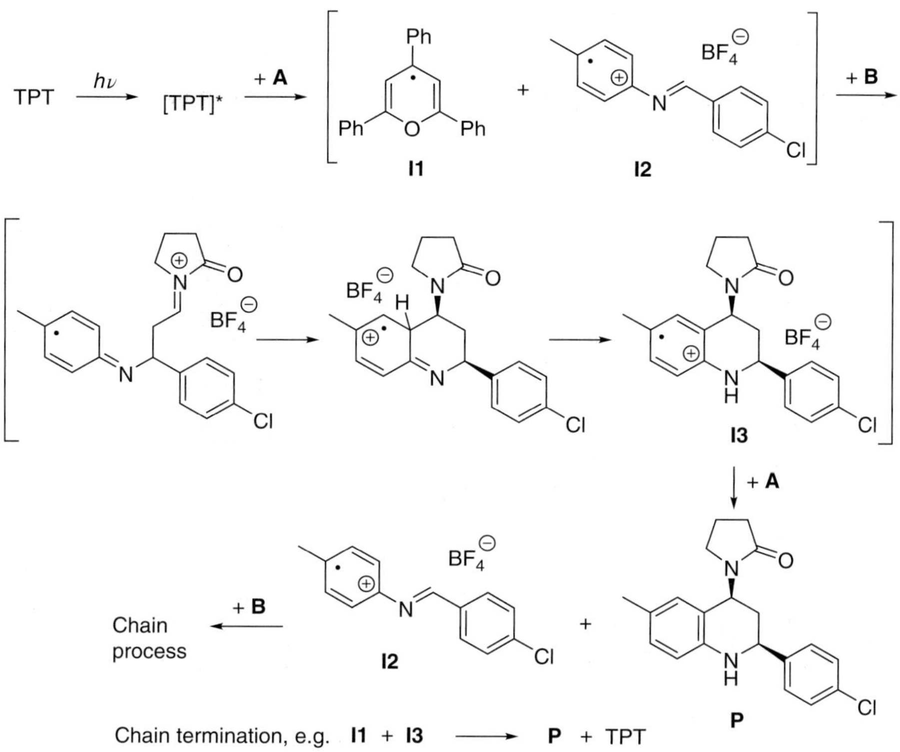

Problem 6.73 Propose a mechanism (5 mol% $Ru(bpy)_3Cl_2$ as a sensitizer, visible light) of the following reaction [80, 81].

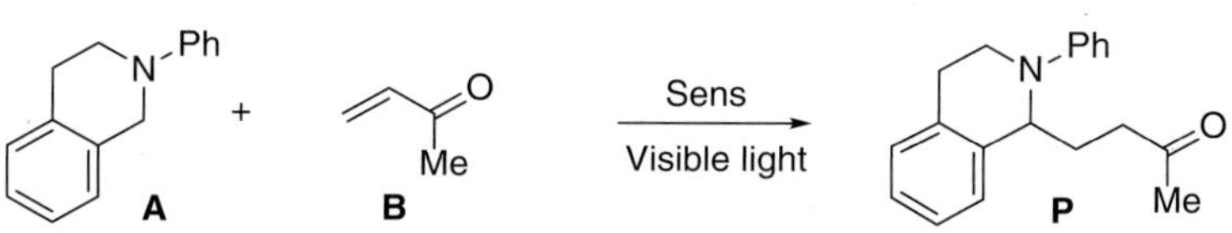

Answer

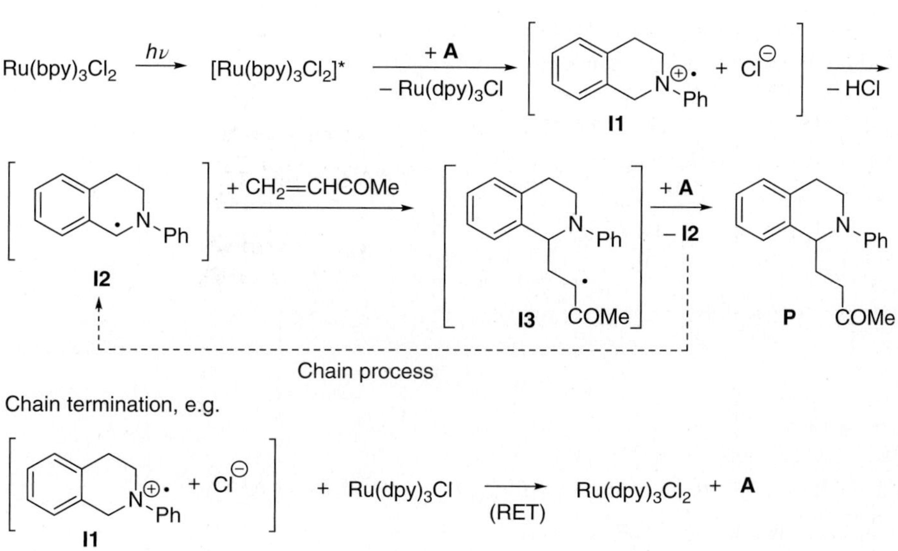

Problem 6.74 What is product **P** of the visible-light-induced cyclization of **A** (sensitizer: 10-mesityl-9-methylacridinium perchlorate)? [82]

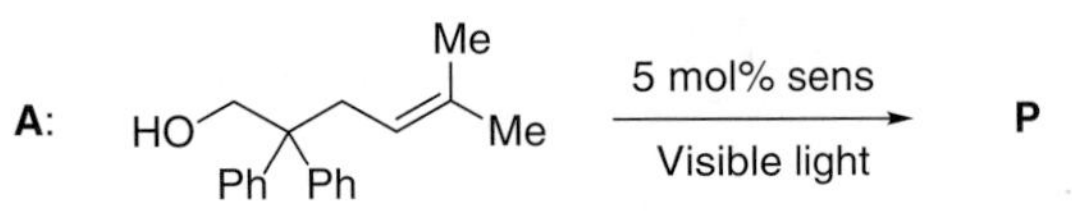

Answer

Under protic acid conditions, alkene **A** adds a proton generating the tertiary carbenium ion intermediate **I1,** which undergoes cyclization into intermediate **I2**. Upon deprotonation **I2** produces tetrahydropyrane **Q**. **Q** corresponds to the Markovnikov product.

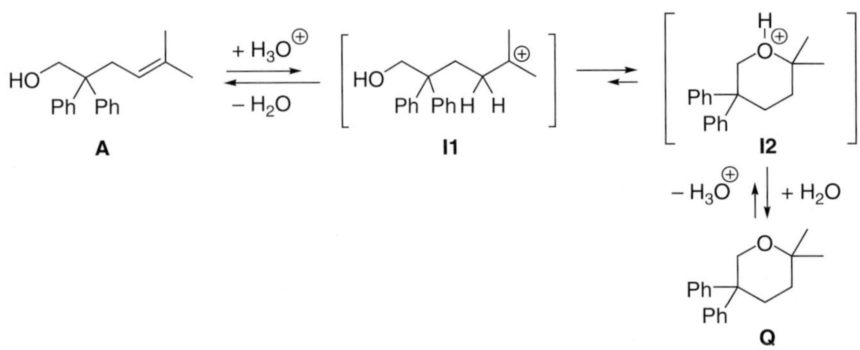

Under visible light irradiation the acridimium perchlorate sensitizer induces a photoinduced electron transfer (PET) from the alkene **A** giving the corresponding radical-cation **I3** in which the positive charge resides on two carbon centers as shown with limiting structures **I3** and **I3′**. This permits an *exo*-trig cyclization, a process favored entropically and which leads to the five-membered ring radical-cation **I4** instead of the six-membered ring isomeric radical-cation **I5** (resulting from an *endo*-trig cyclization; Balwin's rules). Note that the 5–6 kcal mol^{-1} of ring strain in **I4** is compensated by the difference in stability between a tertiary radical (**I4**) and a primary radical (**I5**). In fact, the cyclization is controlled kinetically, and only a part of the stability difference between **I4** and **I5** contributes to the difference of activation enthalpy of these irreversible reactions (oxonium ions are much more stable than carbenium ions, formation of a C—O bond at the expense of a C=C double bond that is converted into a single C—C bond).

Radical-cation **I4** is then isomerized into radical-cation **I6** through hydrogen atom migration. **I6** starts a chain reaction with **A** by ET giving product **P** and radical-cation intermediate **I3**.

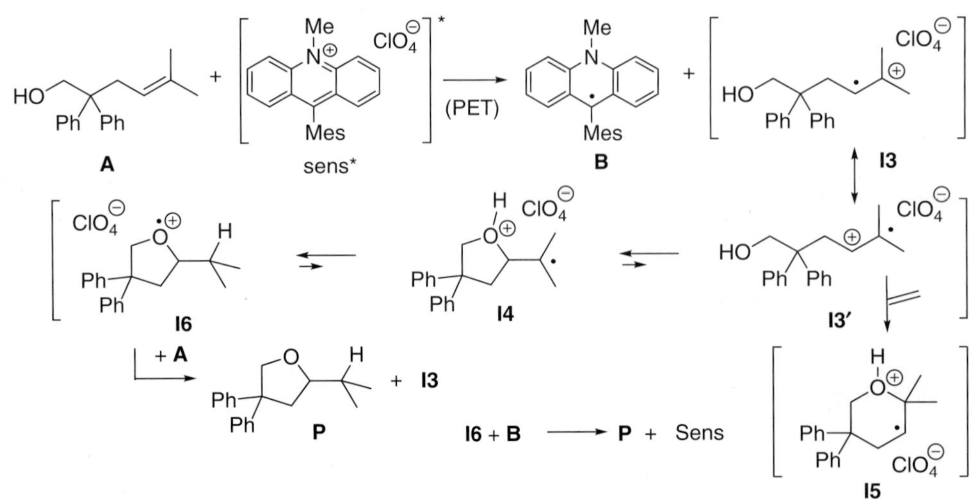

Problem 6.75 Propose a mechanism for the oxidative conversion of thiobenzamide **A** into benzothiazole **P** [83].

Answer

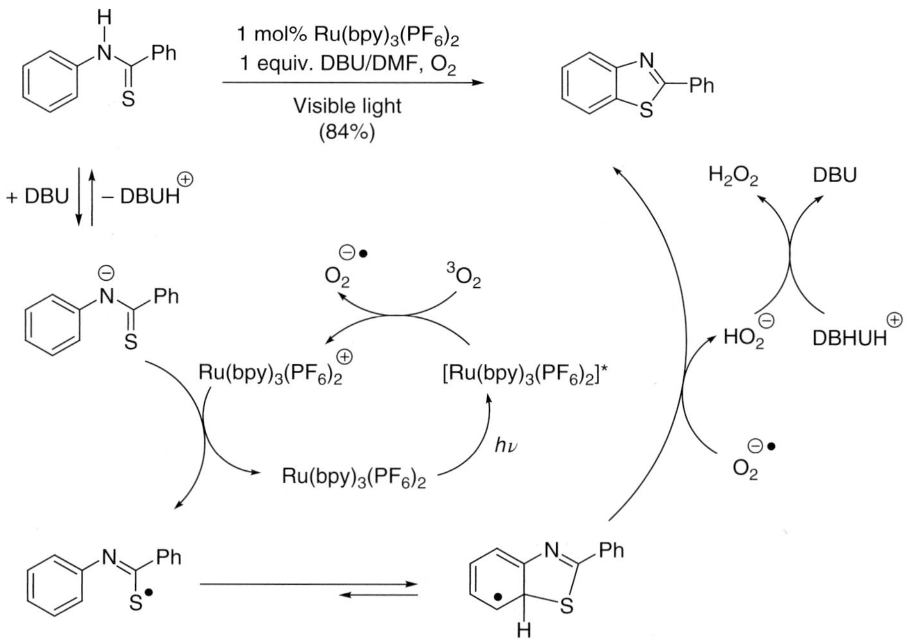

DBU = 1,8-diazabicyclo[5.4.0]undec-7-ene

Problem 6.76 Oxidative 1,1′-biaryl coupling of β-naphthols can be achieved under visible light irradiation in the presence a Ru(II) complex as sensitizer, and $Co(acac)_3$ as oxidant. Propose a mechanism for this reaction [84, 85].

Answer

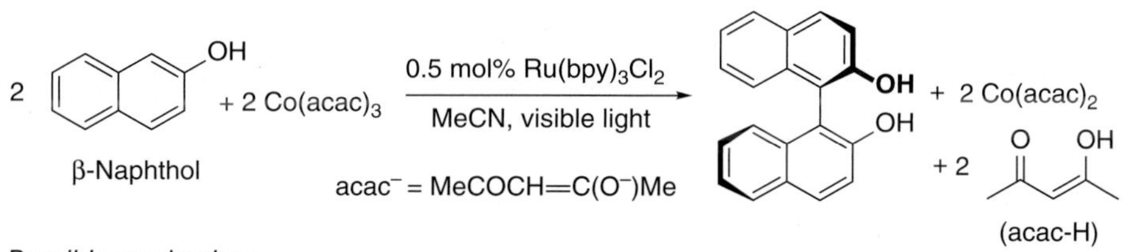

Possible mechanism:

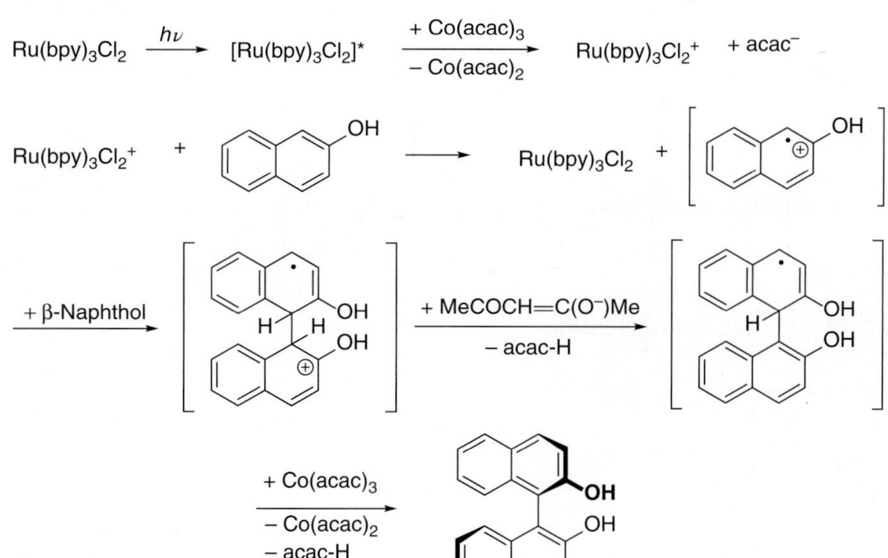

Problem 6.77 UV irradiation of **A** in methanol gives two products **P** and **Q**. What are they? [86]

A: 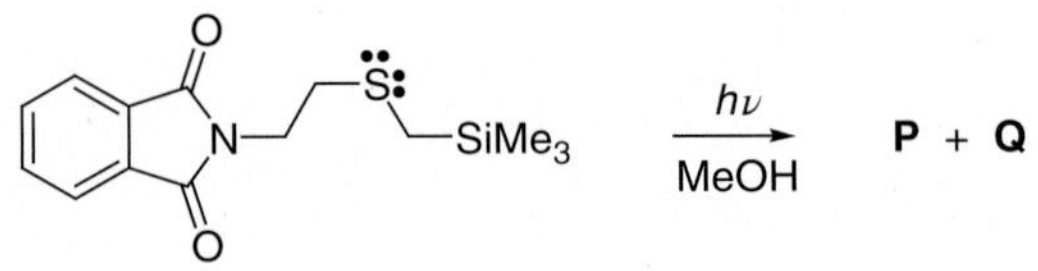

Answer

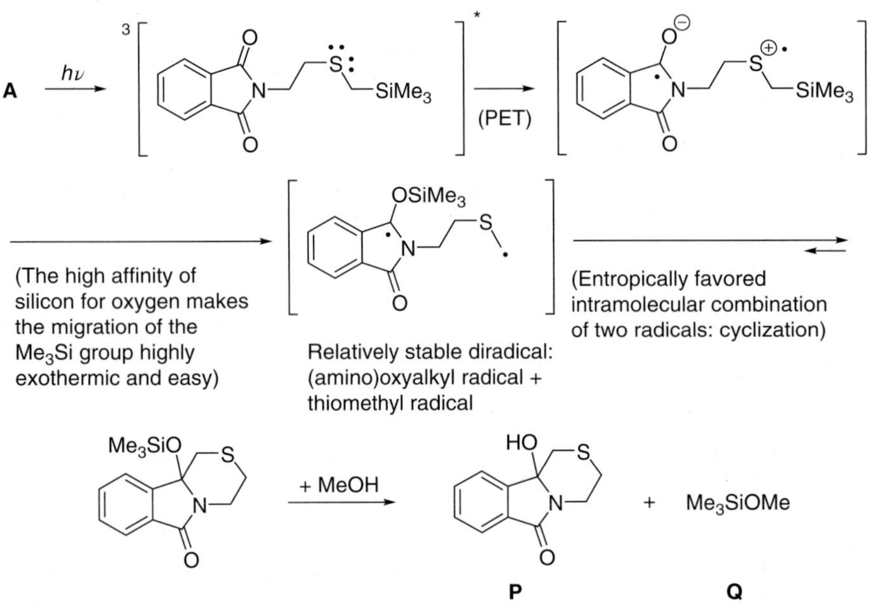

Problem 6.78 What are the main products of the following photoreactions that use visible light, 2.5 mol% *fac*-Ir(ppy)$_3$ as sensitizer, 10 equiv. of Bu$_3$N, and 10 equiv. of formic acid? [87]

Answer

fac-Ir(ppy)$_3$ = [Ir$^{(III)}$]

$$[\mathrm{Ir}^{(III)}] \xrightarrow{h\nu} [\mathrm{Ir}^{(III)}]^* \xrightarrow{+\ \mathrm{Bu_3N:}} [\mathrm{Ir}^{(II)}]^{\ominus} + \mathrm{Bu_3N^{\bullet\oplus}}$$

$$[\mathrm{Ir}^{(II)}]^{\ominus} + \mathrm{R{-}I} \longrightarrow [\mathrm{R}^{\bullet}] + \mathrm{I}^{\ominus} + [\mathrm{Ir}^{(III)}]$$

$$[\mathrm{R}^{\bullet}] + \mathrm{Bu_3N^{\bullet\oplus}} \longrightarrow \mathrm{R{-}H} + \mathrm{MeCH_2CH_2CH{=}\overset{\oplus}{N}Bu_2}$$

↓ (Radical rearrangement)

$$[\mathrm{R'^{\bullet}}] + \mathrm{Bu_3N^{\bullet\oplus}} \longrightarrow \mathrm{R'{-}H} + \mathrm{MeCH_2CH_2CH{=}\overset{\oplus}{N}Bu_2}$$

$$\mathrm{MeCH_2CH_2CH{=}\overset{\oplus}{N}Bu_2\ I^{\ominus}} + \mathrm{HCOOH} \longrightarrow \mathrm{CO_2} + \mathrm{Bu_3\overset{\oplus}{N}H\ I^{\ominus}}$$

(a)

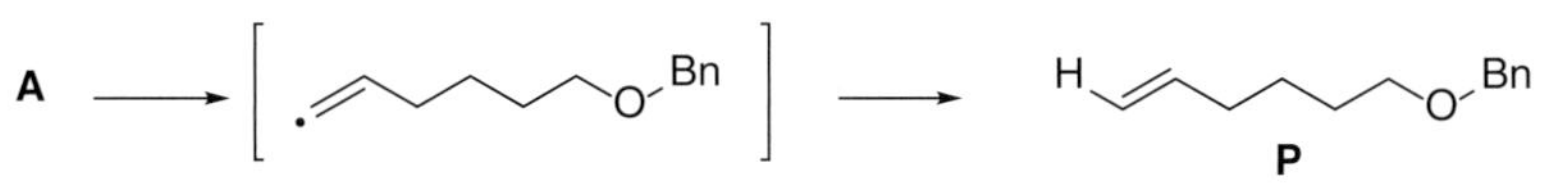

(b)

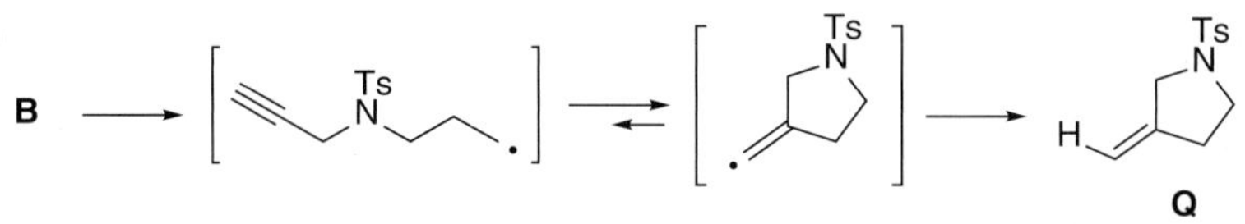

Exo-dig cyclization of the initial alkyl radical: intramolecular addition of the alkyl radical onto the alkyne moiety generating an alkenyl radical intermediate with a five-membered ring. This cyclization is preferred to the alternative cyclization that would produce a six-membered alkenyl radical, for entropic reasons and conformational effect in the transition state of the cyclization.

Problem 6.79 Lucigenin (dimethyl diacridylium salts **A**) and its derivatives react with H_2O_2 in alkaline solution with emission of green to blue light [88, 89]. Propose a mechanism for this reaction.

A:

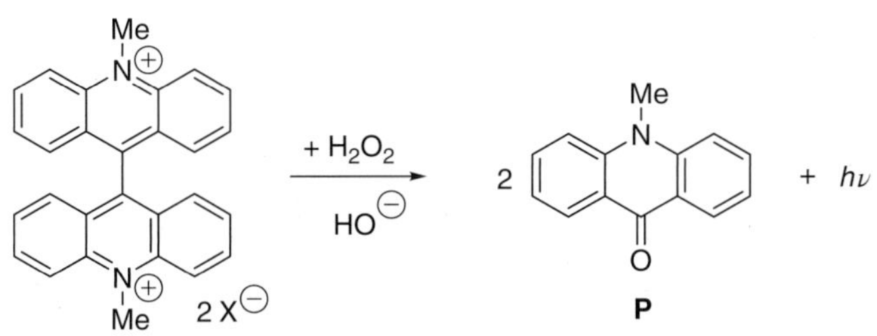

Answer

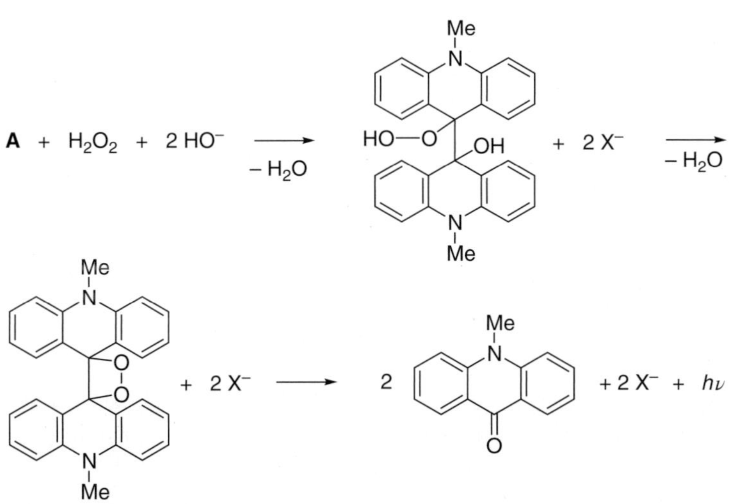

Additional Material

We present an alternative model to interpret the alkyl substituent bathochromic shifts. One can represent oneself the valence electronically excited state of ethylene resulting from the $\pi \rightarrow \pi^*$ transition as a "parallel 1,2-diradical" in which the two electrons repel each other much more (c.168 kcal mol^{-1} above the ground state of ethylene) than in the perpendicular 1,2-diradical, the transition state of the rotation about the π(C=C) bond of ethylene ($\Delta^{\ddagger}H = 64$ kcal mol^{-1}, Section 4.5.1). The electron residing in the highest occupied molecular orbital (HOMO) ($\pi(CH_2{=}CH_2)$) is bonding whereas the electron residing in the lowest unoccupied molecular orbital (LUMO) ($\pi^*(CH_2{=}CH_2)$) is antibonding. The two electrons in perpendicular 1,2-diradical are nonbonding. In $\pi^*(CH_2{=}CH_2)$ of parallel ethylene the electron cannot circulate around the two carbon nuclei. Half of the time it circulates around one of the carbon atoms, and the other half of the time about the other carbon atom. Thus,

$\psi^*_1(CH_2{=}CH_2)$ can be represented by the two limiting structures **7** ↔ **7′**. With this simple model one explains the highly polarizable singlet excited states of alkenes [90] and the bathochromic shifts induced by alkyl and other substituents (stabilizing substituent effects on radicals, carbenium ions, and carbanions) on the $\pi \rightarrow \pi^*$ transition of ethylene (Table 6.1). For instance, the UV absorption spectrum of (*E*)-stilbene shows $\lambda_{max} = 295$ nm ($\Delta E = 97$ kcal mol^{-1}). The UV absorption spectrum of (*Z*)-stilbene shows $\lambda_{max} = 280$ nm ($\Delta E = 102$ kcal mol^{-1}). This is due to the fact that conjugation between the phenyl substituents and the ethylene moiety is reduced in the partially twisted (*Z*)-stilbene (gauche interactions between the Ph rings) compared with planar (*E*)-stilbene. This simple model can be applied to all other chromophores as shown with the Woodward–Fieser rules (Table 6.3).

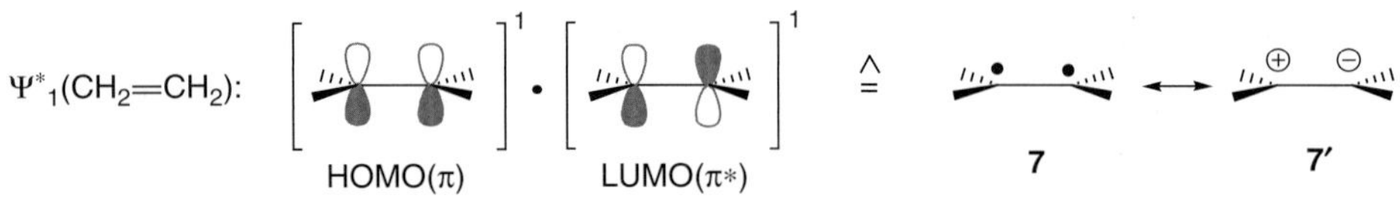

Scheme 6.2 Simple model for the valence electronically excited state of alkenes.

References

1 Turro, N.J., Ramamurphy, V., and Scaiano, J.C. (2009). *Modern Molecular Photochemistry of Organic Molecules*, 1–1084. Sausalito, CA: University Sciene Books.

2 Kropp, P.J. (1995). *CRC Handbook of Organic Photochemistry and Photobiology* (ed. W.M. Horspol and P.-S. Song), 105. Boca Raton, FL: CRC Press.

3 Beccalli, E.M. and Marchesini, A. (1984). Synthesis of strained cycloalkene derivatives. *Gazz. Chim. Ital.* 114: 287–288.

4 Zimmerman, H.E., Givens, R.S., and Pagni, R.M. (1968). Control of photochemical reaction pathways by excited-state multiplicity. Mechanistic and exploratory organic photochemistry. 34. *J. Am. Chem. Soc.* 90: 4191–4193.

5 Zimmerman, H.E., Givens, R.S., and Pagni, R.M. (1968). Photochemistry of benzobarrelene. Mechanistic and exploratory photochemistry. 35. *J. Am. Chem. Soc.* 90: 6096–6108.

6 Bender, C.O., Brooks, D.W., Cheng, W. et al. (1978). Polar substituents in barrelene photochemistry - Mechanistic aspects of photochemistry of 2-cyanobenzobarrelene. *Can. J. Chem.* 56: 3027–3037.

7 Pàquette, L.A., Cottrell, D.M., Snow, R.A. et al. (1975). Substituent control of regiospecific pathways in di-pi-methane photorearrangements which utilize benzo-vinyl bonding schemes. *J. Am. Chem. Soc.* 97: 3275–3276.

8 Paquette, L.A., Cottrell, D.M., and Snow, R.A. (1977). Control by meta substituents of benzo-vinyl bonding options during triplet sensitized photorearrangement of benzonorbornadienes and anti-7,8-benzotricyclo[4.2.2.02,5]deca-3,7,9-trienes. *J. Am. Chem. Soc.* 99: 3723–3733.

9 Santiago, C., Houk, K.N., Snow, R.A., and Paquette, L.A. (1976). Bridging regioselectivity in triplet-sensitized di-pi-methane photorearrangements of ortho-substituted benzonorbornadienes – case for importance of benzene HOMO and LUMO polarization. *J. Am. Chem. Soc.* 98: 7443–7445.

10 Griffin, G.W., Covell, J., Petterso, R.C. et al. (1965). Photoisomerization of cyclopropane derivatives. Photointerconversion of propenes and cyclopropanes. *J. Am. Chem. Soc.* 87: 1410–1411.

11 Doyle, T.D., Filipescu, N., Benson, W.R., and Banes, D. (1970). Photocyclization of α,α,-diethyl-4,4′-stilbenediol – isolation of a stable tautomer of elusive dihydrophenanthrenes. *J. Am. Chem. Soc.* 92: 6371–6372.

12 Pollard, R. and Wan, P. (1993). Synthetic applications of diaryl ether photochemistry – a review. *Org. Prep. Proced. Int.* 25: 1–13.

13 Turro, N.J., Leermake, P.A., Wilson, H.R. et al. (1965). Photochemistry of 1,3-cyclobutanediones. Decomposition modes and chemical intermediates. *J. Am. Chem. Soc.* 87: 2613–2615.

14 Lewis, F.D., Johnson, R.W., and Johnson, D.E. (1974). Conformational control of photochemical behavior – competitive alpha cleavage and gamma-hydrogen abstraction of alkyl phenyl ketones. *J. Am. Chem. Soc.* 96: 6090–6099.

15 Paulson, S.E., Liu, D.L., Orzechowska, G.E. et al. (2006). Photolysis of heptanal. *J. Org. Chem.* 71: 6403–6408.
16 Solomek, T., Stacko, P., Veetil, A.T. et al. (2010). Photoenolization-induced oxirane ring opening in 2,5-dimethylbenzoyl oxiranes to form pharmaceutically promising indanone derivatives. *J. Org. Chem.* 75: 7300–7309.
17 Wolff, S., Smith, A.B., Schreiber, W.L., and Agosta, W.C. (1972). Photochemistry of cyclopentenones – hydrogen abstraction by beta-carbon atom. *J. Am. Chem. Soc.* 94: 7797–7806.
18 Schuster, D.I., Brown, R.H., and Resnick, B.M. (1978). Photochemistry of ketones in solution. 53. Stereospecific triplet-state photorearrangements of chiral 2-cyclohexenones – type-a lumiketone rearrangement and phenyl migrations. *J. Am. Chem. Soc.* 100: 4504–4512.
19 Givens, R.S., Oettle, W.F., Coffin, R.L., and Carlson, R.G. (1971). Mechanistic studies on organic photochemistry. 3. Photochemistry of bicyclo[2.2.2]octenone and benzobicyclo[2.2.2]octadienone. *J. Am. Chem. Soc.* 93: 3957–3962.
20 Schwager, L. and Vogel, P. (1980). Interactions between non-conjugated chromopheres. 12. The photochemistry of 5,6-dimethylidene-2-norbornanone - synthesis and Diels–Alder reactivity of 2,3-dimethylidenebicyclo-[2.1.1]hexane. *Helv. Chim. Acta* 63: 1176–1180.
21 Wilzbach, K.E., Harkness, A.L., and Kaplan, L. (1968). Photochemical rearrangement of benzene-1,3,5-D3. *J. Am. Chem. Soc.* 90: 1116–1118.
22 Kaplan, L., Wilzbach, K.E., and Pavlik, J.W. (1972). Photohydration of pyridinium ions. *J. Am. Chem. Soc.* 94: 3283.
23 Penkett, C.S. and Simpson, I.D. (1999). Photosolvolysis reactions of 3-alkoxypyridinium tetrafluoroborate salts. *Tetrahedron* 55: 6183–6204.
24 Dennis, N., Katritzky, A.R., and Wilde, H. (1976). 1,3-Dipolar character of 6-membered aromatic rings. 27. Photochemically induced valence bond tautomerism and dimerization of 3-oxido-1-phenylpyridinium. *J. Chem. Soc., Perkin Trans.* 1: 2338–2343.
25 Su, M.-D. (2006). Photochemical isomerization reactions of cyanopyrroles: a theoretical study. *J. Phys. Chem. A* 110: 12653–12661.
26 Couture, A. and Lablache, A. (1971). Photochemical study. 3. Photochemistry of thiophene. *Tetrahedron* 27: 1059–1068.
27 Guerrini, G., Taddei, M., and Ponticelli, F. (2011). Synthesis of functionalized quinolines and benzo[2,7]naphthyridines based on a photo-Fries rearrangement. *J. Org. Chem.* 76: 7597–7601.
28 Benn, R., Dreeskamp, H., Schuchmann, H.P., and Vonsonntag, C. (1979). Photolysis of aromatic ethers. *Z. Naturforsch., B* 34B: 1002–1009.
29 Li, Y.Z. and Schuster, G.B. (1987). Photochemistry of 9-diazo-3,6-diazafluorene – through-space or through-bond transmission of electronic effects. *J. Org. Chem.* 52: 3975–3979.
30 Miranda, M.A., PerezPrieto, J., FontSanchis, E. et al. (1997). Laser flash, laser-drop, and lamp photolysis of 1,3-dichloro-1,3-diphenylpropane. One- versus two-photon reaction pathways. *J. Org. Chem.* 62: 5713–5719.
31 Vernin, G., Jauffred, R., Ricard, C. et al. (1972). Homolytic aromatic substitution by heterocyclic free-radicals. 2. Reaction of thiazol-2-yl and benzothiazol-2-yl radicals with aromatic-compounds. *J. Chem. Soc., Perkin Trans. 2* 1145–1150.
32 Kampmeier, J.A. and Hoffmeister, E. (1962). Photolysis of 1,2-diiodobenzene - a photochemical source of benzyne. *J. Am. Chem. Soc.* 84: 3787–3788.
33 Chapyshev, S.V. and Tomioka, H. (2003). Photochemical transformations of quintet *m*-phenylenedinitrenes. *Bull. Chem. Soc. Jpn.* 76: 2075–2089.
34 Yokoyama, M., Hirano, S., Hachiya, T., and Togo, H. (1994). Thermolysis and photolysis of sugar diazides. *Chem. Lett.* 1107–1110.
35 Praly, J.P., Distefano, C., Descotes, G., and Faure, R. (1994). Photolysis of sugar anomeric diazides - sugar-derived tetrazoles as evidences for a major nitrene decomposition pathway. *Tetrahedron Lett.* 35: 89–92.
36 Alawode, O.E., Robinson, C., and Rayat, S. (2011). Clean photodecomposition of 1-methyl-4-phenyl-1H-tetrazole-5(4H)-thiones to carbodiimides proceeds via a biradical. *J. Org. Chem.* 76: 216–222.
37 Bégué, D., Qiao, G.G.-H., and Wentrup, C. (2012). Nitrile imines: matrix isolation, IR spectra, structures, and rearrangement to carbodiimides. *J. Am. Chem. Soc.* 134: 5339–5350.
38 Moon, S. and Ganz, C.R. (1968). Photochemistry of homoconjugated cyclic dienes. I. Photoisomerization of 1,4-cyclooctadiene. *Tetrahedron Lett.* 6275–6277.

39 Gleiter, R., Sander, W., Irngartinger, H., and Lenz, A. (1982). A simple path to tricyclo[5.3.0.02,8]decane and its derivatives. *Tetrahedron Lett.* 23: 2647–2650.

40 Cookson, R.C., Crundwell, E., Hill, R.R., and Hudec, J. (1964). Photochemical cyclisation of Diels–Alder adducts. *J. Chem. Soc.* 3062–3075.

41 Becker, D. and Haddad, N. (1986). About the stereochemistry of intramolecular [2+2] photocycloadditions. *Tetrahedron Lett.* 27: 6393–6396.

42 Shepard, M.S. and Carreira, E.M. (1997). Enantioselective allene/enone photocycloadditions: the use of an inexpensive optically active 1,3-disubstituted allene. *Tetrahedron* 53: 16253–16276.

43 Kwak, Y.S. and Winkler, J.D. (2001). Synthesis of 6-aza-bicyclo[3,2,1]octan-3-ones via vinylogous imide photochemistry: an approach to the synthesis of the hetisine alkaloids. *J. Am. Chem. Soc.* 123: 7429–7430.

44 Dilling, W.L. (1966). Intramolecular photochemical cycloaddition reactions of nonconjugated olefins. *Chem. Rev.* 66: 373–393.

45 Eaton, P.E. (1964). Photocondensation reactions of unsaturated ketones. *Tetrahedron Lett.* 3695–3698.

46 Eaton, P.E. (1968). Photochemical reactions of simple alicyclic enones. *Acc. Chem. Res.* 1: 50–57.

47 Corey, E.J., Bass, J.D., Le Mathieu, R., and Mitra, R.B. (1964). Study of photochemical reactions of 2-cyclohexenones with substituted olefins. *J. Am. Chem. Soc.* 86: 5570–5583.

48 Meyers, A.I. and Fleming, S.A. (1986). Efficient asymmetric (2+2) photocycloaddition leading to chiral cyclobutanes - application to the total synthesis of (−)-grandisol. *J. Am. Chem. Soc.* 108: 306–307.

49 Sulikowski, M.M., Davies, G.E.R.E., and Smith, A.B. (1992). Avermectin milbemycin synthetic studies. 7. An approach to the southern-hemisphere of milbemycin alpha-1. *J. Chem. Soc., Perkin Trans. 1* 979–989.

50 Abad, A., Arno, M., Marin, M.L., and Zaragoza, R.J. (1993). Spongian pentacyclic diterpenes - stereoselective synthesis of aplyroseol-1, aplyroseol-2 and deacetylaplyroseol-2. *J. Chem. Soc., Perkin Trans. 1* 1861–1867.

51 Piet, D., de Bruijn, S., Sum, J., and Lodder, G. (2005). New molecular entities via intermolecular meta photocycloaddition. *J. Heterocyclic Chem.* 42: 227–231.

52 Al-Jalal, N. and Gilbert, A. (1990). Substituent effects on the photocycloaddition reactions of anisoles to acrylonitriles. *Recl. Trav. Chim. Pays-Bas* 109: 21–25.

53 Yang, N.C., Masnovi, J., and Chiang, W. (1979). Chemistry of exciplexes. 8. Role of conservation of orbital symmetry in cycloadditions of olefins to aromatic-hydrocarbons. *J. Am. Chem. Soc.* 101: 6465–6466.

54 Yang, N.C., Masnovi, J., Chiang, W. et al. (1981). Chemistry of exciplexes. 11. Photocycloaddition of 1,3-cyclohexadiene to polynuclear aromatic-hydrocarbons. *Tetrahedron* 37: 3285–3300.

55 Yang, N.C.C., Gan, H., Kim, S.S. et al. (1990). Stereoselectivity in 4-Pi/S+4Pi/S photocycloaddition of 1,3-cyclohexadiene to arenes. *Tetrahedron Lett.* 31: 3825–3828.

56 Friedrich, L.E. and Bower, J.D. (1973). Detection of an oxetene intermediate in photoreaction of benzaldehyde with 2-butyne. *J. Am. Chem. Soc.* 95: 6869–6870.

57 Toki, S., Shima, K., and Sakurai, H. (1965). Organic photochemical reactions. I. Synthesis of substituted oxetanes by photoaddition of aldehydes to furans. *Bull. Chem. Soc. Jpn.* 38: 760–762.

58 Shima, K. and Sakurai, H. (1966). Organic photochemical reactions. 4. Photoaddition reactions of various carbonyl compounds to furan. *Bull. Chem. Soc. Jpn.* 39: 1806–1808.

59 D'Auria, M., Emanuele, L., and Racioppi, R. (2004). Regio- and stereoselectivity in the Paterno–Buchi reaction on furan derivatives. *J. Photochem. Photobiol., A* 163: 103–111.

60 D'Auria, M., Racioppi, R., and Viggiani, L. (2010). Paterno-Buchi reaction between furan and heterocyclic aldehydes: oxetane formation vs. metathesis. *Photochem. Photobiol. Sci.* 9: 1134–1138.

61 Griesbeck, A.G., Franke, M., Neudoerfl, J., and Kotaka, H. (2011). Photocycloaddition of aromatic and aliphatic aldehydes to isoxazoles: cycloaddition reactivity and stability studies. *Beilstein J. Org. Chem.* 7: 127–134.

62 Fischer, G., Fritz, H., and Prinzbach, H. (1986). Photochemical-transformations. 66. An intramolecular imine ene photo-[2+2]-cycloaddition reaction. *Tetrahedron Lett.* 27: 1269–1272.

63 Hagenbuch, J.P., Birbaum, J.L., Metral, J.L., and Vogel, P. (1982). Interaction between non-conjugated chromophores. 17. Effects of remote substitution on the photooxidation of exocyclic *S-cis*-butadienes grafted onto bicyclo[2.2.1]heptanes - thermal and rhodium catalyzed rearrangements of 3,6-dihydro-1,2-dioxines. *Helv. Chim. Acta* 65: 887–895.

64 Montagnon, T., Tofi, M., and Vassilikogiannakis, G. (2008). Using singlet oxygen to synthesize polyoxygenated natural products from Furans. *Acc. Chem. Res.* 41: 1001–1011.

65 Wasserman, H.H. and Doumaux, A.R. (1962). An unusual intramolecular Diels–Alder reaction in photooxidation of A (Furano) heterocyclophane. *J. Am. Chem. Soc.* 84: 4611–4612.
66 Schaap, A.P., Recher, S.G., Faler, G.R., and Villasenor, S.R. (1983). Evidence for a perepoxide intermediate in the 1,2-cycloaddition of singlet oxygen to adamantylideneadamantane – nucleophilic oxygen atom transfer to sulfoxides. *J. Am. Chem. Soc.* 105: 1691–1693.
67 Schenck, G.O. and Schulte-Elte, K.H. (1958). Photosensibilisierte synthese tertiarer allylhydroperoxyde aus symmetrisch tetrasubstituierten athylenen. *Liebigs Ann. Chem.* 618: 185–193.
68 Alberti, M.N. and Orfanopoulos, M. (2006). Stereoelectronic and solvent effects on the allylic oxyfunctionalization of alkenes with singlet oxygen. *Tetrahedron* 62: 10660–10675.
69 Adam, W. and Delfierro, J. (1978). Cyclic peroxide series. 61. Synthesis of alpha-hydroperoxy esters by singlet oxygenation of ketene acetals. *J. Org. Chem.* 43: 1159–1161.
70 Wasserman, H.H. and Pickett, J.E. (1985). The fluoride-ion effect in the reactions of singlet oxygen with enols. *Tetrahedron* 41: 2155–2162.
71 Adam, W., Saha-Moller, C.R., and Schambony, S.B. (1999). A highly diastereoselective dioxetane formation by the hydroxy-directed [2+2] cycloaddition of singlet oxygen to a chiral allylic alcohol. *J. Am. Chem. Soc.* 121: 1834–1838.
72 Celaje, J.A., Zhang, D., Guerrero, A.M., and Selke, M. (2011). Chemistry of *trans*-resveratrol with singlet oxygen: [2+2] addition, [4+2] addition, and formation of the phytoalexin moracin M. *Org. Lett.* 13: 4846–4849.
73 Lu, Z., Shen, M., and Yoon, T.P. (2011). [3+2] Cycloadditions of aryl cyclopropyl ketones by visible light photocatalysis. *J. Am. Chem. Soc.* 133: 1162–1164.
74 Cossy, J., Ranaivosata, J.L., Bellosta, V. et al. (1995). Reductive oxa ring opening of 7-oxabicyclo[2.2.1]heptan-2-ones. Synthesis of C-alpha-galactosides of carbapentopyranoses. *J. Org. Chem.* 60: 8351–8359.
75 Lange, G.L., Decicco, C.P., Willson, J., and Strickland, L.A. (1989). Ring expansions of [2+2] photoadducts - potential applications in the synthesis of triquinane and taxane skeletons. *J. Org. Chem.* 54: 1805–1810.
76 Larraufie, M.H., Pellet, R., Fensterbank, L. et al. (2011). Visible-light-induced photoreductive generation of radicals from epoxides and aziridines. *Angew. Chem. Int. Ed.* 50: 4463–4466.
77 Tucker, J.W., Nguyen, J.D., Narayanam, J.M. et al. (2010). Tin-free radical cyclization reactions initiated by visible light photoredox catalysis. *Chem. Commun.* 46: 4985–4987.
78 Edson, J.B., Spencer, L.P., and Boncella, J.M. (2011). Photorelease of primary aliphatic and aromatic amines by visible-light-induced electron transfer. *Org. Lett.* 13: 6156–6159.
79 Zhang, W., Guo, Y.P., Liu, Z.G. et al. (2005). Photochemically catalyzed Diels–Alder reaction of arylimines with *N*-vinylpyrrolidinone and *N*-vinylcarbazole by 2,4,6-triphenylpyrylium salt: synthesis of 4-heterocycle-substituted tetrahydroquinoline derivatives. *Tetrahedron* 61: 1325–1333.
80 Kohls, P., Jadhav, D., Pandey, G., and Reiser, O. (2012). Visible light photoredox catalysis: generation and addition of *N*-aryltetrahydroisoquinoline-derived alpha-amino radicals to Michael acceptors. *Org. Lett.* 14: 672–675.
81 Espelt, L.R., Wiensch, E.M., and Yoon, T.P. (2013). Bronsted acid cocatalysts in photocatalytic radical addition of alpha-amino C–H bonds across Michael acceptors. *J. Org. Chem.* 78: 4107–4114.
82 Hamilton, D.S. and Nicewicz, D.A. (2012). Direct catalytic anti-Markovnikov hydroetherification of alkenols. *J. Am. Chem. Soc.* 134: 18577–18580.
83 Cheng, Y., Yang, J., Qu, Y., and Li, P. (2012). Aerobic visible-light photoredox radical C–H functionalization: catalytic synthesis of 2-substituted benzothiazoles. *Org. Lett.* 14: 98–101.
84 Hamada, T., Ishida, H., Usui, S. et al. (1993). A novel photocatalytic asymmetric-synthesis of (*R*)-(+)-1,1′-bi-2-naphthol derivatives by oxidative coupling of 3-substituted-2-naphthol with delta-[Ru(Menbpy)3]2+ [Menbpy = 4,4′-Di(1R,2S,5R)-(−)-menthoxycarbonyl-2,2′-bipyridine], which possesses molecular helicity. *J. Chem. Soc., Chem. Commun.* 909–911.
85 Hamada, T., Ishida, H., Usui, S. et al. (1994). Enantioselective and photocatalytic oxidation of 1,1′-bi-2-naphthol with a chiral ruthenium complex which includes molecular helicity. *J. Mol. Catal.* 88: L1–L5.
86 Cho, D.W., Yoon, U.C., and Mariano, P.S. (2011). Studies leading to the development of a single-electron transfer (SET) photochemical strategy for syntheses of macrocyclic polyethers, polythioethers, and polyamides. *Acc. Chem. Res.* 44: 204–215.

87 Nguyen, J.D., D'Amato, E.M., Narayanam, J.M., and Stephenson, C.R.J. (2012). Engaging unactivated alkyl, alkenyl and aryl iodides in visible-light-mediated free radical reactions. *Nat. Chem.* 4: 854–859.

88 Gleu, K. and Petsch, W. (1935). Chemo-luminescence of dimethyl-diacridylium salts. *Angew. Chem. Int. Ed. Engl.* 48: 0057–0059.

89 Mccapra, F. and Richardson, D.G. (1964). The mechanism of chemiluminescence – a new chemiluminescent reaction. *Tetrahedron Lett.* 3167–3172.

90 Wulfman, C.E. and Kumei, S. (1971). Highly polarizable singlet excited states of alkenes. *Science* 172: 1061.

7

Catalytic Reactions

Answers to Problems

Problem 7.1 Give a possible mechanism for the thiazolium-catalyzed acylsilane additions to α,β-unsaturated ketones that produce 1,4-diketones [1].

Answer

The carbene catalyst **A** adds to the acylsilane and promotes a Brook rearrangement that affords the silylated intermediate **B**. The alcohol additive present in the medium desilylates **B**, which produces intermediate **C**. Under normal thiazolium-catalyzed benzoin or Stetter reaction conditions, this nucleophilic intermediate can intercept another aldehyde. However, due to the attenuated electrophilicity of the acylsilane (relative to an aldehyde), the reaction proceeds via the preferred 1,4-addition manifold producing intermediate **D**, which expels the carbene catalyst **A** with delivery of the final dicarbonyl compound **P**.

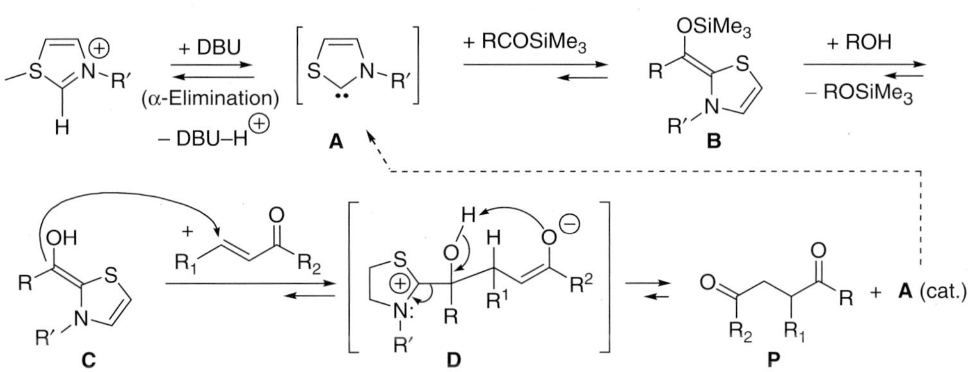

Problem 7.2 What are the main products **P** of the following reactions? [2]

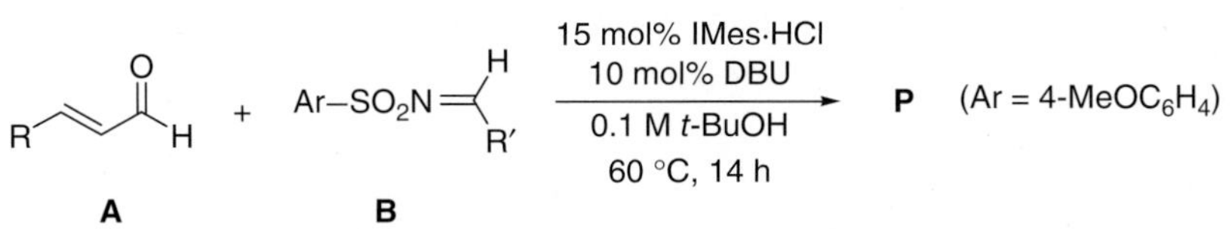

Answer

The catalyst **C** (NHC) adds to enal **A** forming homoenolate **D**. Homoenolate **D** adds then to imine **B** generating the zwitterion **E**. The latter undergoes irreversible intramolecular acyl transfer producing lactam **P** and liberating the catalyst **C**.

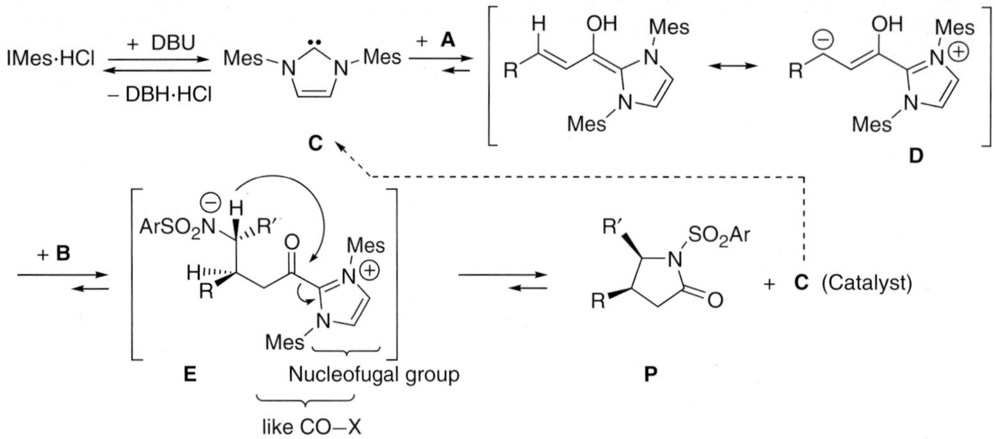

Problem 7.3 Explain the following reaction catalyzed by IMes [3].

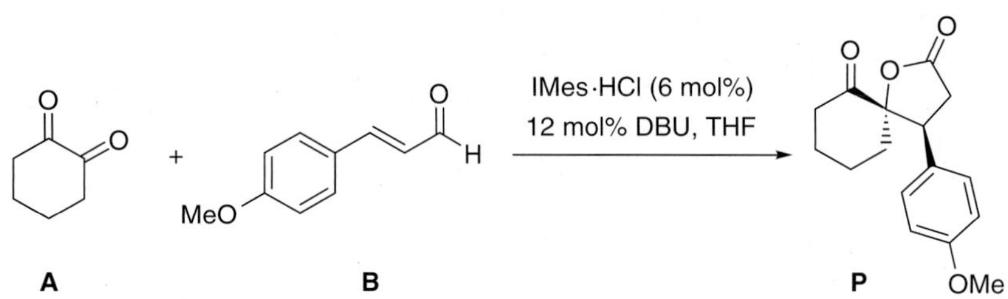

Answer

Enal **B** reacts with the diaminocarbene IMes generating the nucleophilic dienol **C** that adds to the highly electrophilic carbonyl group of cyclohexa-1,2-dione (**A**) producing the zwitterion **D** that subsequently undergoes isomerization into **E**. Then, an intramolecular acyl transfer regenerates the diaminocarbene IMes (catalyst) and produces the spiro γ-butyrolacetone **P**.

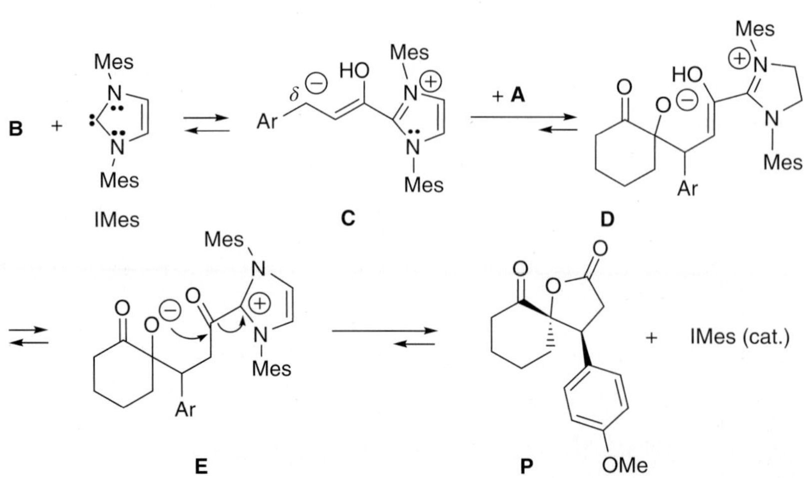

Problem 7.4 What is product **P** when **A** is treated with Me_3P and *t*-amyl-OH? [4]

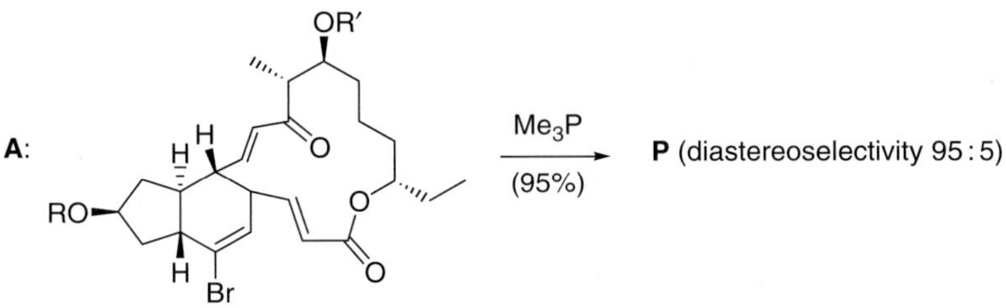

Answer

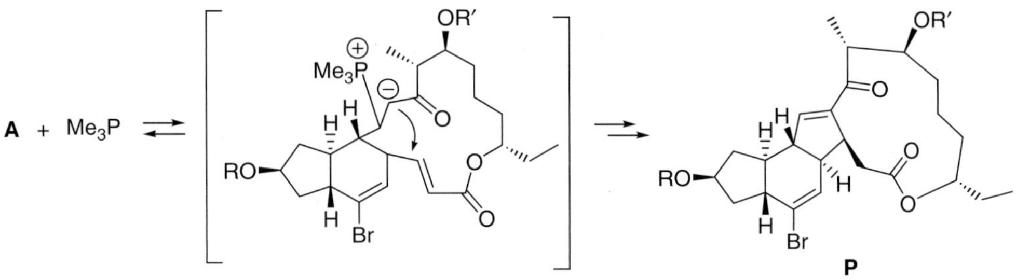

Problem 7.5 Propose a mechanism for the following condensations catalyzed by pyridine [5].

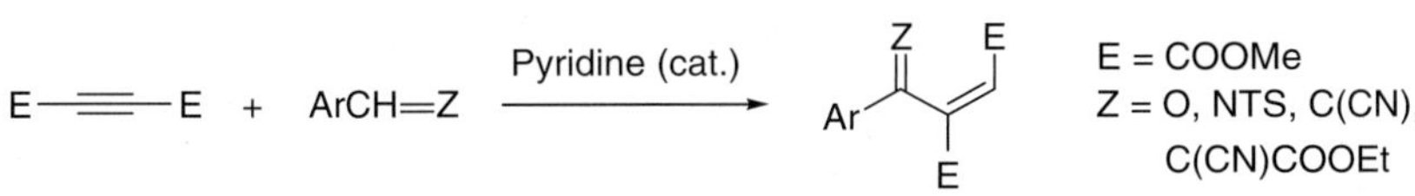

Answer

The nucleophilic pyridine adds to dimethyl acetylenedicarboxylate giving the zwitterionic intermediate **A**. The latter adds to ArCH=Z giving **B**. Protonation and deprotonation convert **B** into **C**. Rotation gives **D**. 1,4-Elimination of pyridine from **D** furnishes product **P** and liberates pyridine (catalyst).

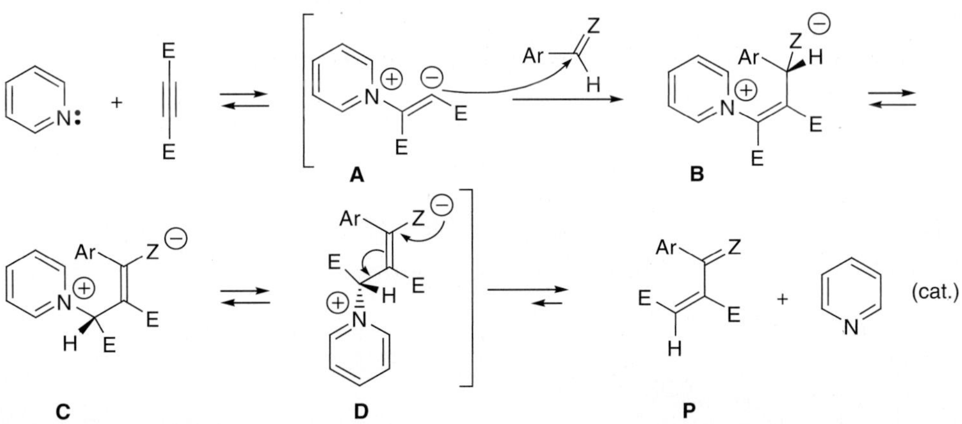

Problem 7.6 Propose a mechanism for the DABCO (1,4-diazabicyclo[2.2.2]octane) catalyzed chromene synthesis **A** + **B** → **C** [6].

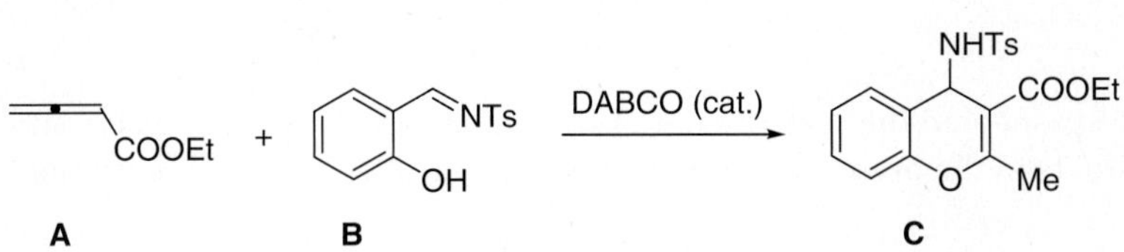

Answer

DABCO adds to the ethyl but-2,3-dienoate **A** giving the zwitterionic intermediate **D**. Protonation of the latter with **B** furnishes ion pair **E**, which then undergoes 1,4-addition of the phenolate moiety, followed by cyclization providing **F**. Protonation/deprotonation (formal 1,3-transfer of a proton) and β-elimination of DABCO (catalyst) gives **C**.

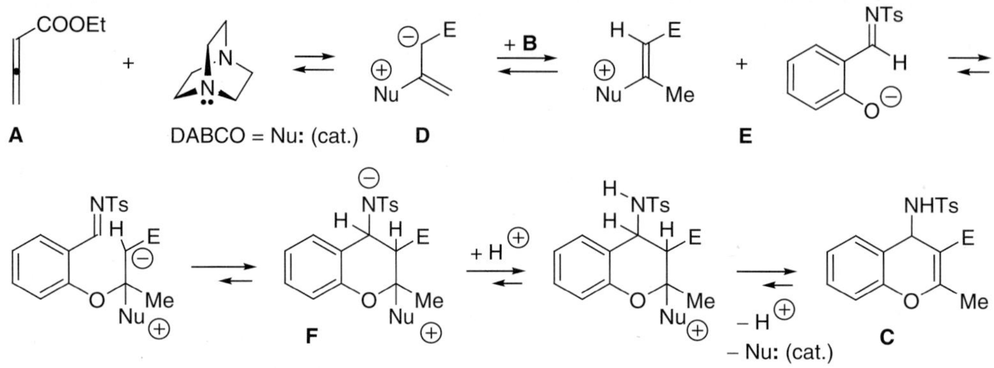

Problem 7.7 Propose a mechanism for the Ph_3P-catalyzed double Michael addition **A** + **B** → **C** [7, 8].

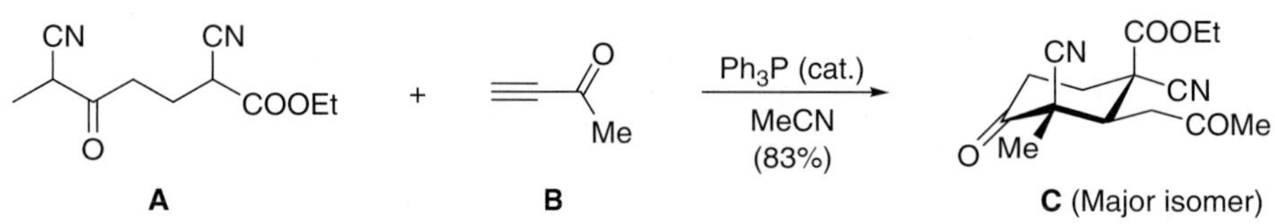

Answer

The zwitterion **D** resulting from the addition of Ph_3P to butynone (**B**) deprotonates **A** giving phosphonium ion **I** and enolate **E**. The latter adds to **B** providing equilibrating intermediates **F** ⇄ **G**. Enolate **G** undergoes intramolecular Michael addition giving **C** next to **D** after proton transfer from phosphonium ion **I**.

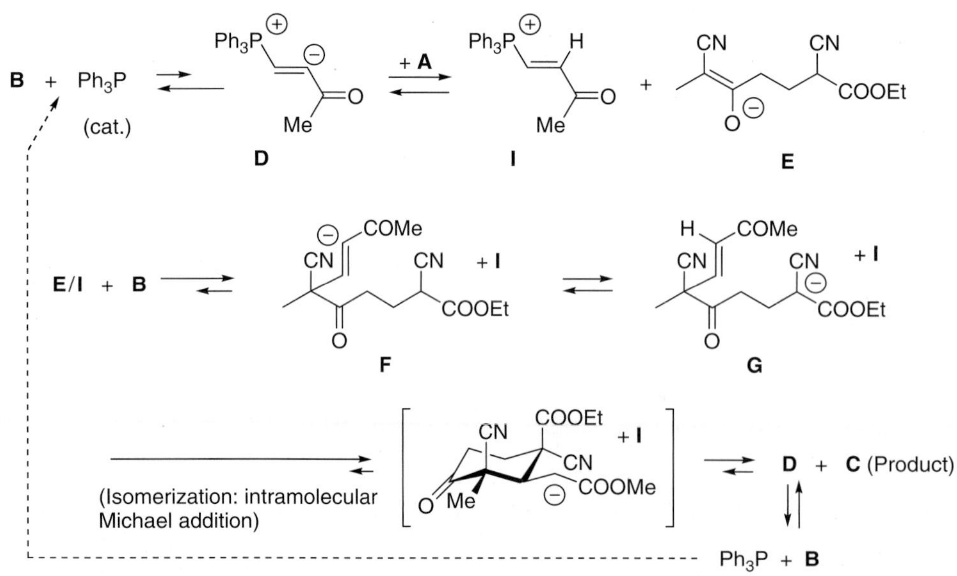

Problem 7.8 Lu and coworkers [9, 10] have pioneered the catalytic (3+2)-cycloaddition of allenoate esters and conjugated enones to give cyclopentane adducts in the presence of Ph_3P. Cowen and Miller [11] have developed

an enantioselective version of the reaction using phosphine-containing α-amino acid derivatives as catalysts. Propose a mechanism for the (3+2)-cycloaddition **A** + **B** → **P** in the presence of catalyst **C**.

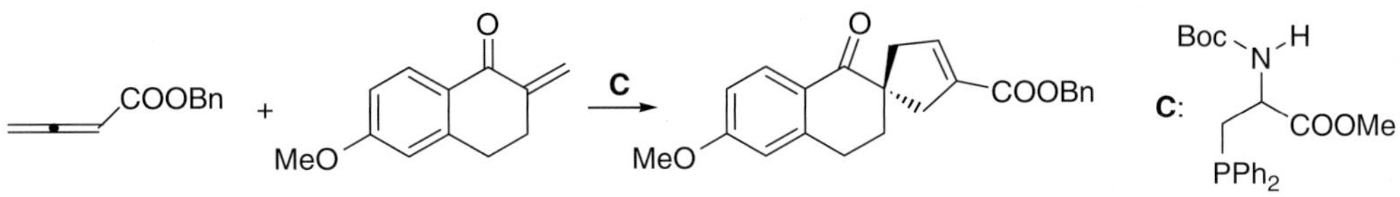

A **B** **P**

Answer

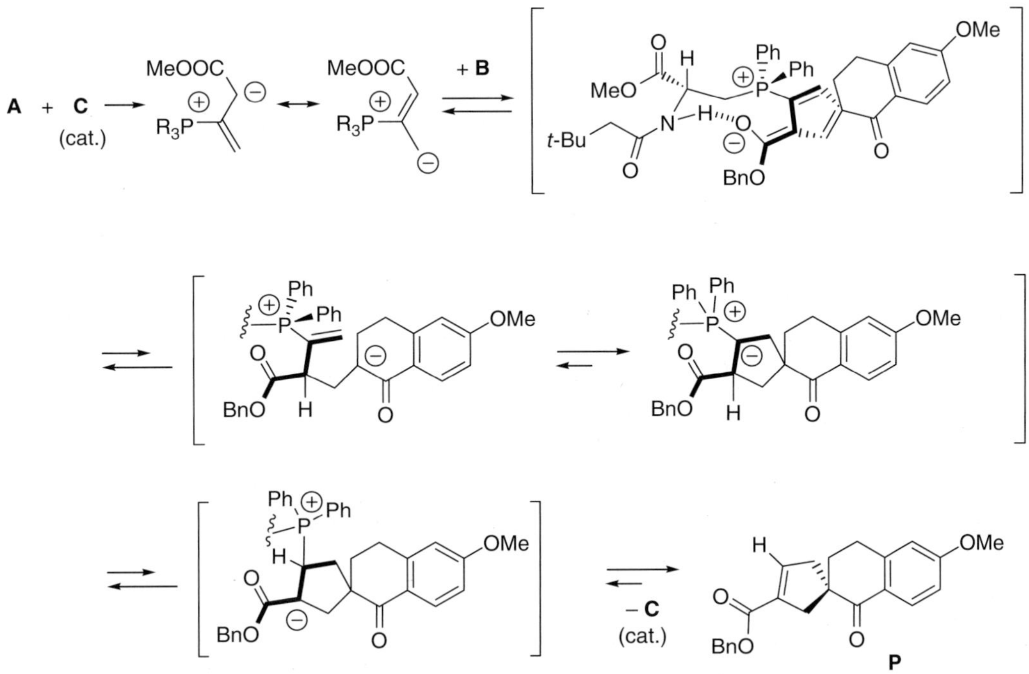

Problem 7.9 Propose a mechanism for the reaction **A** + **B** → **C** [12].

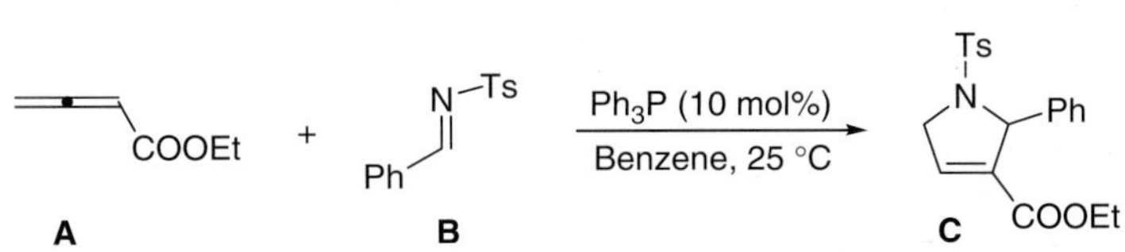

Answer

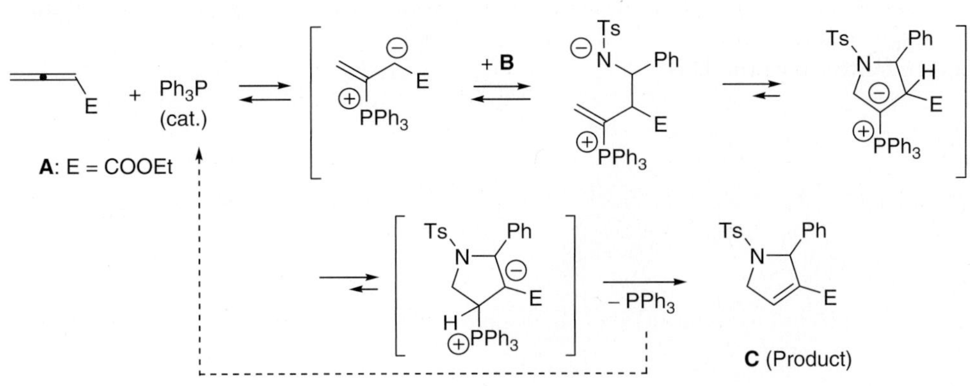

Problem 7.10 Propose a mechanism for the (8+3)-annulations **A** + **B** → **C** [13].

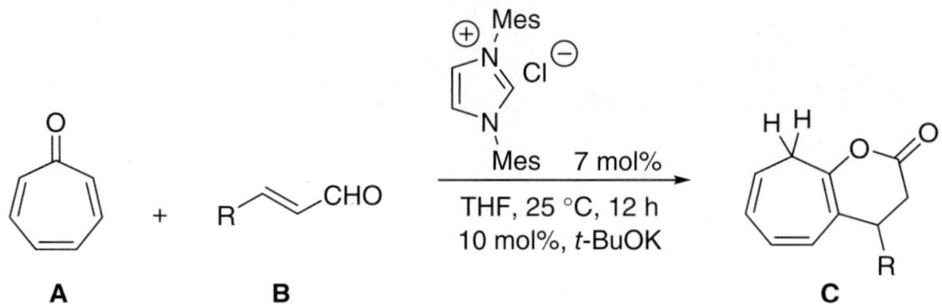

Answer

α-Elimination from 1,3-dimesitylimidazolium chloride (IMes·HCl) induced by *t*-BuOK generates the corresponding diaminocarbene **D**. It adds to aldehyde **B** and forms dienol **E**. The latter is a nucleophile that adds to tropone (**A**) giving the zwitterionic intermediate **F**, which equilibrates with its keto form **G**. Trienolate **G** is isomerized into **H** through an intramolecular nucleophilic addition. Elimination of the diaminocarbene **D** (catalyst) form **H** provides enol ester **I** that undergoes tautomerization into product **C**.

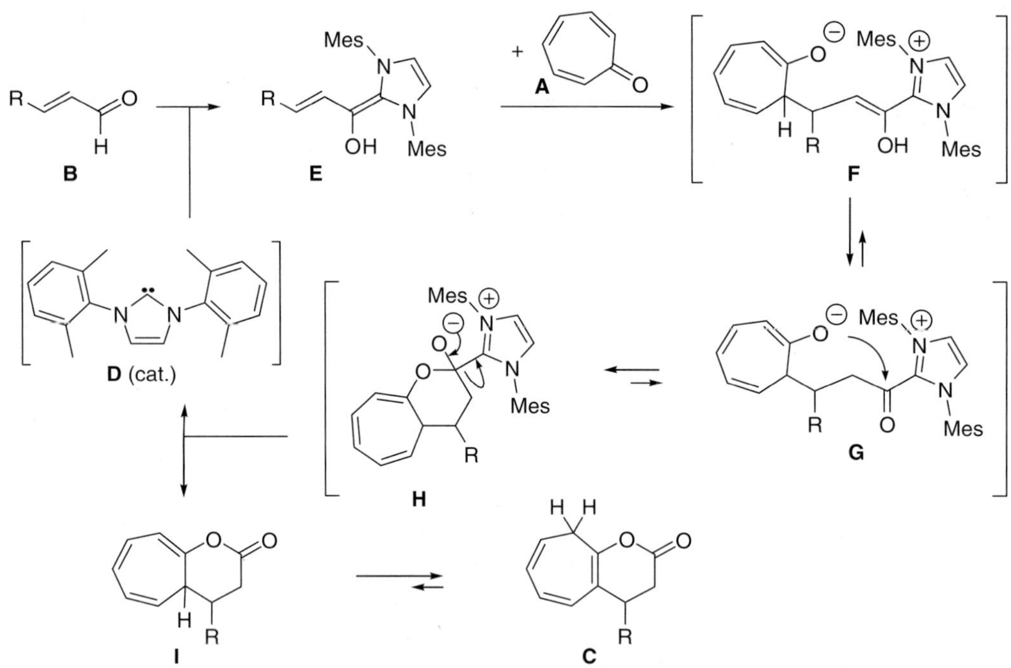

Problem 7.11 $Fe(ClO_4)_3$ catalyzes the (2+2)-cyclodimerization of *trans*-anethole and gives a *trans,anti,trans*-cyclobutane derivative stereospecifically in good yield. Alumina supported $Fe(ClO_4)_3$ is an especially efficient catalyst. Propose a mechanism for this reaction [14].

Answer

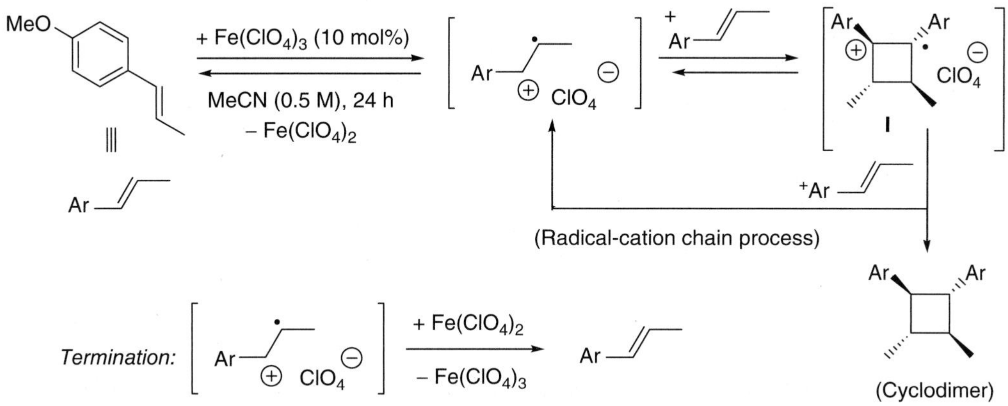

The *trans,trans,trans*-cyclodimer is the major stereomer for steric reasons (reversible reaction up to **I**)

The 1,2-diaryl disubstitution regioselectivity arises from the fact that radical-cation **I** is the most stable regioisomeric intermediate that can be formed. The *p*-methoxyphenyl group stabilizes carbenium ion and radical much better than a methyl group.

Problem 7.12 What is the major product **P** of the following visible-light-induced reaction (bpz, 2,2-bipyrazine)? [15]

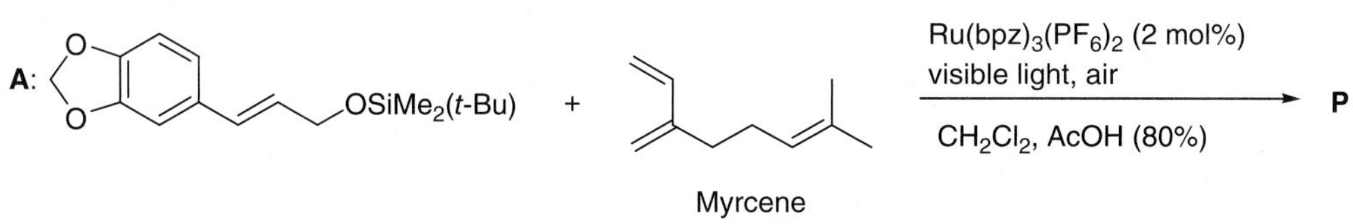

Answer

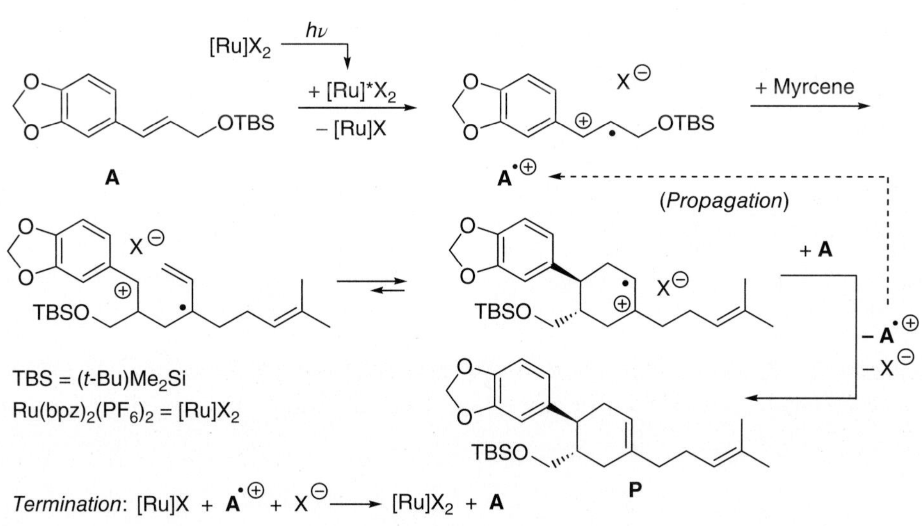

Problem 7.13 Propose a mechanism for the $(p\text{-BrC}_6H_4)_3N^{\cdot+}/SbCl_6^-$ catalyzed epoxidation of 1-vinylcyclohexene with benzeneselenic anhydride. What is the main product or reaction? [16]

Answer

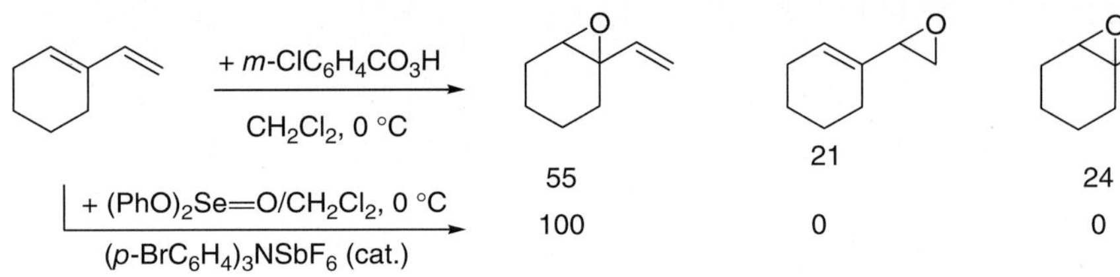

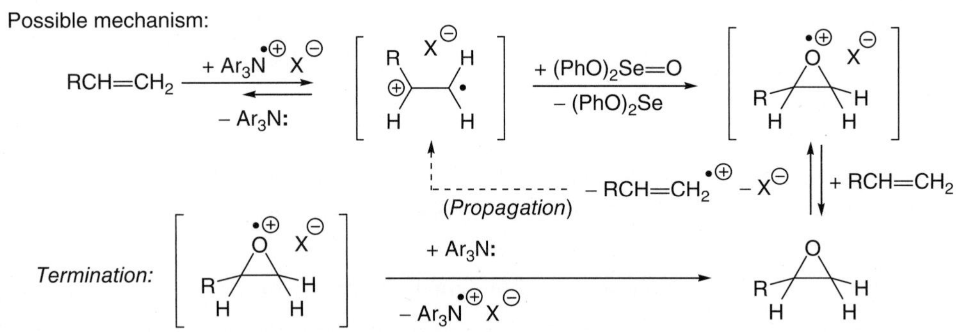

Problem 7.14 Explain the high *syn*-diastereoselectivity of reaction **232** + CH_2=CHOSi$(SiMe_3)_3$ → **233**.

Answer

The *β*-silyloxy substituent in **232** is voluminous and non-chelating. Thus the preferred approach of the nucleophile onto **232** when coordinated to $TTMSSNTf_2$ follows a Felkin–Ahn trajectory (steric factors govern the diastereoselectivity; the most stable rotamer reacts with the enoxysilane).

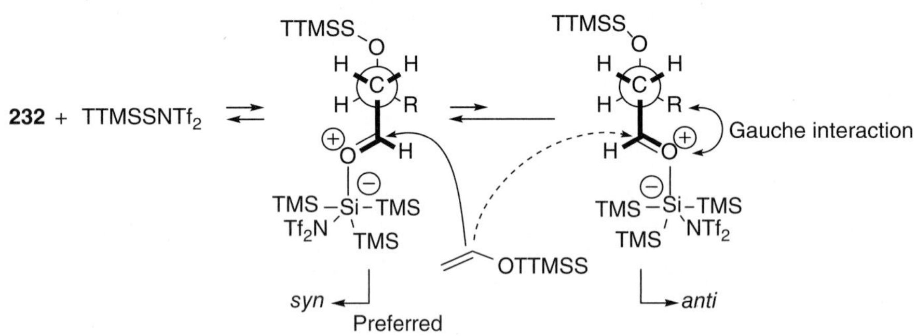

Problem 7.15 What is the major product of reaction of dimethyl oxomalonate with 2-methylhepta-2,6-diene in the presence of 0.2 equiv. of $SnCl_4$?

Answer

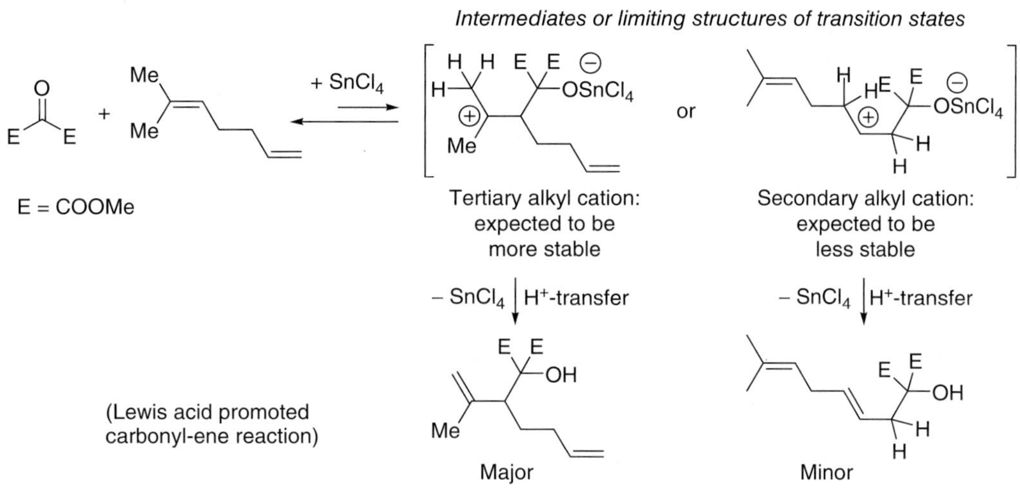

Problem 7.16 Ene-reaction of formaldehyde with methyl hept-6-ynoate using $Me_2AlCl + MeAlCl_2$ as catalyst produces an antibiotic isolated from *Sapium japonicum*. What is this product? [17]

Answer

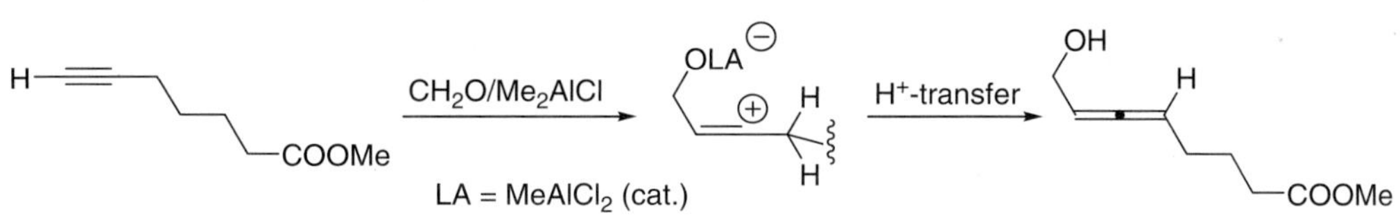

Problem 7.17 What is the major product of reaction of isoprene with 3-methylbutanal in the presence of Me_2AlCl? [18]

Answer

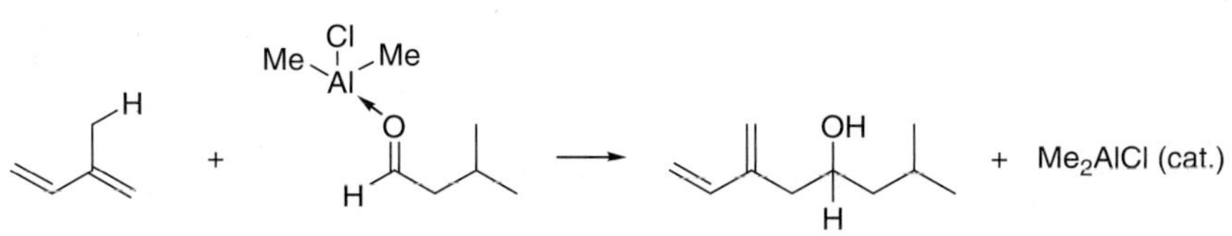

Problem 7.18 Explain the regioselectivity of Lewis acid catalyzed cycloadditions (7.65) and (7.66).

Answer

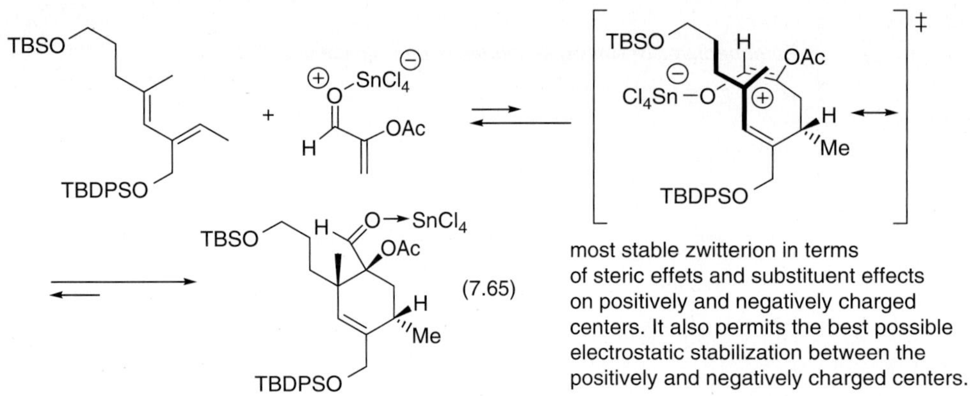

(7.65)

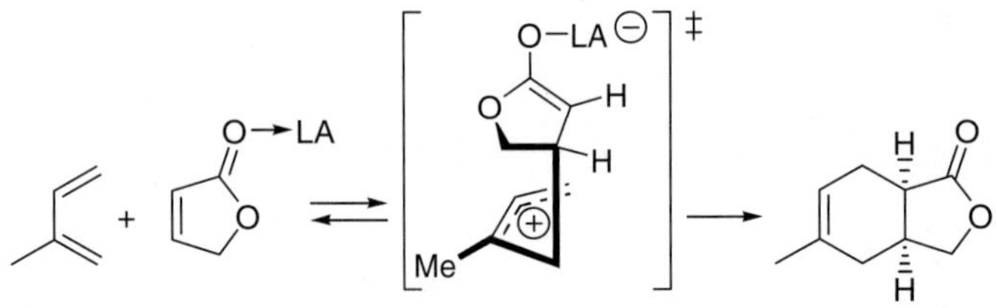

(7.66)

Problem 7.19 What is the main product **P** of reaction **A** + **B** [19].

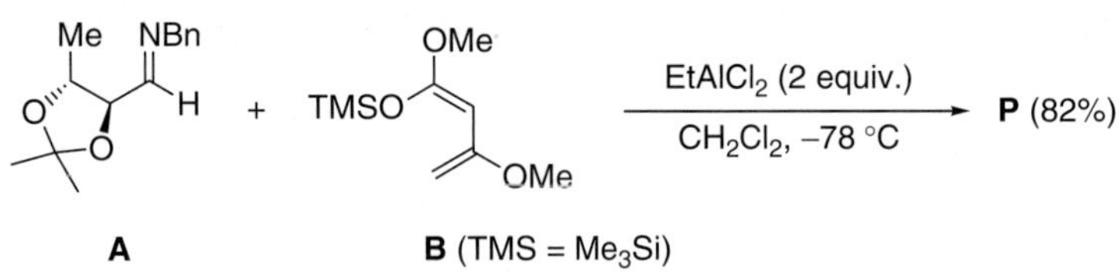

Answer

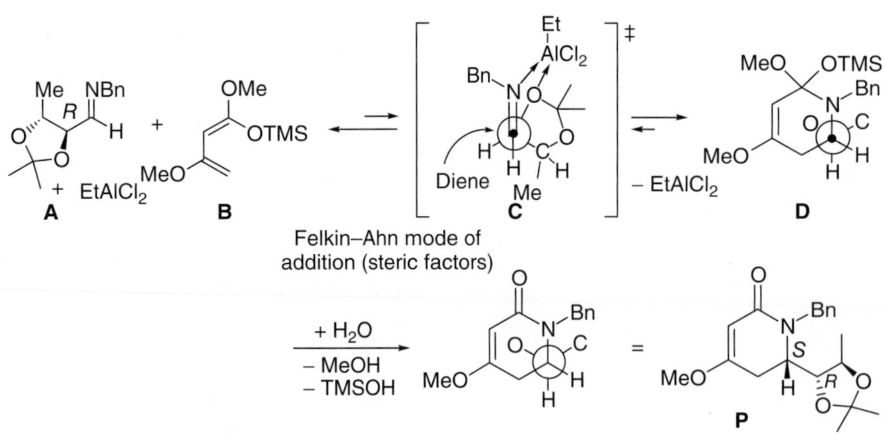

Problem 7.20 Give a possible mechanism for reaction (7.72).

Answer

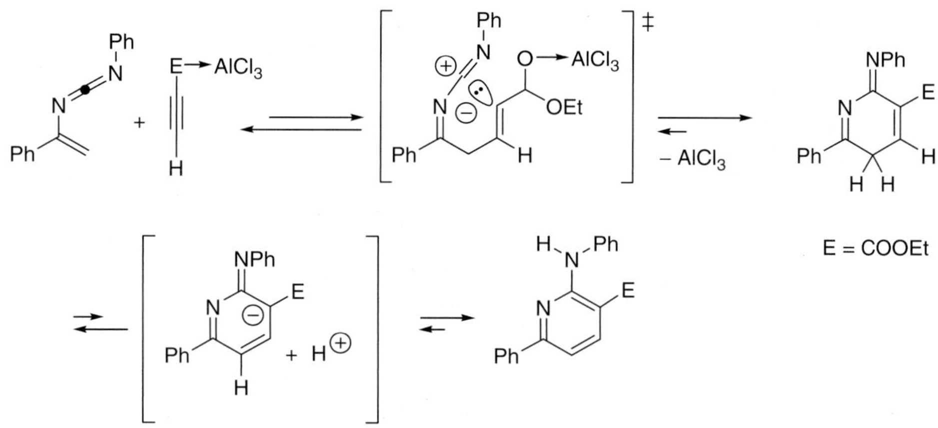

Problem 7.21 What are the successive products **B** and **C** produced in the following reaction? [20, 21].

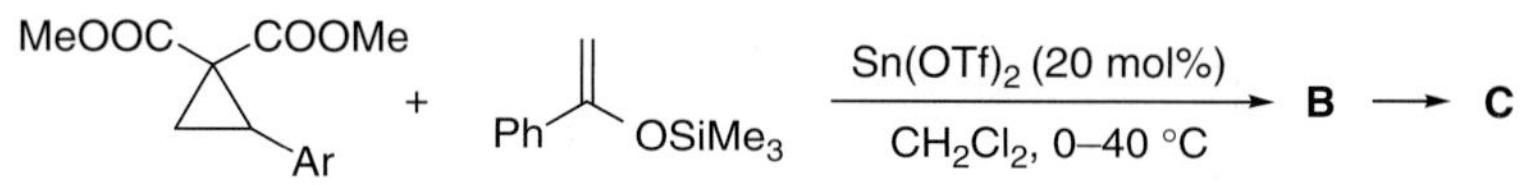

Answer

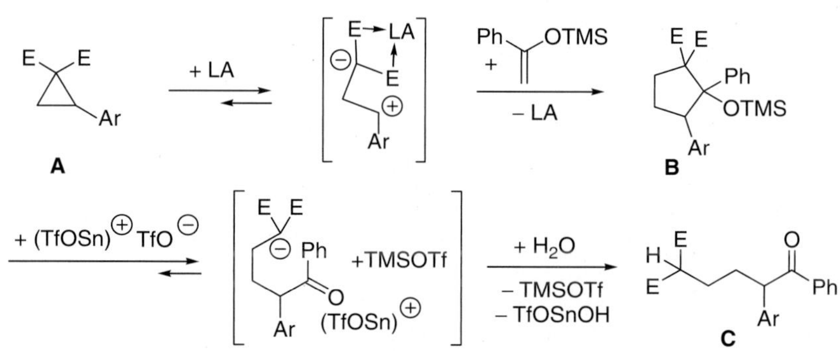

Problem 7.22 Explain the following reaction [22].

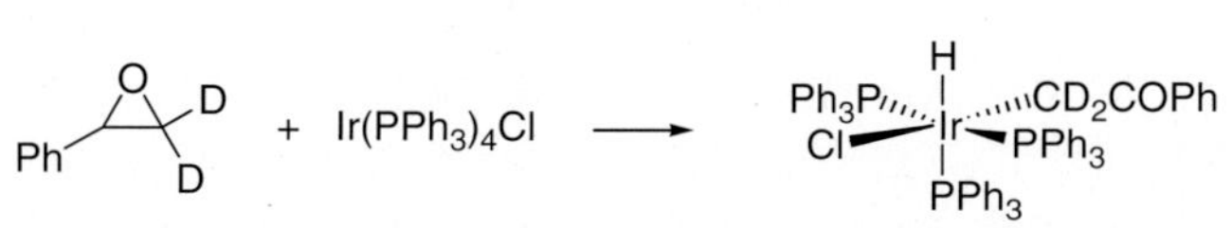

Answer

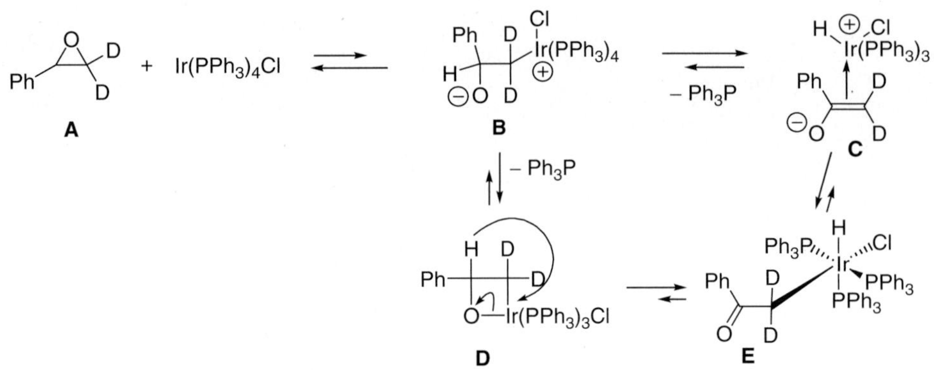

Problem 7.23 What is the structure of product **P** in the following reaction?

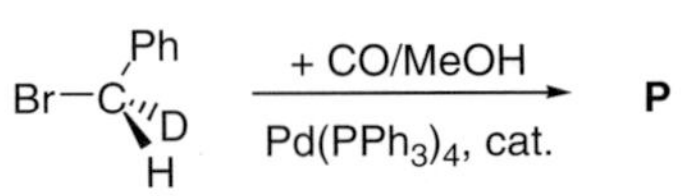

Answer

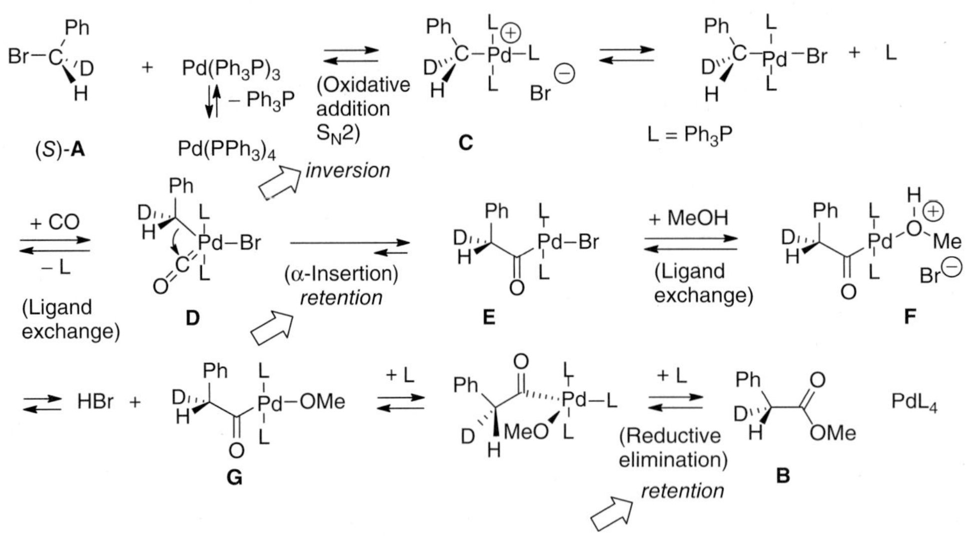

Problem 7.24 Explain the following transformation: $Cp_2Mo(PR_3)Me^+/PF_6^- \rightarrow Cp_2Mo(H)CH_2P^+R_3/PF_6^-$ [23].

Answer

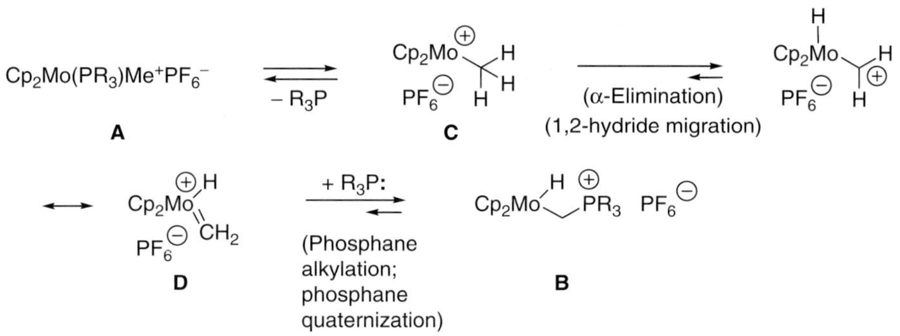

Problem 7.25 $Ir(PPh_3)_2(Cl)_2(COCH_2Ph)$ ($\nu_{C=O} = 1670\,cm^{-1}$) rearranges into **P** ($\nu_{C=O} = 2040\,cm^{-1}$) at 30 °C in benzene. What is the structure of **P**?

Answer

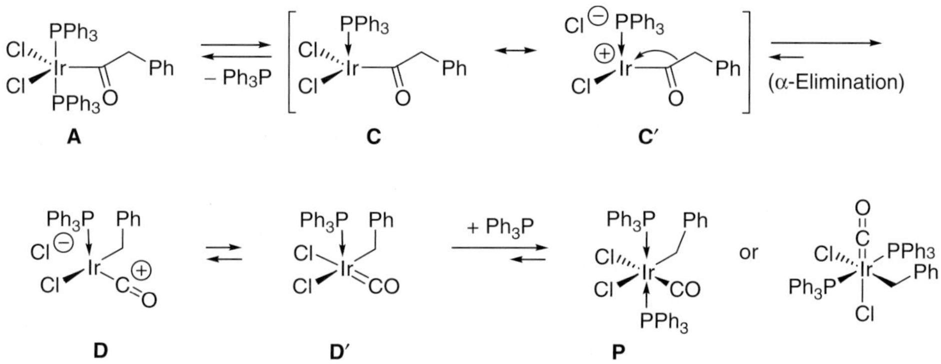

Problem 7.26 What is product **P** of the reaction $C_4H_9—C{\equiv}C—D + Cp_2Zr(H)(Cl) \rightarrow$ **P**?

Answer

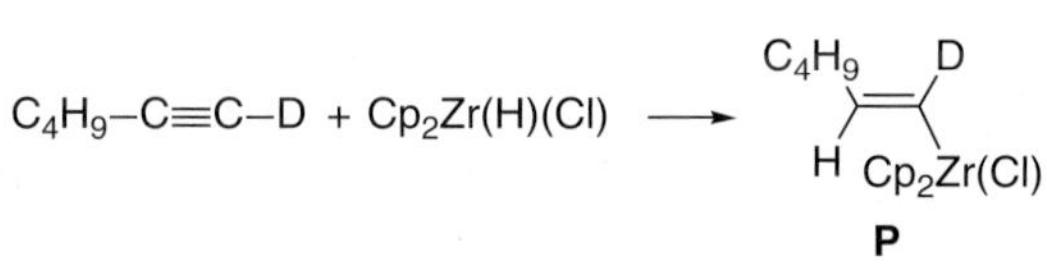

Problem 7.27 Internal alkenes and alkynes isomerize to 1-alkyl and alk-1-en-1-yl complexes, respectively, when reacted with the Schwartz's reagent. Explain.

Answer

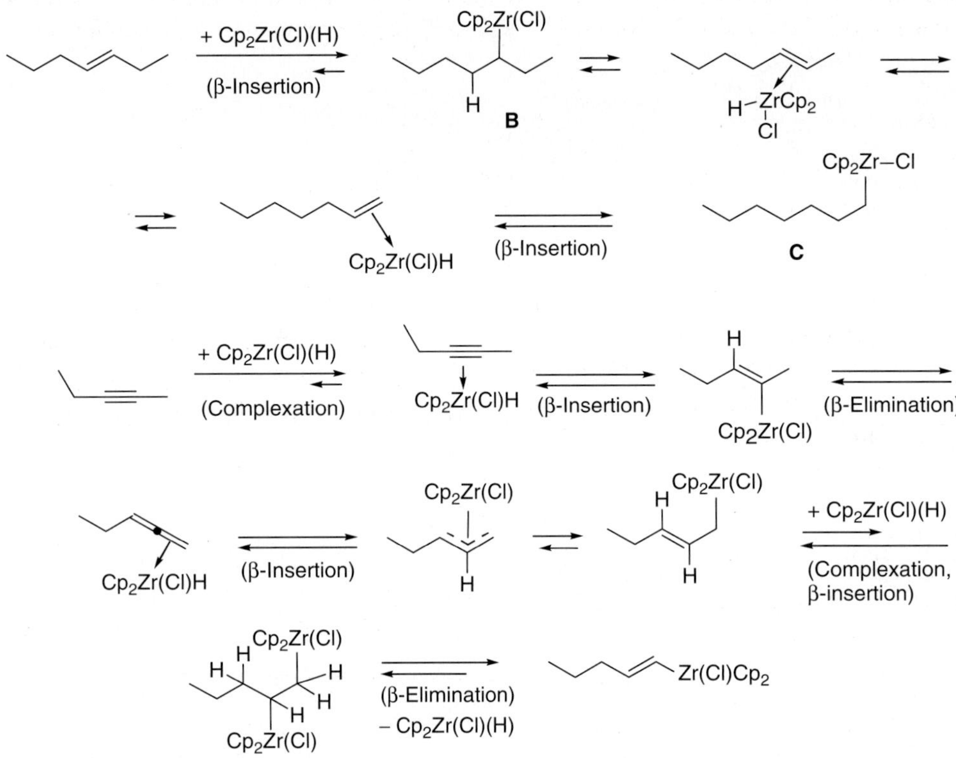

Back-strain (steric repulsions between the ligands about the Zr center) renders all branched alkyl and alkenyl–Zr(Cl)Cp_2 isomeric complexes less stable than alk-1-yl and alk-1-en-1-yl complexes.

Problem 7.28 Explain the stereoselectivity of the following reactions [24].

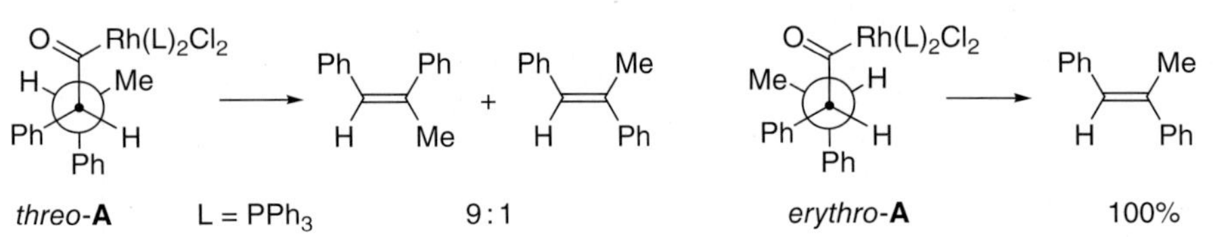

Answer

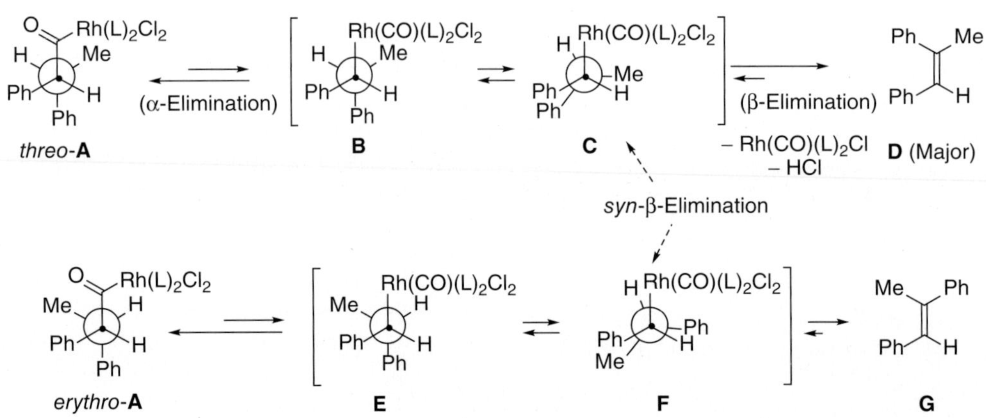

Problem 7.29 Snyder and Hours reported the $PtCl_2$-catalyzed rearrangement **A** → **P** [25]. With the deuterated derivatives **B**, **C**, and **D**, products **Q**, **T**, and **W**, respectively were obtained with the following kinetic isotopic effects: $k(\mathbf{A})/k(\mathbf{B}) = 1.54$, $k(\mathbf{A})/k(\mathbf{C}) = 5.44$ and $k(\mathbf{A})/k(\mathbf{D}) = 11.3$ at 140 °C [26]. Propose a mechanism for these catalyzed isomerizations.

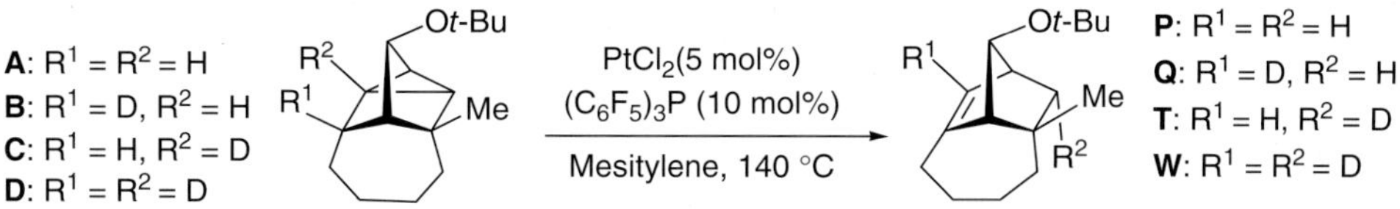

Answer

The first step is the oxidative addition of the platinum catalyst [Pt] to one of the cyclopropane σ(C—C) bonds. This is followed either by α- or β-elimination of hydrogen or deuterium. The data are consistent only with α-elimination, which generates a carbenemetal intermediate in the rate-determining step. R^2 migrates in the α-elimination; R^1 would migrate in the β-elimination.

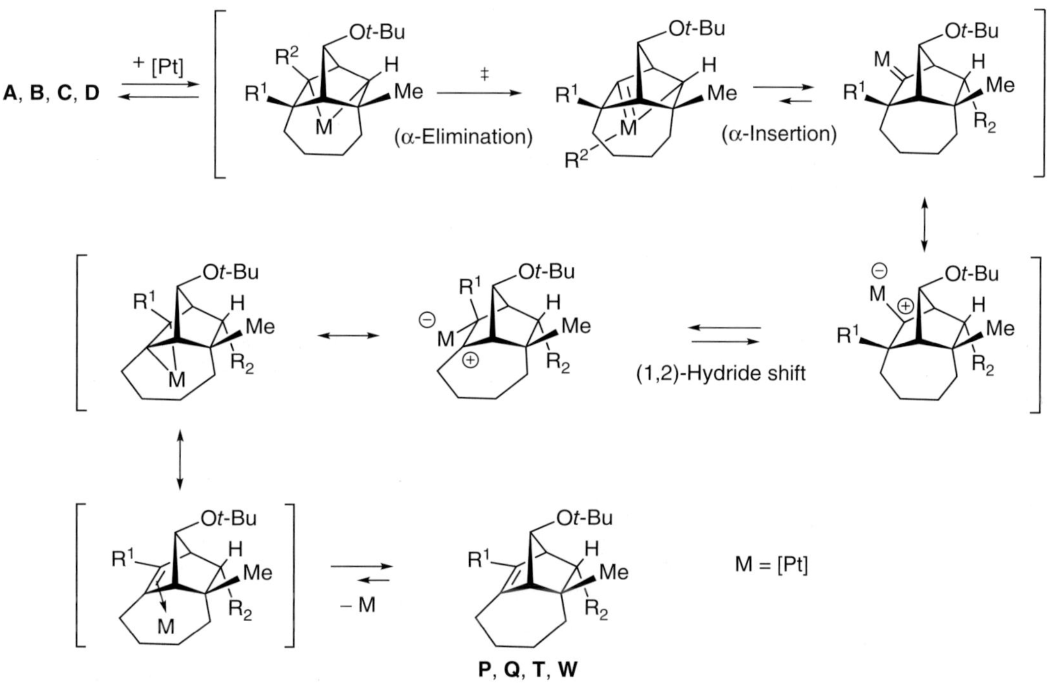

An alternative mechanism involving β-elimination instead of α-elimination of hydrogen (deuterium) is ruled out. For instance:

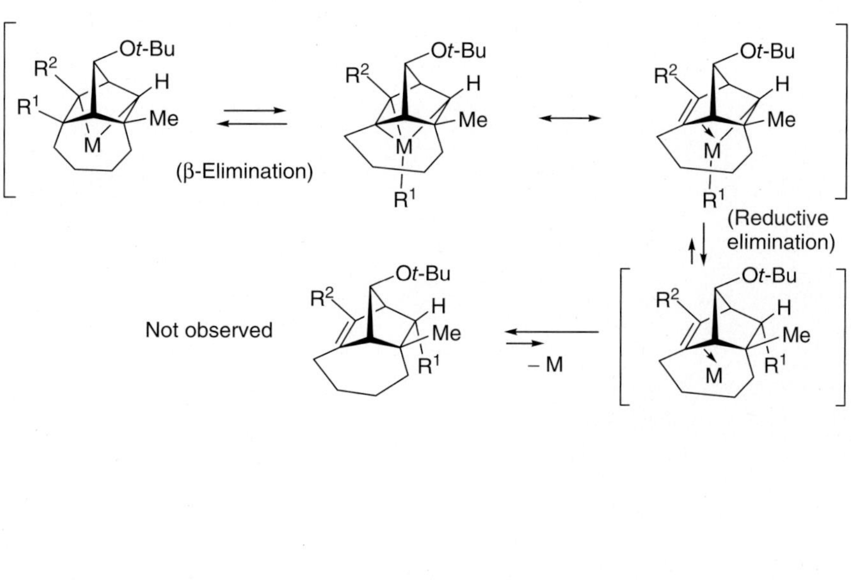

Problem 7.30 Propose a mechanism for the reaction of **A** with the (alkylidene)chromium complex **B** that gives **P**. How can one prepare **B**? [27]

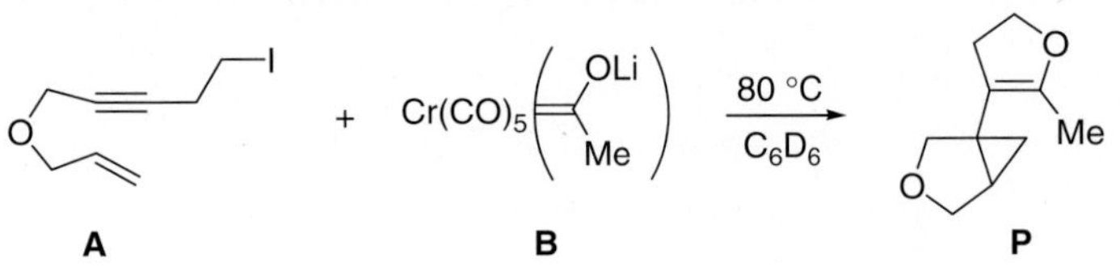

Answer

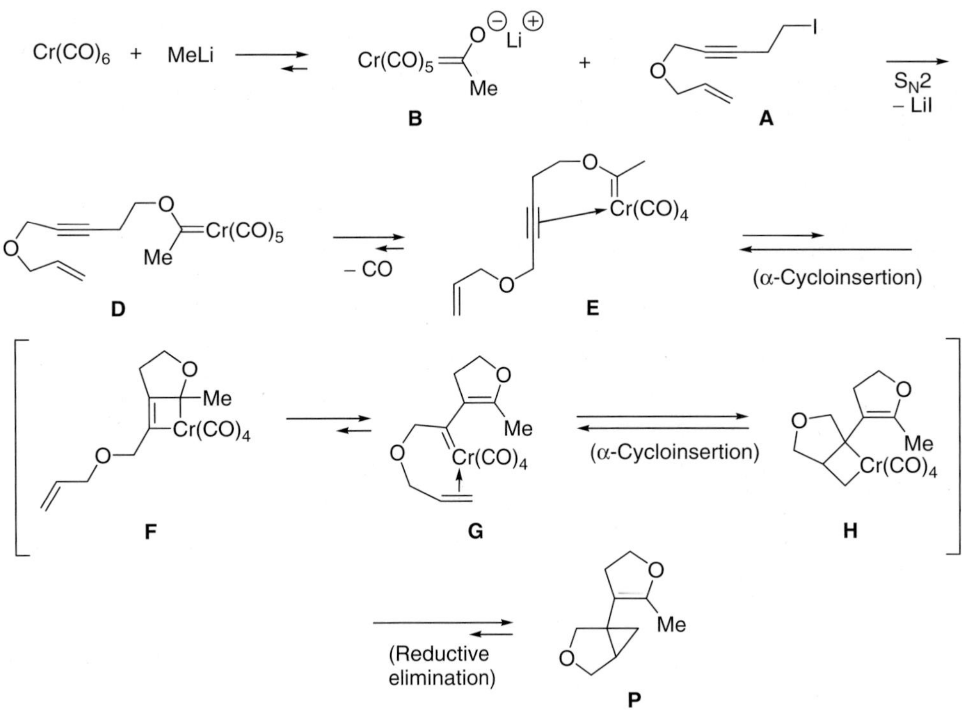

Problem 7.31 Propose mechanisms for all the synthetic steps given in Scheme 7.63.

Answer

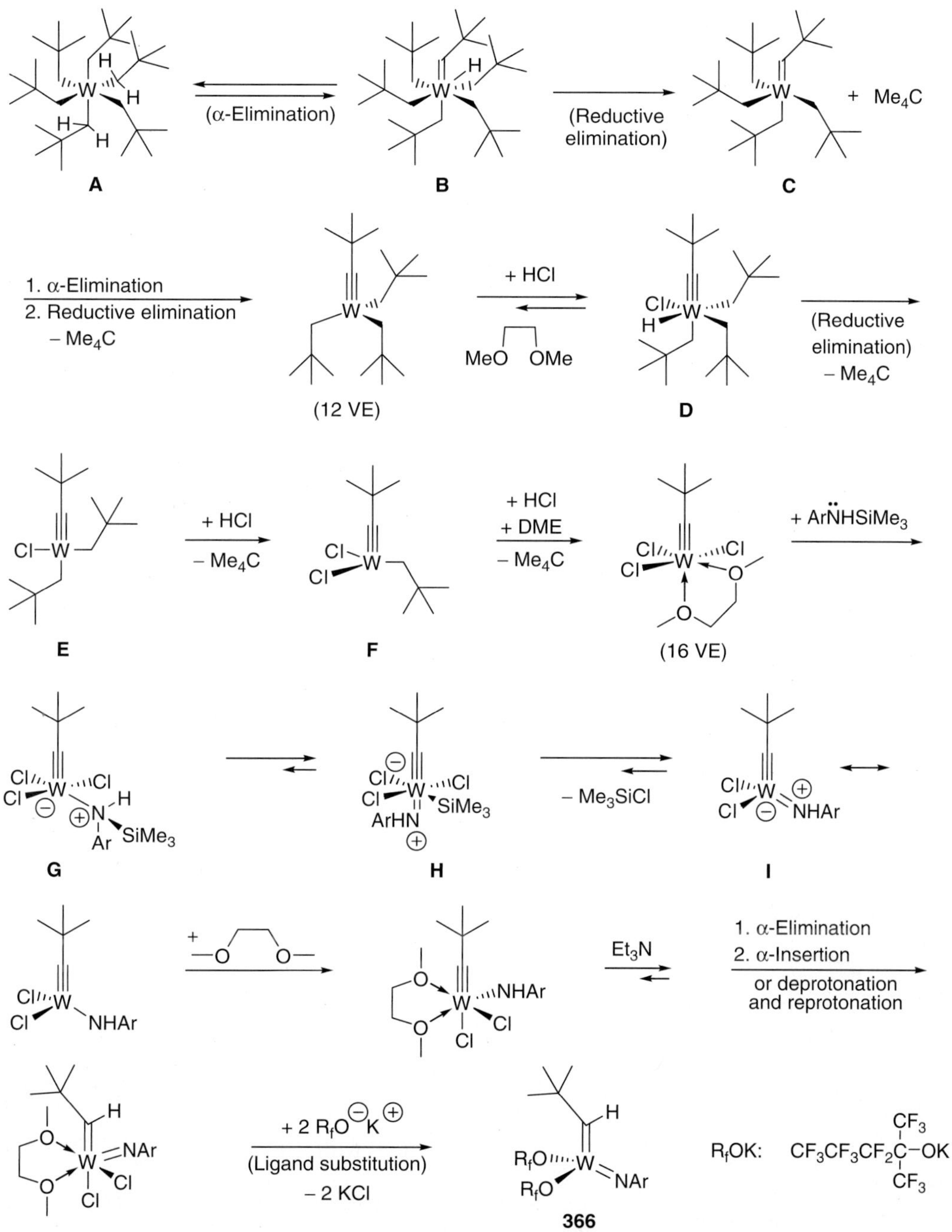

Problem 7.32 Propose a mechanism for the oligomerization and polymerization of terminal alkynes induced by an (alkylidyne)metal catalyst.

Answer

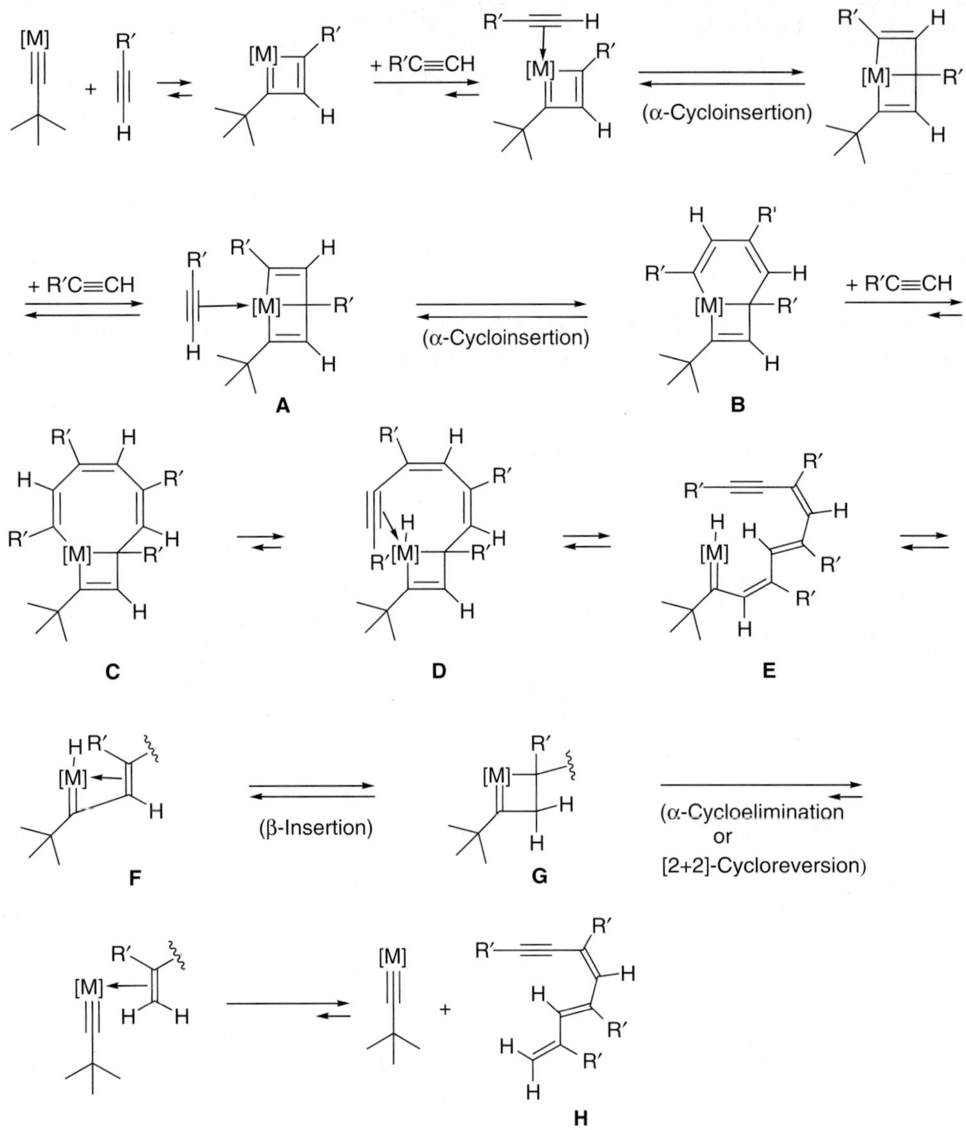

Problem 7.33 What are the products of reaction of Cp_2Zr(propene) with (i) deca-2,8-diyne; (ii) with $Me_3SiCH{=}CH_2 + PhC{\equiv}CPh$? [28]

Answer

(a)

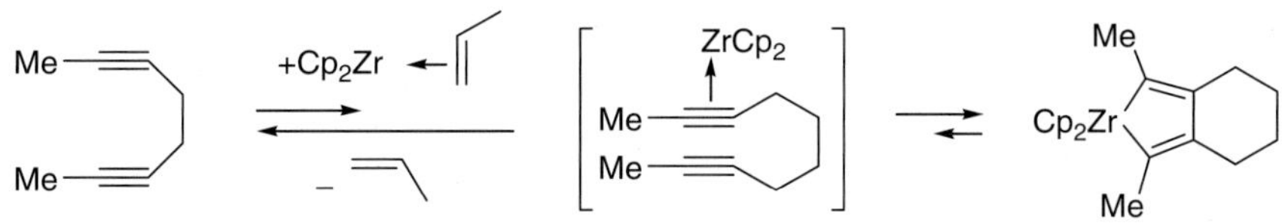

(b)

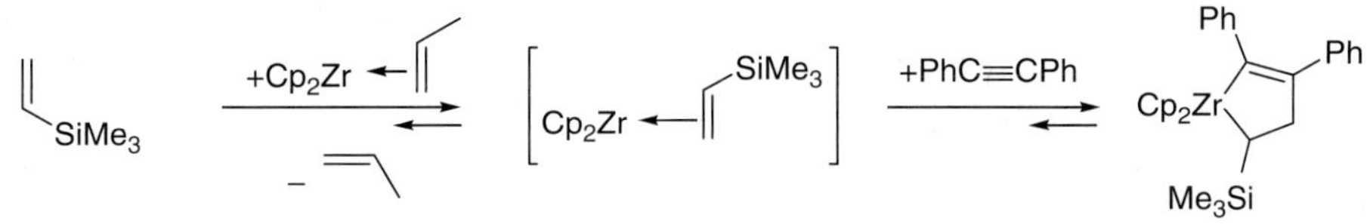

Problem 7.34 Give a possible mechanism for the chemoselective hydrogenation of cinnamaldehyde into (*E*)-3-phenylprop-2-en-1-ol catalyzed by the cationic iron complex $[P(o\text{-}Ph_2PC_6H_4)_3FeH]^+/BF_4^-$ in *i*-PrOH (CF_3COOH at 140 °C and under H_2 (20 bar)) [29].

Answer

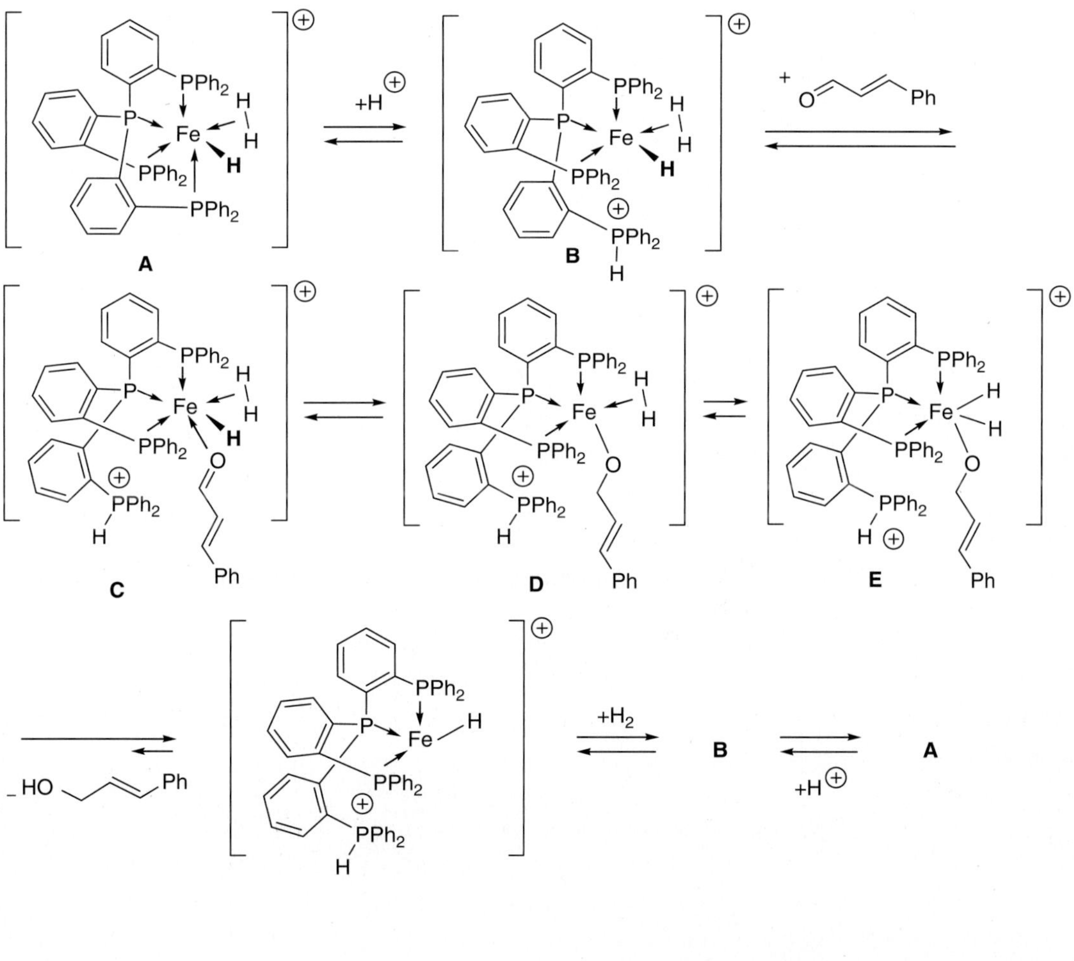

Problem 7.35 Give a plausible mechanism for the hydrogenation of CO_2 into HCOOH catalyzed by the iron complex **428** [30].

Answer

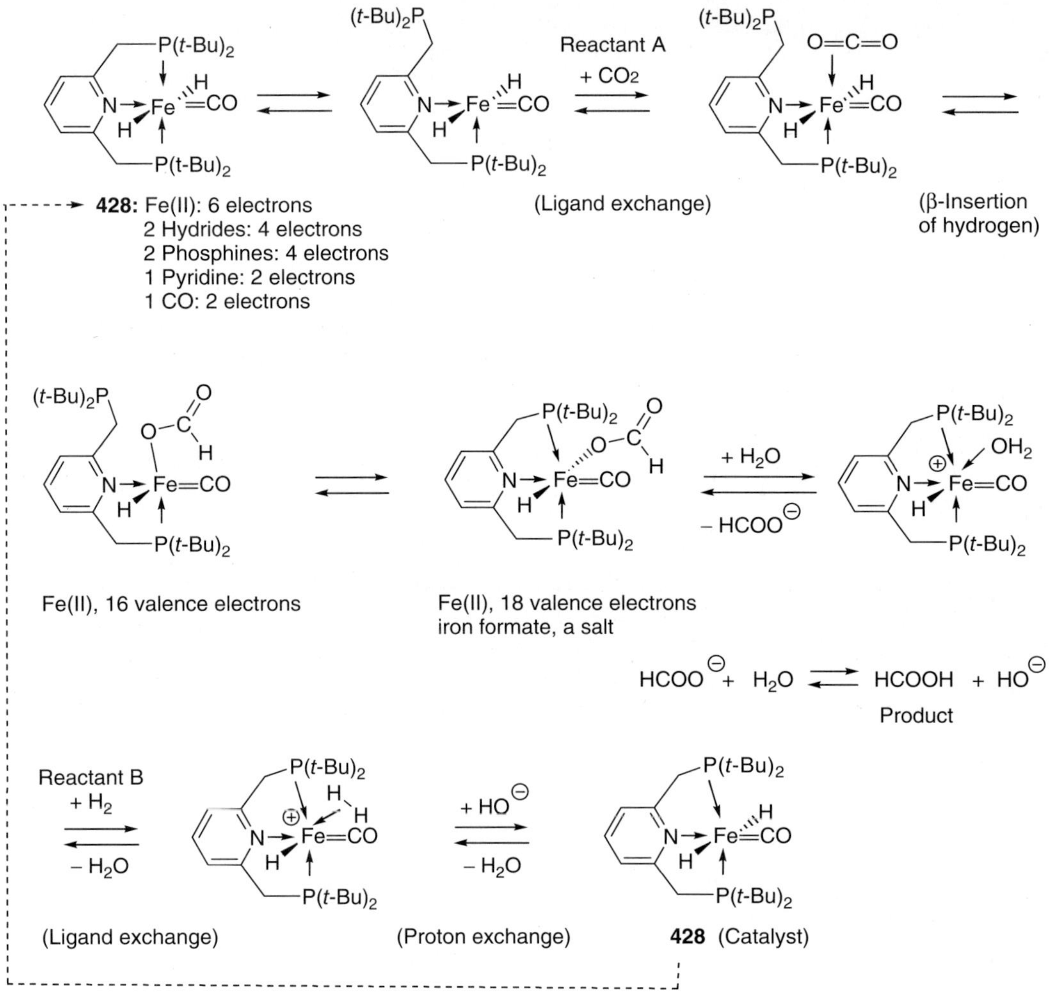

Problem 7.36 Propose a mechanism for the hydrosilylation of ketones with $(EtO)_2MeSiH$ in toluene at 65 °C catalyzed by iron(II) complex of type **452** [31, 32].

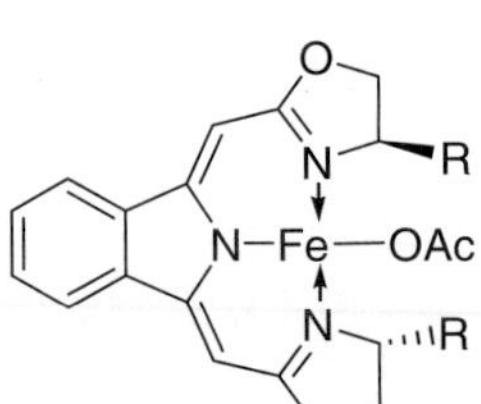

452

Answer

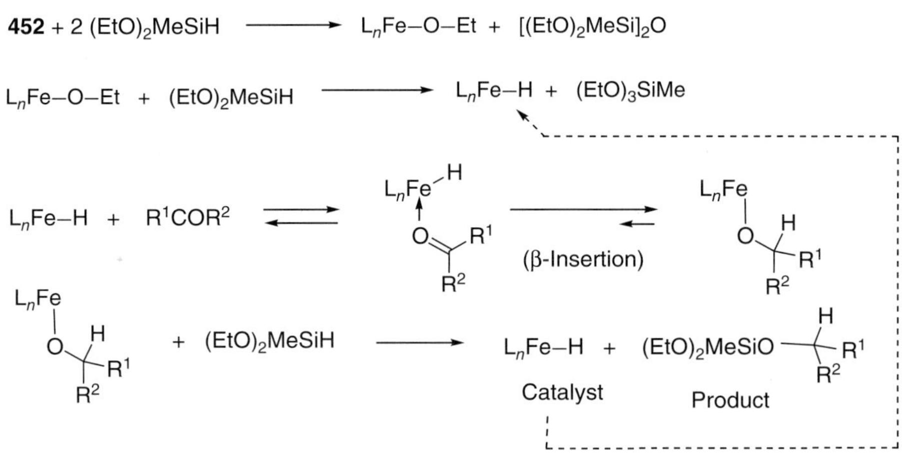

Problem 7.37 Propose a mechanism for the bis(tributylphosphine)dichloroplatinum(II)-catalyzed alkene hydrosilylation by Et_3SiH [33].

Answer

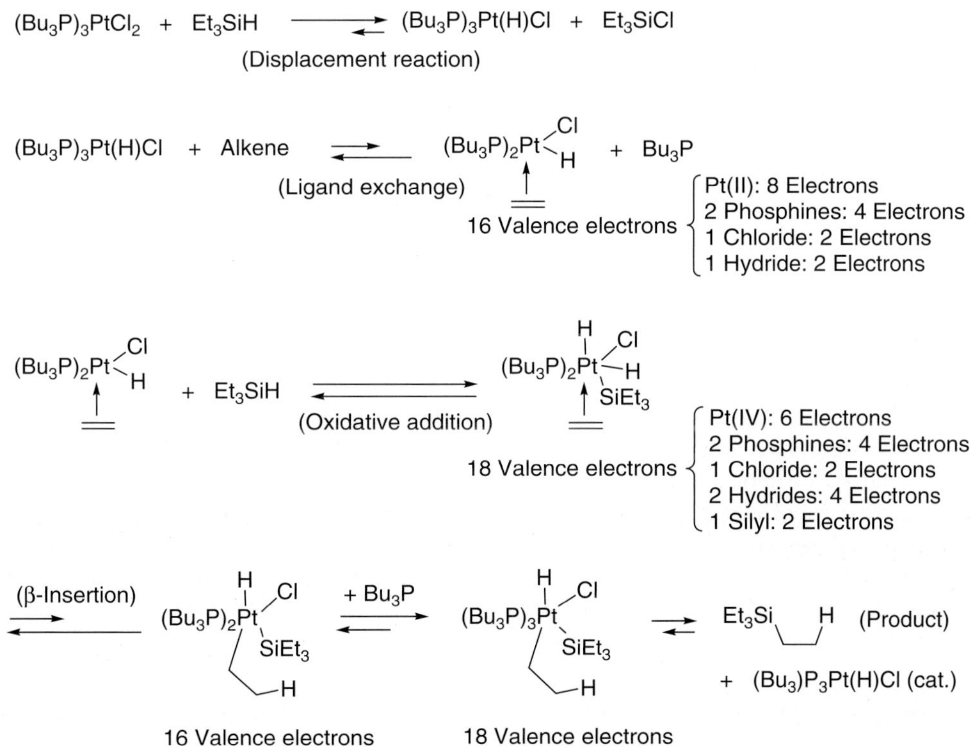

Problem 7.38 Propose a mechanism for the hydrosilylation of alkenes with $PhSiH_3$ catalyzed by the cationic ruthenium complex $[Cp^*(i\text{-}Pr_3P)(H)_2Ru{=}Si(H)Ph{\cdot}Et_2O]^+/[B(C_6F_5)_4]^-$ [34].

Answer

Et_2O is loosely bound in the catalyst and it is exchanged by the alkene giving intermediate **A**.

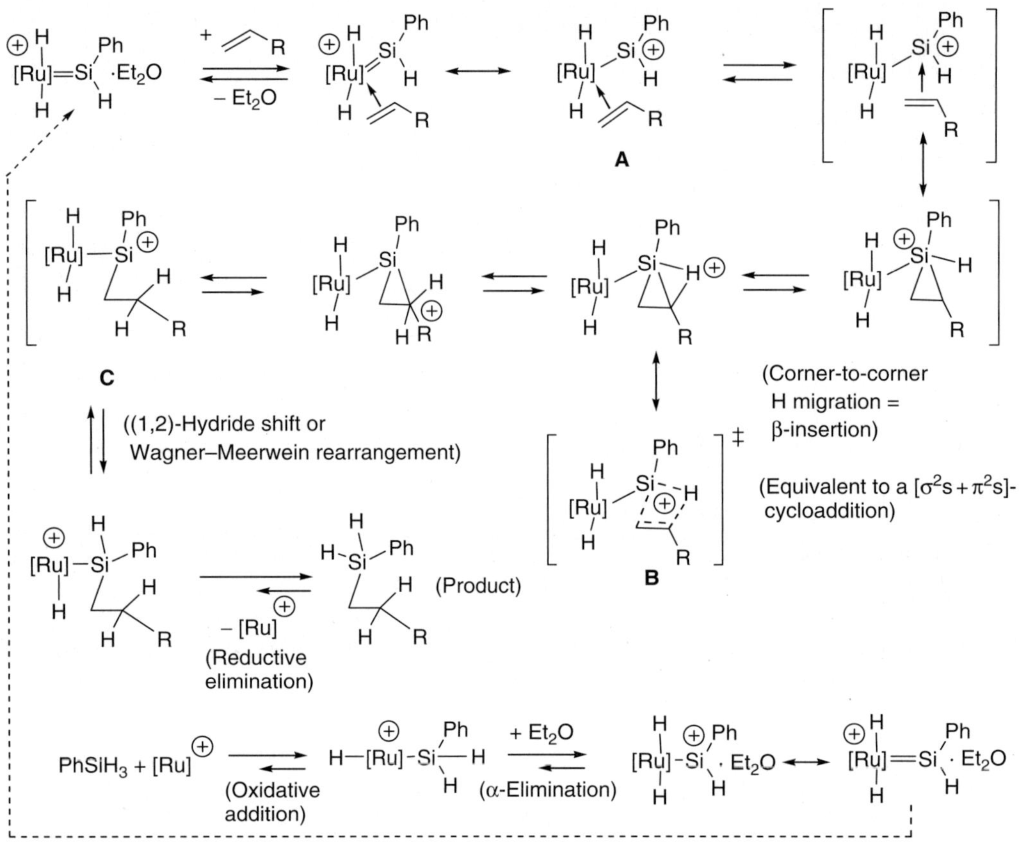

A can be seen as a ruthenasilicenium ion that adds to the alkene, probably via a four-membered cyclic transition state **B** equilibrating with another ruthenasilicenium ion **C**. The latter undergoes reductive elimination providing the product of hydrosilylation and a 16 valence electron complex that adds across the Si—H bond of the silane reagent. An α-elimination and subsequent alkene coordination follow, generating intermediate **A**, which starts another catalytic cycle. This is the Glaser–Tilley mechanism. The π-complex of the silicenium cation (sextet of electron about Si atom) with the alkene is a corner-protonated silacyclopropane, analog of corner-protonated cyclopropanes that are intermediates in Wagner–Meerwein rearrangements. Migration of hydrogen from a corner to another corner in corner-protonated cyclopropanes is an easy process.

Problem 7.39 Propose a mechanism for the formation of methyl 3,3-dimethoxypropanoate by oxidation of methyl acrylate by O_2 in MeOH catalyzed $PdCl_2/CuCl_2$ [35].

Answer

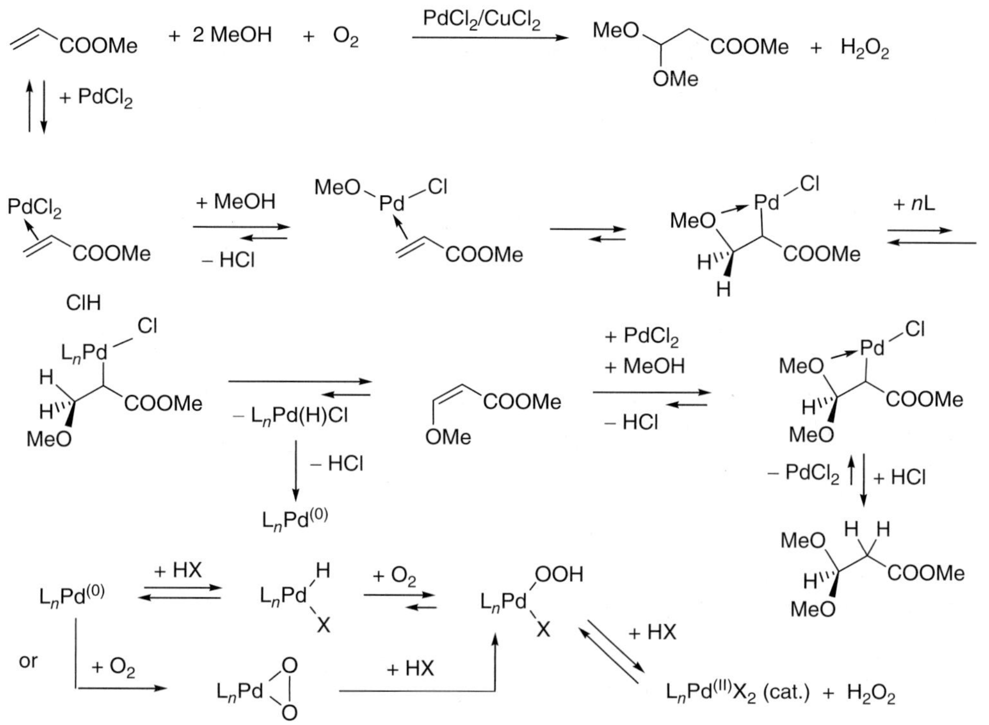

Problem 7.40 Propose a mechanism for the following three-component synthesis of 1,4-dihydopyridines.

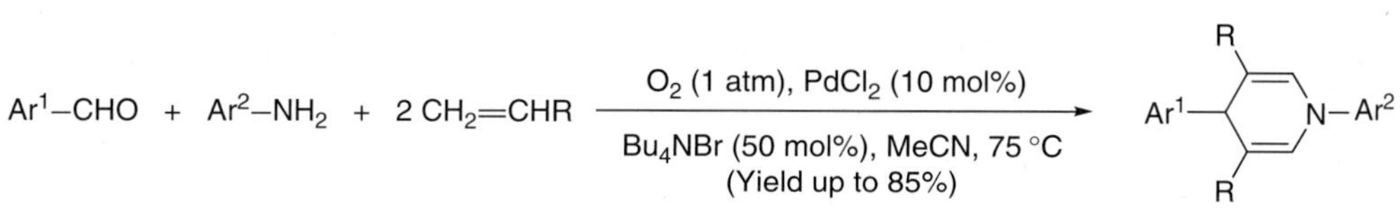

Answer

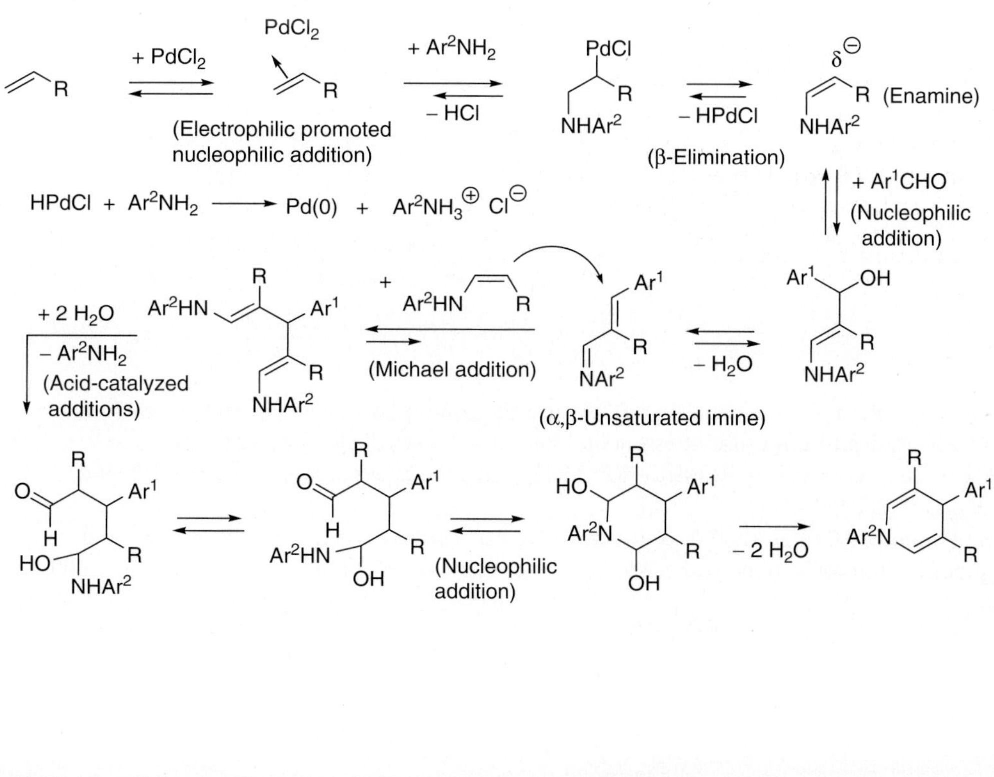

Problem 7.41 Propose a Cu-catalyzed synthesis of aryl thioethers [36, 37].

Answer

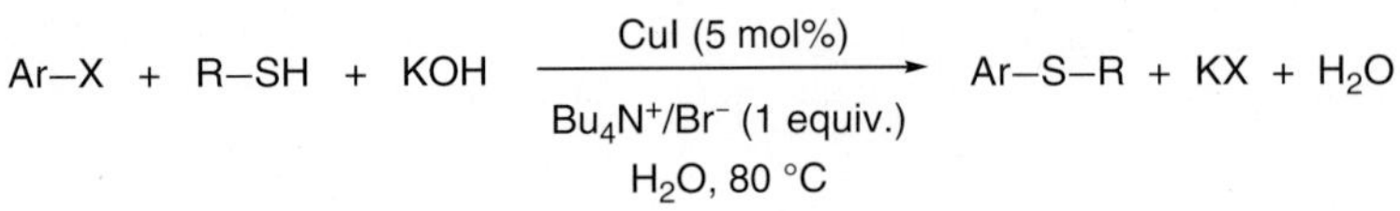

Proposed mechanism:

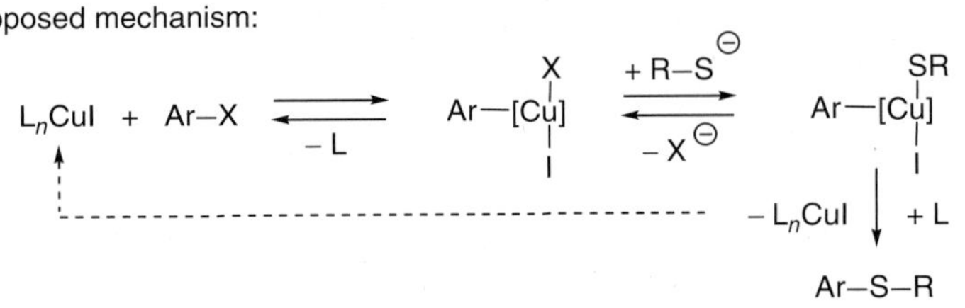

Problem 7.42 What is the product **P** of the following reaction? [38]

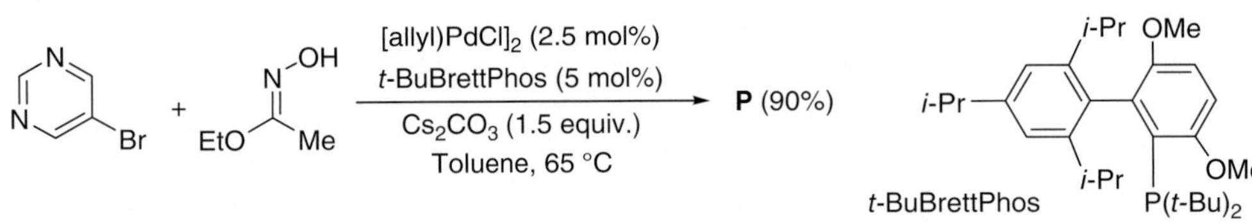

Answer

P:

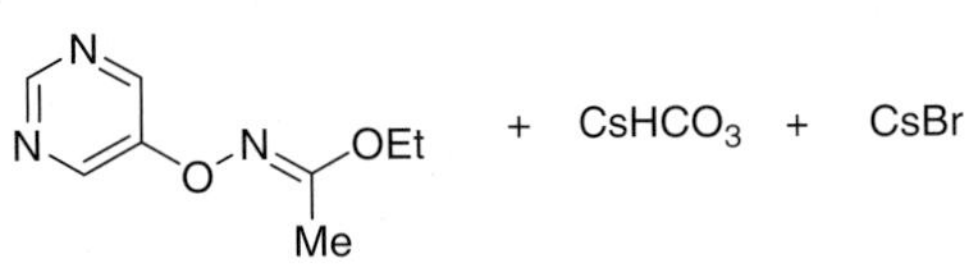

Possible mechanism:

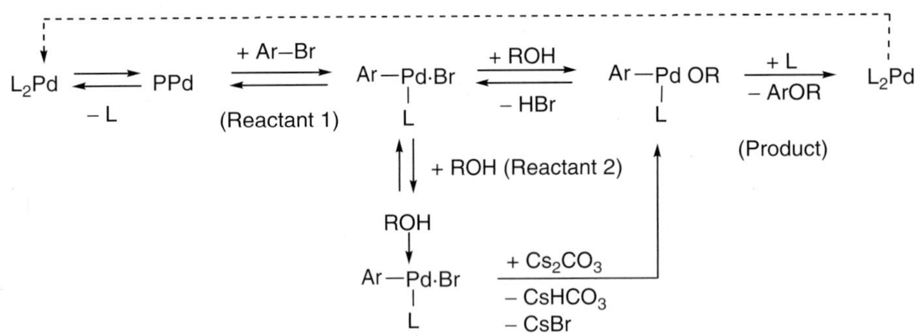

References

1 Mattson, A.E., Bharadwaj, A.R., and Scheidt, K.A. (2004). The thiazolium-catalyzed sila-Stetter reaction: conjugate addition of acylsilanes to unsaturated esters and ketones. *J. Am. Chem. Soc.* 126: 2314–2315.

2 He, M. and Bode, J.W. (2005). Catalytic synthesis of gamma-lactams via direct annulations of enals and *N*-sulfonylimines. *Org. Lett.* 7: 3131–3134.

3 Nair, V., Vellalath, S., Poonoth, M. et al. (2006). *N*-heterocyclic carbene catalyzed reaction of enals and 1,2-dicarbonyl compounds: stereoselective synthesis of spiro gamma-butyrolactones. *Org. Lett.* 8: 507–509.

4 Mergott, D.J., Frank, S.A., and Roush, W.R. (2004). Total synthesis of (−)-spinosyn A. *Proc. Natl. Acad. Sci. U.S.A.* 101: 11955–11959.

5 Nair, V., Menon, R.S., Sreekanth, A.R. et al. (2006). Engaging zwitterions in carbon–carbon and carbon–nitrogen bond-forming reactions: a promising synthetic strategy. *Acc. Chem. Res.* 39: 520–530.

6 Shi, Y.L. and Shi, M. (2005). DABCO-catalyzed reaction of allenic esters and ketones with salicyl *N*-tosylimines: synthesis of highly functionalized chromenes. *Org. Lett.* 7: 3057–3060.

7 Grossman, R.B., Pendharkar, D.S., and Patrick, B.O. (1999). [n+1]-Annulation route to highly substituted cyclic ketones with pendant ketone, nitrile, and ester functionality. *J. Org. Chem.* 64: 7178–7183.

8 Grossman, R.B., Comesse, S., Rasne, R.M. et al. (2003). Phosphoramidites are efficient, green organocatalysts for the Michael reaction. Mechanistic insights into the phosphorus-catalyzed Michael reaction of alkynones and implications for asymmetric catalysis. *J. Org. Chem.* 68: 871–874.

9 Zhang, C.M. and Lu, X.Y. (1995). Phosphine-catalyzed cycloaddition of 2,3-butadienoates or 2-butynoates with electron-deficient olefins – a novel [3+2]-annulation approach to cyclopentenes. *J. Org. Chem.* 60: 2906–2908.

10 Du, Y.S., Lu, X.Y., and Yu, Y.H. (2002). Highly regioselective construction of spirocycles via phosphine-catalyzed [3+2]-cycloaddition. *J. Org. Chem.* 67: 8901–8905.

11 Cowen, B.J. and Miller, S.J. (2007). Enantioselective [3+2]-cycloadditions catalyzed by a protected, multifunctional phosphine-containing alpha-amino acid. *J. Am. Chem. Soc.* 129: 10988–10989.

12 Xu, Z.R. and Lu, X.Y. (1998). A novel [3+2] cycloaddition approach to nitrogen heterocycles via phosphine-catalyzed reactions of 2,3-butadienoates or 2-butynoates and dimethyl acetylenedicarboxylate with imines: a convenient synthesis of pentabromopseudilin. *J. Org. Chem.* 63: 5031–5041.

13 Nair, V., Poonoth, M., Vellalath, S. et al. (2006). An *N*-heterocyclic carbene-catalyzed [8+3] annulation of tropone and enals via homoenolate. *J. Org. Chem.* 71: 8964–8965.

14 Ohara, H., Itoh, T., Nakamura, M., and Nakamura, E. (2001). [2+2]-Cycloaddition reaction of styrene derivatives using an Fe(III) salt catalyst. *Chem. Lett.* 624–625.

15 Yoon, T.P. (2013). Visible light photocatalysis: the development of photocatalytic radical ion cycloadditions. *ACS Catal.* 3: 895–902.

16 Bauld, N.L. and Mirafzal, G.A. (1991). Hole-catalyzed epoxidation. *J. Am. Chem. Soc.* 113: 3613–3614.

17 Rodini, D.J. and Snider, B.B. (1980). Dimethylaluminum halide induced reactions of formaldehyde with alkynes – synthesis of alpha-allenic alcohols and zeta-3-chloroallylic alcohols. *Tetrahedron Lett.* 21: 3857–3860.

18 Snider, B.B. and Rodini, D.J. (1980). Dialkylaluminum chloride catalyzed ene reactions of aldehydes – synthesis of ipsenol. *Tetrahedron Lett.* 21: 1815–1818.

19 Bennett, D.M., Okamoto, I., and Danheiser, R.L. (1999). Hetero [4+2] cycloadditions of (trialkylsilyl) vinylketenes. Synthesis of α,β-unsaturated δ-valerolactones and -lactams. *Org. Lett.* 1: 641–644.

20 Fang, J., Ren, J., and Wang, Z.W. (2008). $Sc(OTf)_3$-catalyzed smooth tandem [3+2] cycloaddition/ring opening of donor-acceptor cyclopropane 1,1-diesters with enol silyl ethers. *Tetrahedron Lett.* 49: 6659–6662.

21 Takasu, K., Nagao, S., and Ihara, M. (2006). Construction of highly-functionalized cyclopentanes from silyl enol ethers and activated cyclopropanes by [3+2] cycloaddition catalyzed by triflic imide. *Adv. Synth. Catal.* 348: 2376–2380.

22 Milstein, D. and Calabrese, J.C. (1982). Oxidative addition of unactivated epoxides to iridium(i) complexes – formation of stable *cis*-hydridoformylmethyl and hydridacylmethyl complexes. *J. Am. Chem. Soc.* 104: 3773–3774.

23 Green, M.L.H. (1978). Studies on synthesis, mechanism and reactivity of some organo-molybdenum and organo-tungsten compounds. *Pure Appl. Chem.* 50: 27–35.

24 Stille, J.K., Huang, F., and Regan, M.T. (1974). Mechanism of acid chloride decarbonylation with chlorotris(triphenylphosphine)rhodium(I) – stereochemistry and direction of elimination. *J. Am. Chem. Soc.* 96: 1518–1522.

25 Hours, A.E. and Snyder, J.K. (2006). $PtCl_2$-promoted cyclopropane opening in [4+2+2] homo Diels–Alder cycloadducts. *Tetrahedron Lett.* 47: 675–678.

26 Hours, A.E. and Snyder, J.K. (2008). $PtCl_2$-mediated cyclopropane opening: a mechanistic study. *Organometallics* 27: 410–417.

27 Hoye, T.R., Chen, K.J., and Vyvyan, J.R. (1993). Preparation of Fischer carbene complexes by alkylation of acylmetalates with alkyl iodides. *Organometallics* 12: 2806–2809.

28 Negishi, E. and Takahashi, T. (1994). Patterns of stoichiometric and catalytic reactions of organozirconium and related complexes of synthetic interest. *Acc. Chem. Res.* 27: 124–130.

29 Wienhofer, G., Westerhaus, F.A., Junge, K. et al. (2013). A molecularly defined iron-catalyst for the selective hydrogenation of α,β-unsaturated aldehydes. *Chem. Eur. J.* 19: 7701–7707.

30 Langer, R., skin-Posner, Y., Leitus, G. et al. (2011). Low-pressure hydrogenation of carbon dioxide catalyzed by an iron pincer complex exhibiting noble metal activity. *Angew. Chem. Int. Ed.* 50: 9948–9952.

31 Bleith, T., Wadepohl, H., and Gade, L.H. (2015). Iron achieves noble metal reactivity and selectivity: highly reactive and enantioselective iron complexes as catalysts in the hydrosilylation of ketones. *J. Am. Chem. Soc.* 137: 2456–2459.

32 Bleith, T. and Gade, L.H. (2016). Mechanism of the iron(II)-catalyzed hydrosilylation of ketones: activation of iron carboxylate precatalysts and reaction pathways of the active catalyst. *J. Am. Chem. Soc.* 138: 4972–4983.

33 Chalk, A.J. and Harrod, J.F. (1965). Homogeneous catalysis. 2. Mechanism of hydrosilation of olefins catalyzed by group 8 metal complexes. *J. Am. Chem. Soc.* 87: 16–21.

34 Glaser, P.B. and Tilley, T.D. (2003). Catalytic hydrosilylation of alkenes by a ruthenium silylene complex. Evidence for new hydrosilylation mechanism. *J. Am. Chem. Soc.* 125: 13640–13641.

35 Wu, W.Q. and Jiang, H.F. (2012). Palladium-catalyzed oxidation of unsaturated hydrocarbons using molecular oxygen. *Acc. Chem. Res.* 45: 1736–1748.

36 Rout, L., Saha, P., Jammi, S., and Punniyamurthy, T. (2008). Efficient copper(I)-catalyzed C–S cross coupling of thiols with aryl halides in water. *Eur. J. Org. Chem.* 640–643.

37 Sperotto, E., van Klink, G.P.M., de Vries, J.G., and Van Koten, G. (2008). Ligand-free copper-catalyzed C–S coupling of aryl iodides and thiols. *J. Org. Chem.* 73: 5625–5628.

38 Maimone, T.J. and Buchwald, S.L. (2010). Pd-catalyzed *O*-arylation of ethyl acetohydroximate: synthesis of *O*-arylhydroxylamines and substituted benzofurans. *J. Am. Chem. Soc.* 132: 9990–9991.

8

Transition-Metal-Catalyzed C—C Bond Forming Reactions

Answers to Problems

Problem 8.1 What products can form upon carbonylation of ethanol catalyzed by a Rh catalyst in the presence of ethyl iodide? [1]

Answer

Because of possible β-elimination in (ethyl)rhodium intermediates ethylene is expected to be formed next to the expected product of ethoxycarbonylation: ethyl propionate. HI is a co-product of the reaction, a very strong acid that catalyzes the formation of diethyl ether from ethanol.

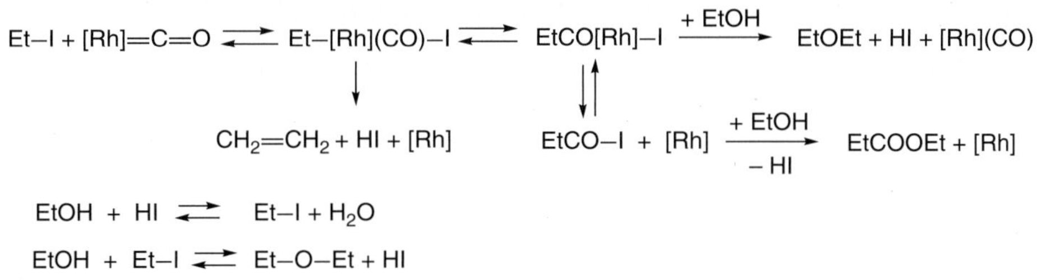

Problem 8.2 Propose a mechanism for the hydrocarbonylation of acetic acid into propionic acid catalyzed by $[Rh(CO)_2Cl]_2$ and LiI [2].

Answer

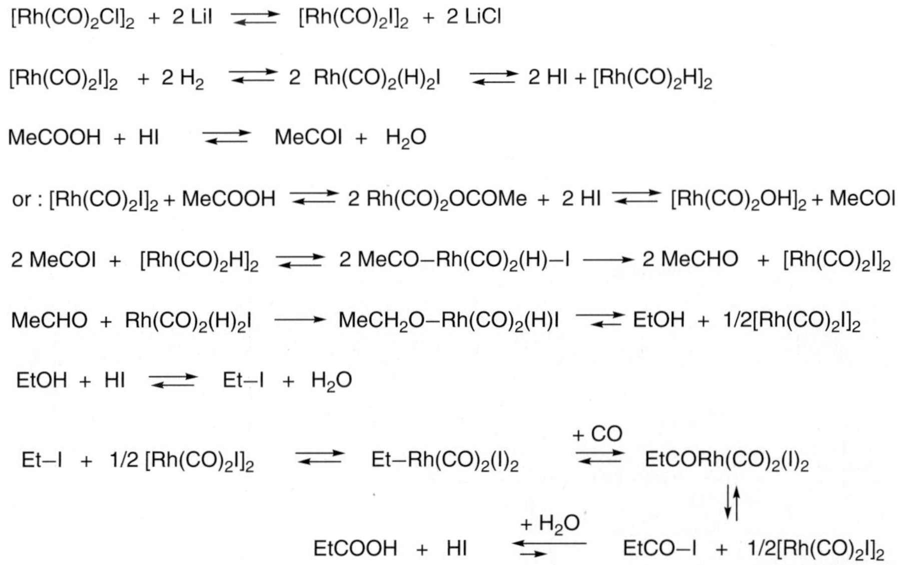

Organic Chemistry: Theory, Reactivity and Mechanisms in Modern Synthesis, Workbook, First Edition. Pierre Vogel and Kendall N. Houk.

Problem 8.3 Explain the following reaction [3].

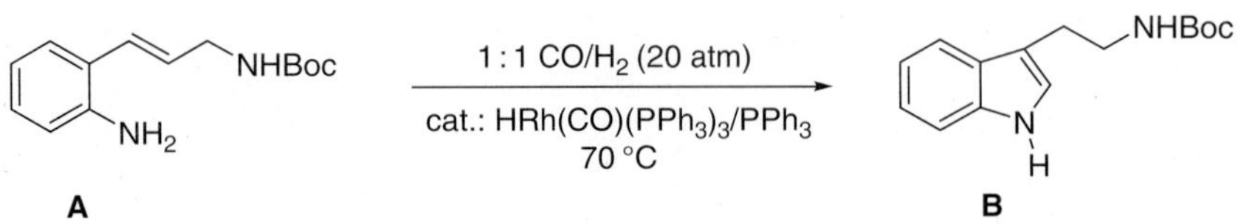

Answer

A priori, hydrocarbonylation of **A** can give aldehydes **B** and **D** competitively, irrespective of whether the amino or BocHN groups play lateral controls or not. Aldehyde **B** undergoes fast intramolecular addition of the amino moiety producing a α-aminoalcohol **C** that loses a mole of water irreversibly (due to the gain of indole aromaticity), leading to the final product **P**. If formed, the regioisomeric aldehyde **D** eliminates H_2O giving the less stable imine **Q,** which is not stabilized by aromaticity as is **P**. This raises the question whether hydroformylation is a reversible or irreversible reaction under the conditions applied.

If one takes as model reaction the hydroformylation of propene at 25 °C and under 1 atm, and using thermochemical data given by the NIST Chemistry WebBook, one estimates a standard heat of reaction $\Delta_r H°(CH_2 = CHCH_3 + H_2 + CO \rightleftarrows MeCH(CHO)Me)) = \Delta_f H°(MeCH(CHO)Me) - \Delta_f H°(\text{propene}) - \Delta_f H°(H_2) - \Delta_f H°(CO) = -57.6 - 4.9 - 0 + 26.0 = -30.1\ \text{kcal mol}^{-1}$ (Section 1.4), and an entropy variation $\Delta_r S°(CH_2{=}CHCH_3 + H_2 + CO \rightleftarrows MeCH(CHO)Me)) = S°(MeCH(CHO)Me) - S°(\text{propene}) - S°(H_2) - S°(CO) = 82.6 - 63.8 - 31.2 - 47.2 = -59.6\ \text{eu}$. The reaction under study is run at 70 °C instead of 25 °C. Gibbs energy at 25 °C(298 K) $\Delta_f G° = \Delta_f H° - T\Delta_r S° = -30\,100 - 298(-59.6)\ \text{cal mol}^{-1}$ (under 1 atm). Using this expression for the same reaction run at $T = 70\,°C$ gives $\Delta_r G^{343\,K} = 30\,100 - 343(-59.6) = -9658\ \text{cal mol}^{-1} \approx -9.7\ \text{kcal mol}^{-1}$. The error done on the heat of reaction when using standard values for a reaction at 343 K instead of 298 K is compensated to a large extent by the error done on the $-T\Delta_r S°$ term. Thus, $\Delta_r G^{343K} = -9.7\ \text{kcal mol}^{-1}$ is a good estimate for the Gibbs energy of the hydroformylation of propene at 70 °C (Section 1.3). This is an exergonic reaction: it can be considered as irreversible. This is also the case for hydroformylation of **A** into **B** or **D**. Considering the fact that the reaction is run under 20 atm instead of 1 atm (standard conditions), reversibility is further decreased by the mass law effect. The non-observation of imine **Q** (that can form reversibly or not from **D**) suggests that aldehyde **B** is formed more rapidly than aldehyde **D** (kinetic control on the regioselectivity of the hydroformylation of **A**).

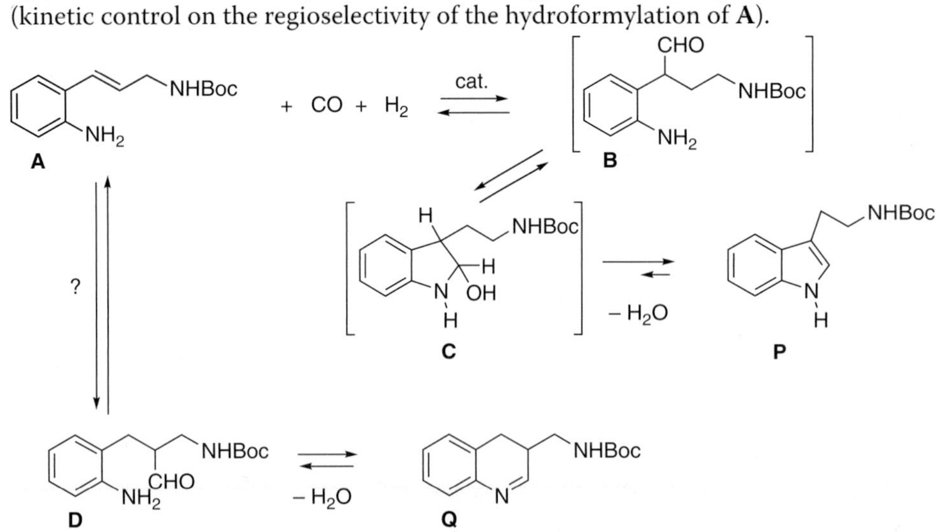

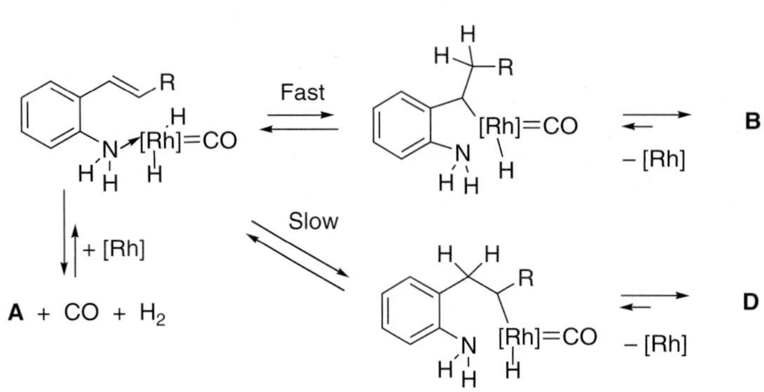

Problem 8.4 Explain the following indole synthesis [4].

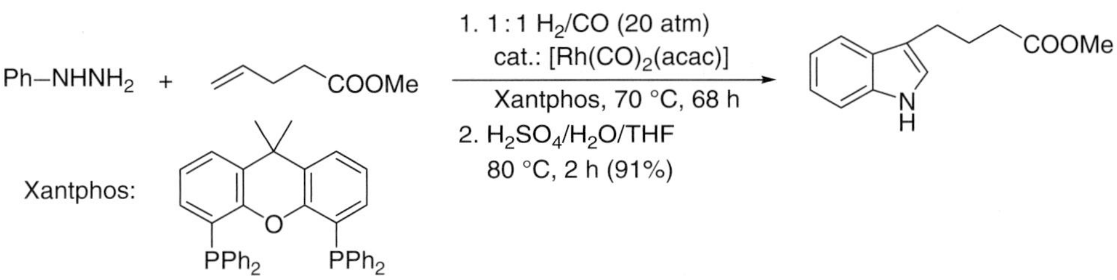

Answer

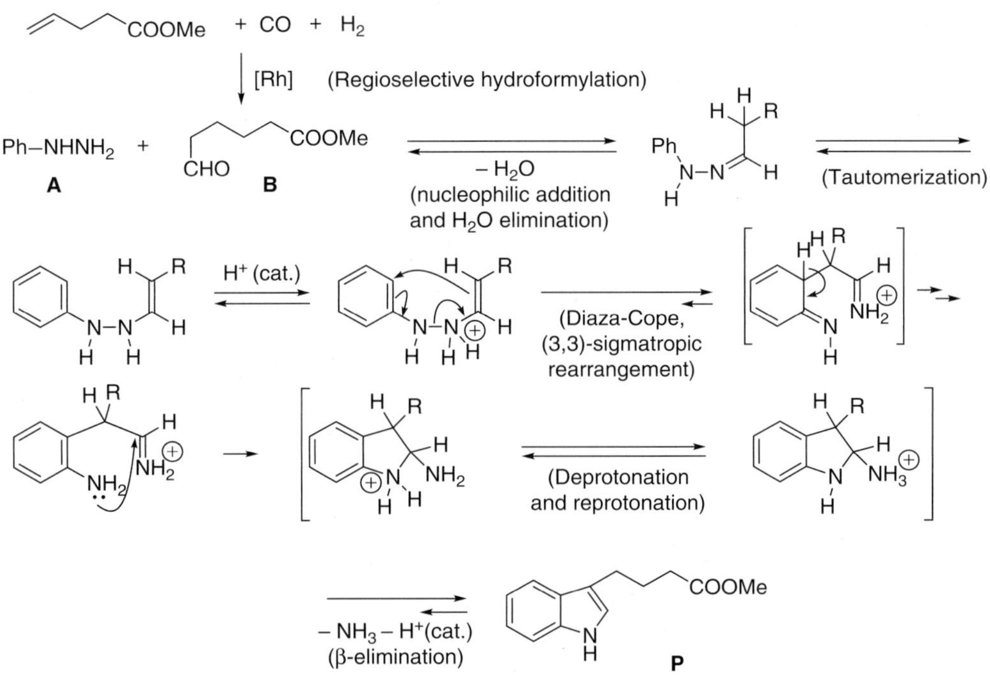

The conversion of **A** + **B** into **P** is an example of the Fischer indole synthesis (Section 5.5.9.1, Scheme 5.97).

Problem 8.5 Propose a mechanism for the silylformylation reaction: $PhCH_2CHO + PhMe_2SiH + CO \rightarrow PhCH_2CH(CHO)\text{—}OSiMe_2Ph$ catalyzed by $[Rh(cod)Cl]_2$ in tetrahydrofuran (THF) [5].

Answer

Possible mechanism

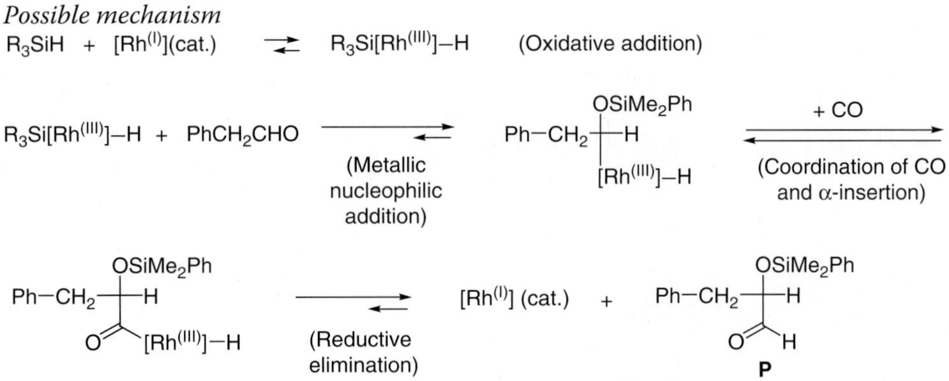

Problem 8.6 What is a possible product of reaction $HC{\equiv}C(CH_2)_3CHO + Et_3SiH + CO$ (1 atm) catalyzed by $Rh_2Co_2(CO)_{12}$ at 25 °C? [6].

Answer

Oxidative addition of an unsaturated rhodium(I) complex onto the Si—H bond of the silane gives a silylrhodium(III) hydride that reacts with the terminal alkyne **A** equilibrating with **B** (regioselective anti-Markovnikov *syn*-addition). Intermediate **B** undergoes metal carbonylation and α-insertion producing **P** competitively with an intramolecular hydride addition onto its carbaldehyde moiety leading to alcoholate **D**. The latter gives the second possible product **Q** via a reductive elimination.

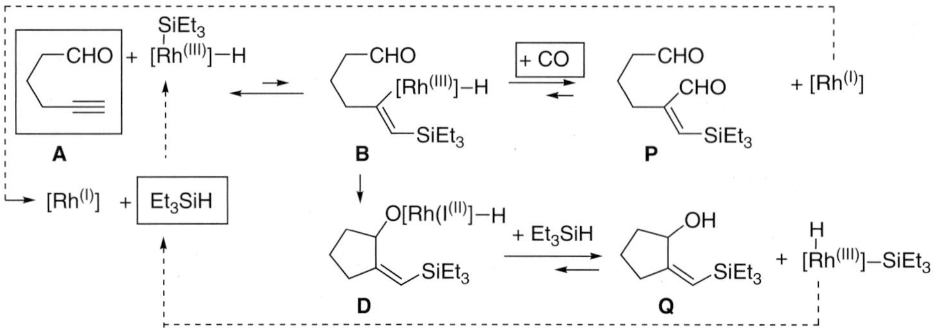

Problem 8.7 (A) Tsuji et al. found that acetylene reacts with CO in the presence of one equivalent of $PdCl_2$ affording maleyl and fumaryl chloride [7]. (B) They found also that diphenylacetylene in ethanol reacts with CO in the presence of a catalytic amount of $PdCl_2$ giving α,β-diphenyl-γ-crotonolactone (66%) and ethyl α,β-diphenylmaleate (26%) in the presence of HCl [8]. Propose mechanisms for these reactions [9, 10].

Answer

Possible mechanism

(a)

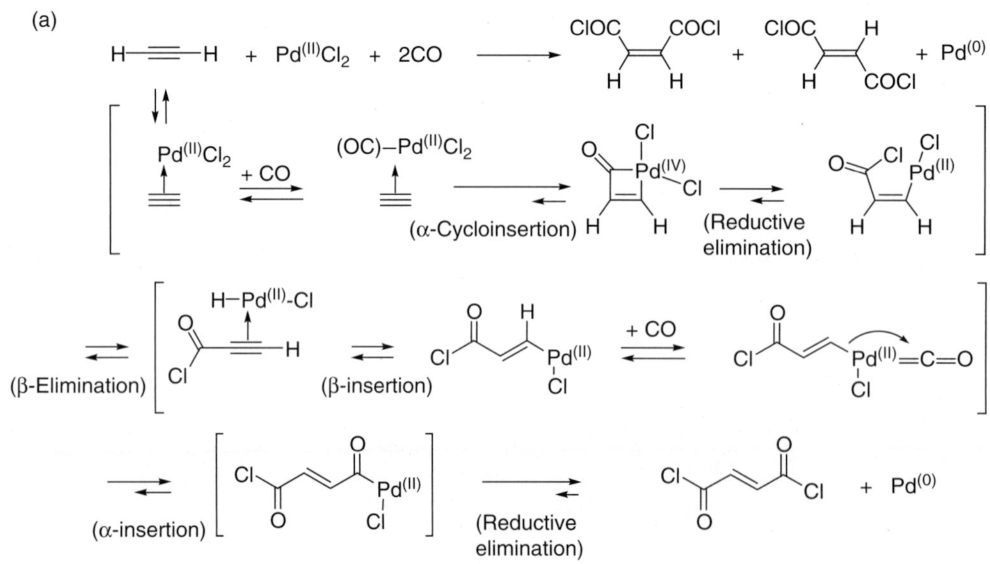

(b)

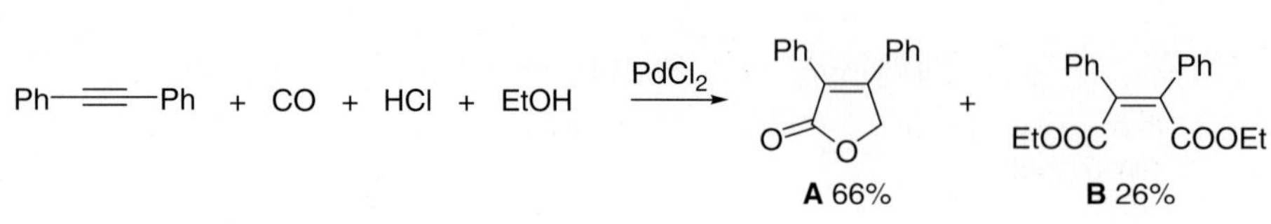

Possible mechanism

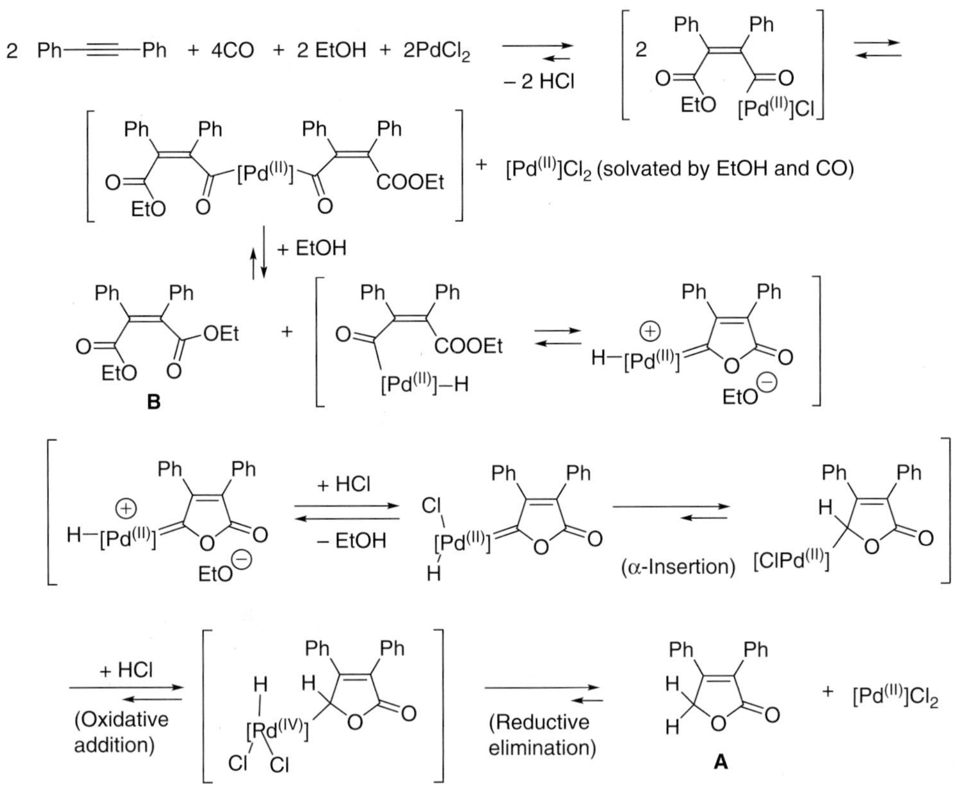

Problem 8.8 What is the major product **P** of the following reaction? [11]

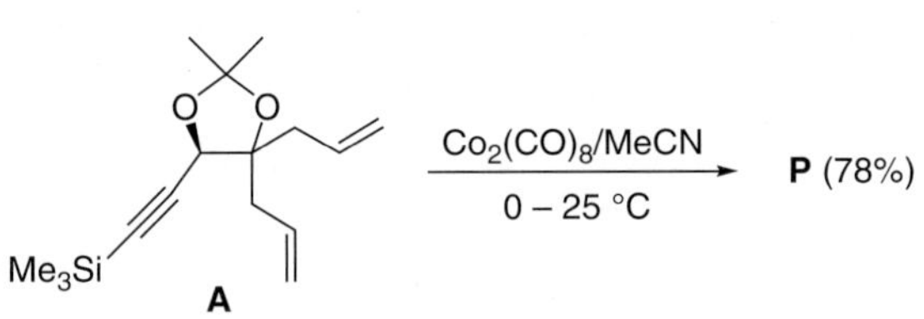

Answer

An example of diastereoselective intramolecular Pauson–Khand reaction catalyzed by $Co_2(CO)_8$.

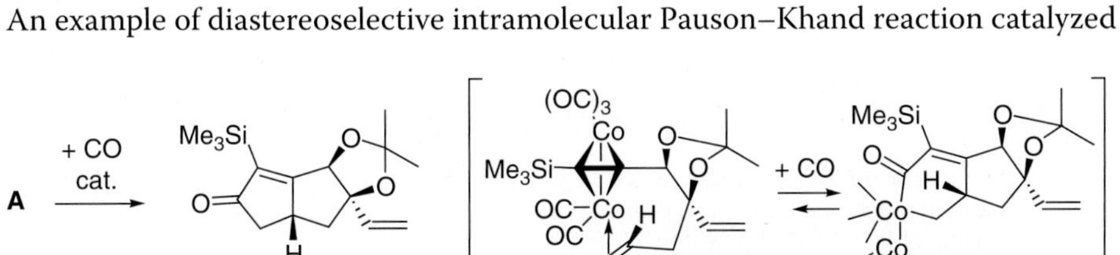

The large metallic moiety occupies the least hindered face of the diene-yne

Problem 8.9 What is product **P** of the following reaction? [12]

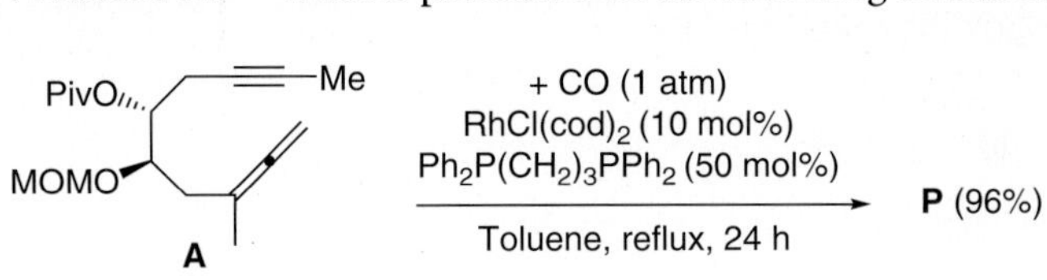

Answer

This is another example of intramolecular Pauson–Khand reaction.

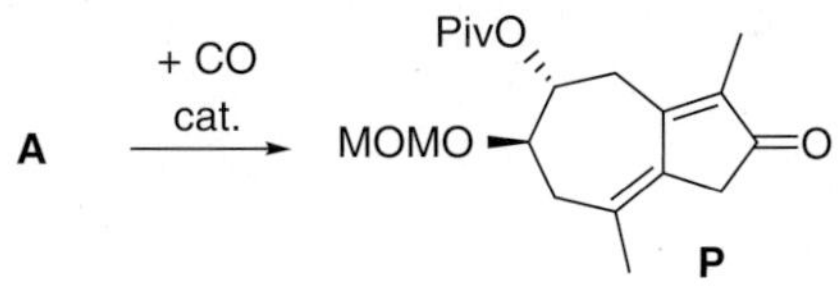

Problem 8.10 Under atmospheric pressure of CO the following reaction gives mostly product **B**, whereas under 100 atm of CO, the ketocarboxamide **D** becomes the dominant product [13]. Explain these results.

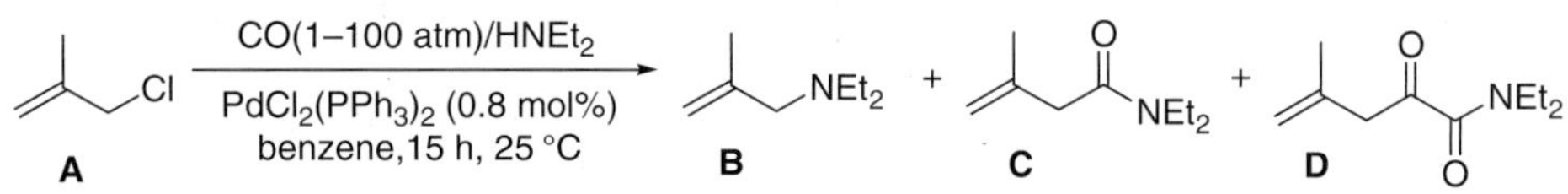

Answer

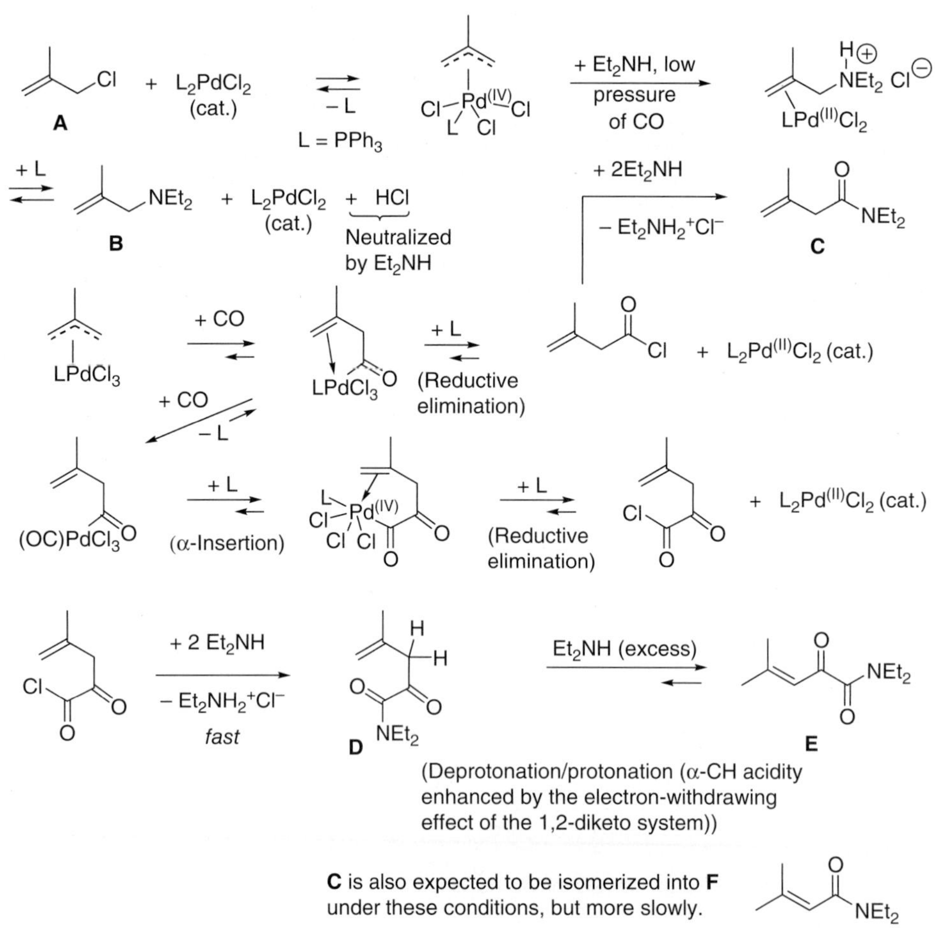

Problem 8.11 What is the major product **P** of the following reaction? [14].

$$\text{PhI} + n\text{-PrI} + \text{CO} \xrightarrow[\text{Pd(Ph}_3\text{P)}_4\text{ (2 mol\%), THF, 50 °C}]{+\text{ CO (1 atm) + Zn/Cu (2 equiv.)}} \textbf{P}$$

Answer

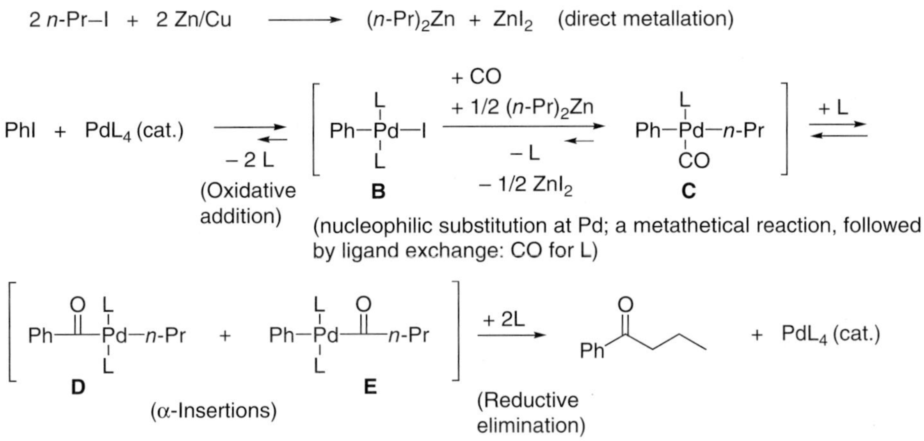

Problem 8.12 Estimate the heat of isomerization of hep-6-en-2-yl radical *trans*-(2-methylcyclopentyl)methyl radical (see reaction (8.43)) and the equilibrium constant of this isomerization under standard conditions (1 atm, 25 °C, gas phase).

Answer

Isomerization of hept-1-ene into *trans*-1,2-dimethylcyclopentane is an example of intramolecular hydrocarbation. The heat of this isomerization is given by the difference in standard heats of formation of these compounds. Table 1.2 gives $\Delta_f H°$(hept-1-ene) = −15.1 kcal mol^{-1}, but does not give $\Delta_f H°$(*trans*-1,2-dimethylcyclopentane) but it gives $\Delta_f H°$(methylcyclopentane) = −25.3 kcal mol^{-1}. One estimates $\Delta_f H°$(*trans*-1,2-dimethylcyclopentane) from $\Delta_f H°$(methylcyclopentane) by considering the exchange of a $CH_2(sp^3)$ moiety for a CHMe group. Table 2.1 gives for these groups [C—$(C)_2(H)_2$], [C—$(C)_3$(H), and [C—(C)$(H)_3$] −4.92, −1.75, and −10.12 kcal mol^{-1}, respectively. Thus the exchange of CH_2 for CHMe is associated with +4.92 − 1.75 − 10.12 = −6.95 ≈ −7 kcal mol^{-1}. Table 1.2 gives also $\Delta_f H°$(methylcyclopentane) − $\Delta_f H°$(cyclopentane) = −25.3 − (−18.3) = −7 kcal mol^{-1}. This shows that methyl substitution of cyclopentane does not introduce significant destabilization due to gauche interactions. One can thus estimate $\Delta_f H°$(*trans*-1,2-dimethylcyclopentane) = $\Delta_f H°$(methylcyclopentane) − 7 = −25.3 − 7 = −32.3 kcal mol^{-1}. The NIST Webbook of Chemistry gives $\Delta_f H°$(*trans*-1,2-dimethylcyclopentane) = −32.7 kcal mol^{-1}. For the isomerization hept-6-en-2-yl radical into *trans*-(2-methylcyclopentyl)methyl radical one can write:

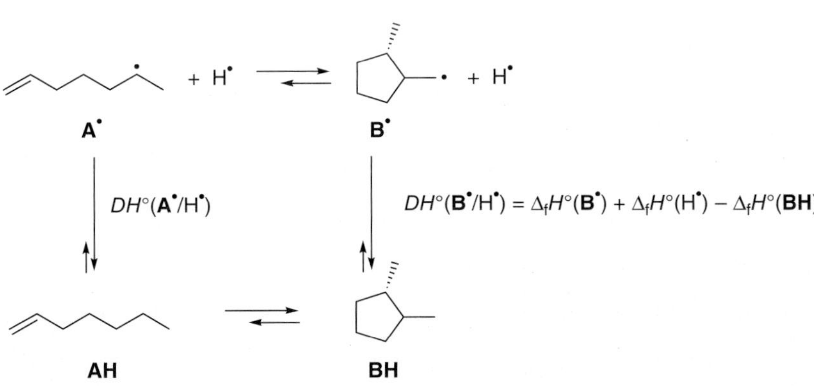

$$\Delta_r H°(\mathbf{A^\bullet} \rightleftarrows \mathbf{B^\bullet}) = \Delta_f H°(\mathbf{B^\bullet}) - \Delta_f H°(\mathbf{A^\bullet}) = \Delta_f H°(\mathbf{B^\bullet}) + \Delta_f H°(H^\bullet) - \Delta_f H°(\mathbf{A^\bullet}) - \Delta_f H°(H^\bullet) =$$

$$= \Delta_f H°(\mathbf{BH}) - \Delta_f H°(\mathbf{AH}) + DH°(\mathbf{B^\bullet}/H^\bullet) - DH°(\mathbf{A^\bullet}/H^\bullet) =$$

$$= \Delta_f H°(\mathbf{BH}) - \Delta_f H°(\mathbf{AH}) + \Delta_f H°(\mathbf{B^\bullet}) + \Delta_f H°(H^\bullet) - \Delta_f H°(\mathbf{BH}) - \Delta_f H°(\mathbf{AH}) - \Delta_f H°(\mathbf{A^\bullet}) - \Delta_f H°(H^\bullet)$$

$$= \Delta_r H°(\mathbf{AH} \rightleftarrows \mathbf{BH}) + DH°(\mathbf{B^\bullet}/H^\bullet) - DH°(\mathbf{A^\bullet}/H^\bullet)$$

Table 7.1 does not give $DH°(\mathbf{B}^{\cdot}/\mathrm{H}^{\cdot})$ and $DH°(\mathbf{A}^{\cdot}/\mathrm{H}^{\cdot})$. They can be estimated as $DH°(\mathbf{B}^{\cdot}/\mathrm{H}^{\cdot}) = DH°$(*trans*-(2-methylcyclopentyl)methyl$^{\cdot}$/H$^{\cdot}$) $\approx DH°$ ($Me_2CHCH_2^{\cdot}/H^{\cdot}$) = 100.1 kcal mol^{-1} and $DH°(\mathbf{A}^{\cdot}/\mathrm{H}^{\cdot}) = DH°$(hex-6-en-2-yl$^{\cdot}$/H$^{\cdot}$) $\approx DH°$ (but-2-yl$^{\cdot}$/H$^{\cdot}$) = 99.1 kcal mol^{-1}.

This gives $\Delta_r H°(\mathbf{A}^{\cdot} \rightleftarrows \mathbf{B}^{\cdot}) = \Delta_r H°(\mathbf{AH} \rightleftarrows \mathbf{BH}) + DH_o(\mathbf{B}^{\cdot}/\mathrm{H}^{\cdot}) - DH°(\mathbf{A}^{\cdot}/\mathrm{H}^{\cdot}) = -32.3 - (-15.1) + 100.1 - 99.1 = -16.2\ \mathrm{kcal\ mol^{-1}}$ (*the heat of the radical isomerization corresponds to that of the heat of the corresponding hydrocarbation corrected for the difference in radical stability*).

Entropy variation $\Delta_r S°(\mathbf{A}^{\cdot} \rightleftarrows \mathbf{B}^{\cdot}) = \Delta S°(\mathbf{B}^{\cdot}) - \Delta S°(\mathbf{A}^{\cdot})$ corresponds to a first approximation to the entropy decrease due to the blockade of rotation about four σ(C—C) bonds when going from the open radical $\mathbf{A}^{\cdot}$ to the cyclized radical $\mathbf{B}^{\cdot}$. This amounts to 4 (−5 eu) = −20 eu (Section 2.10). No correction is necessary for symmetry change since $\mathbf{B}^{\cdot}$ and $\mathbf{A}^{\cdot}$ maintain the same symmetry and have both a free rotating methyl group. Thus, $\Delta_r G° = \Delta_r H° - T\Delta_r S° = -16\,200 - 298(-20) = -16\,200 + 5960 = 10\,240\ \mathrm{cal\ mol^{-1}}$

Equilibrium constant $K^{298\mathrm{K}}(\mathbf{A}^{\cdot} \rightleftarrows \mathbf{B}^{\cdot}) = \exp(-\Delta_r G°/RT) = \exp(10\,240/298 \times 1.987) = 3.2 \times 10^7$.

Problem 8.13 What is product **P** of carbonylation of **A** catalyzed by **60**? [15]

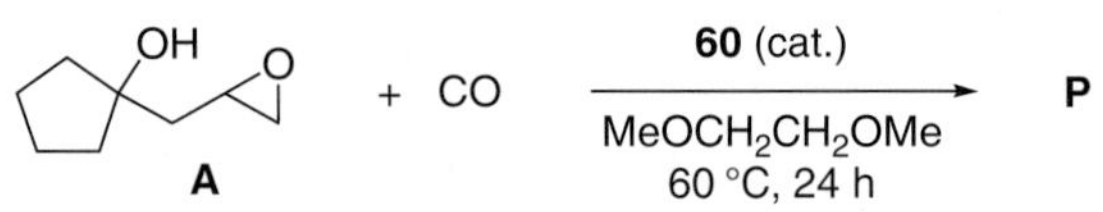

Answer

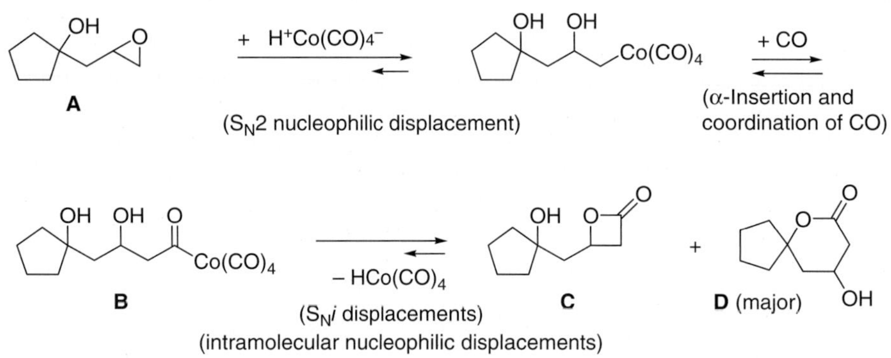

Problem 8.14 What is product **P** of the carbonylation of **A** (60 atm of CO, 2 days)? [16]

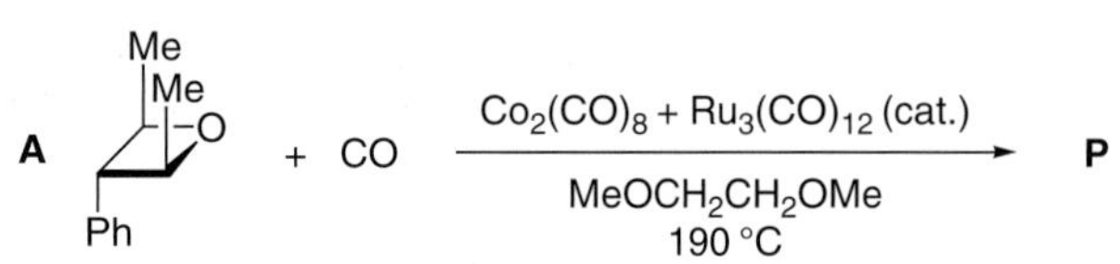

Answer

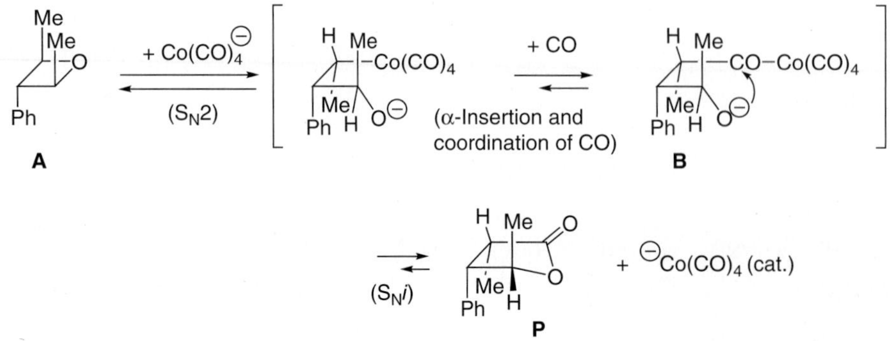

Problem 8.15 What is the major product reaction of $CO + H_2$ with (2*S*)-2-benzyloxymethyloxirane catalyzed by $Co_2(CO)_8$ and ligand **63** in $HC(OMe)_3$? [17]

Answer

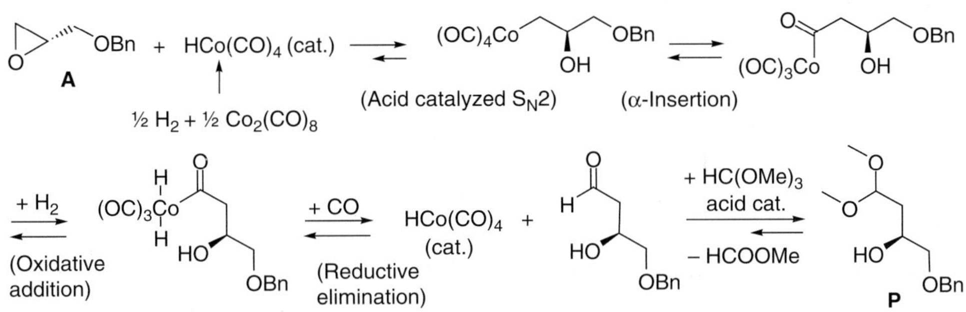

Problem 8.16 What is the major product of the reaction of $CO + PhMe_2SiH$ with 7-oxa bicyclo[4.1.0]heptane catalyzed by $[Rh(CO)_2Cl]_2$ in CH_2Cl_2 containing 1-methylpyrazole? [18]

Answer

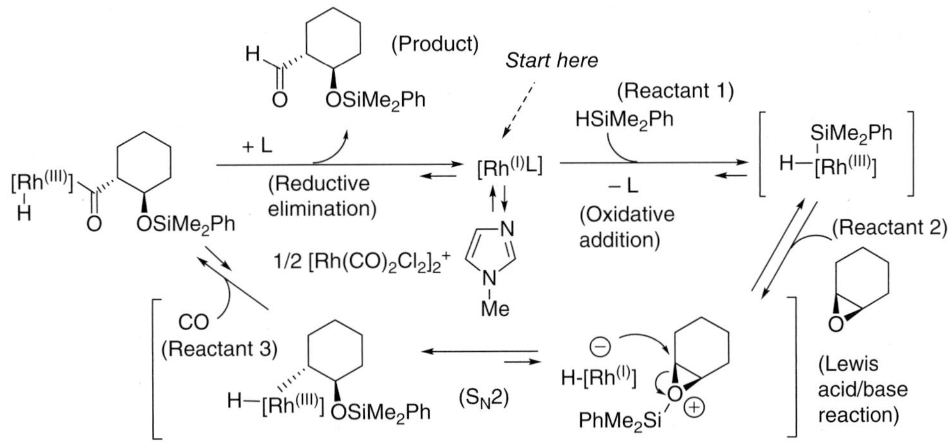

Problem 8.17 What are the major products **P**, **Q** of reactions **A** + **B** and **C** + **B** catalyzed by $CpRu(MeCN)_3PF_6$ (acetone, 25 °C)? [19]

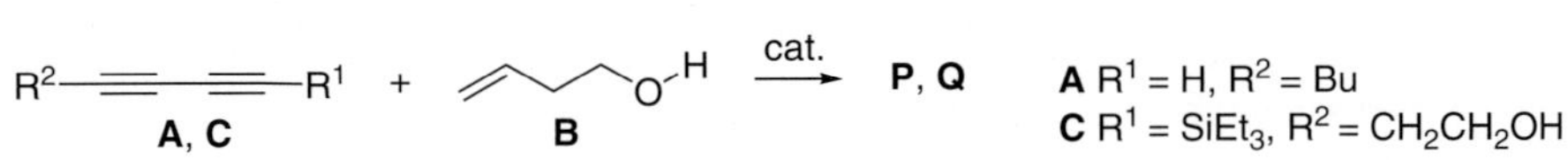

Answer

Examples of Ru^+-catalyzed Alder ene-reactions (hydrocarbations).

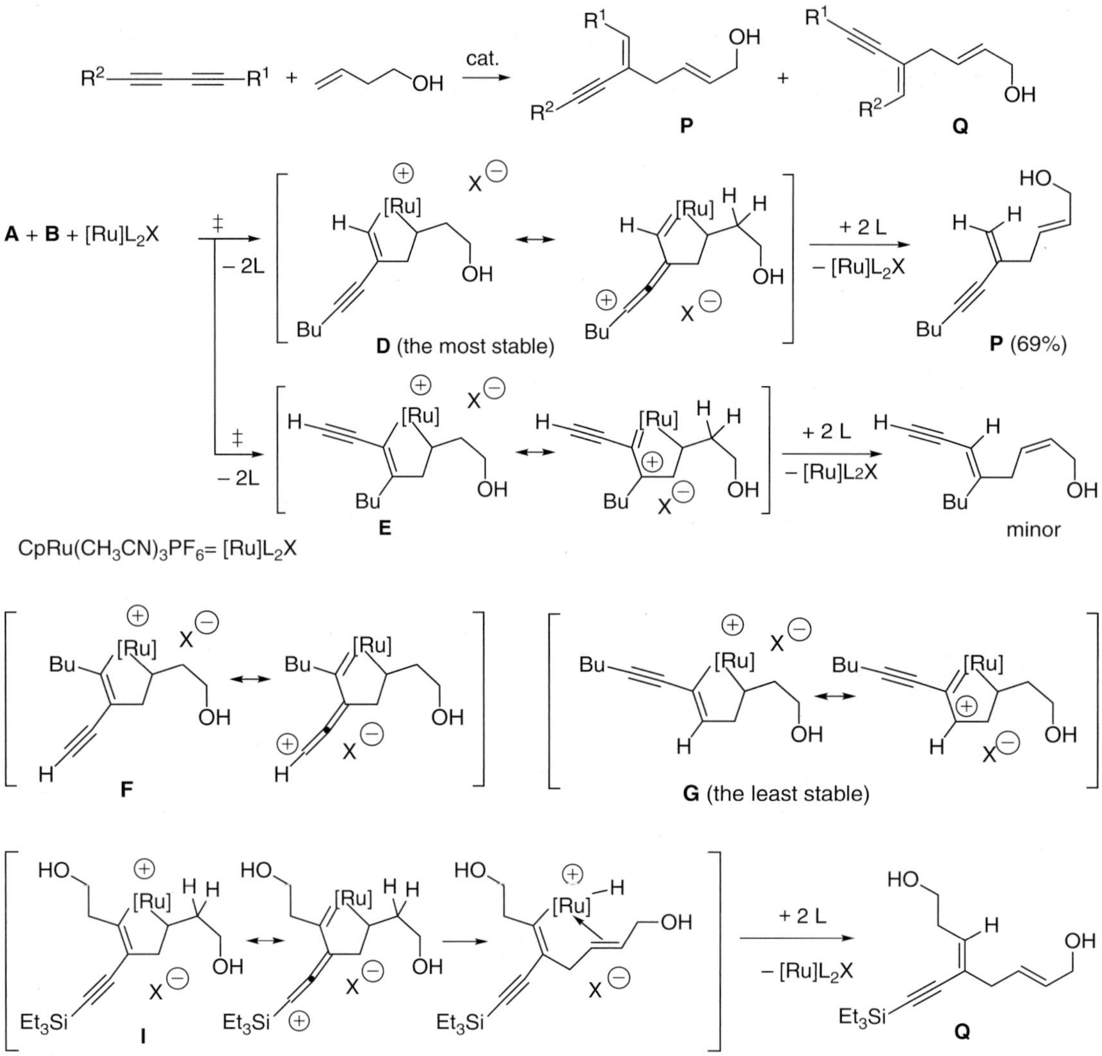

Problem 8.18 What is the main product **P** of the following reaction? [20].

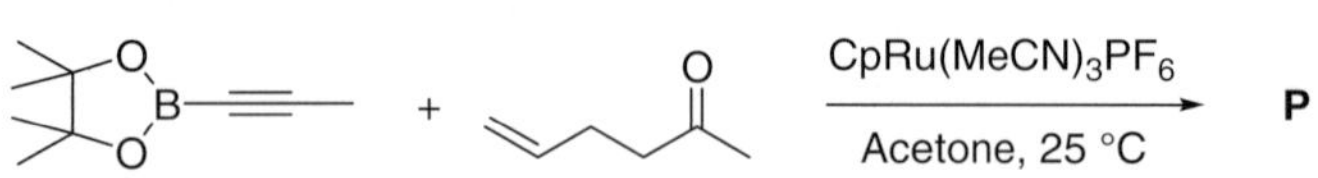

Answer

$[Ru]^{\oplus}$ = CpRu$^{(II)}$(MeCN)$_n^{\oplus}$ n = 1,2 (MeCN can be substituted by solvent)

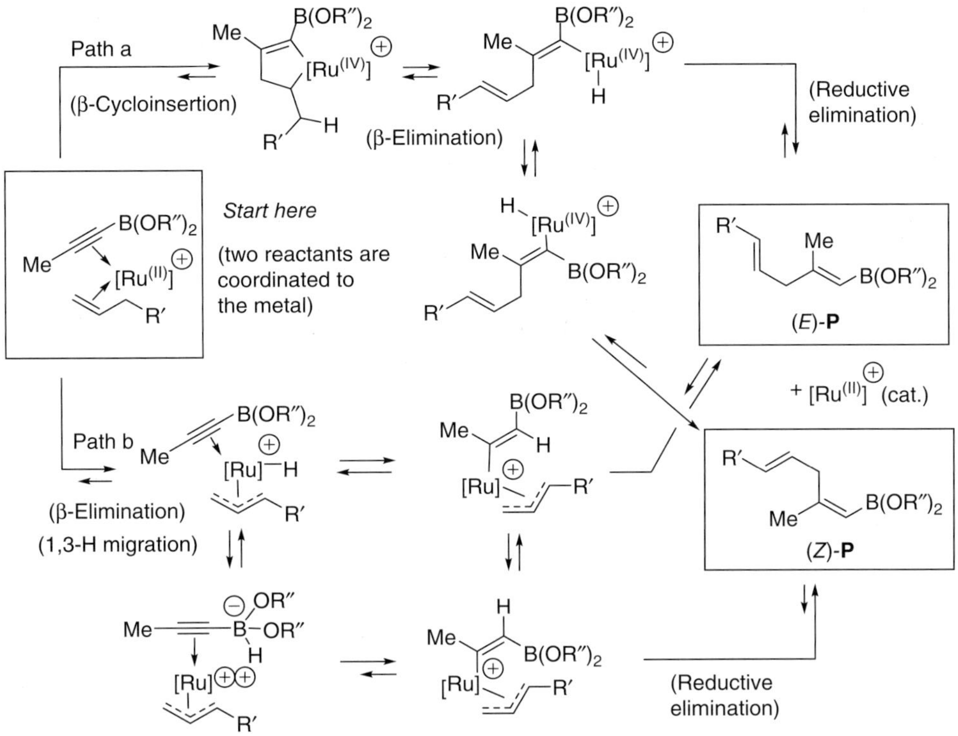

Problem 8.19 What is the main product of the reaction of benzonitrile with CH_2=CHSi(OEt)$_3$ catalyzed by $RuH_2(CO)(PPh_3)_3$ in toluene heated under reflux? [21]

Answer

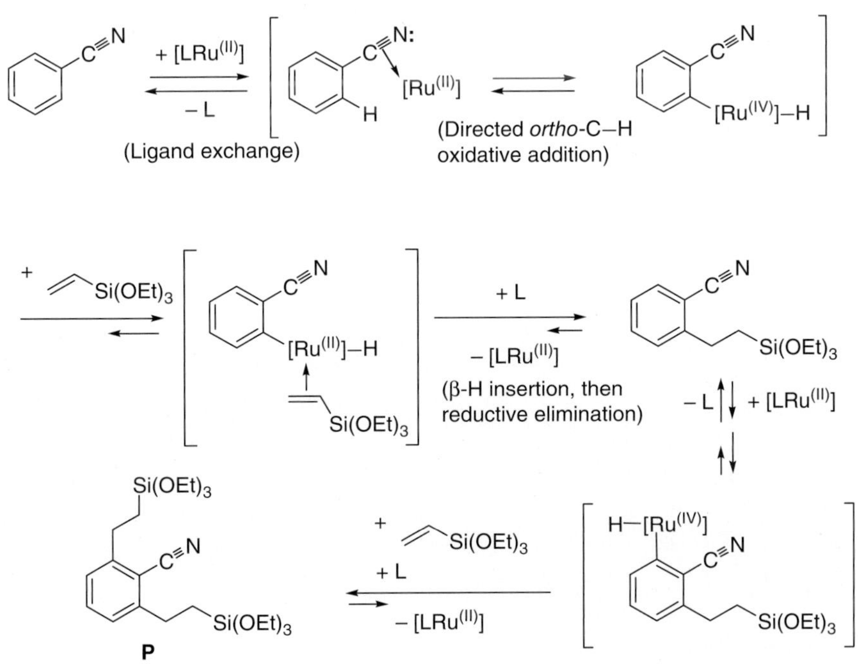

Problem 8.20 What is the main product of the reaction of 1-tetralone with MeC≡CSiMe$_3$ catalyzed by RuH$_2$(CO)(PPh$_3$)$_3$ in toluene heated under reflux? [22]

Answer

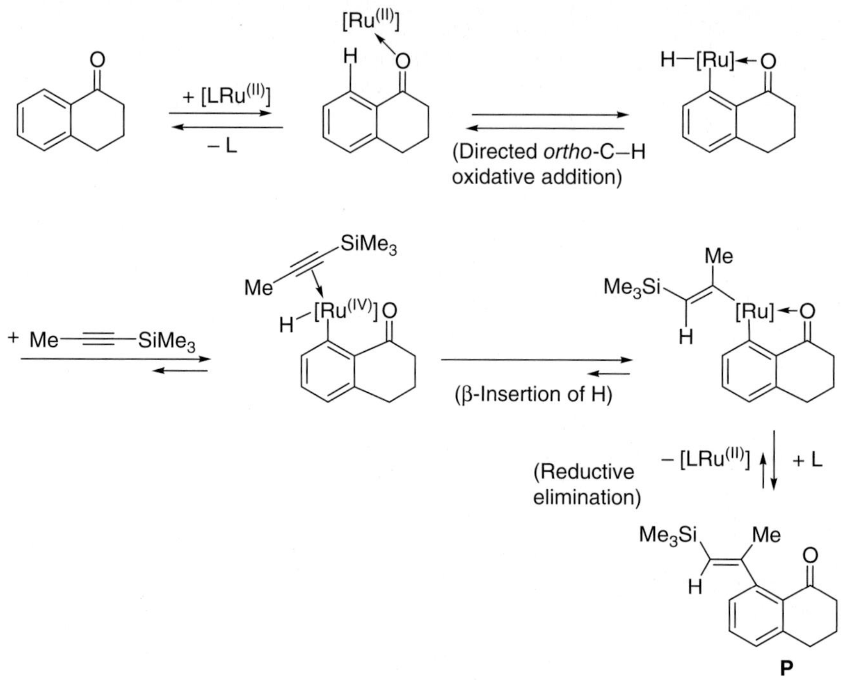

Problem 8.21 What is product **P** of the following reaction? [23]

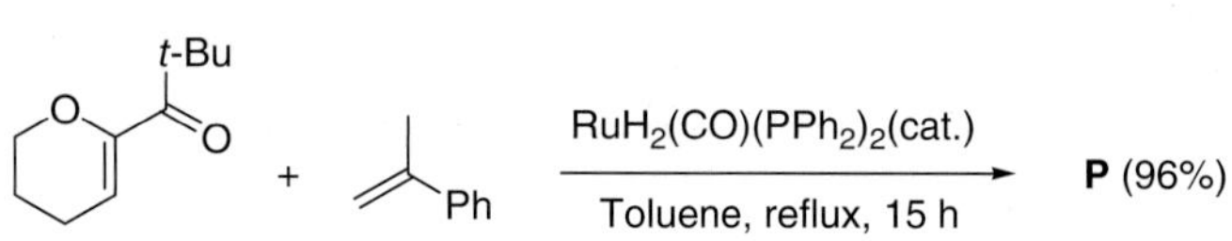

Answer

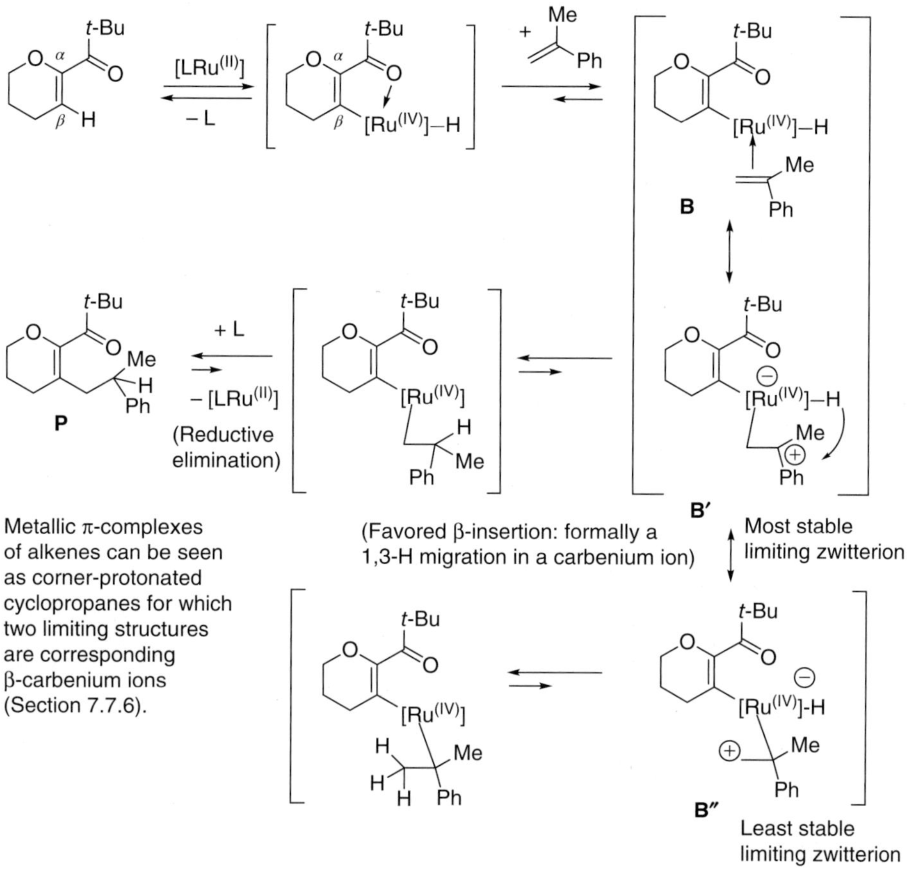

Problem 8.22 What is the main product of reaction of PhC(=NOMe)Et with Ph—N=C=O catalyzed by $[Cp^*Rh(MeCN)_3][SbF_6]_2$ in 1,2-dichloroethane at 100 °C [24].

Answer

Oxidative addition onto the *ortho* C—H bond; lateral control by the *O*-methyloxime, followed by a nucleophilic metal-aryl addition to phenyl isocyanate:

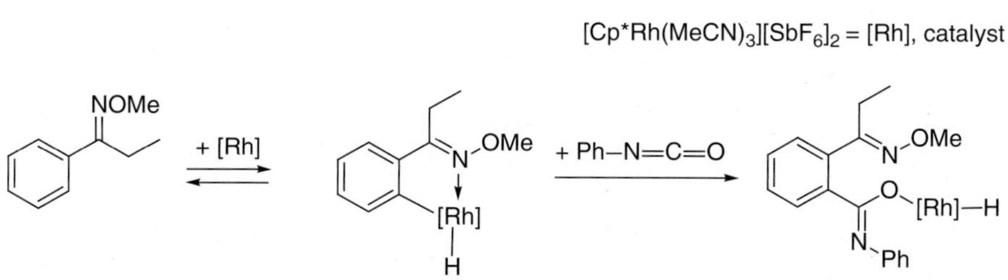

A reductive elimination produces the corresponding benzamide **B**:

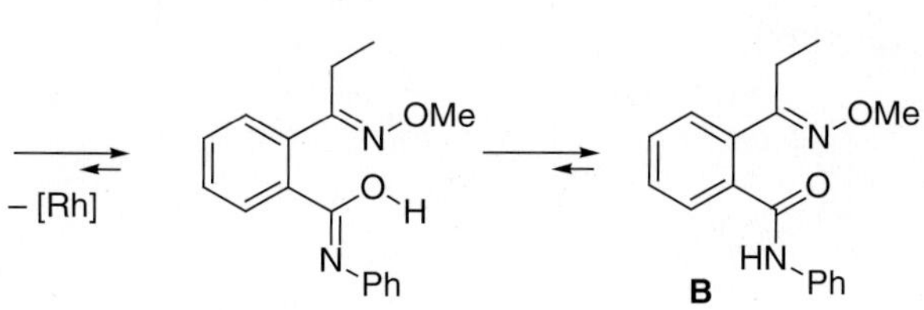

On heating **B** to 100 °C, intramolecular nucleophilic addition of the amide moiety onto the *O*-methyloxime group gives intermediate **D**, which eliminates one equivalent of $MeONH_2$ providing the final product **P**.

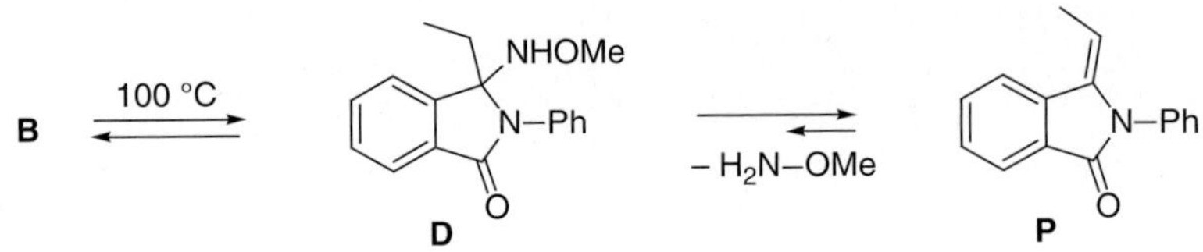

Problem 8.23 The following cycloisomerizations of enynes **A** (intramolecular hydroalkenylations of alkynes) have been catalyzed [25] by the Jonas' iron complex [26, 27]. Propose a mechanism for these isomerizations.

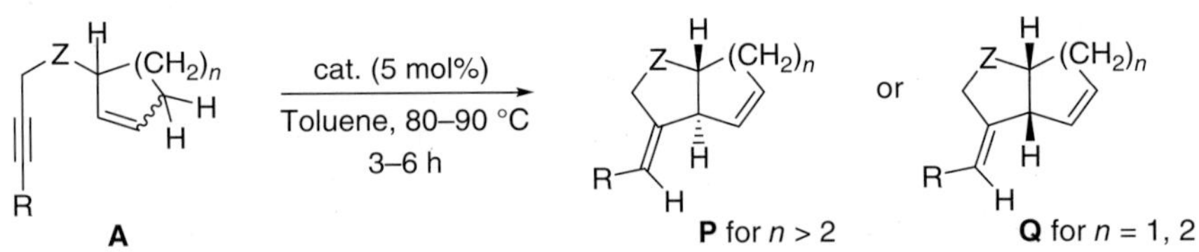

Z = $C(COOEt)_2$, $C(CH_2OR)_2$, O, NBn, NTs
R = Ar, H, Me, cyclopropyl, PhCO

Jonas complex: $CpFe(C_2H_4)_2Li(TMEDA)$
TMEDA = $Me_2NCH_2CH_2NMe_2$

Answer

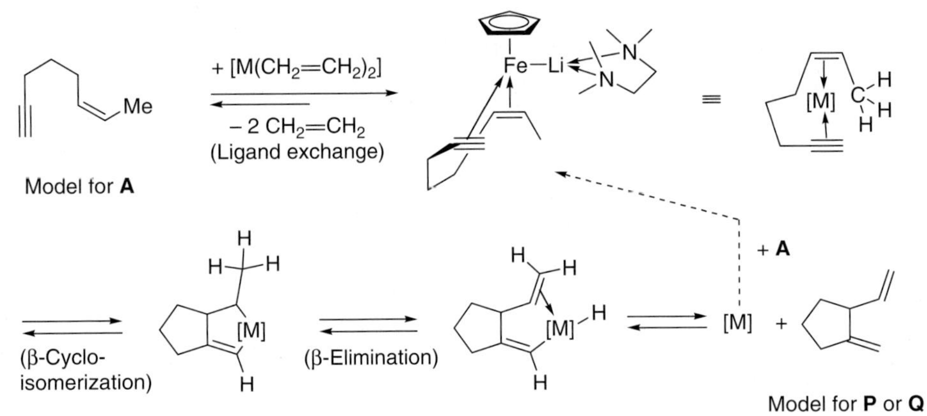

Synthesis of the catalyst $[M(CH_2{=}CH_2)_2]$:

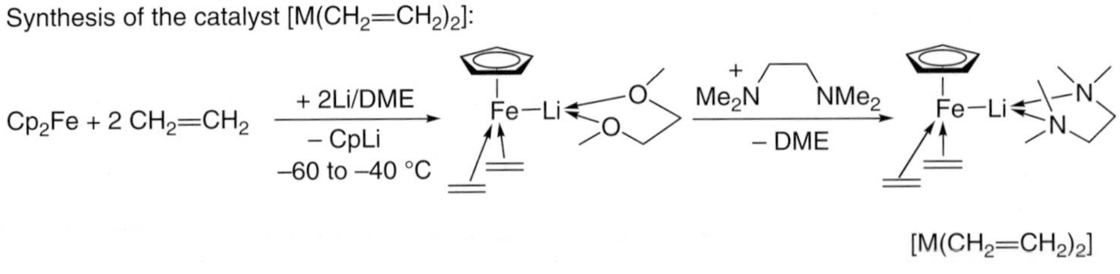

Problem 8.24 Propose a mechanism for the following reaction [28].

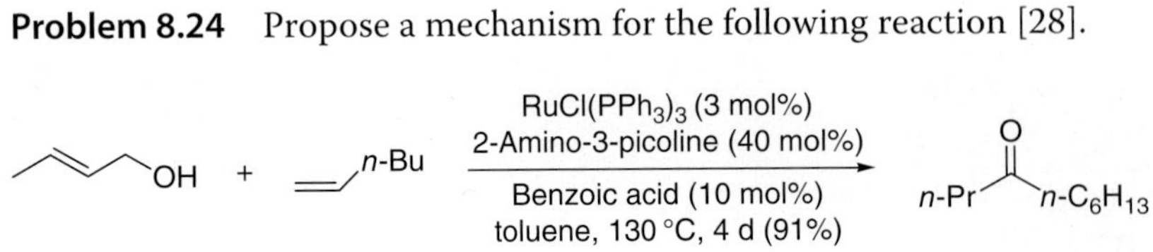

Answer

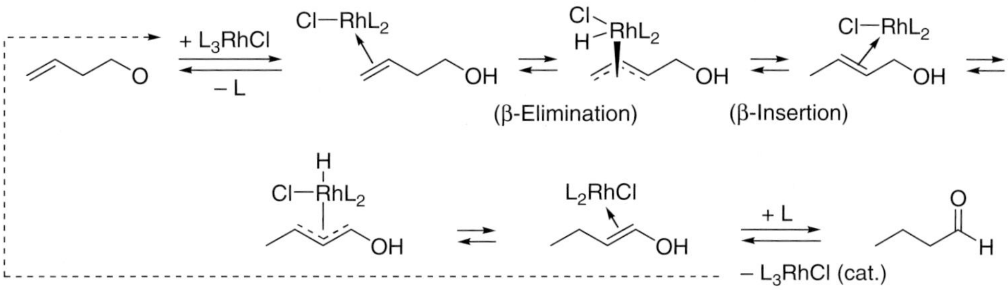

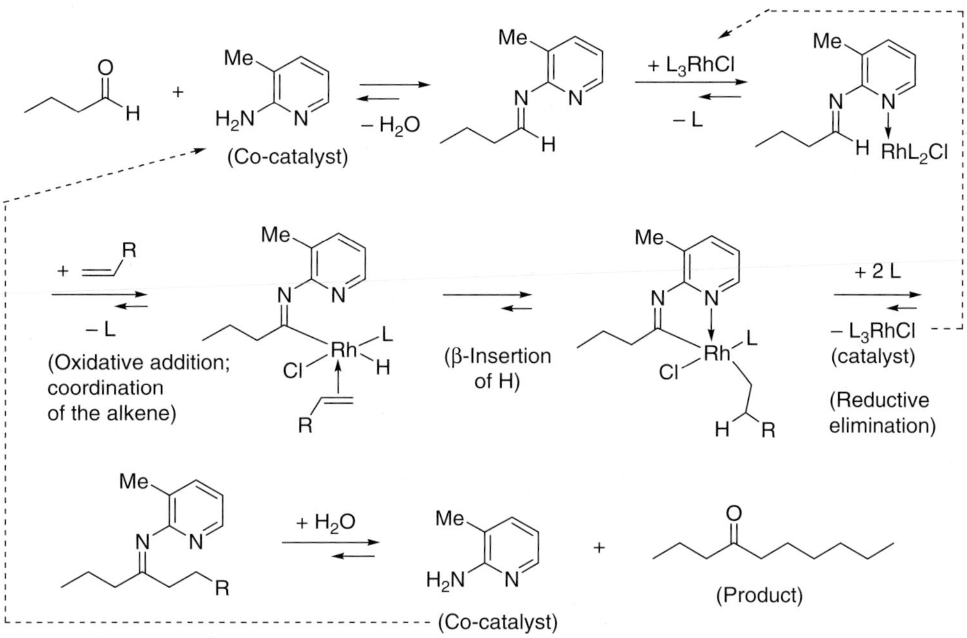

Problem 8.25 What is the main product **P** of reaction of 2-methylthiobenzaldehyde and (2-nitrophenyl) acetylene catalyzed by Rh(norbornadiene)$_2$BF$_4$ and bis(dicyclohexylphosphino)methane in acetone at 55 °C? What is the product **Q** resulting from the treatment of **P** with Sn and aqueous HCl in EtOH? [29, 30]

Answer

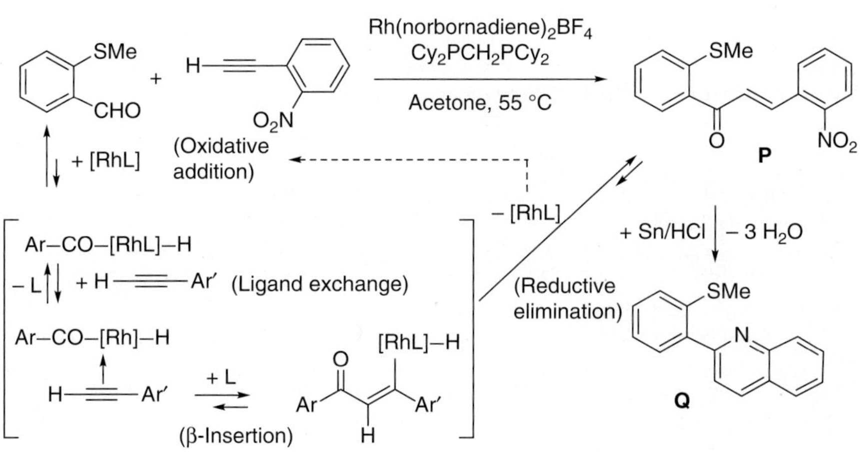

Problem 8.26 Propose a mechanism for the nickel-catalyzed hydrocyanation of **93** into adiponitrile [31].

Answer

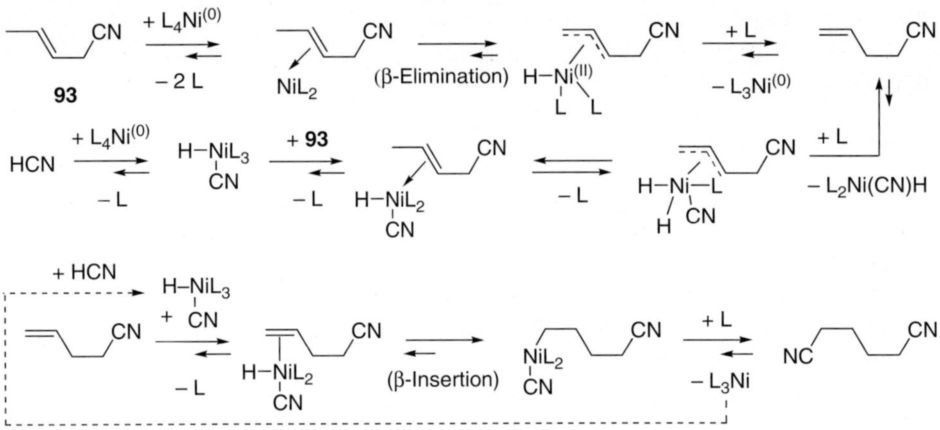

Problem 8.27 In the presence of an excess of isopropanol, $Ru(CO)(PPh_3)_3(H)Cl$ and a diphosphine ligand as catalyst, allenes **A** react with formaldimines forming homoallylamines with complete regioselectivity. Propose a mechanism for this reaction [32].

A:

Answer

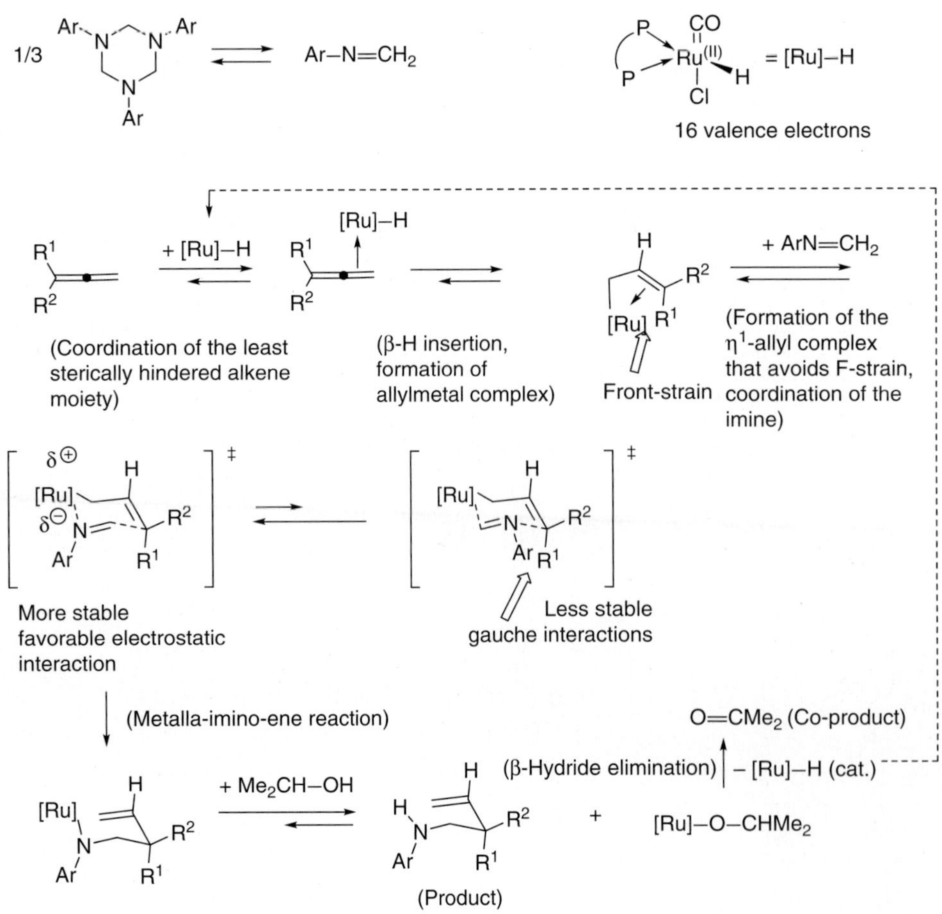

Problem 8.28 Propose a mechanism for the following reaction [33].

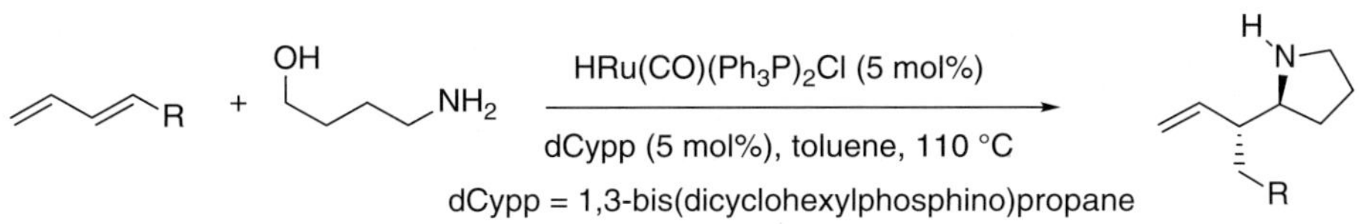

Answer

Ligand exchange generates HRu(dCypp)Cl, which coordinates the diene. Hydrogen β-insertion produces allylruthenium complex **A**. The latter is displaced by the alcohol with liberation of HCl forming the complex intermediate **B**. Hydride transfer to the metallic center (β-elimination) gives intermediate **C**, which equilibrates with **D** and aldehyde **E**. HCl catalyzes the intramolecular formation of imine **I** that equilibrates with **D** to produce intermediate **F**. The latter undergoes β-insertion of the allyl group leading finally to the product of imine *anti*-crotylation with high enantioselectivity.

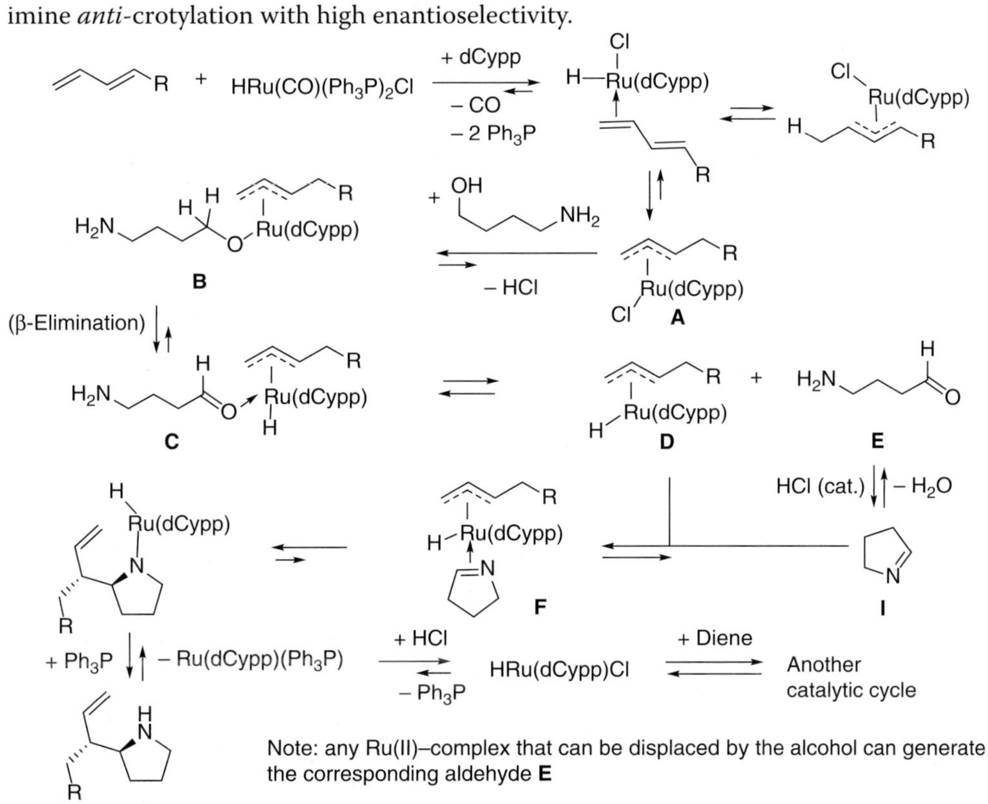

Problem 8.29 What is the product **P** formed by reaction of bicyclo[2.2.1]hept-2-ene with biphenylene in the presence of 10 mol% of [(dtbpm)RhCl]$_2$ (dtbpm = (*t*-Bu)$_2$PCH$_2$P(*t*-Bu)$_2$]? [34]

Answer

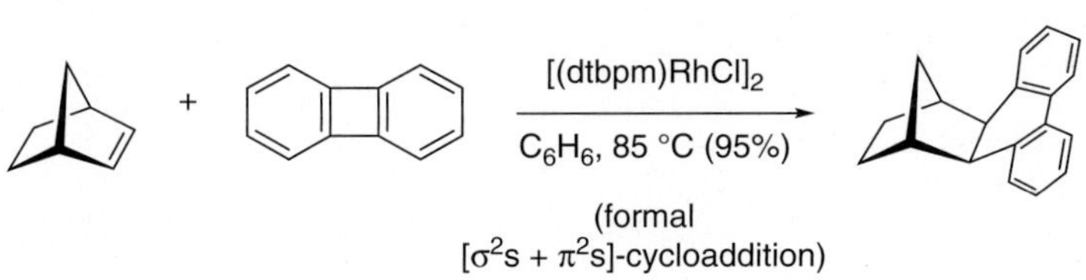

Problem 8.30 What are the successive products **B** and **C** produced in the following reaction? [35, 36]

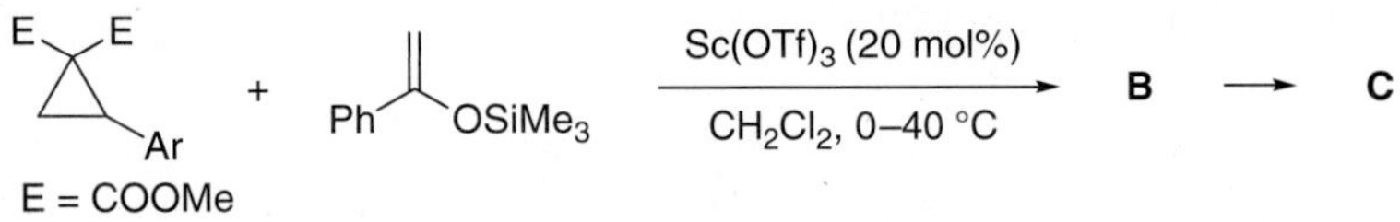

Answer

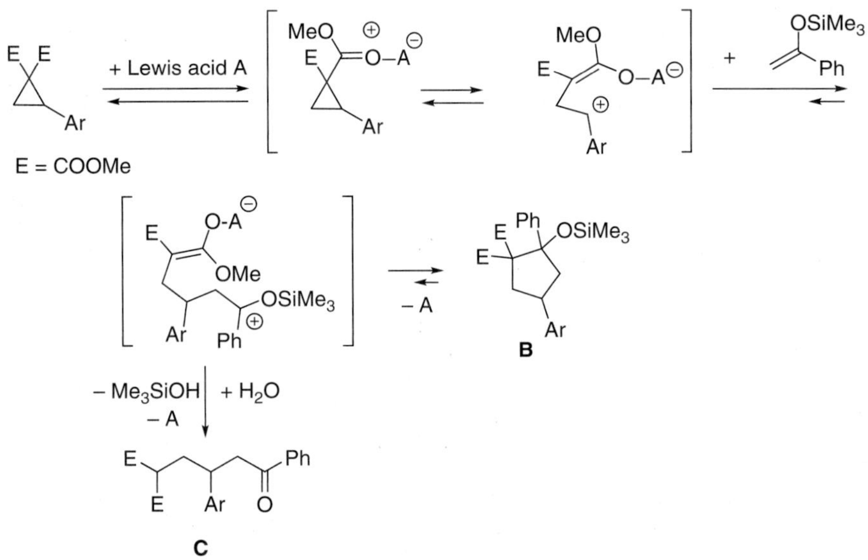

Problem 8.31 What are the two products **B** (73%) and **C** (7%) formed in the following reaction? [37, 38]

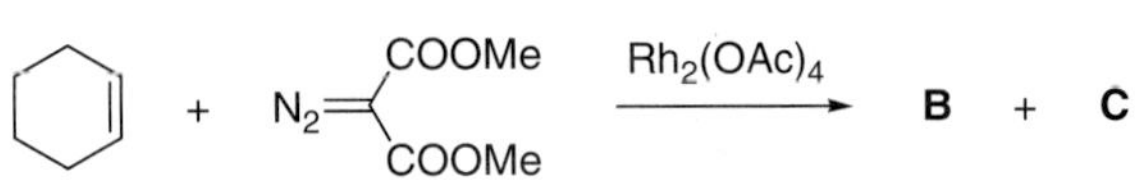

Answer

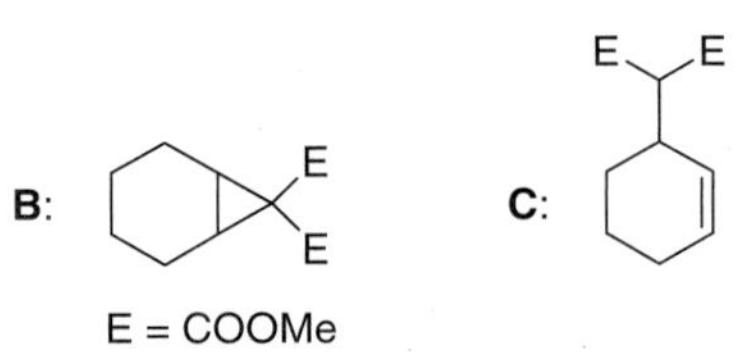

Problem 8.32 What are the major products **P** and **Q** of the reactions of *N*-phenylmaleimide (**C**) with dimethyl norbornadiene-2,3-dicarboxylate (**A**) and with 2-methoxynorbornadiene (**B**) catalyzed by $Ni(cod)_2$ and PPh_3 in toluene at 80–110 °C? [39].

Answer

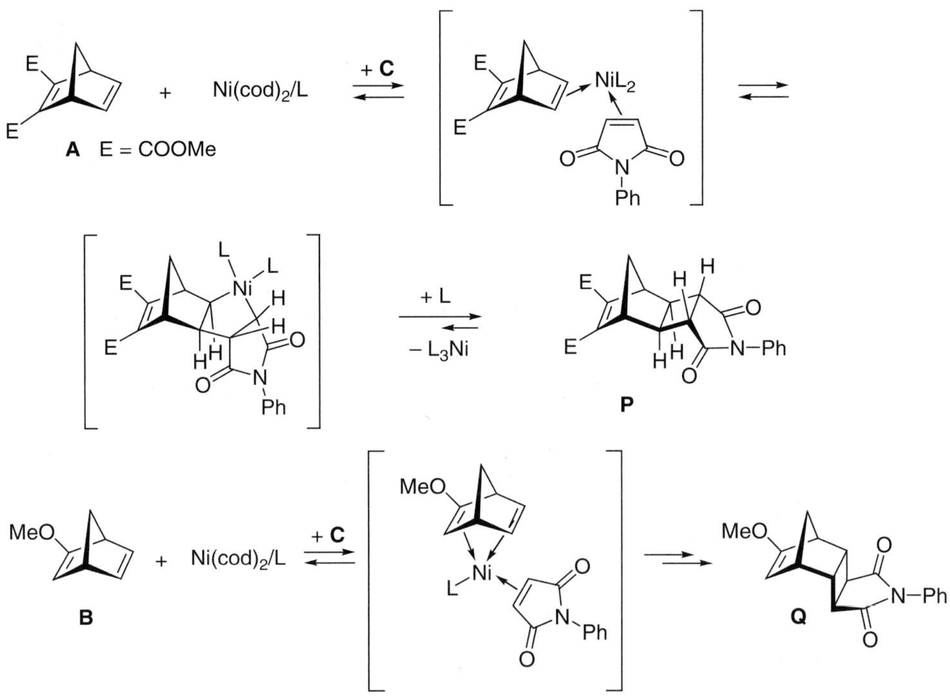

The maleate moiety of **A** (electron poor) cannot coordinate nickel as well as the methyl vinyl ether moiety of **B** (electron rich).

Problem 8.33 Estimate the heat of the $[\pi^2+\pi^2]$-cycloadditions of norbornadiene + PhC≡CCOOMe.

Answer

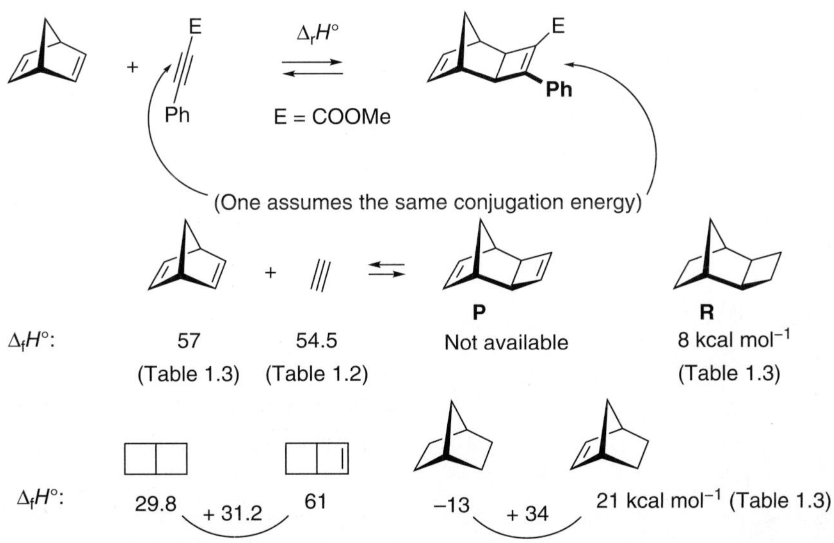

Approximation: $\Delta_f H°(\mathbf{P}) = \Delta_f H°(\mathbf{R}) + 31.2 + 34 = 73.2\,\text{kcal mol}^{-1}$
Estimation: $\Delta_r H° = 73.2 - 57 - 54.5 = -38.3\,\text{kcal mol}^{-1}$
Compare with $\Delta_r H°$(acetylene + ethylene = cyclobutene) $= 37.5 - 54.2 - 12.5 = -29.2\,\text{kcal mol}^{-1}$
Compared with ethylene, norbornadiene is a strained alkene. Part of this strain is released during the (2+2)-cycloaddition. The standard heat of formation of **R** (Table 1.3) is nearly the same as that estimated from

norbornane to which one exchanges two CH_2 fragments for two CH fragments, adds two CH_2 fragments, and considers the ring strain of cyclobutane: $-13 + 10 - 4 - 10 + 26 = 9\,\text{kcal mol}^{-1}$.

Problem 8.34 What is the main product of the following reaction? [40, 41]

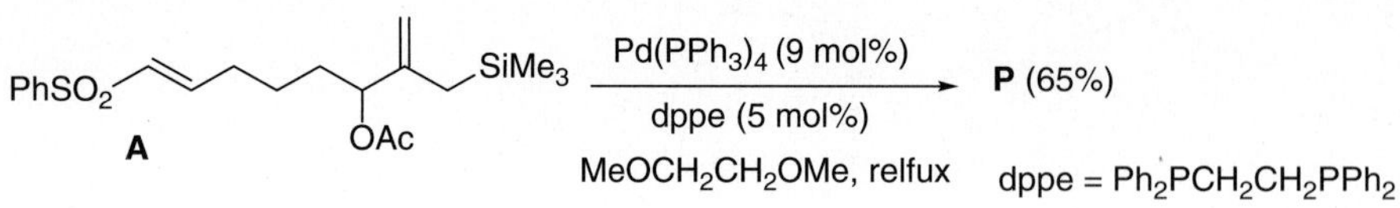

Answer

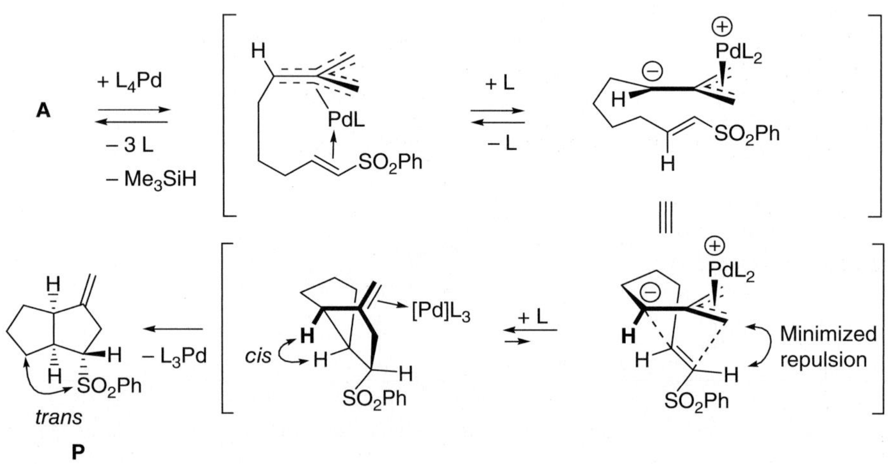

Problem 8.35 Propose a mechanism for reaction (8.140) using (alkyl)CH=C=CHCOOMe.

Answer

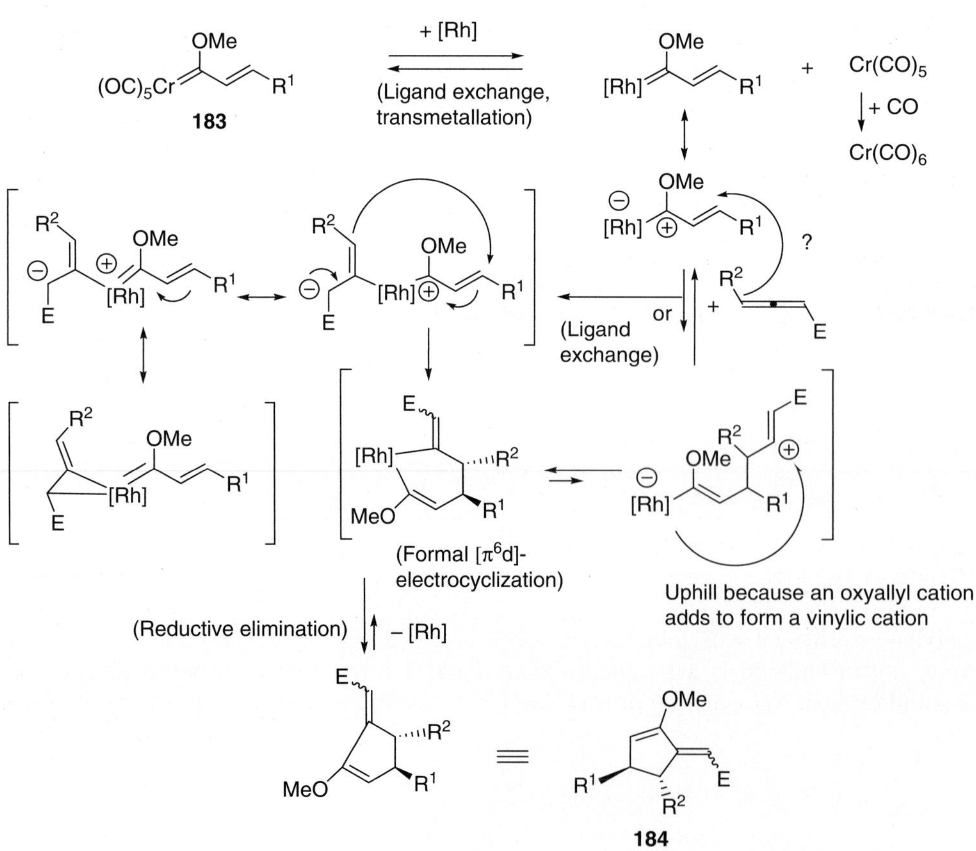

Problem 8.36 Propose a mechanism for reaction (8.143) [42].

Answer

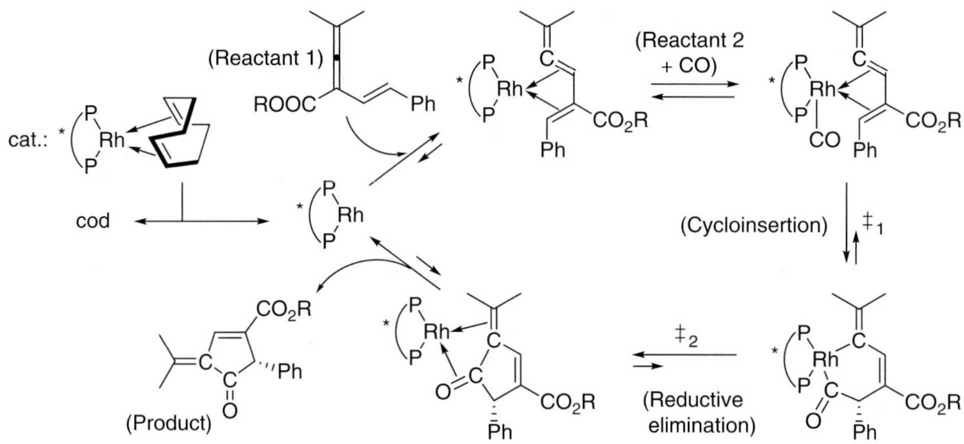

Problem 8.37 Propose a mechanism for reaction (8.144).

Answer

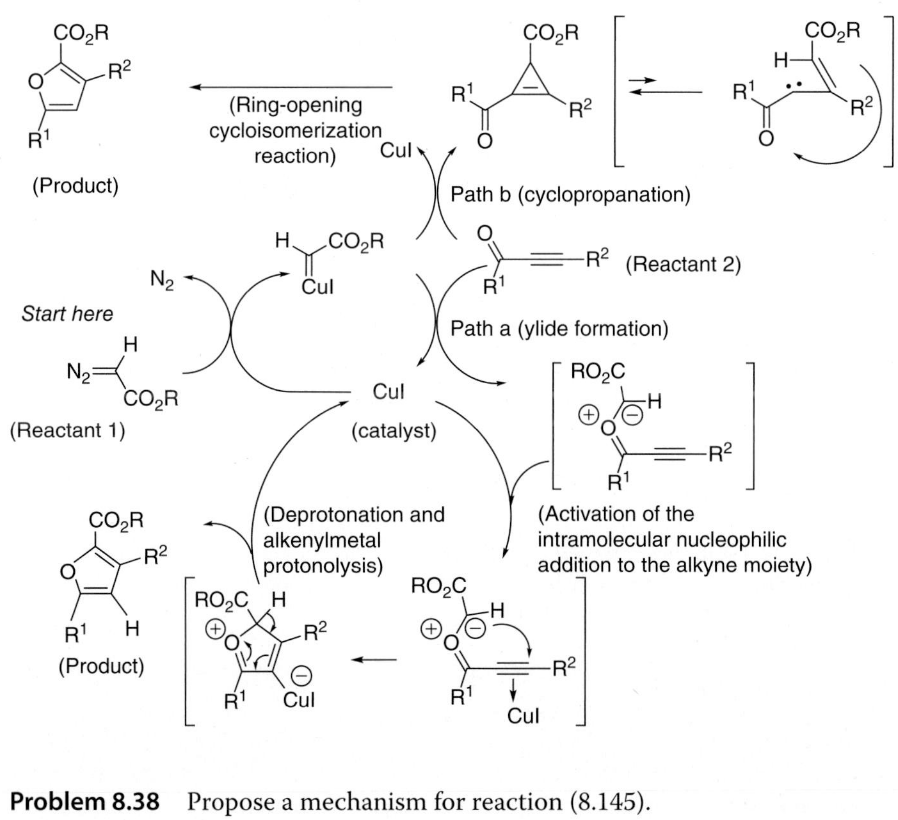

Problem 8.38 Propose a mechanism for reaction (8.145).

Answer

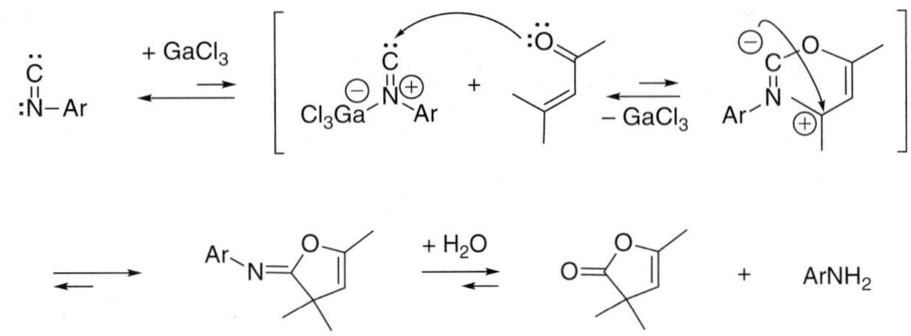

Problem 8.39 $Fe(CO)_5$-mediated carbonylative (5+1)-cycloaddition of vinylcyclopropane generates cyclohexanone [43, 44]. Ir-catalyzed carbonylative (5+1)-cycloaddition of allenylcyclopropanes gives 2-alkyline-3-cyclohexen-1-ones [45]. Propose a reaction mechanism for the $Ru_3(CO)_{12}$-catalyzed carbonylation of cyclopropyl-C(=N-*t*-Bu)-Ph into 3,4-dihydro-1-(*t*-butyl)-6-phehyl-2(1*H*)-pyridinone [46].

Answer

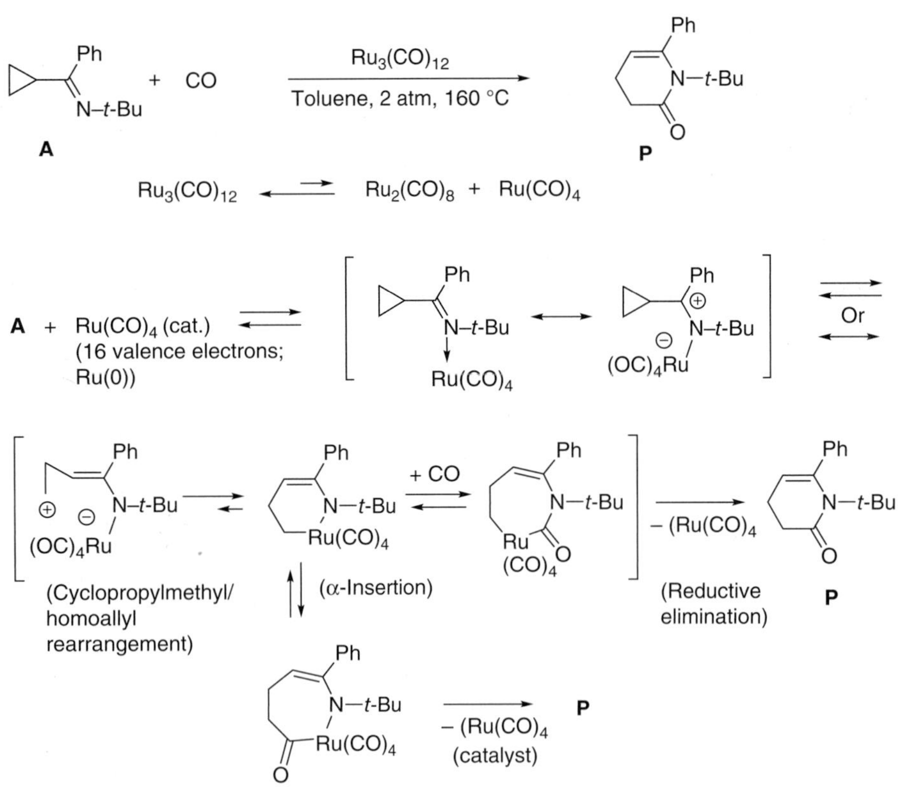

Note: an alternative alkyl group α-insertion cannot be excluded

Problem 8.40 Hilt et al. [47] have developed an enantioselective version of the Co-catalyzed intermolecular Diels–Alder addition, for instance **A** → **P**. Give a mechanism for the two successive reactions.

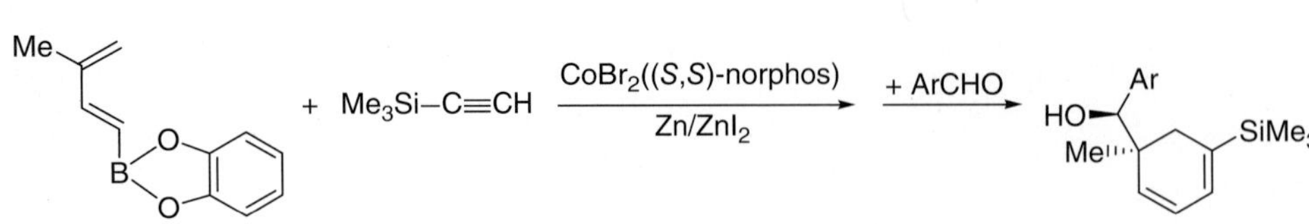

A for Ar = 4-$NO_2C_6H_4$: 87% yield, 71% ee **P**

(*S*,*S*)-norphos: (*S*,*S*)-2,3-bis(diphenylphospanyl)bicyclo[2.2.1]hept-5-ene

Answer

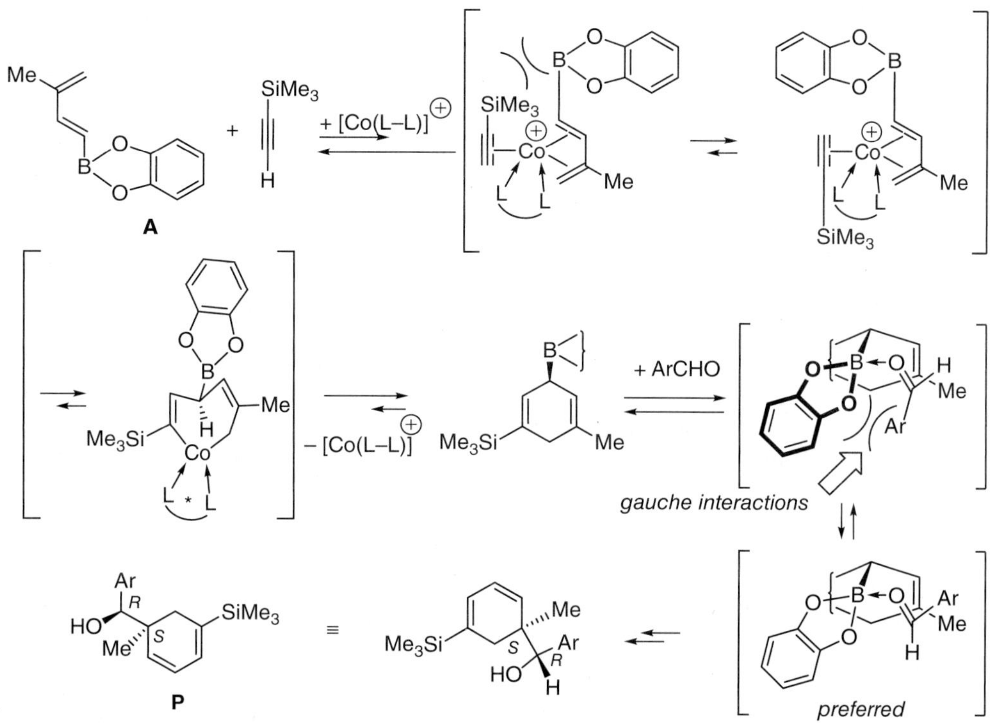

Problem 8.41 Propose a total synthesis of (±)-alloyohimban (**P**) and (±)-yohimban (**Q**) starting from tryptamine (**A**), 3,5-hexadienoyl chloride (**B**), and propargyl bromide (**C**) [48].

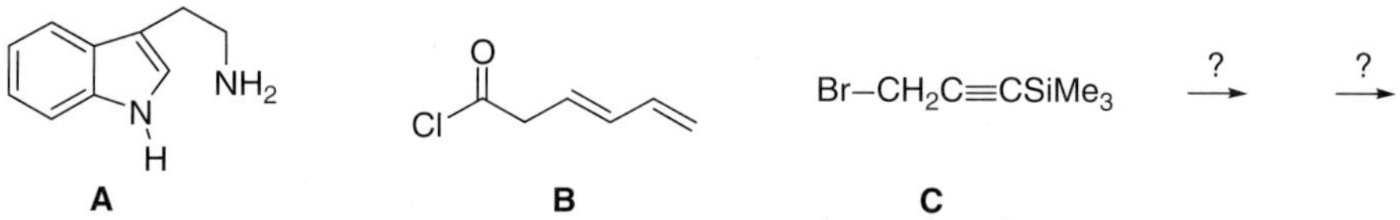

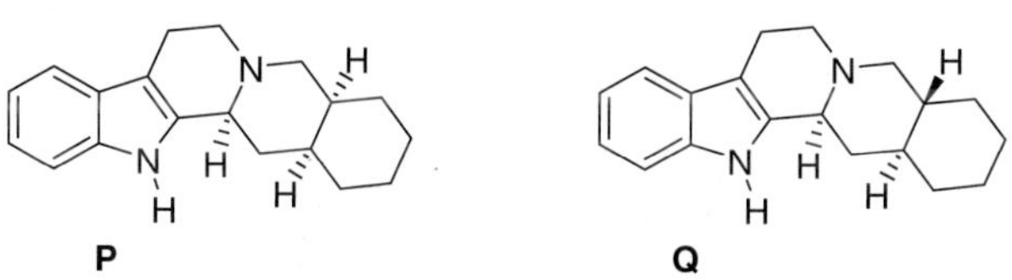

Answer

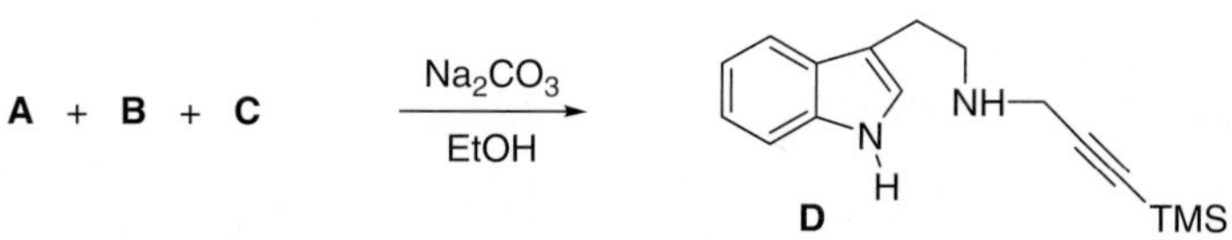

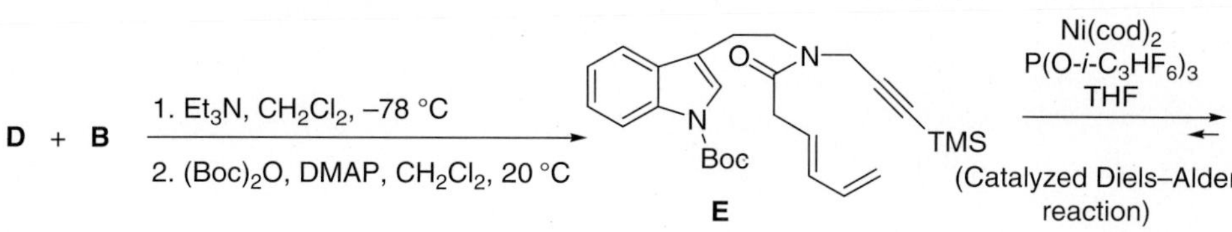

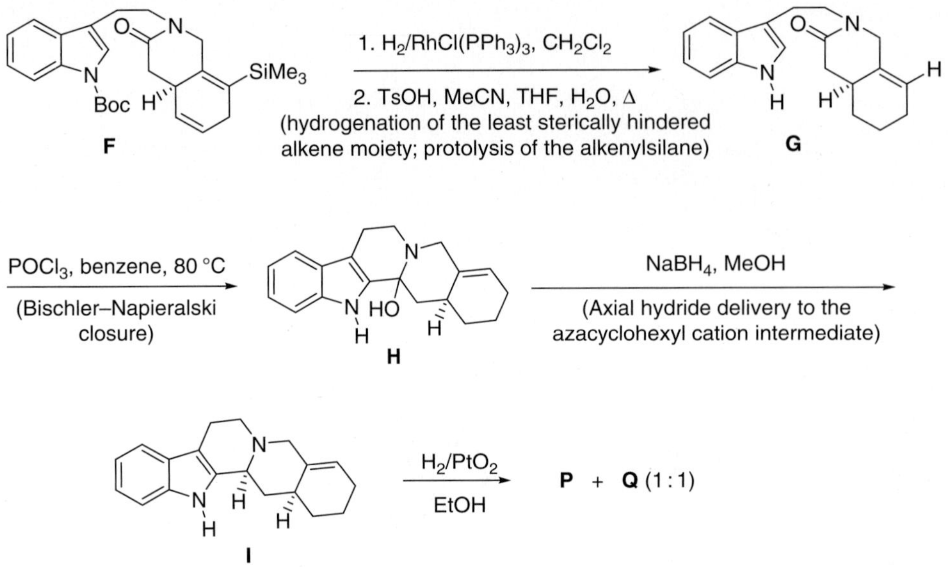

Problem 8.42 Kesenheimer and Groth have utilized the $CpCo(C_2H_4)_2$-catalyzed (2+2+2)-cycloaddition in the synthesis of the core of several antibiotics, especially the angucyclines [49]. Propose a reaction sequence converting **A** into (−)-8-*O*-methyltetrangomycin (**P**).

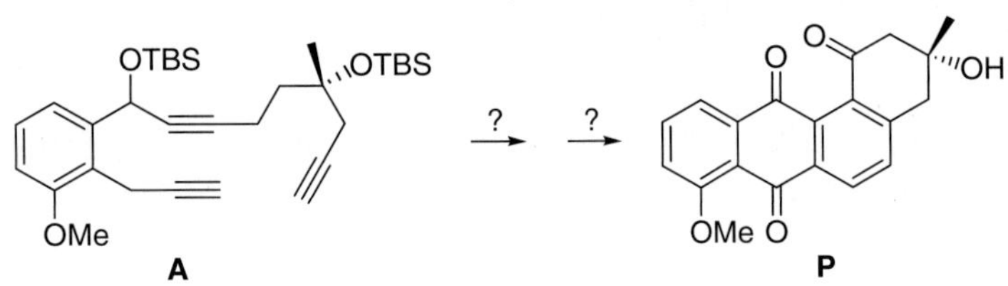

Answer

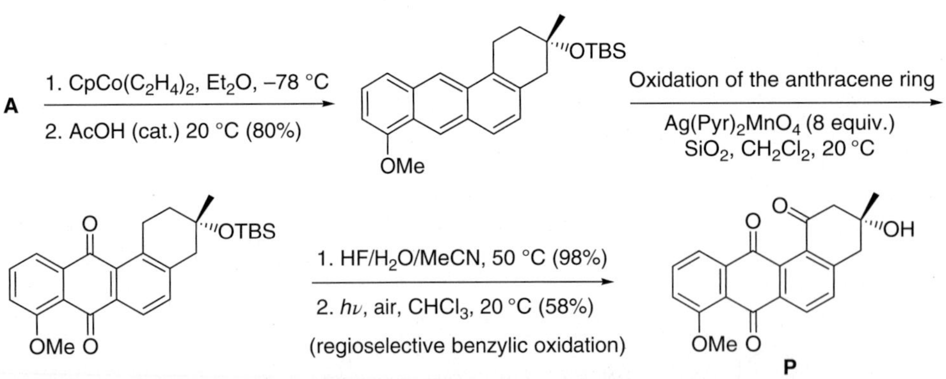

Problem 8.43 Propose a mechanism for the Ru-catalyzed (2+2+2)-cycloaddition (8.166) [50].

Answer

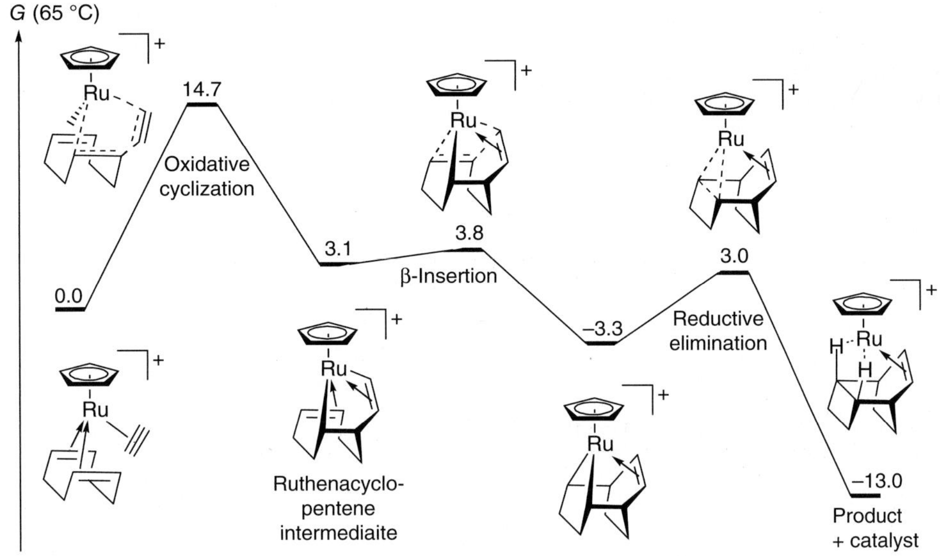

Problem 8.44 Propose a mechanism for the Rh-catalyzed carbonylative (3+2+1)-cycloaddition (8.171)

Answer

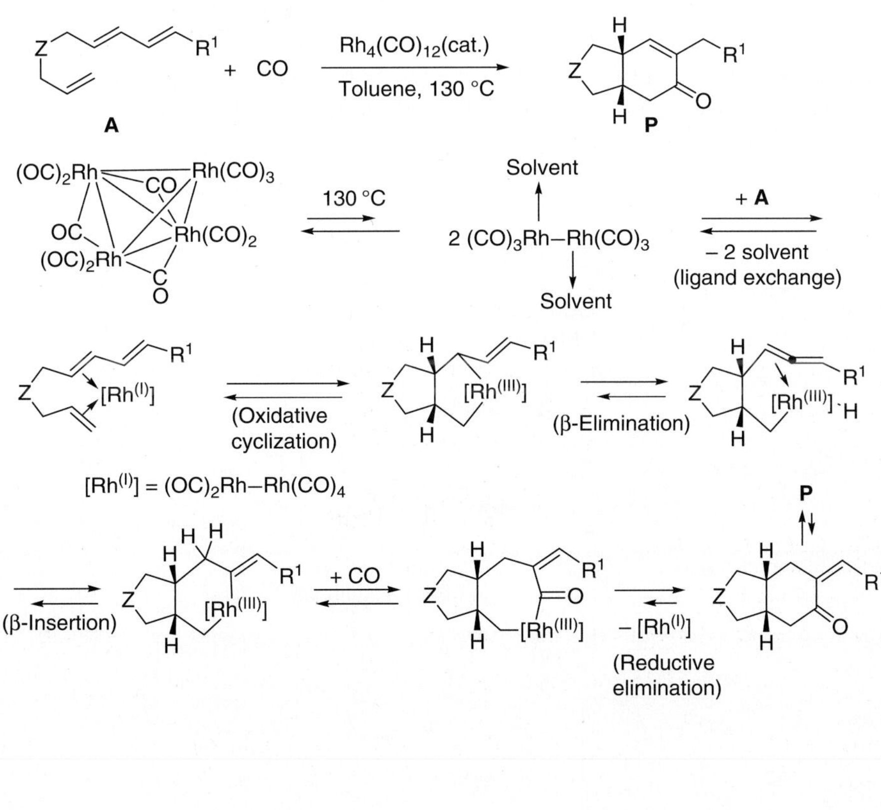

Problem 8.45 Propose a mechanism for the following [$HCo(CO)_4$]-catalyzed three-component synthesis of 1,3-oxazinan-4-ones [51].

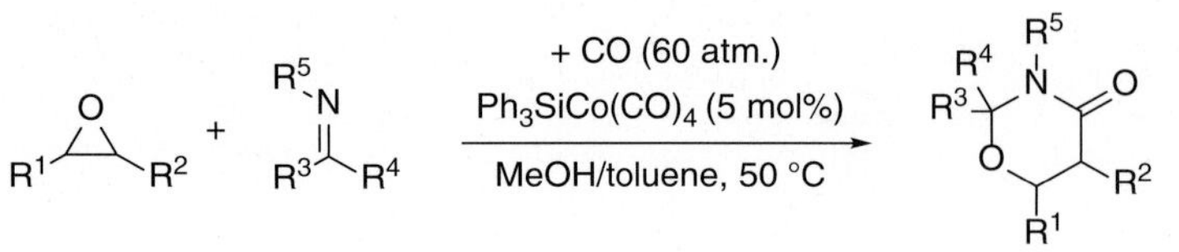

Answer

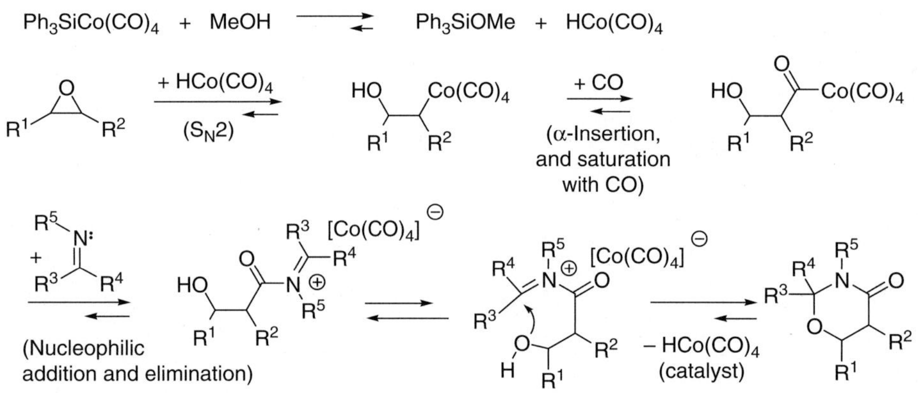

Problem 8.46 Saito et al. have reported on the Ni(cod)$_2$/PPh$_3$-catalyzed (3+2+2)-cycloaddition of ethyl cyclopropylideneacetate to terminal alkynes [52] and the Ni(cod)$_2$/topp (tri-*O*-biphenylyl phosphite)-catalyzed (4+3)-cycloaddition of ethyl cyclopropylideneacetate to 1,3-diene [53]. Propose a mechanism for the latter reaction.

Answer

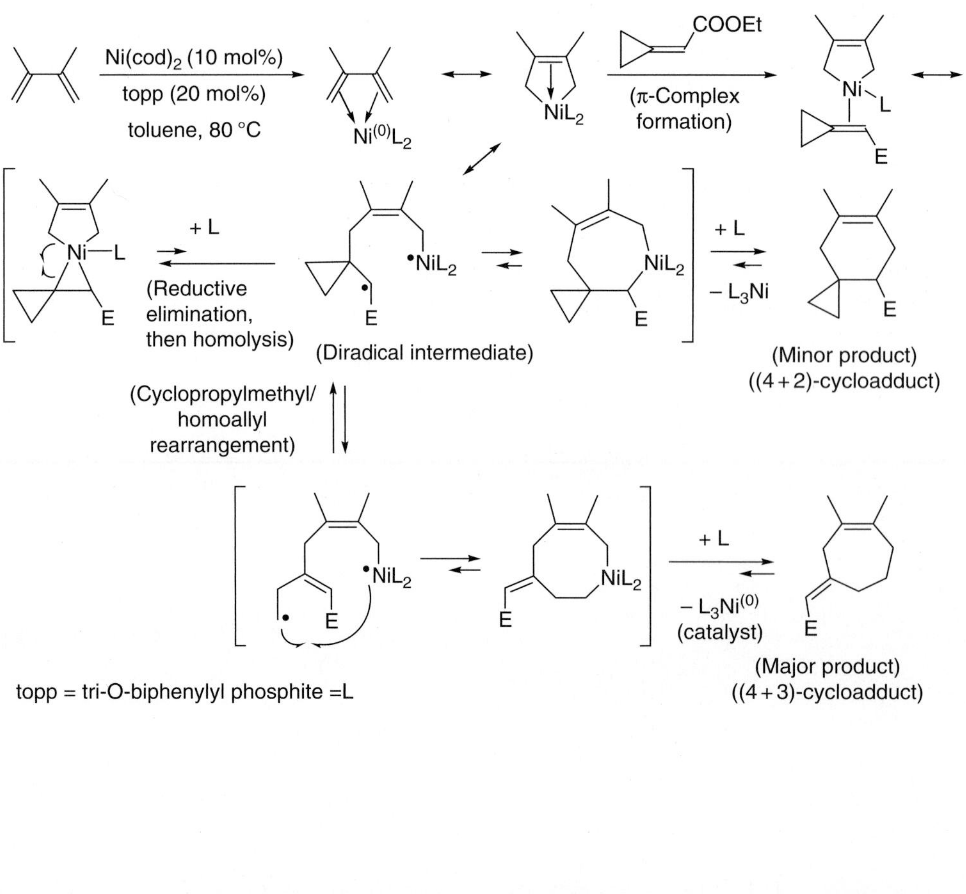

topp = tri-O-biphenylyl phosphite =L

Problem 8.47 What is the major product **P** of reaction **A** + **B**? [54]; see also [55]:

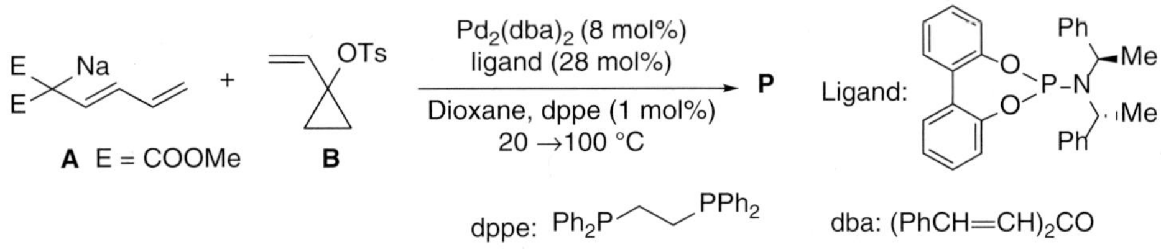

Answer

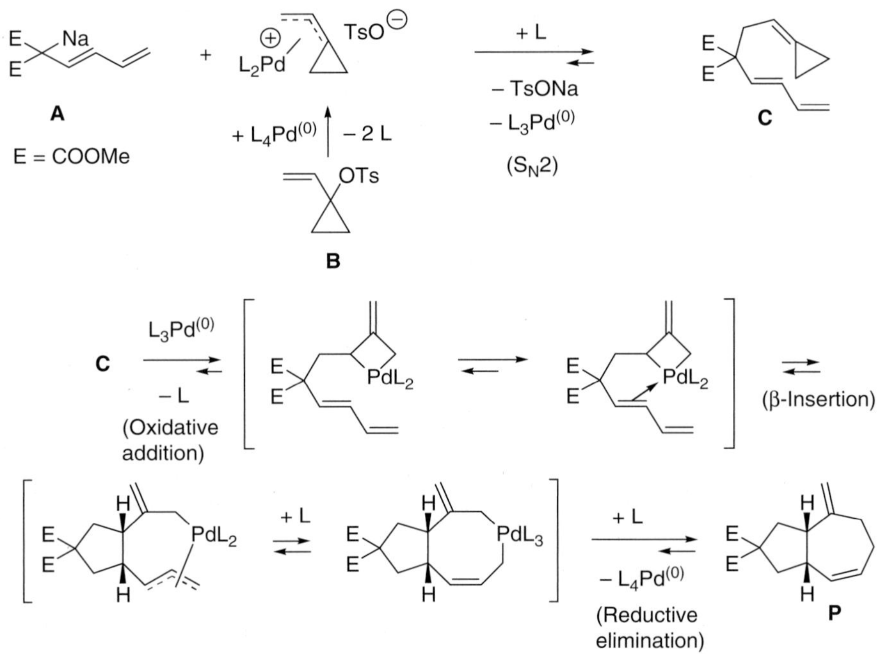

Problem 8.48 The reaction of **A** with $Me_3SiCH{=}N_2$ in the presence of $Ni(cod)_2$ as catalyst at 60 °C produces **B** (69%) with high stereoselectivity. On heating **B** in toluene at 105 °C, product **P** is obtained, which realizes a formal (4+2+1)-cycloaddition of **A** with $Me_3SiCH{=}N_2$. What is product **P**? Propose a mechanism for the formation of **B** [56].

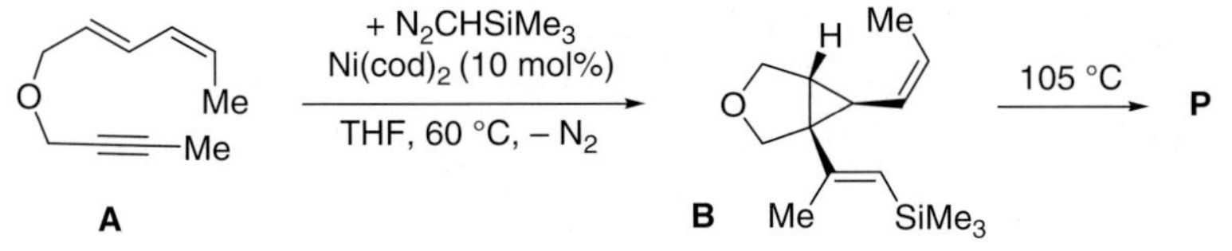

Answer

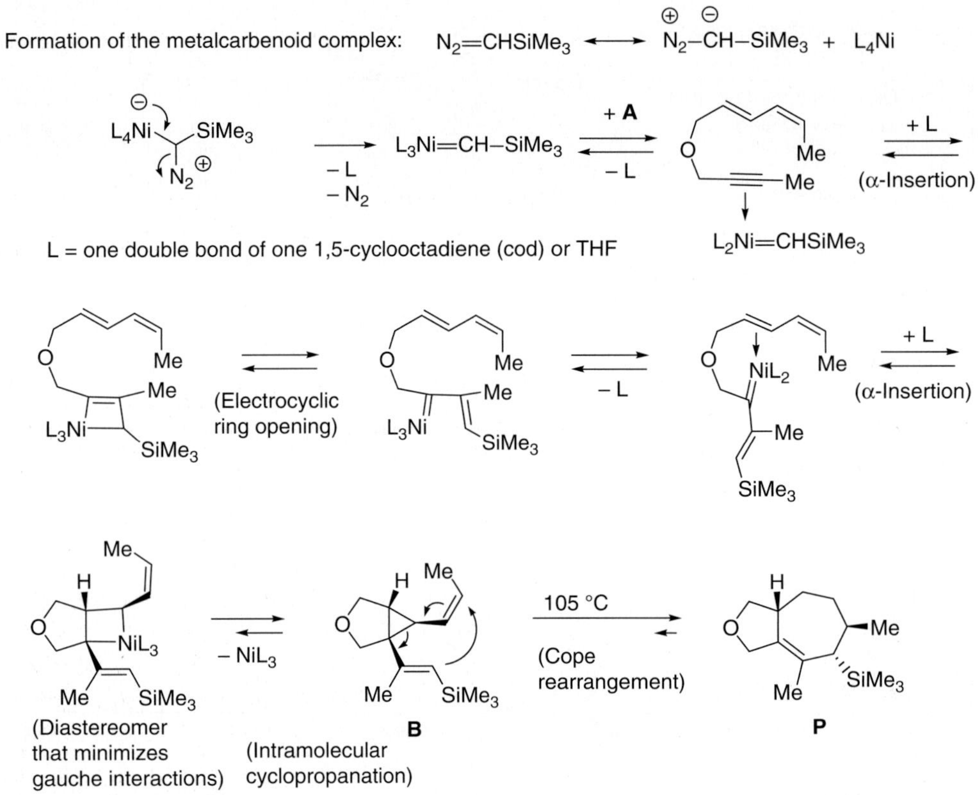

Problem 8.49 What is the main product of the following reaction? [57]

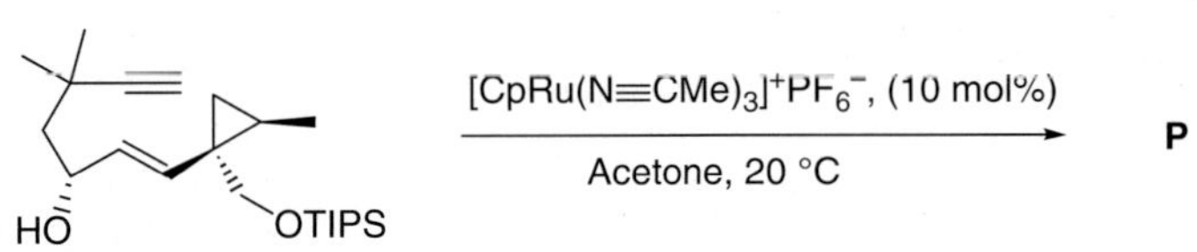

Answer

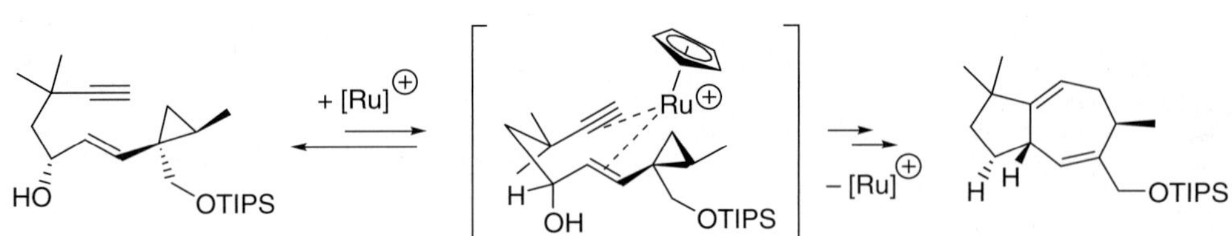

Problem 8.50 Using three transition-metal-catalyzed reactions convert **A** into **P** [58, 59].

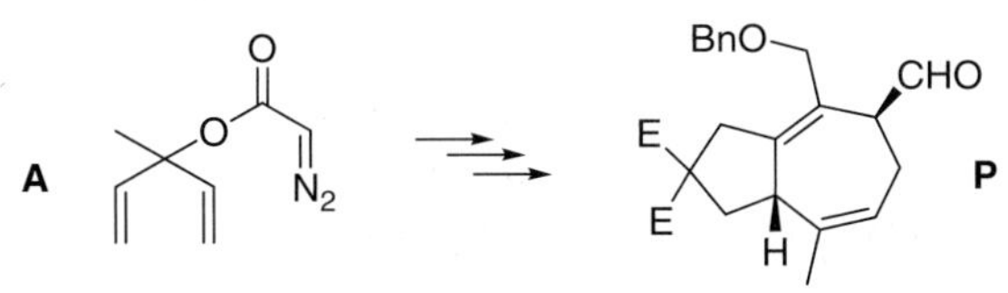

Answer

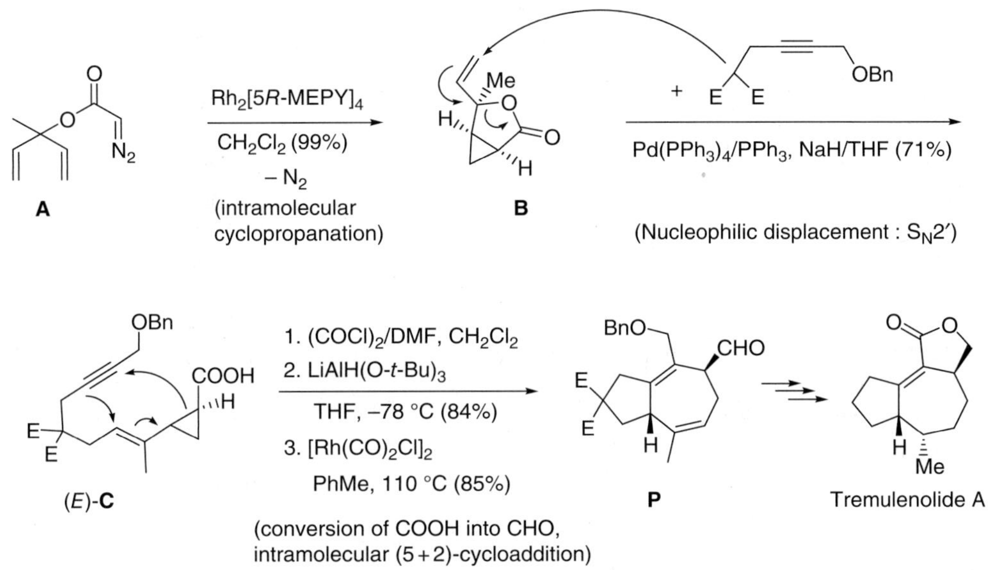

Problem 8.51 Propose a mechanism for the $[Rh(CO)_2Cl]_2$-catalyzed (5+2)-cycloaddition of vinylcyclopropane + ethyne [60].

Answer

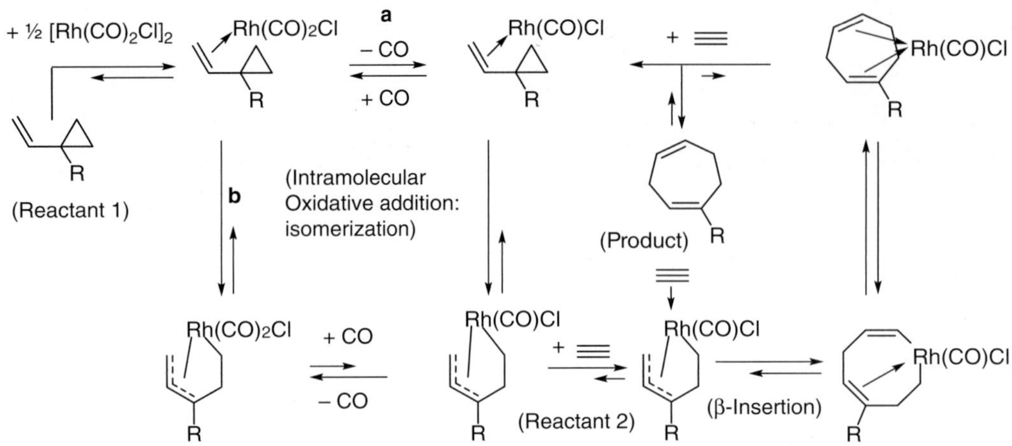

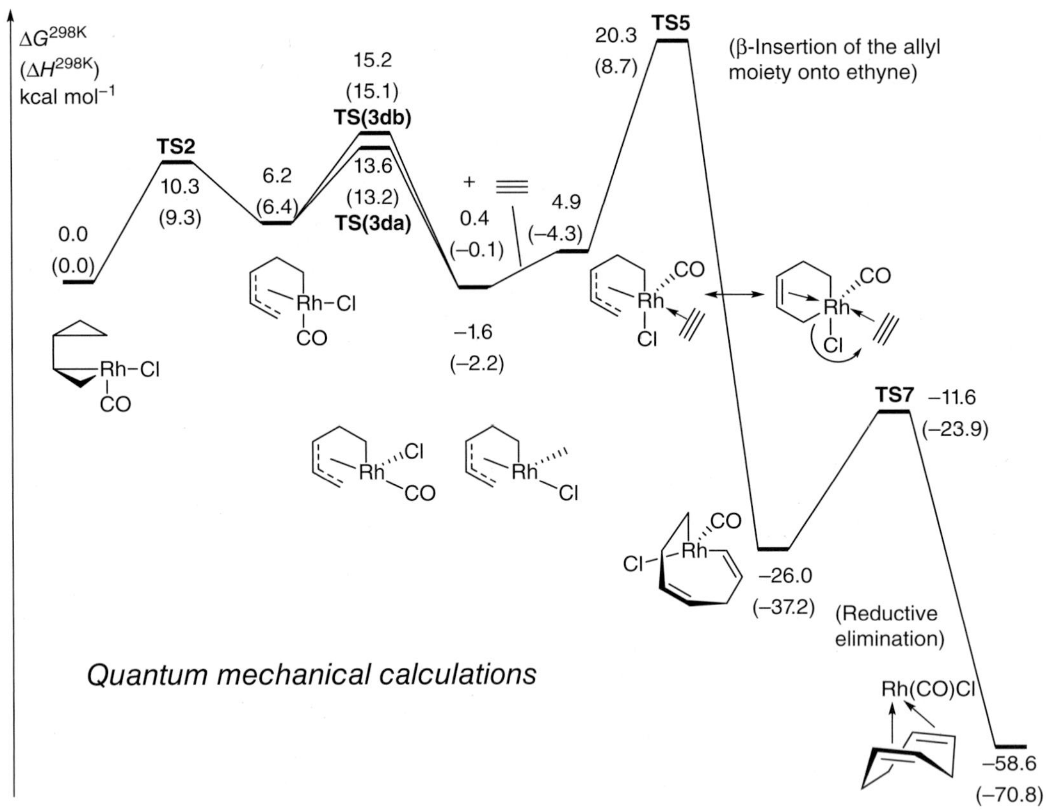

Problem 8.52 What is the product **P** of the following reaction?

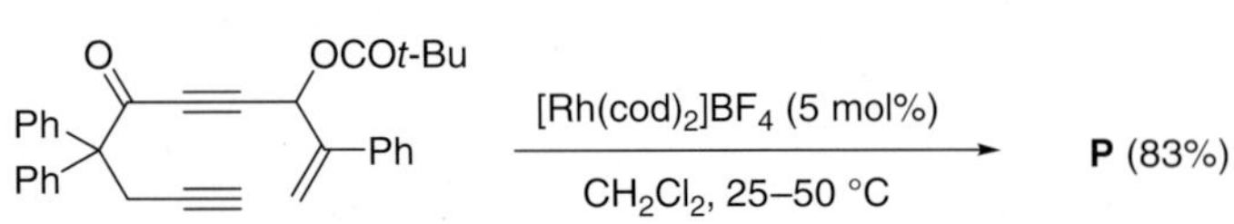

Answer

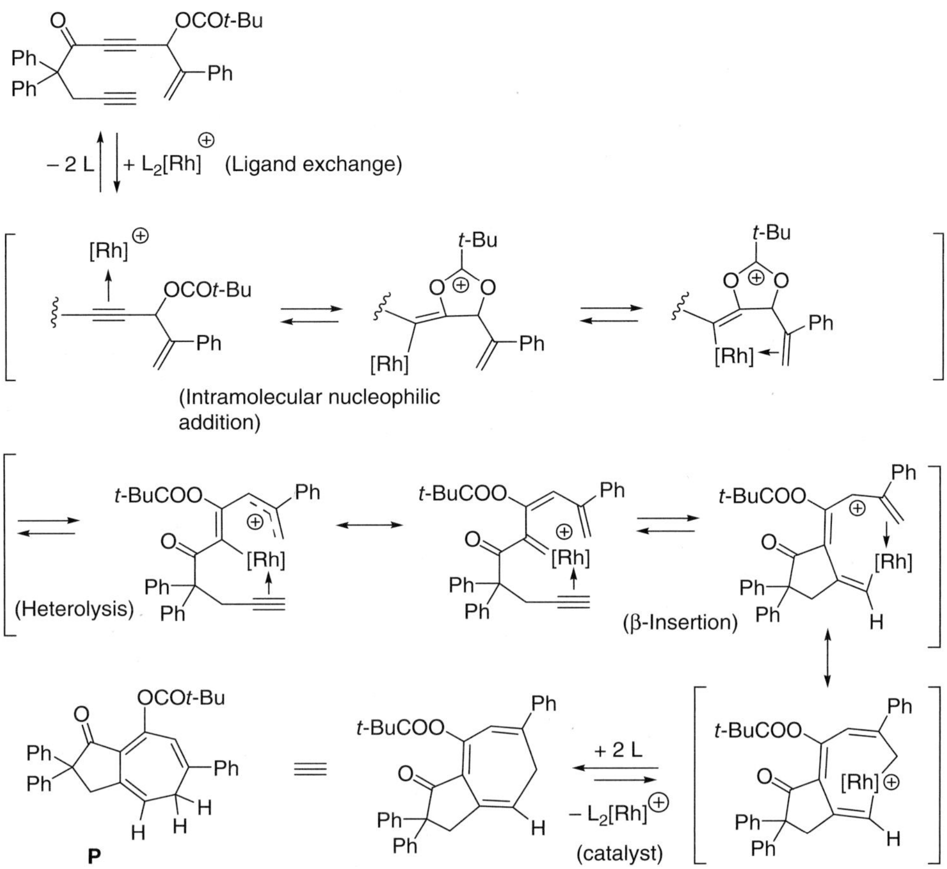

Problem 8.53 Propose a mechanism for the following (5+2)-cycloannulation [61].

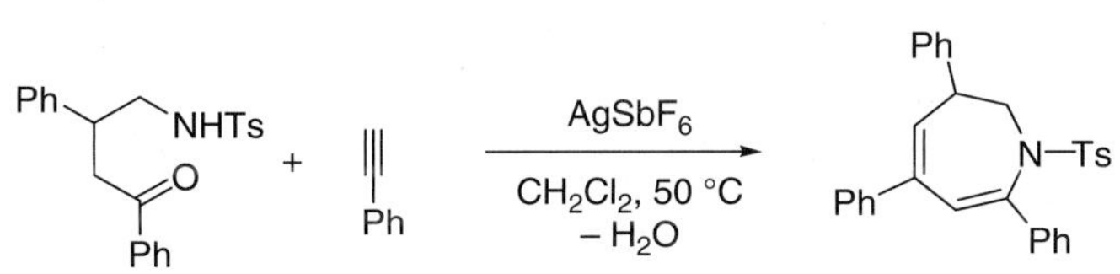

Answer

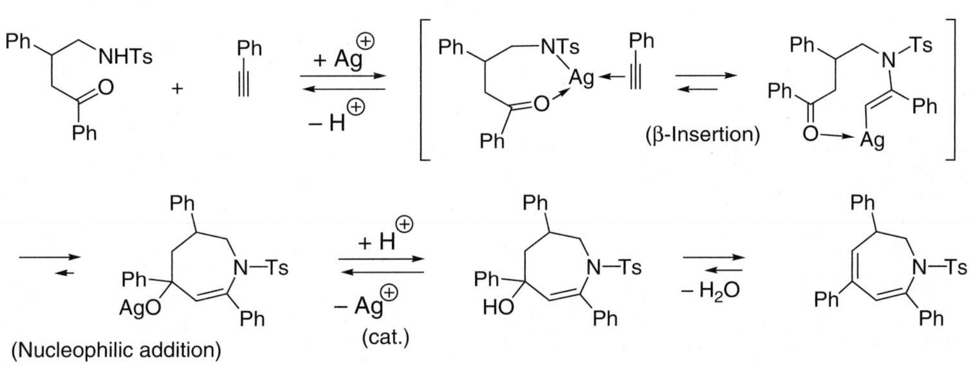

Problem 8.54 In the presence of CO transition-metal-catalyzed (2+2+2)-cycloadditions can become (2+2+2+1)-cycloadditions that permit the three component construction of seven-membered ring systems. What is the main product **P** of the following reaction? Propose a reaction mechanism [62].

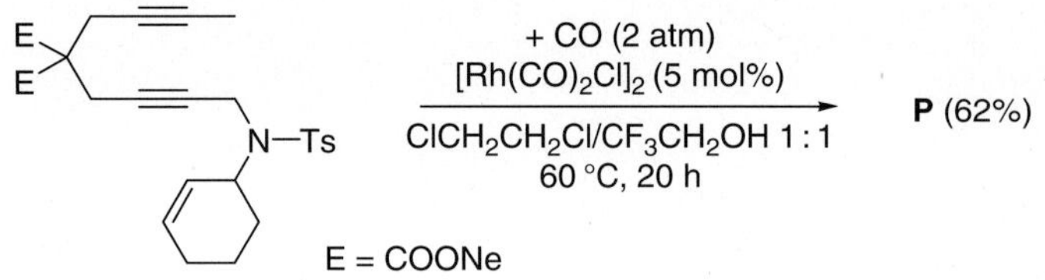

Answer

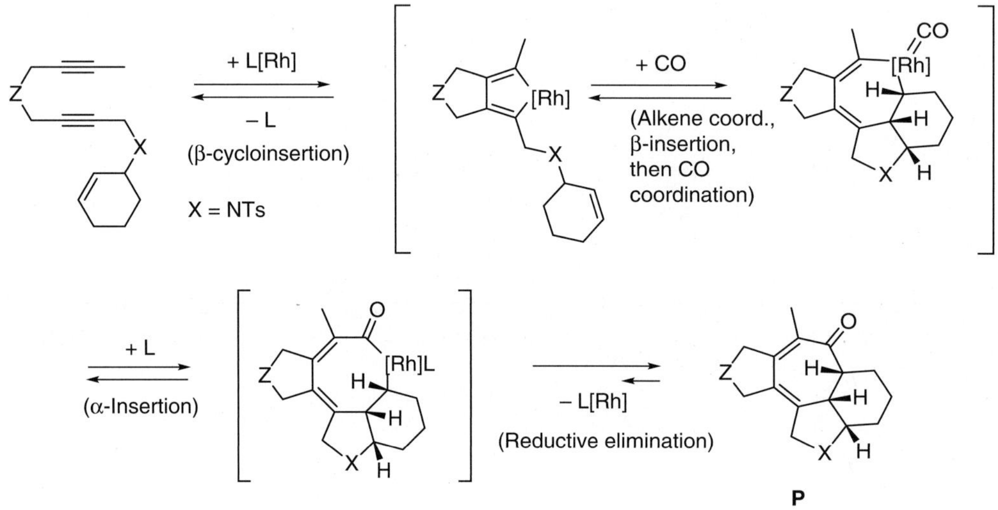

Problem 8.55 In the presence of two equivalents of $TiCl_4$ in CH_2Cl_2 **A** is converted into compound **P**. Propose a structure for **P** [63].

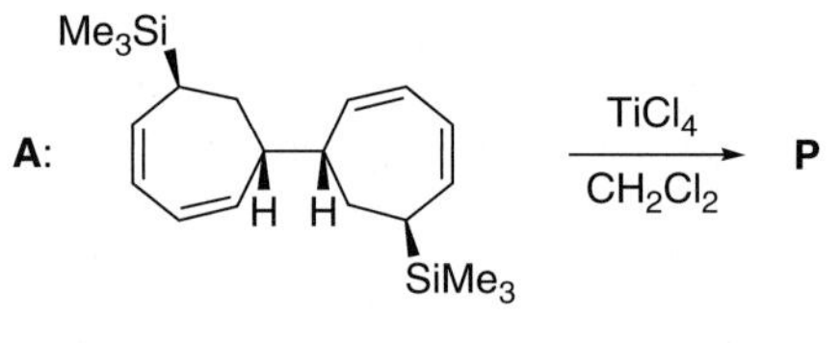

Answer

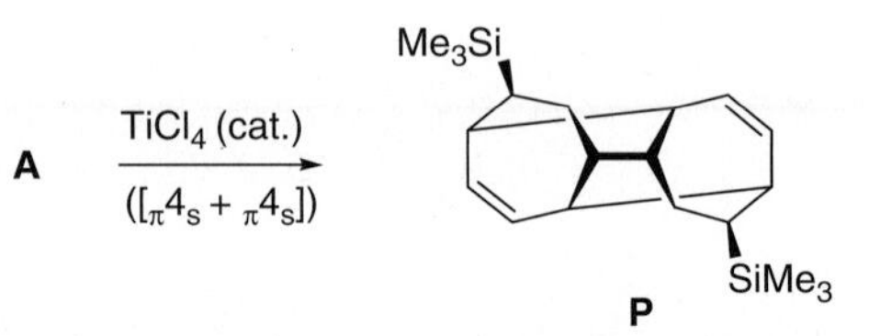

Problem 8.56 In the presence of $[P(t\text{-Bu})_2(ortho\text{-biphenyl})]AuCl/AgSbF_6$ (2.5 mol%) in 1,2-dichloroethane at 35 °C 2-(4-chlorophenyl)oxetane reacts with *t*-butyl 3-phenylprop-2-yne-1-carboxylic ester giving a product **P** of (4+4)-annulation. What is **P** and propose a reaction mechanism [64].

Answer

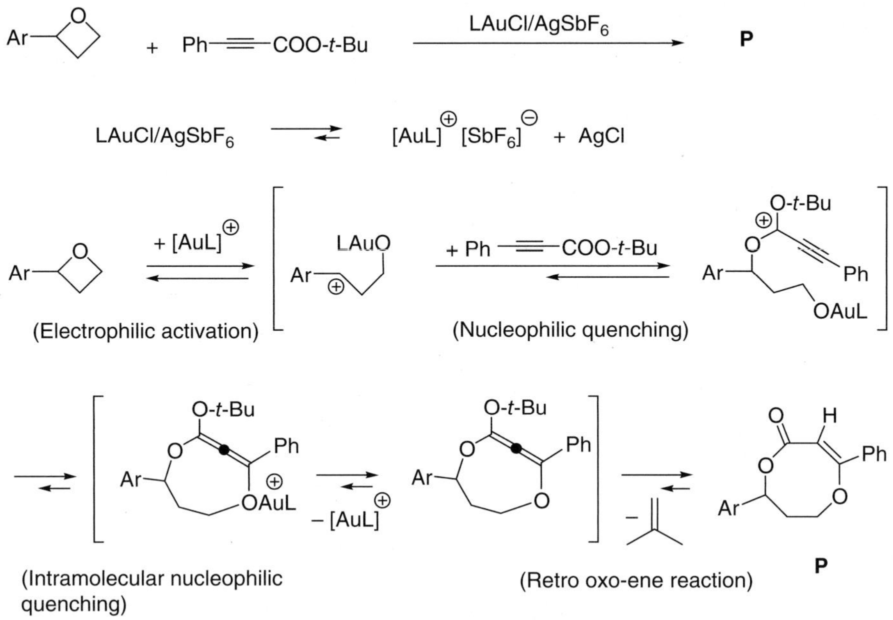

Problem 8.57 In the presence of Co(acac)$_2$, Et$_2$AlCl, and of an enantiomerically pure diphosphine ((*R*)-1,2-bis(diphenylphosphino)propane), norbornadiene and isoprene undergo a (4+2+2)-cycloaddition. What is the main product of this reaction [65].

Answer

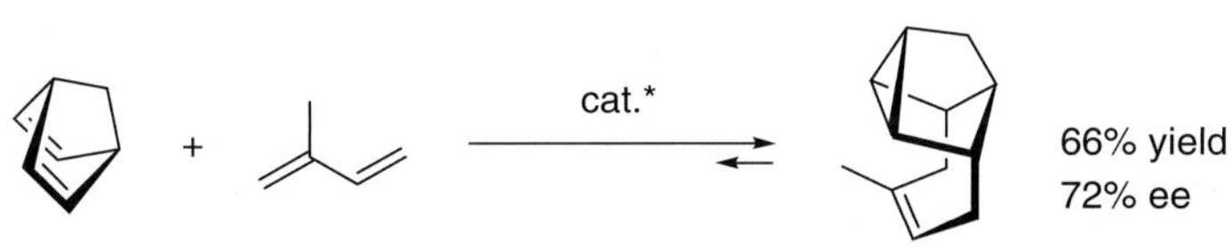

P

Problem 8.58 Isoprene and 2,2-bis(propargyl)malonic diesters undergo [Cp*Ru(MeCN)$_3$]PF$_6$-catalyzed (2+2+2)- and (4+2+2)-cycloadditions. What are the possible products and the mechanism of their formation? [66]

Answer

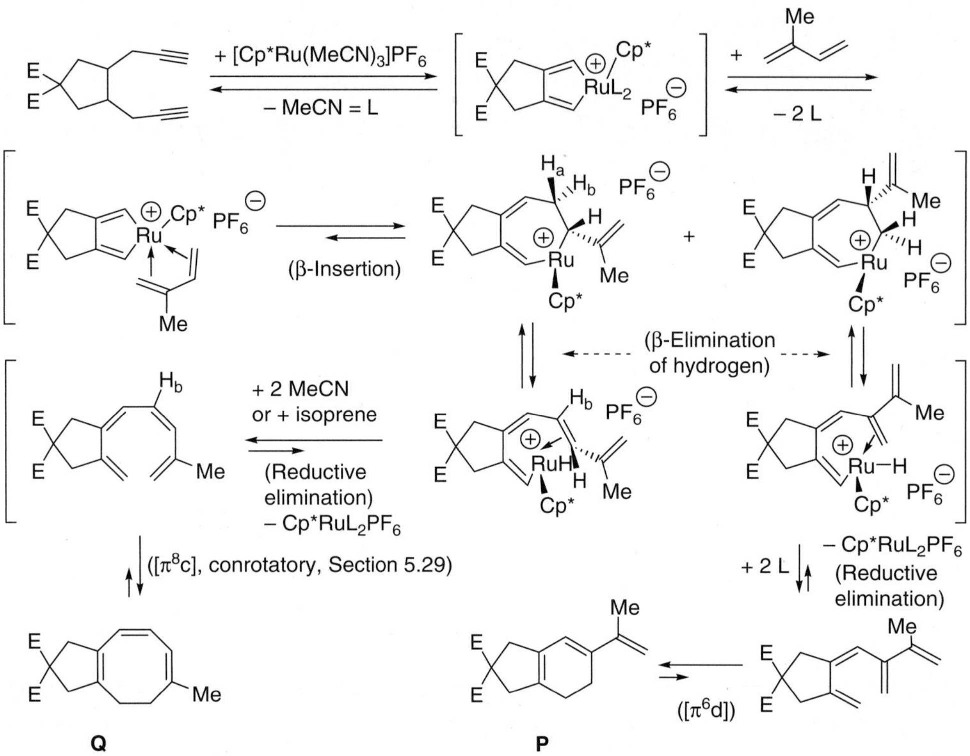

Problem 8.59 In the presence of [Rh(cod)Cl]$_2$ (2.5 mol%) and AgOTf (5 mol%) in toluene at 20 °C 2 equiv. of **A** add to 2,3-dimethylbutadiene giving product **P** in 98% yield. What is **P**? [67]

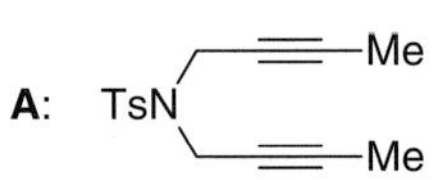

Answer

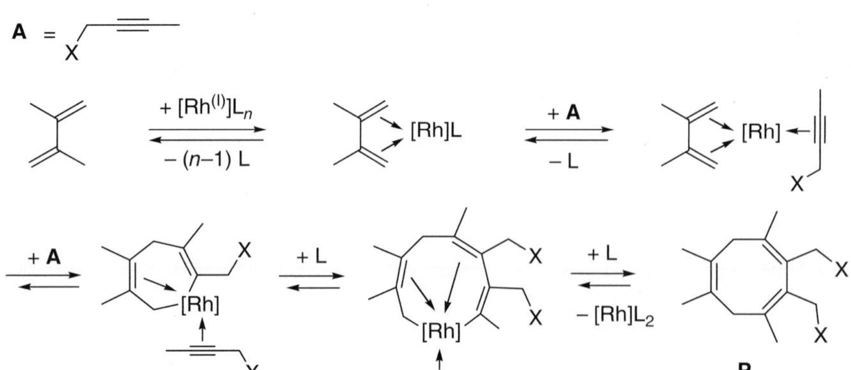

Problem 8.60 The Co(I)-catalyzed (6+2)-cycloaddition of cyclooctatetraene to alkynes affords bicyclo[4.2.2] deca-2,4,7,9-tetraenes **P**. Another type of cycloadducts **Q** forms concurrently. What is it? Propose a mechanism for these reactions [68, 69].

Answer

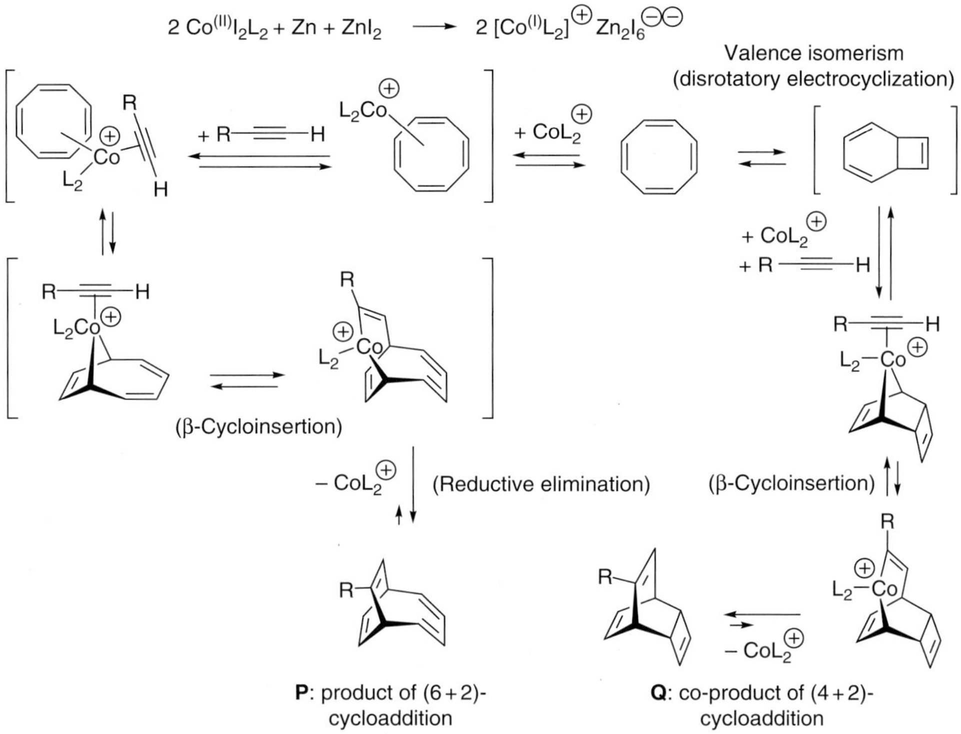

Problem 8.61 The Rh(I)-catalyzed cyclization **A** → **P** represents a formal intramolecular (6+2)-cycloaddition of a 4-allenal with an alkyne moiety (IMes, 1,3-bis(2,4,6-trimethylphenyl)imidazole-2-ylidene, cod, cycloocta-1, 5-diene). Propose a mechanism for this reaction [70, 71].

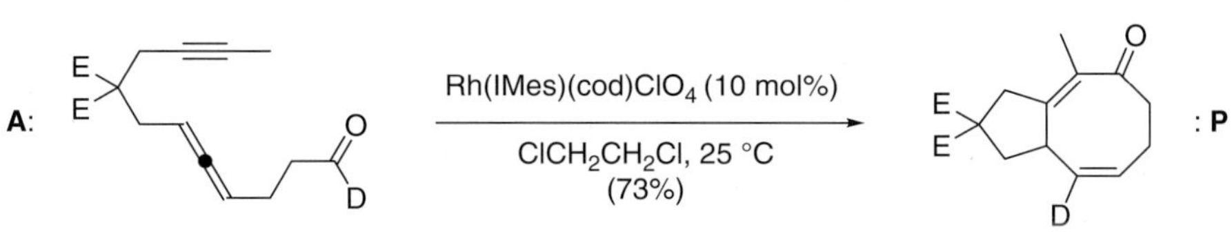

Answer

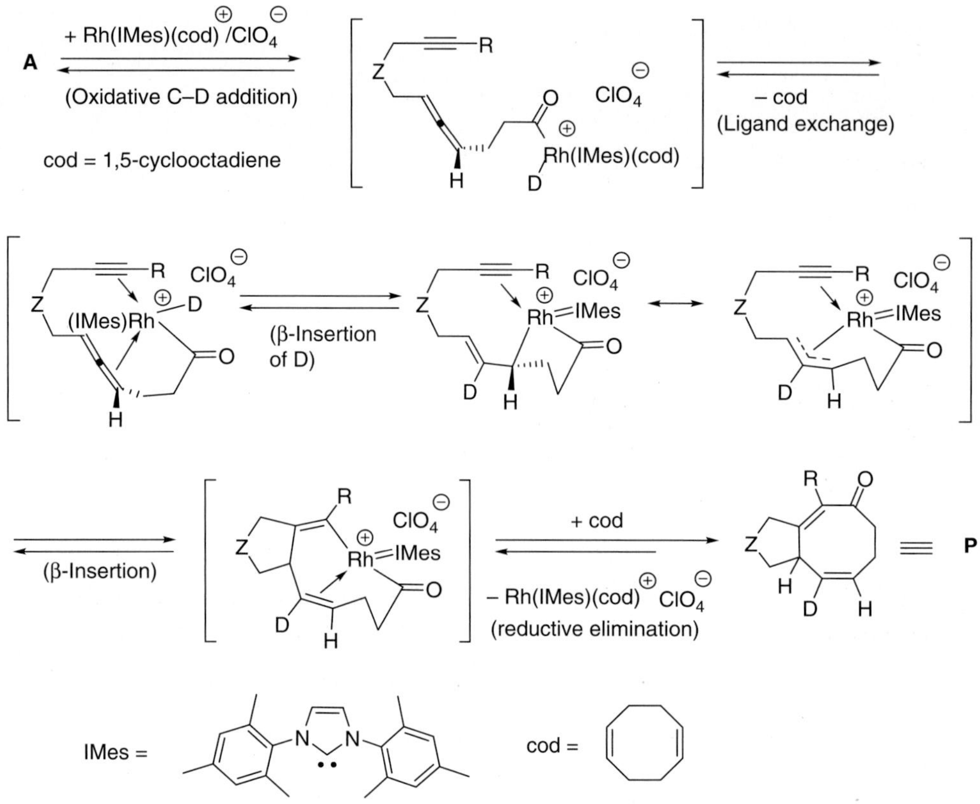

Problem 8.62 Propose a mechanism for reaction (8.203).

Answer

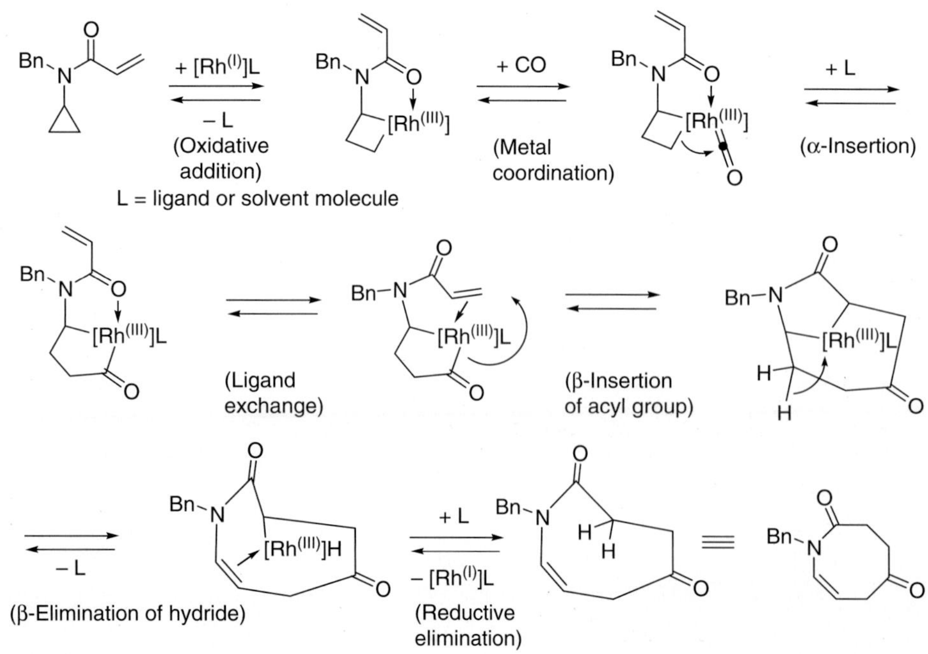

Problem 8.63 What is the main product of the reaction of tropone and cyclohexa-1,3-diene in the presence of $FeCl_3$ (25 mol%)? [72]

Answer

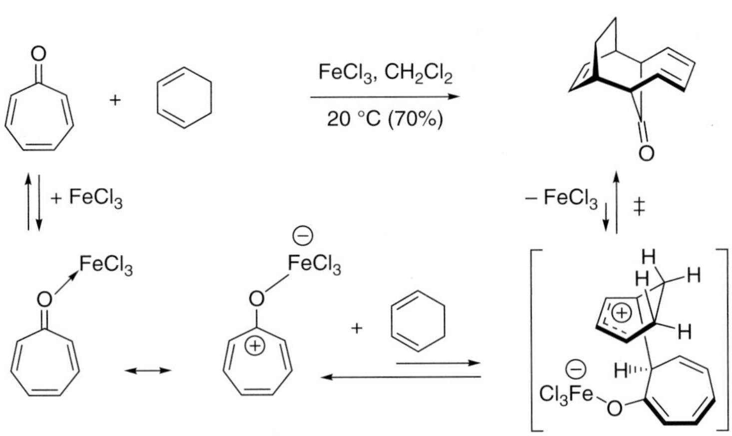

Problem 8.64 Werner and coworkers have found that cationic $[Rh(I)(\eta^4\text{-butadiene})_2(PR_3)]^+X^-$ ($X = CF_3SO_3$, $C_6F_5)_4B$; R = *I*-Pr. cyclohexyl) catalyzes the cyclotetramerization of butadiene giving predominantly (1*Z*,5*Z*,9*E*,13*E*)-cyclohexadecatetraene [73]. Propose a mechanism for this reaction [74].

Answer

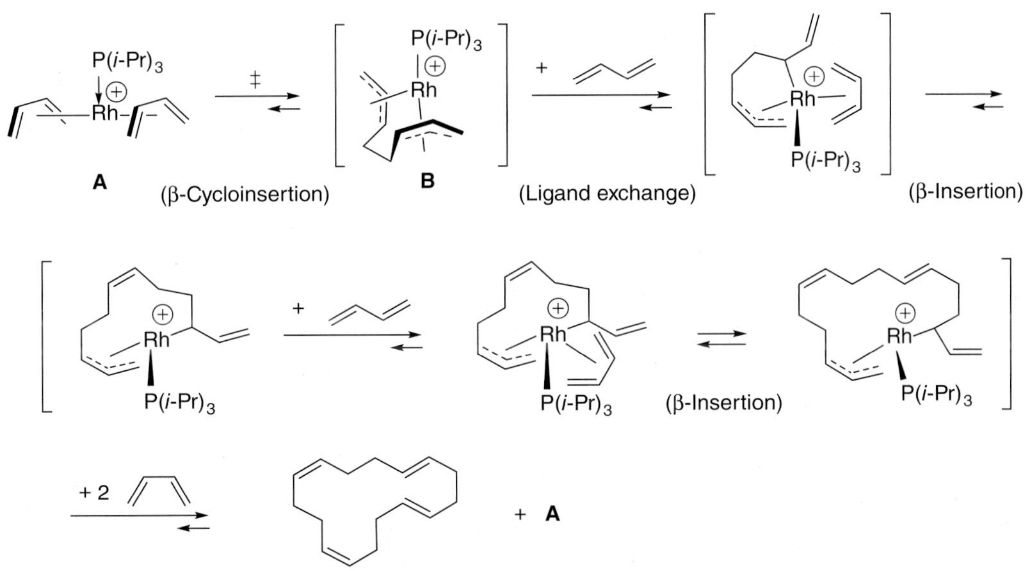

Problem 8.65 Propose a reaction mechanism for the Rh-catalyzed isomerization (8.213).

Answer

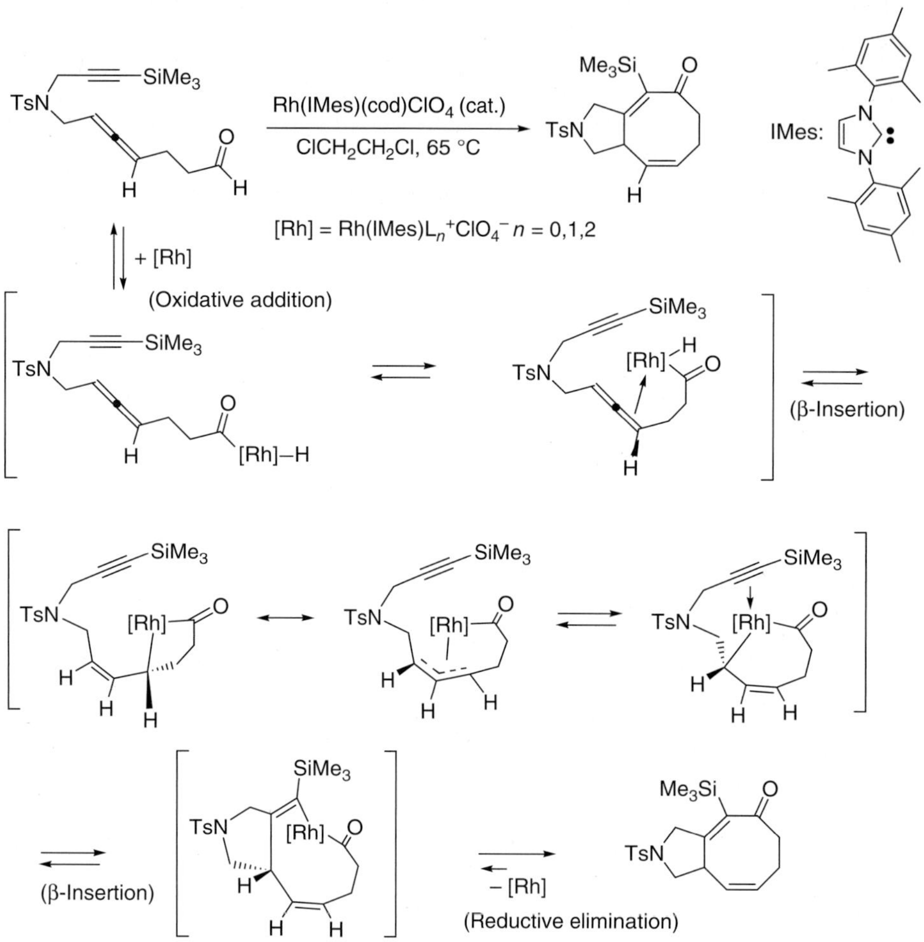

Problem 8.66 Benzooxepines are obtained by reaction of *o*-vinylphenol with internal acetylenes in the presence of $[Cp^*RhCl_2]_2$ (2.5 mol%), $Cu(OAc)_2/H_2O$ (50 mol%), and air in MeCN at 55 °C [75]. Propose a mechanism for this oxidative (5+2)-heterocarba-annulation.

Answer

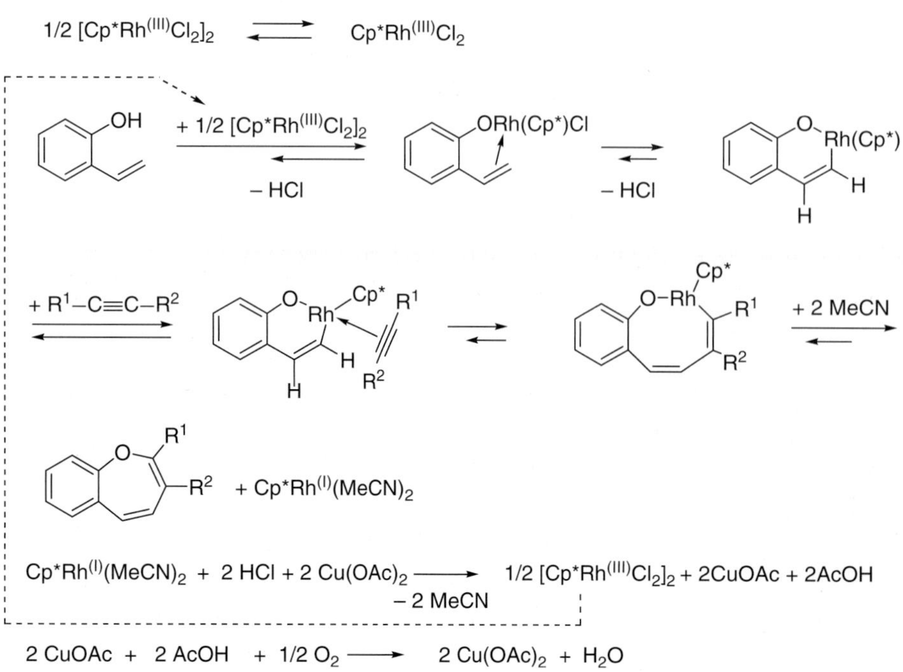

Problem 8.67 In the presence of 5 mol% of CuBr/3,3′-bis(*t*-butyl)-2,2′-bipyridine/ketoABNO and 30 mol% of *p*-(dimethylamino)pyridine molecular oxygen (1 atm) promotes the oxidative Strecker reaction between HCN and *N*-(*p*-methoxyphenyl) cyclohexylmethyl amine at 50 °C [76]. What is the main product of this reaction? Propose a mechanism for it.

Answer

Possible Mechanisms

Secondary amine oxidation is expected to produce the corresponding imines that add HCN. Several mechanisms can be envisioned for this reaction. In a **first hypothesis**, single electron transfer (SET) from the amine to the aminyloxyl radical generates an amininium radical cation, which then reacts with a second equivalent of the aminyloxy radical:

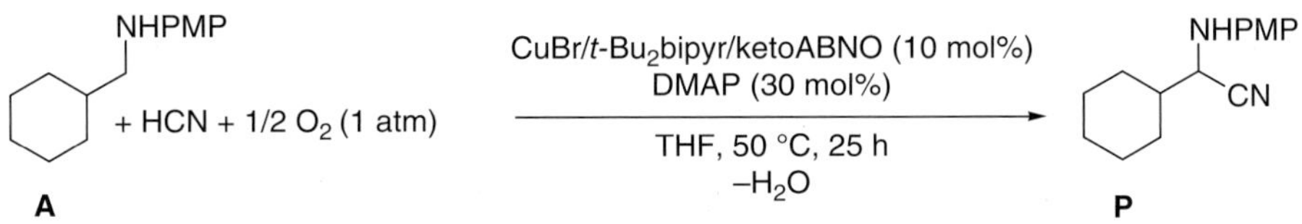

PMP = 4-MeOC$_6$H$_4$

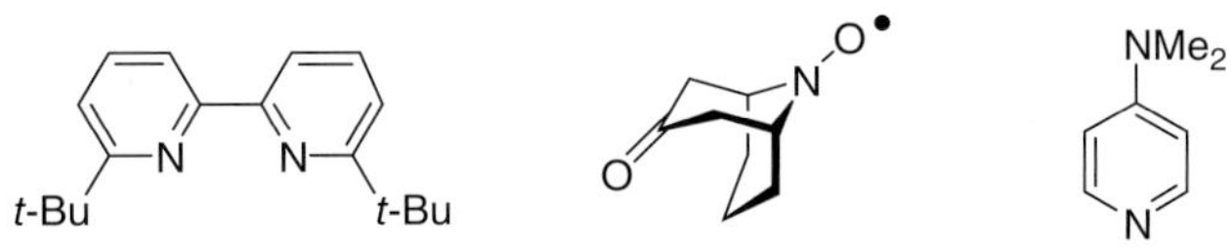

t-Bu$_2$bipyr = L$_2$ ketoABNO = R$_2$NO$^\bullet$ DMAP

Possible Mechanisms

Secondary amine oxidation is expected to produce the corresponding imines that add HCN. Several mechanisms can be envisioned for this reaction. In a **first hypothesis**, SET from the amine to the aminyloxyl radical generates an amininium radical cation, which then reacts with a second equivalent of the aminyloxy radical:

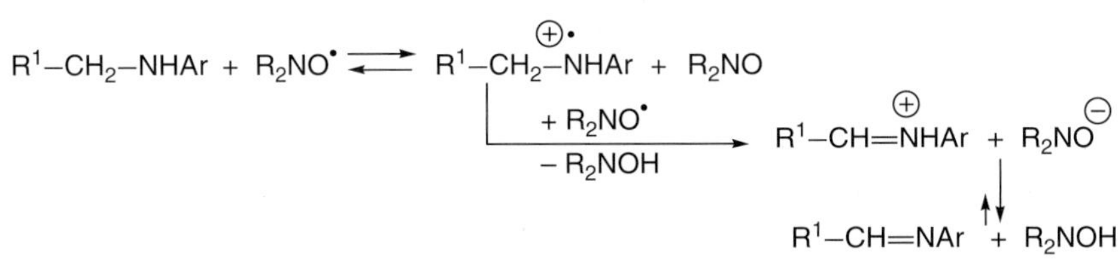

Scheme 7.93 (Section 7.11.4) proposes a mechanism for the aerobic Cu(I)-catalyzed oxidation of TEMPO-H into (2,2,6,6-tetramethylpiperidin-1-yl)oxyl (TEMPO).

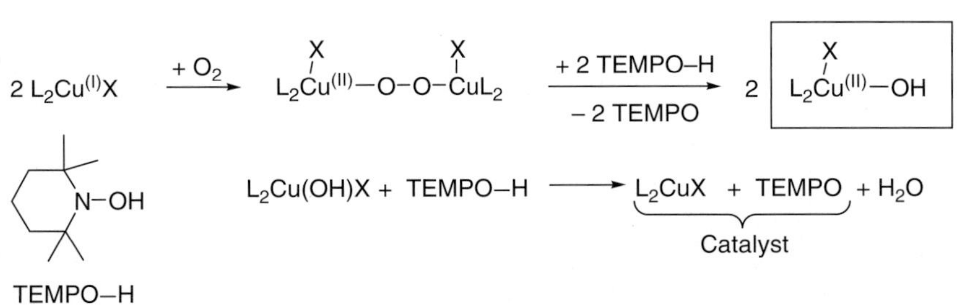

In a **second hypothesis** one envisions the formation of a quaternary complex in which the hydrogen abstraction is facilitated:

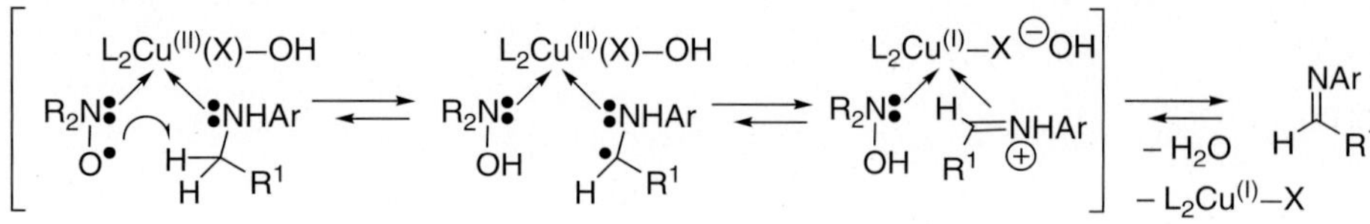

Problem 8.68 Give a mechanism for the reaction:

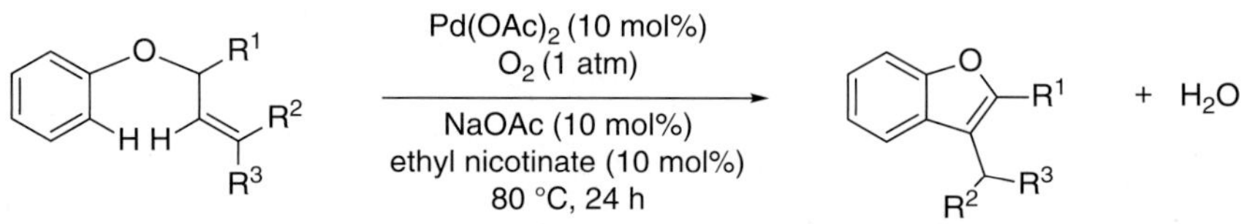

Answer

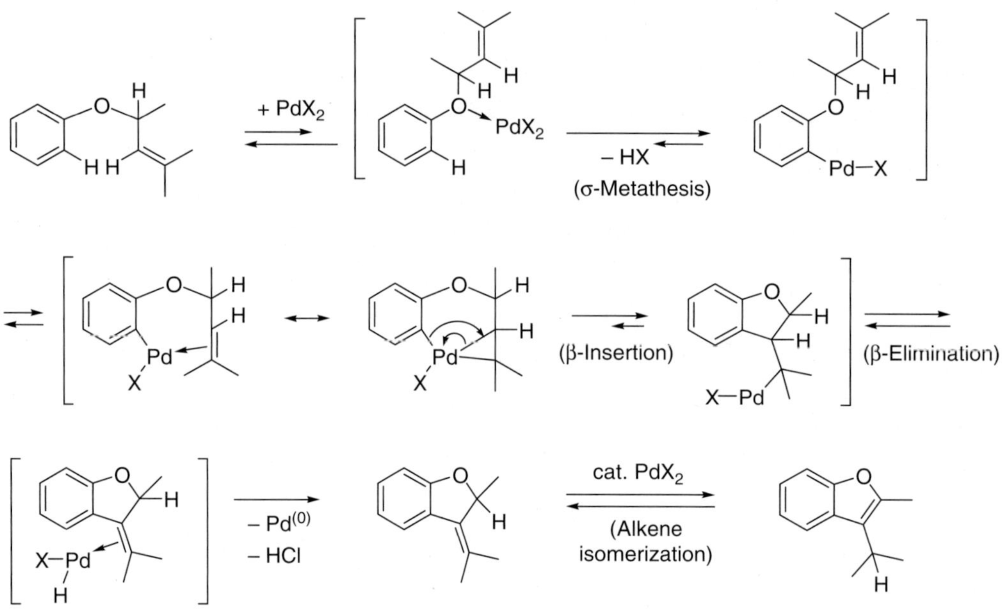

$$Pd^{(0)} + 2\,HX + \tfrac{1}{2}\,O_2 \longrightarrow Pd^{(II)}X_2 + H_2O$$

Problem 8.69 What is the main product **P** for the oxidative cross-coupling of **A** and **B**?

$$\underset{\mathbf{A}}{PhCH_2{-}NMe_2} + \underset{\mathbf{B}}{HC{\equiv}CPh} \xrightarrow[(t\text{-BuO})_2\ (2\ \text{equiv.}),\ 100\ °C,\ 24\ h]{\text{Air, FeCl}_2\ (10\ \text{mol\%})} \mathbf{P}$$

Answer

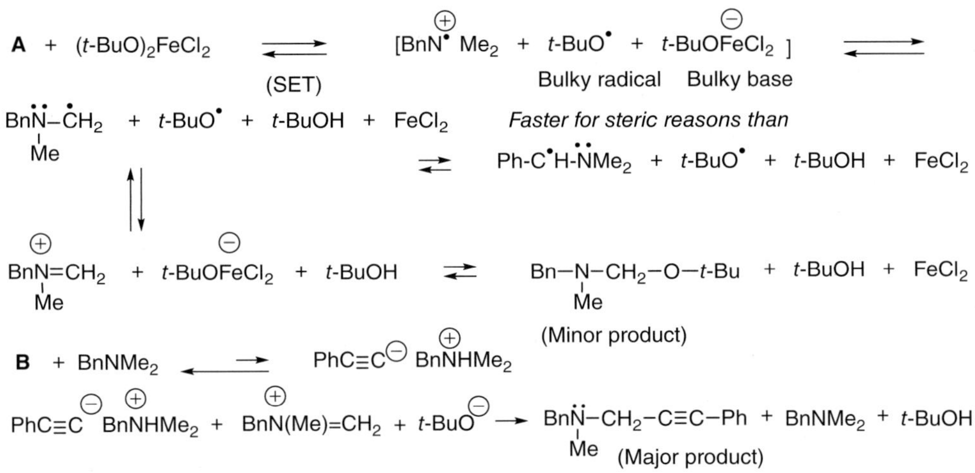

Problem 8.70 Propose a mechanism for the oxidative cross-coupling **A** + **B** → **P** [77].

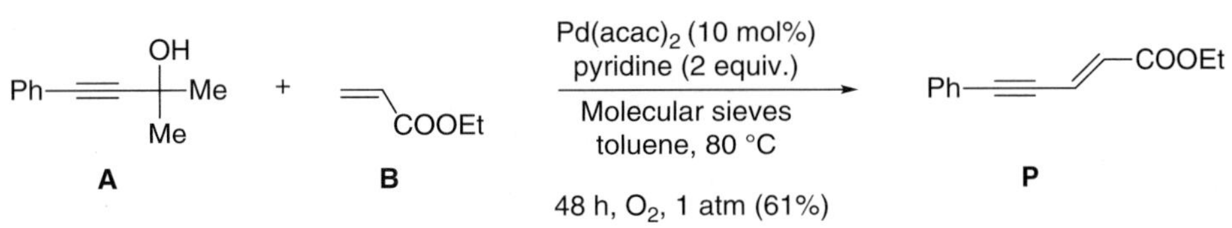

Answer

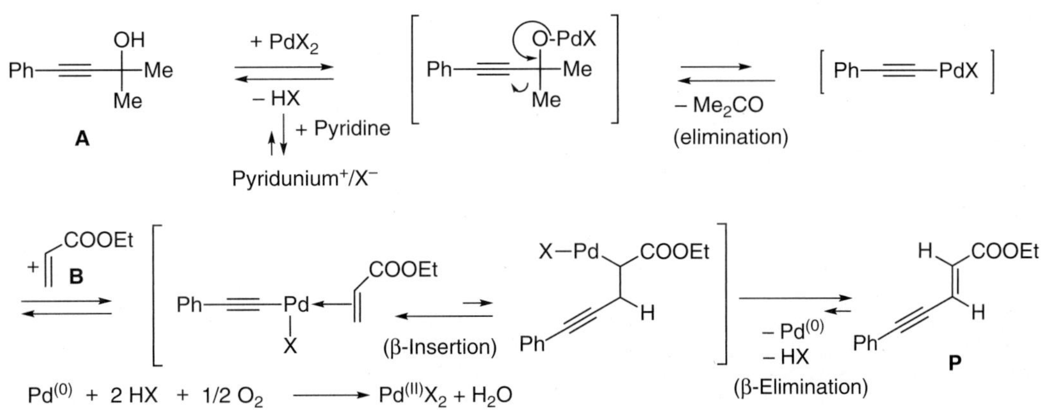

Problem 8.71 What are the possible products of the following reactions? [78].

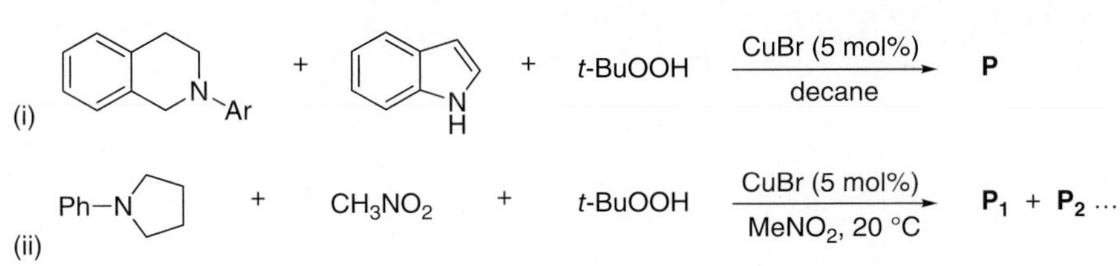

Answer

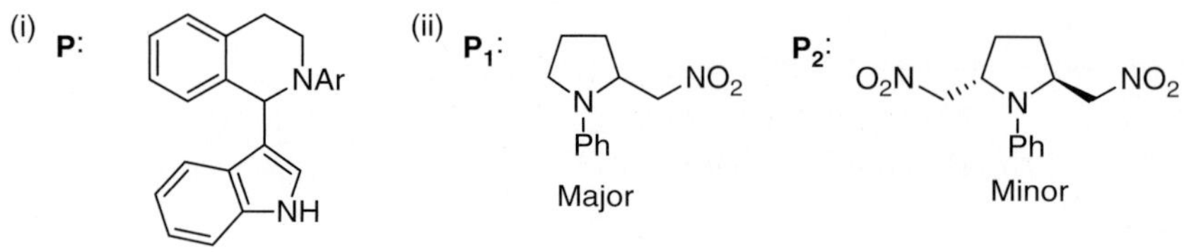

Problem 8.72 Styrene (in excess) and 8-oxabicyclo[3.3.1]oct-7-en-3-one equilibrate with a compound **P** in the presence of 1 mol% of Grubbs II catalyst ([**Ru**]-**2**). What is **P** and give a possible mechanism [79].

Answer

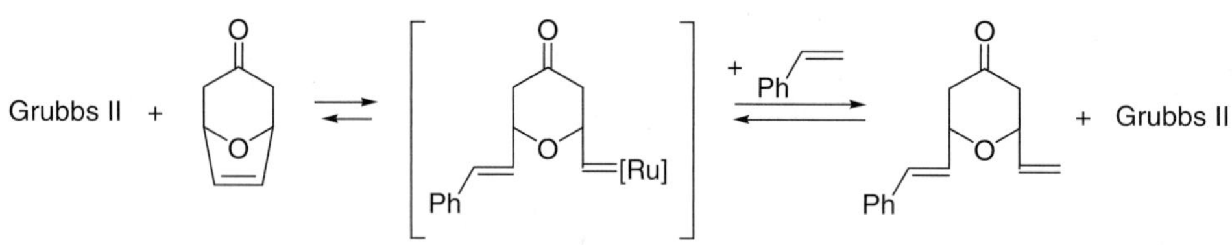

Problem 8.73 What is product **P** of the two following successive reactions? [80]

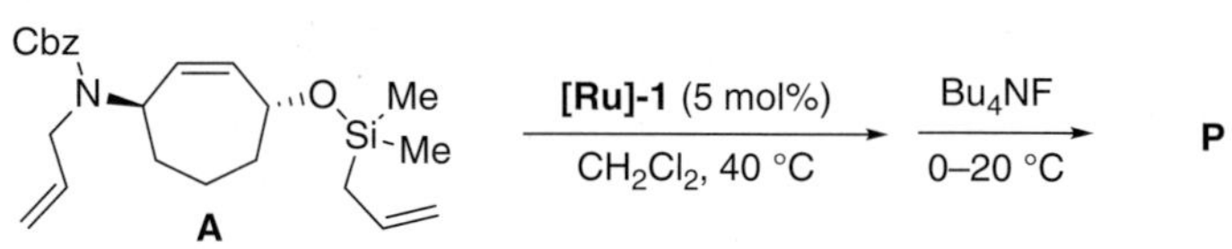

Answer

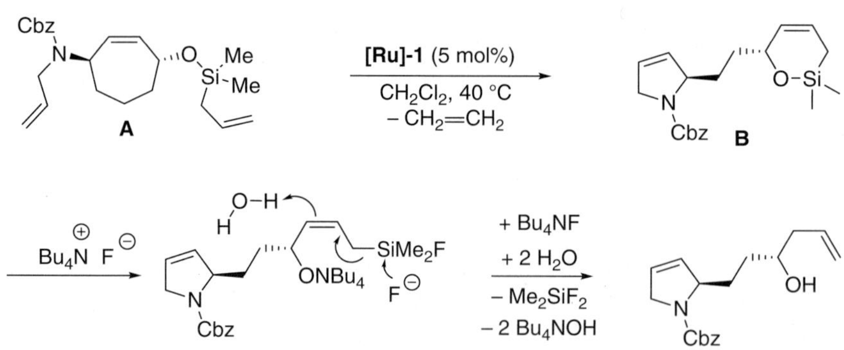

Problem 8.74 What is the major product **P** of the following reaction? [81].

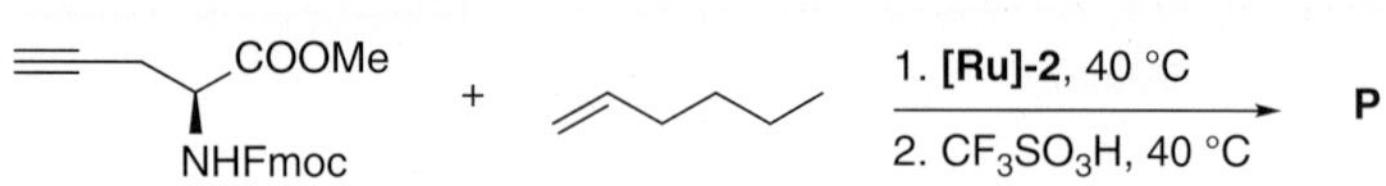

Answer

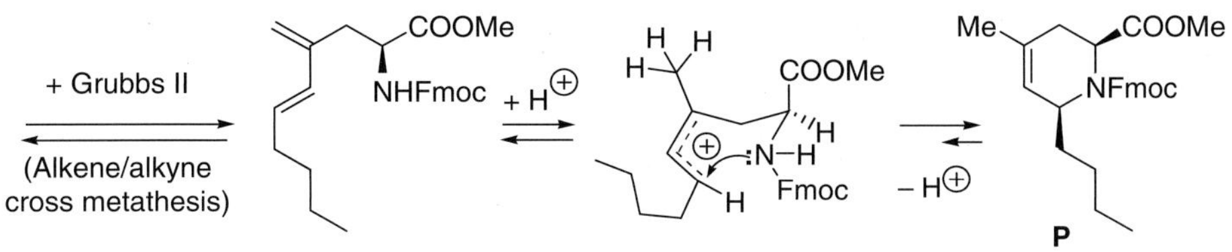

Problem 8.75 What is the major product **P** of the following reaction? [82]

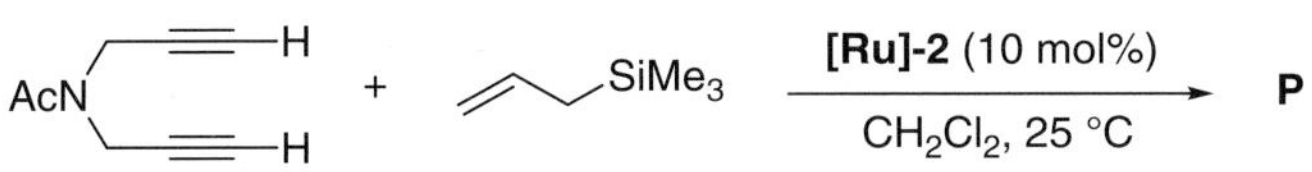

Answer

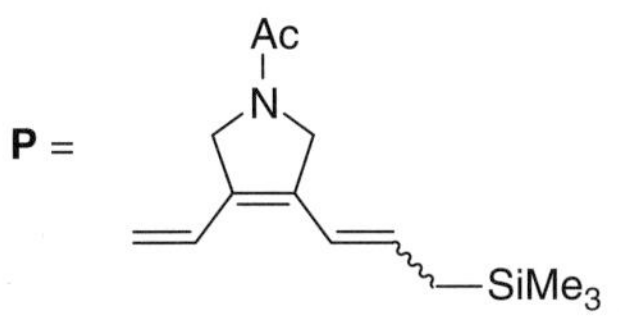

Problem 8.76 Give a mechanism for the following reaction [83].

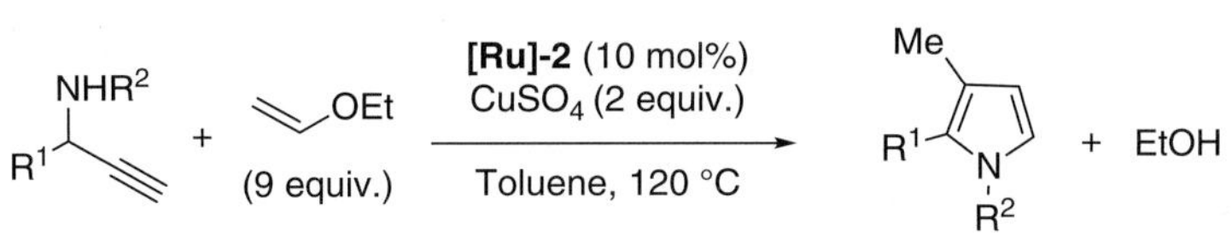

Answer

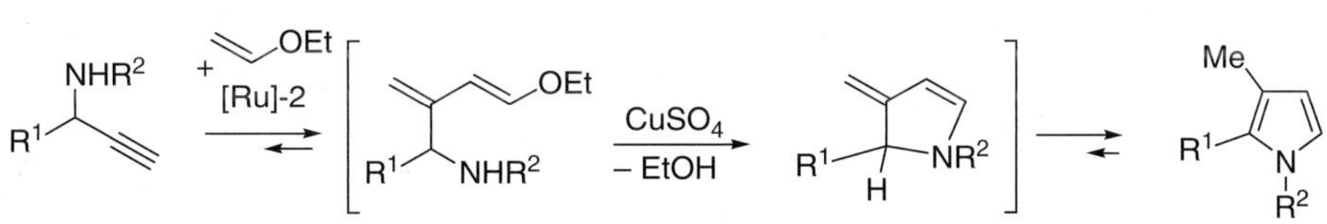

Problem 8.77 What are the main products **P** + **Q** of the following reactions? [84]

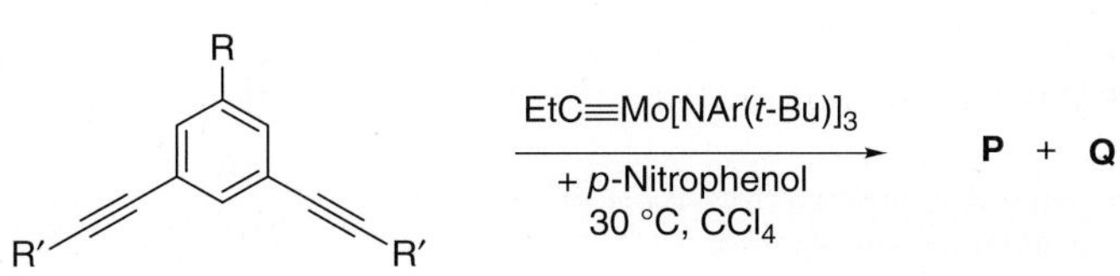

Answer

P: R′—≡—R′ **Q**:

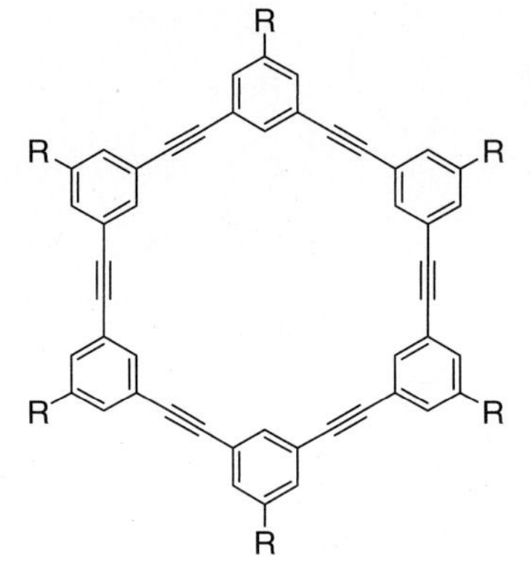

Problem 8.78 What products form in the following reaction? [85].

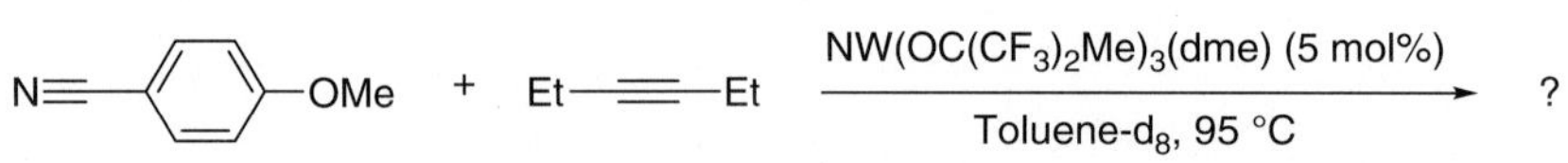

Answer

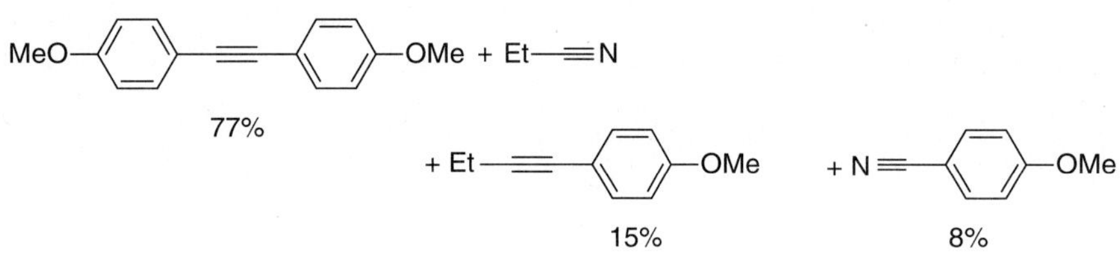

Problem 8.79 Propose a synthesis of compound **P** [86].

P:

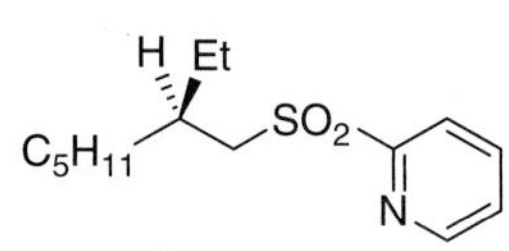

Answer

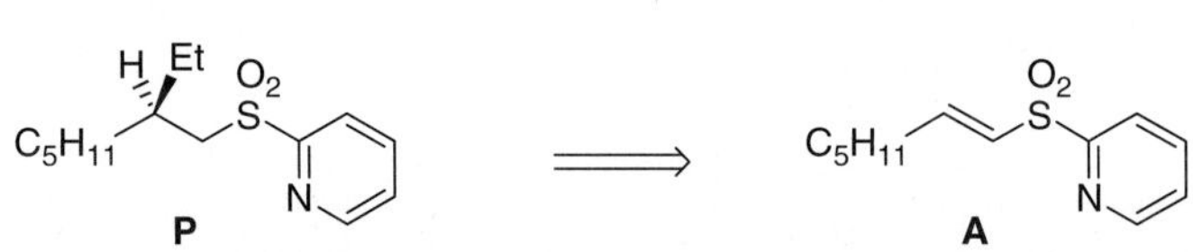

Asymmetric nucleophilic addition of EtMgBr to the Michael acceptor **A** applying a Cu(I)-catalyst bearing an enantiopure ligand should be found. Feringa and coworkers have reported:

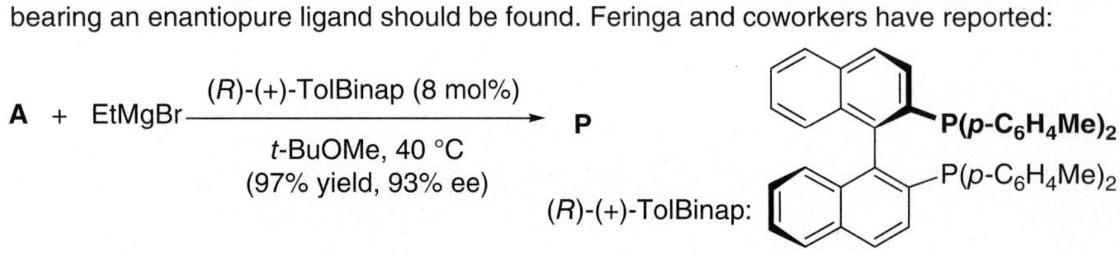

Problem 8.80 Propose a synthesis of 2-hexyl-1-phenyloct-1-ene starting with hex-1-ene and 1-phenyloct-1-yne [87].

Answer

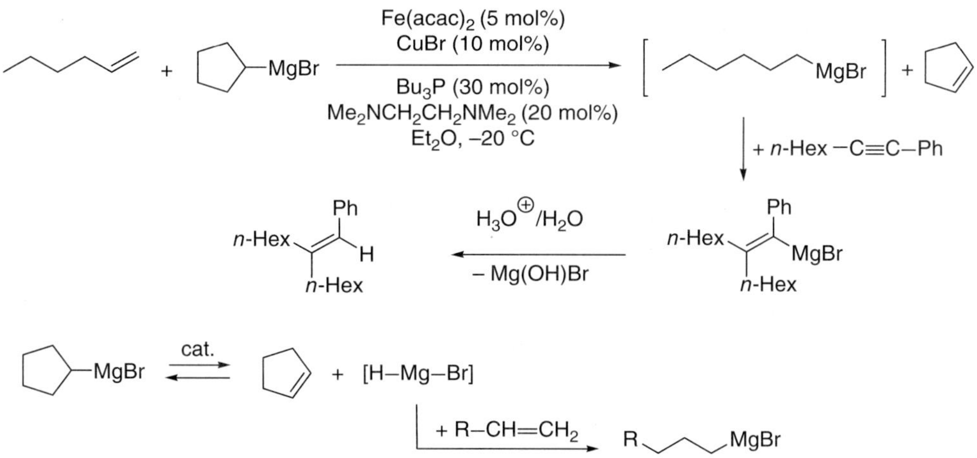

Problem 8.81 What are the products of reaction of benzoxanorbornadiene with Ph_2Zn in the presence of $FeCl_3$ (1 mol%) and 1,2-[(4-$FC_6H_4)_2P]_2C_6H_4$ (2 mol%) in THF/toluene at 0 °C and subsequent acidic aqueous work-up? [88]

Answer

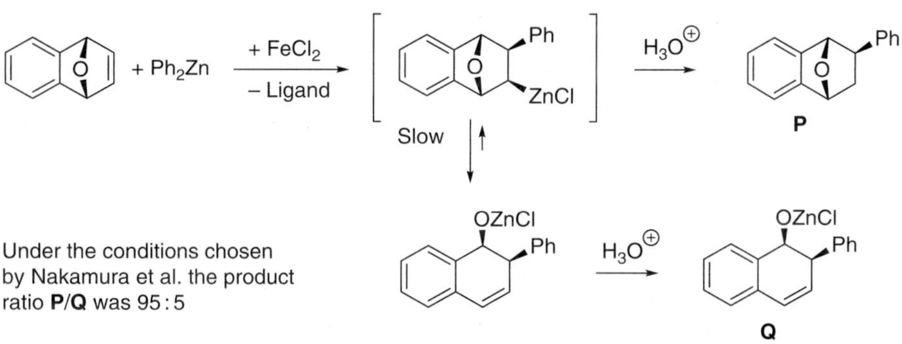

Problem 8.82 What is the product of the following two-step reaction? [89]

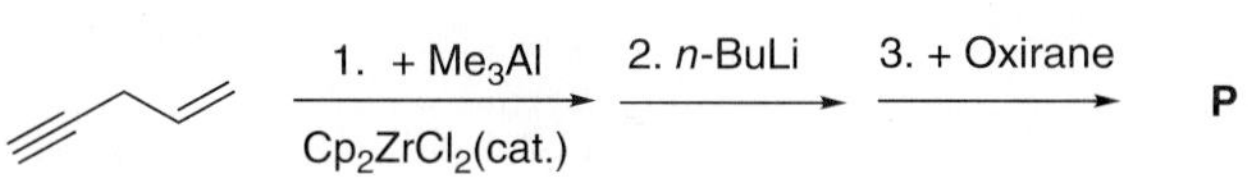

Answer

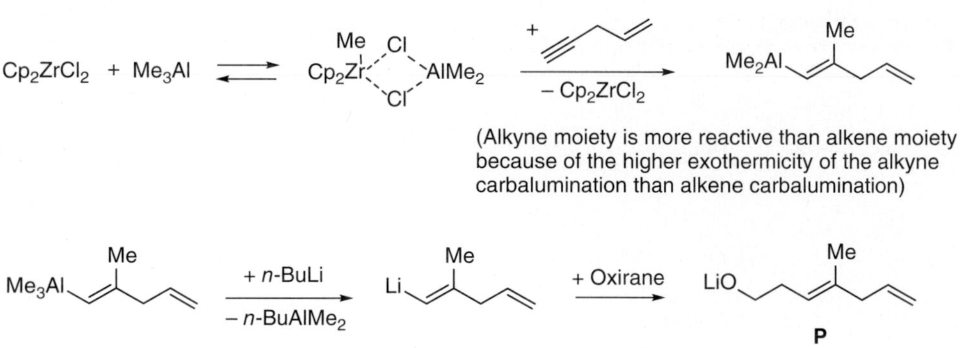

Problem 8.83 Propose a method to convert bis(allyl)dimethylsilane into **P** [90, 91].

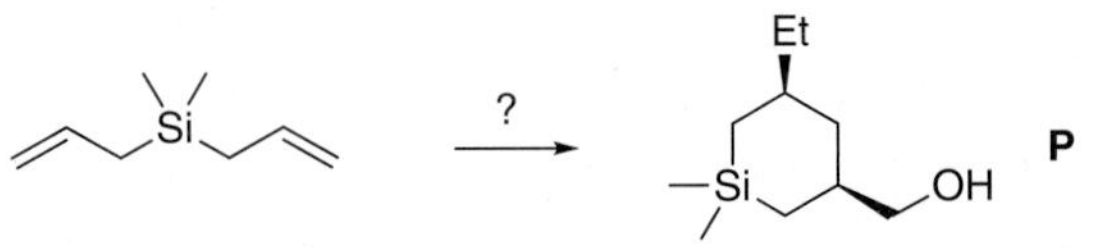

Answer

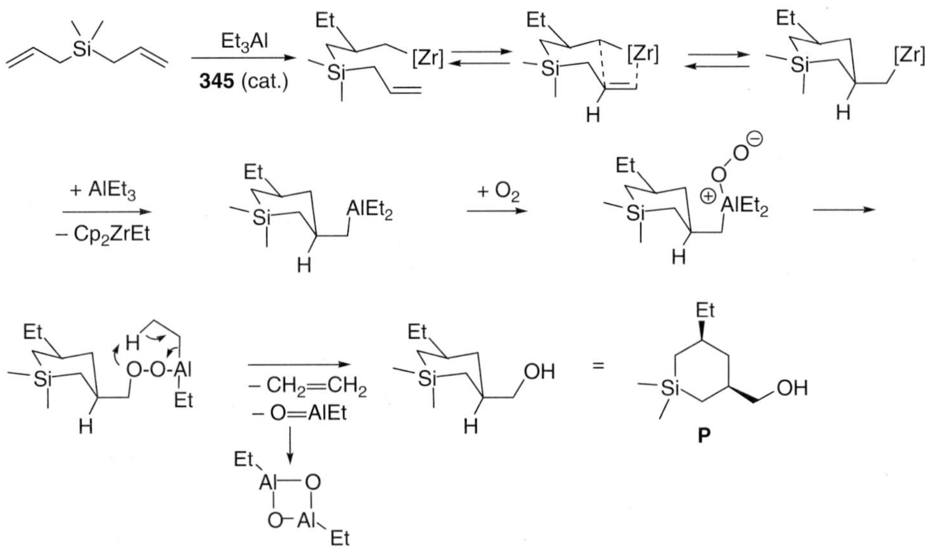

Problem 8.84 Propose a mechanism for the Pd(0)-catalyzed carbosilylation shown in Scheme 8.86a [92].

Answer

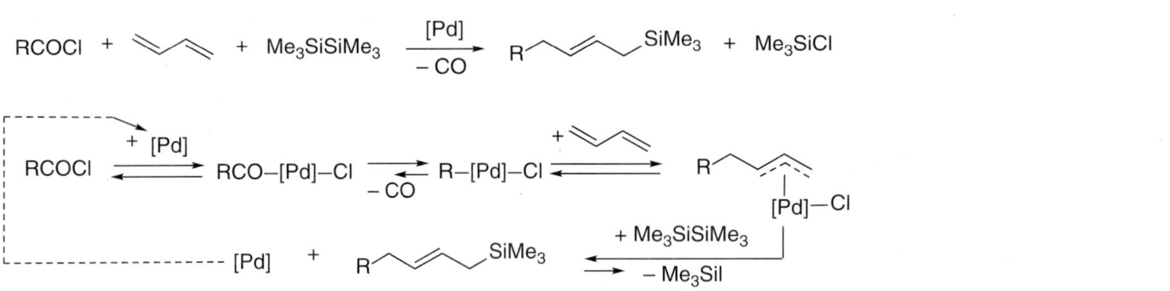

Problem 8.85 What are the four hydrocarbons formed in the reaction of *n*-BuMgBr + *n*-decyl-Br in the presence of $Fe(OAc)_2$ (1 mol%) and $[o\text{-}(Ph_2P)C_6H_4]_2O$ (1 mol%)? [93]

Answer

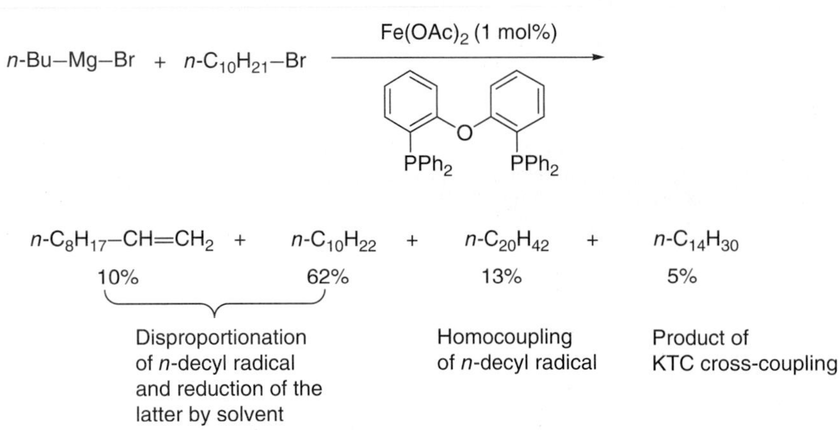

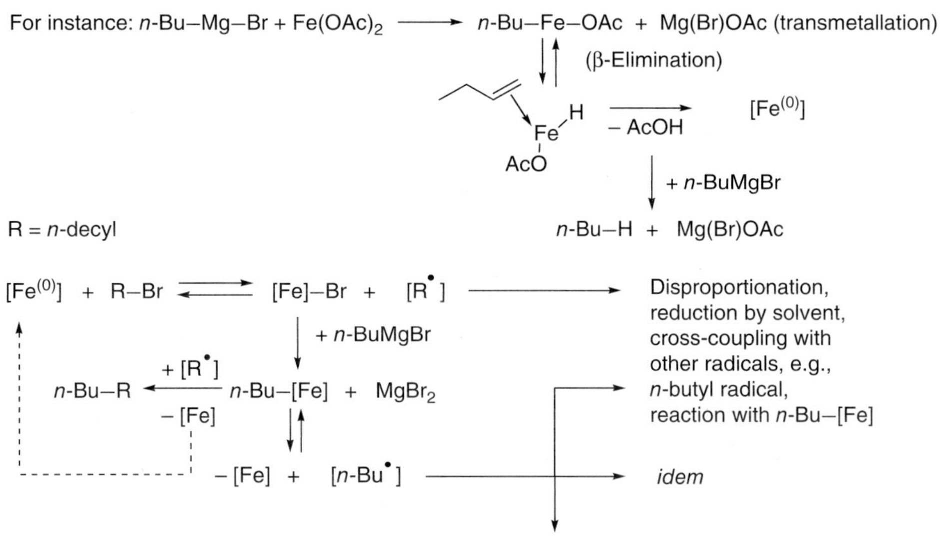

Problem 8.86 What is the major product of the following reaction? [94]

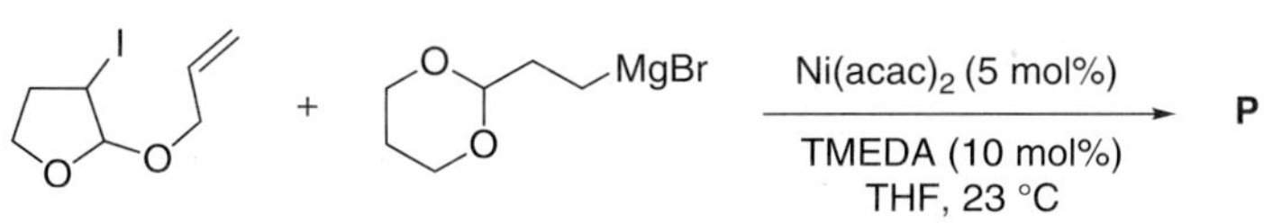

Answer

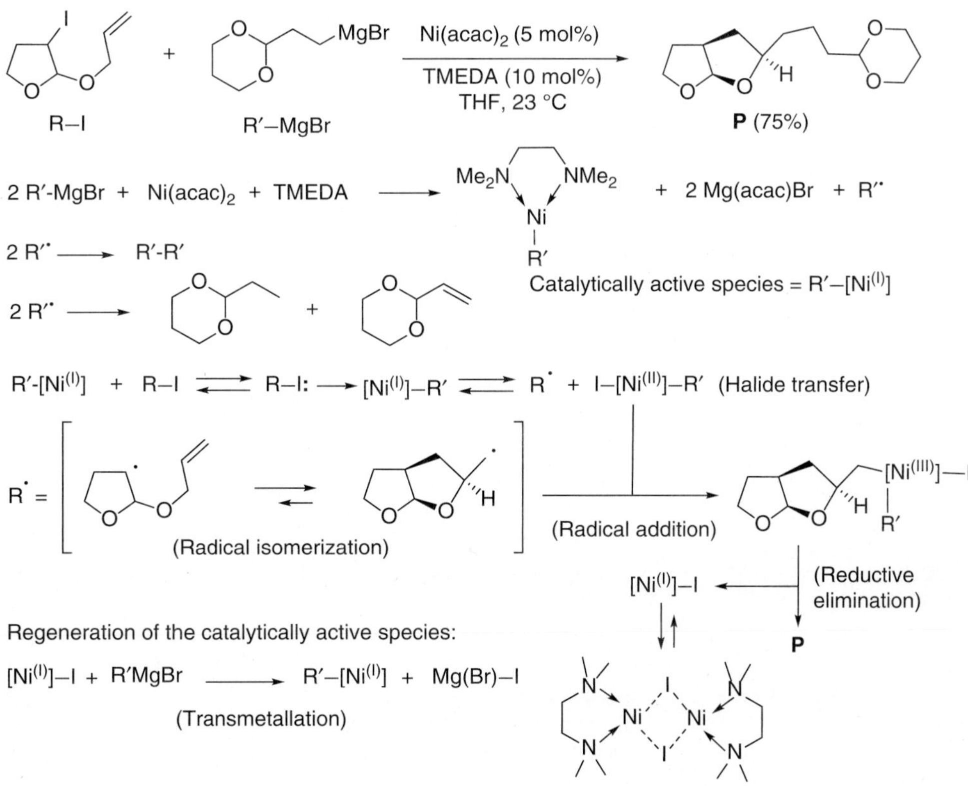

Problem 8.87 Propose syntheses for the protected α-amino acids **A**, **B**, and **C** (Bn = $PhCH_2$; Boc = *t*-BuOCO) [95–98].

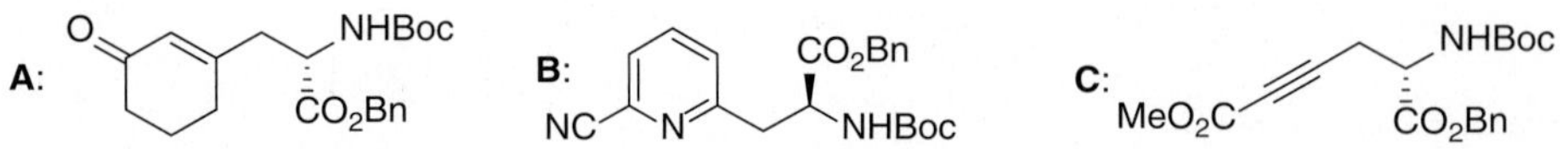

Answer

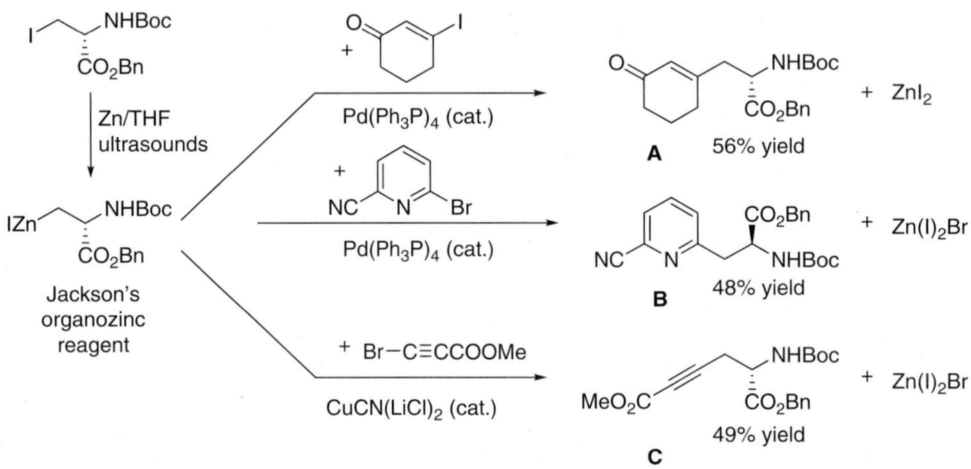

Problem 8.88 Propose a mechanism for the carbonylative Stille reaction (8.296).

Answer

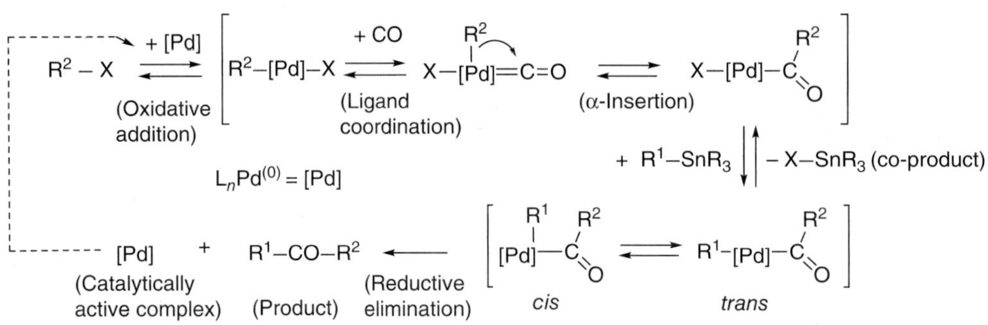

Problem 8.89 What is the product formed on treating 2-iodophenyl (trimethylstannyl)methyl ether with $Pd(Ph_3P)_4$ in toluene at 40 °C [99].

Answer

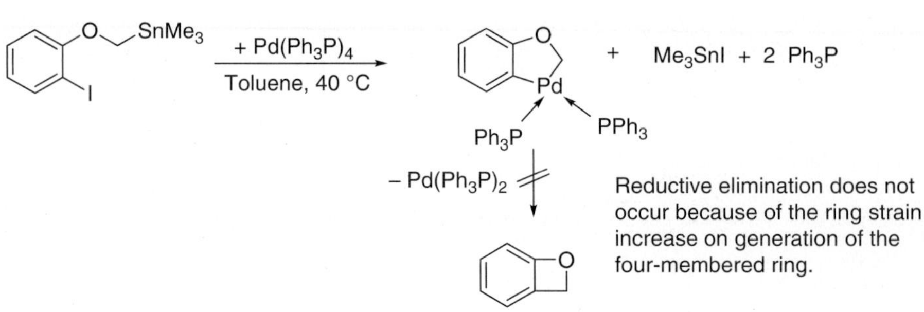

Problem 8.90 What is the main product **P** of the following reaction in 1 : 1 methoxycyclopentane/H_2O at 105 °C? [100]

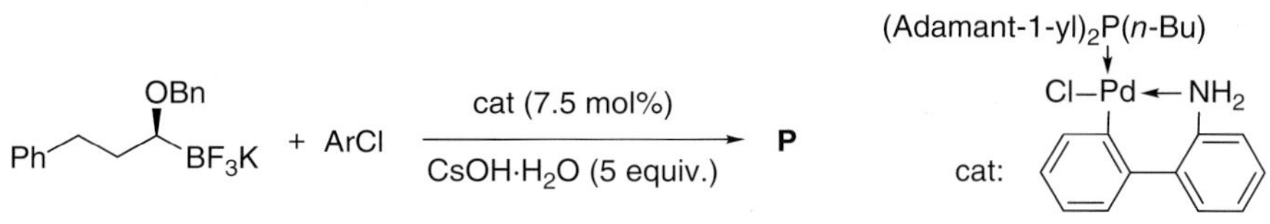

Answer

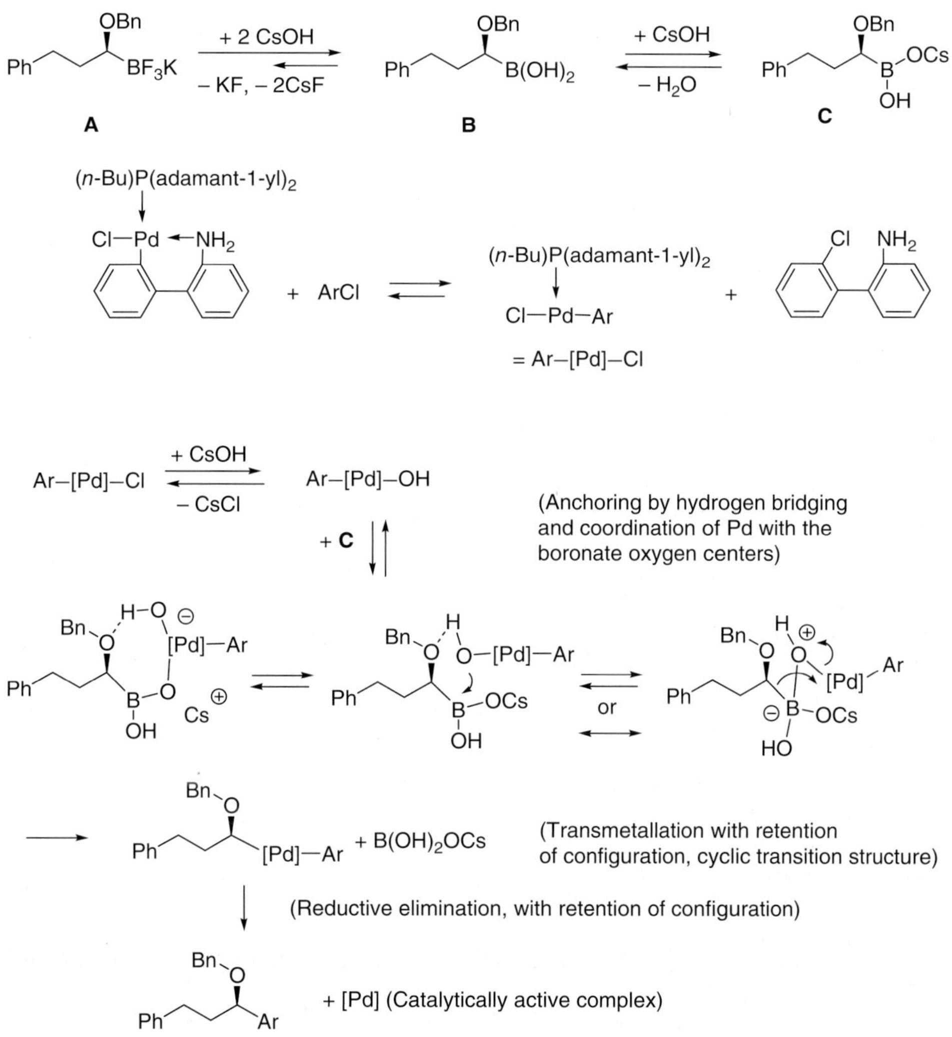

Problem 8.91 Aryltriazenes in combination with a Lewis acid undergo Pd-catalyzed cross-coupling reactions with arylboronic acids. Propose a mechanism for the following reaction and give the co-products of the reaction [101].

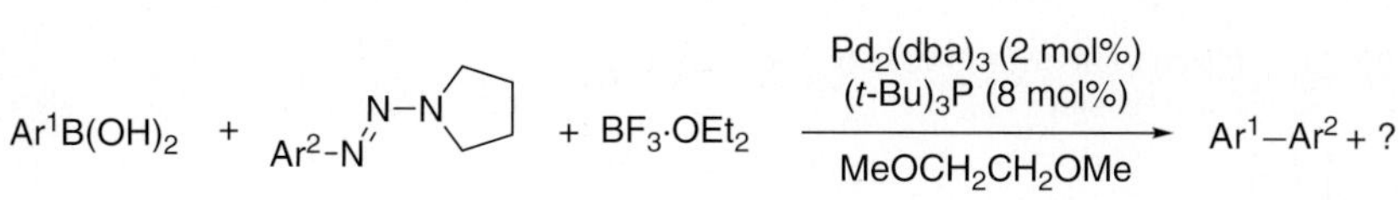

Answer

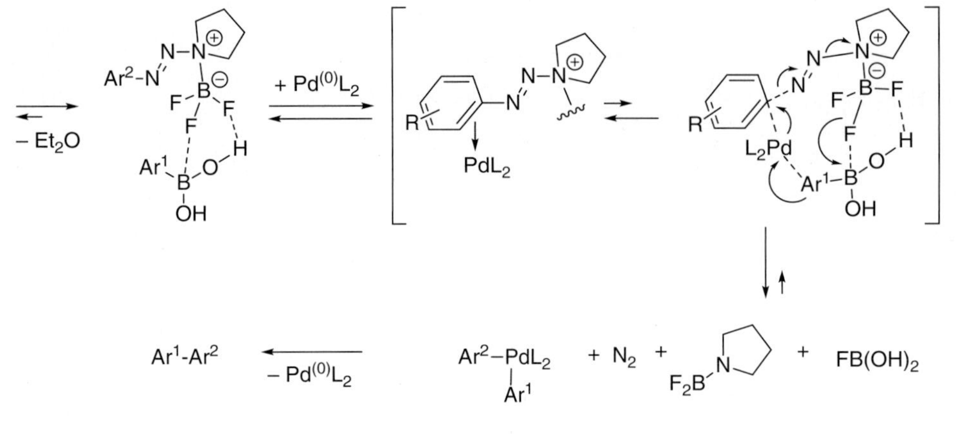

Problem 8.92 Starting with allylbenzene, and using potassium (4-bromophenyl)trifluoroborate and 4-bromoanisole as reagents, propose reaction conditions to obtain product **P** in three steps [102].

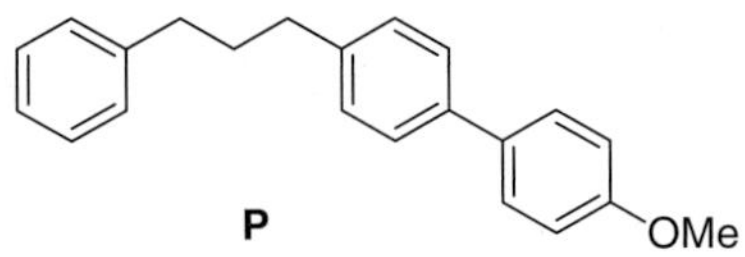

Answer

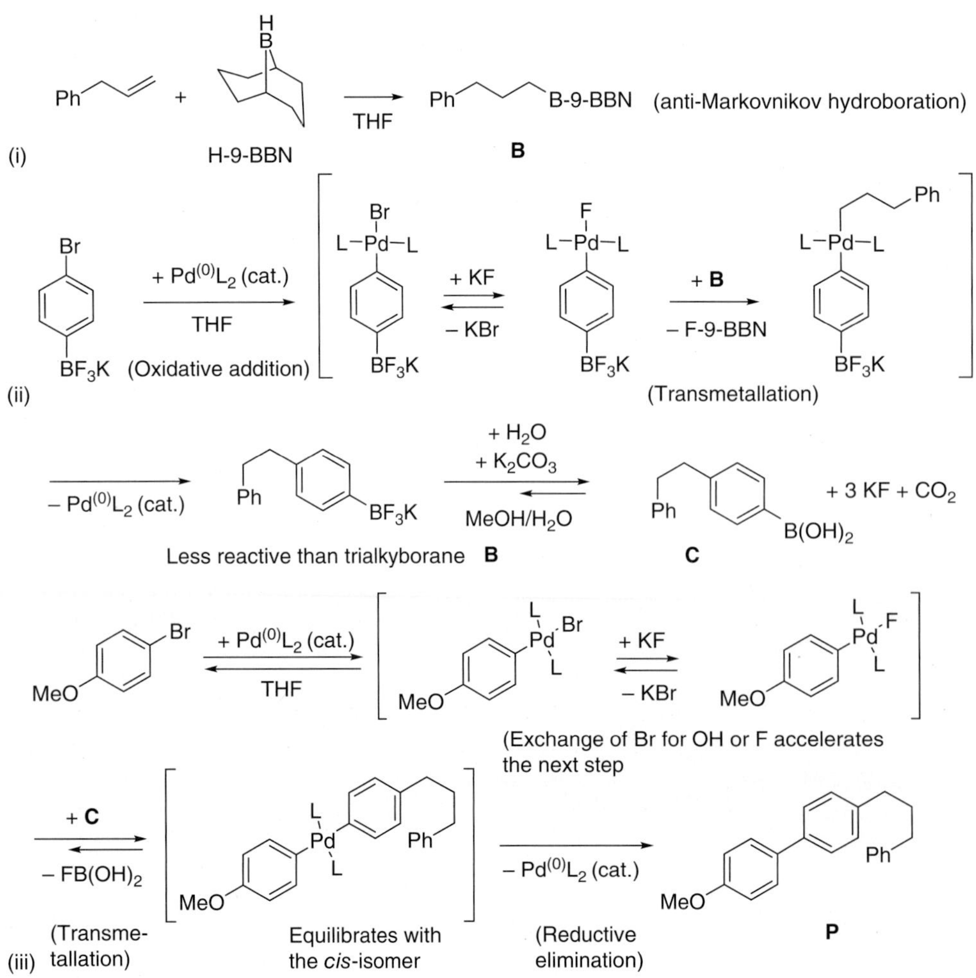

Problem 8.93 Give the product and co-products of the following cross-coupling reaction and propose a reaction mechanism.

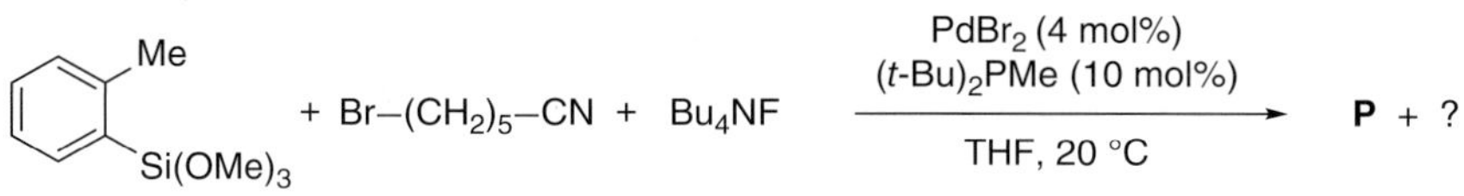

Answer

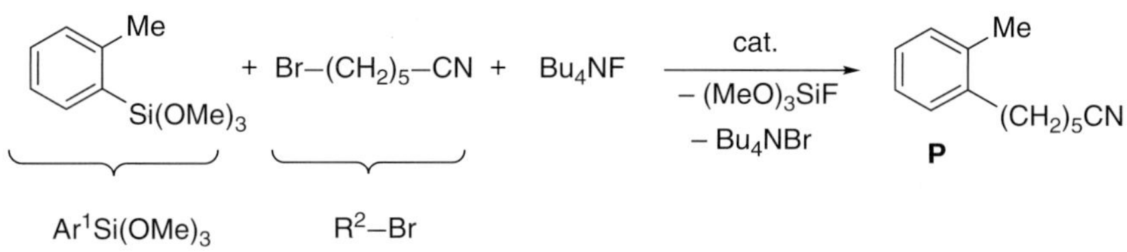

$Ar^1Si(OMe)_3 + Bu_4NF \rightleftharpoons Ar^1Si(OMe)_3F^{\ominus}\ Bu_4N^{\oplus}$

$PdBr_2 + 3\ R_3P + H_2O \longrightarrow Pd^{(0)}(R_3P)_2 + R_3P{=}O + 2\ HBr$

(Formation of the catalytically active species: $[Pd^{(0)}]$)

$HBr + Bu_4NF \rightleftharpoons Bu_4NBr + HF$

$[Pd^{(0)}] + R^2{-}Br \rightleftharpoons R^2{-}[Pd^{(II)}]{-}Br$ (Oxidative addition of the electrophile)

$R^2{-}[Pd^{(II)}]{-}Br + Ar^1Si(OMe)_3F^{\ominus}\ Bu_4N^{\oplus} \rightleftharpoons R^2{-}[Pd^{(II)}]{-}Ar^1 + (MeO)_3SiF + Bu_4NBr$

(Transmetallation)

(Reductive elimination)

$R^2{-}[Pd^{(II)}]{-}Ar^1 \longrightarrow [Pd^{(0)}]$ (cat.) $+ Ar^1{-}R^2 = \mathbf{P}$

Problem 8.94 What is the major product **P** of the following reaction? [103]

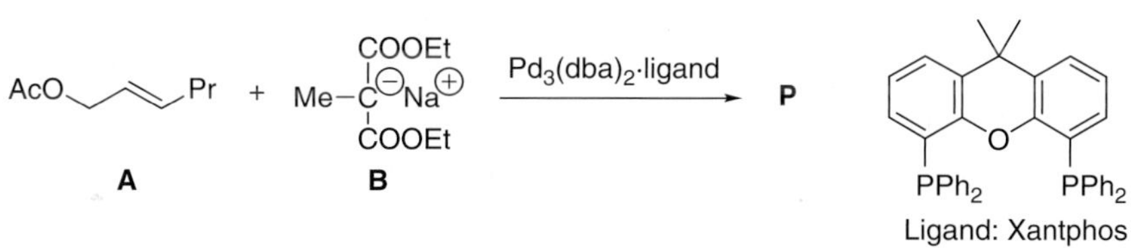

Answer

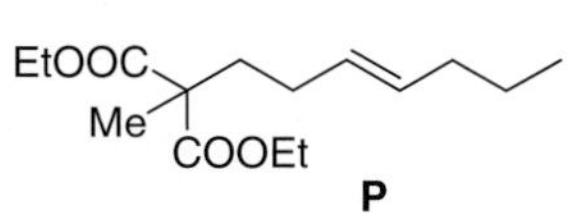

Problem 8.95 Propose a mechanism for reaction (8.336) [104].

Answer

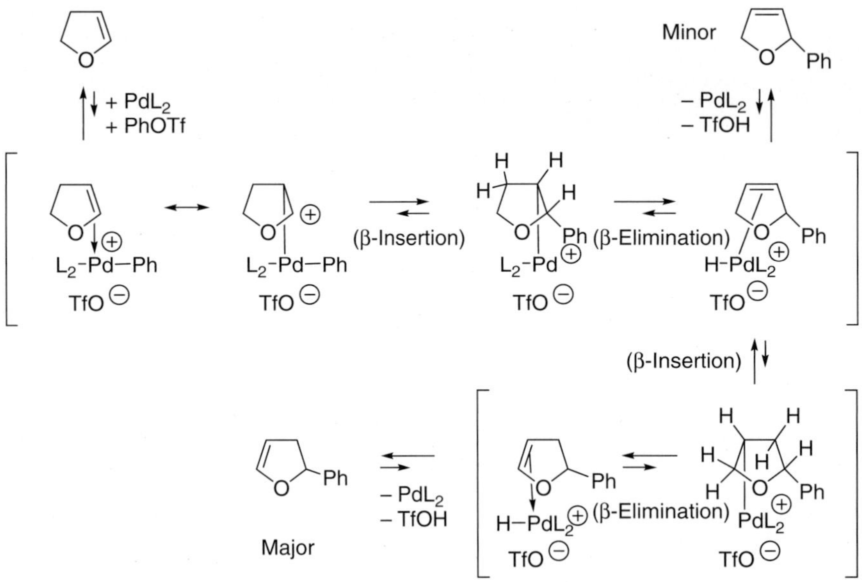

Problem 8.96 Give a possible mechanism for the Rh-catalyzed decarbonylative arylation of styrene [105].

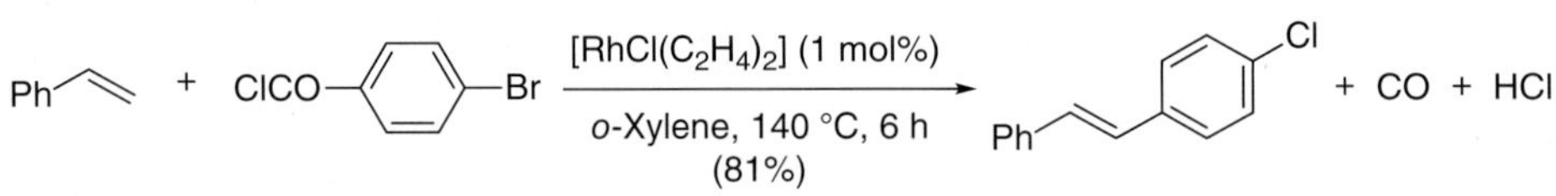

Answer

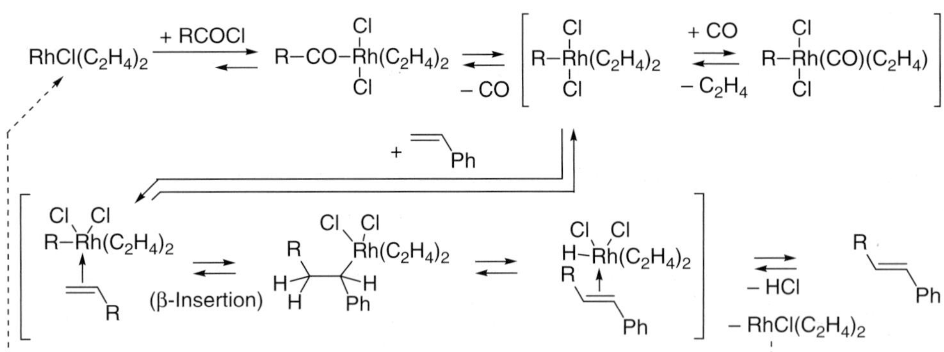

Problem 8.97 What are the possible products of the following reaction **A** → **P**? [106]

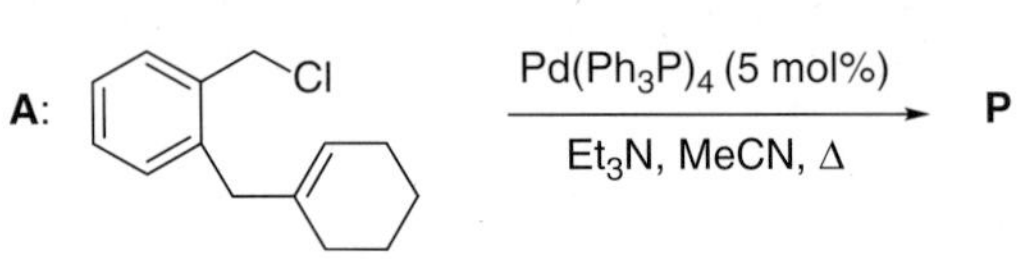

Answer

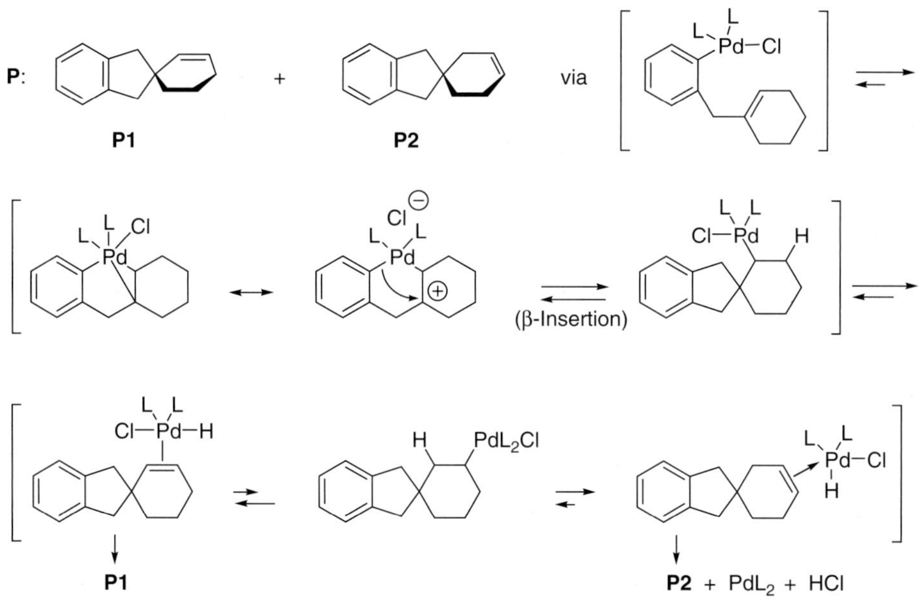

Problem 8.98 What are the major products **P** of the Ni-catalyzed cross-couplings of **A** + **B**? [107]

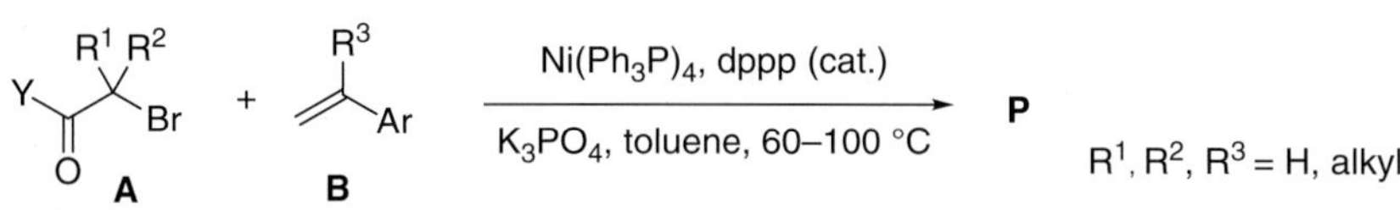

Answer

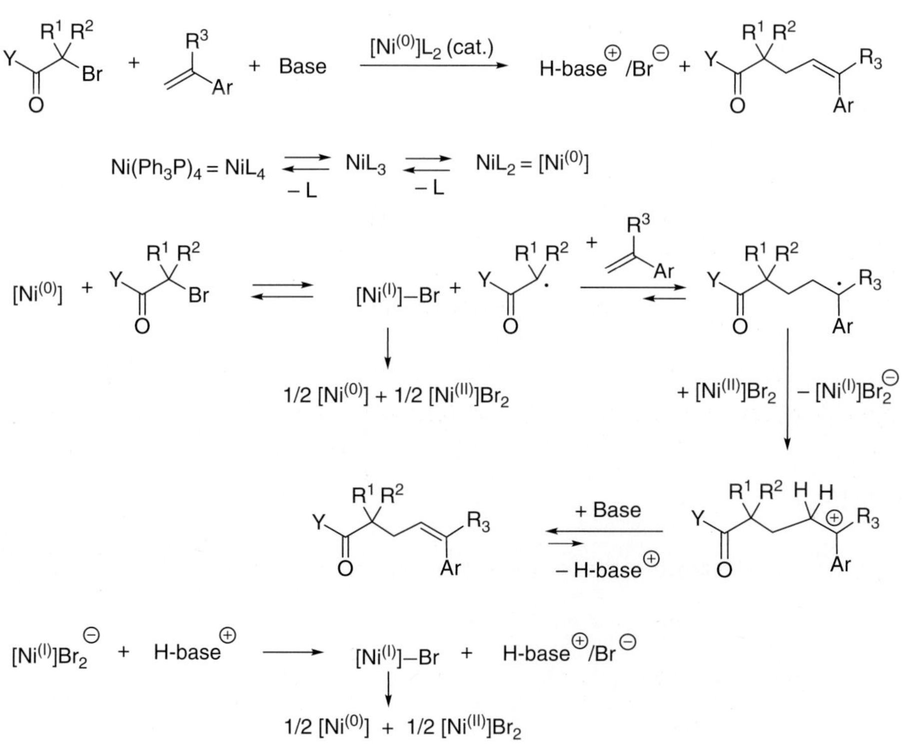

Problem 8.99 Propose a reaction mechanism for the Cu-catalyzed couplings (8.344) [108].

Answer

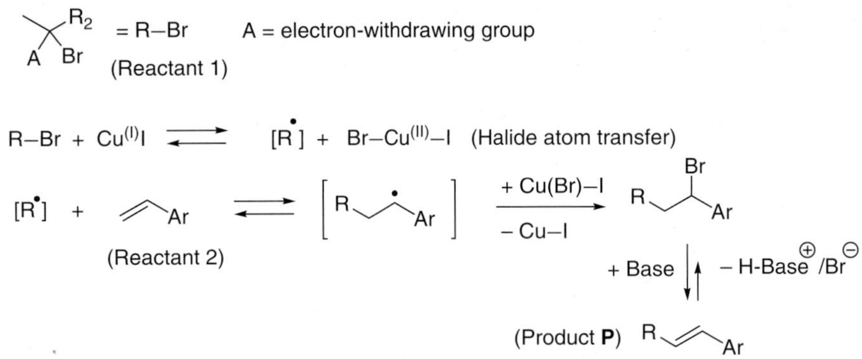

Problem 8.100 Propose a reaction mechanism for the following decarboxylative Sonogashira–Hagihara coupling [109].

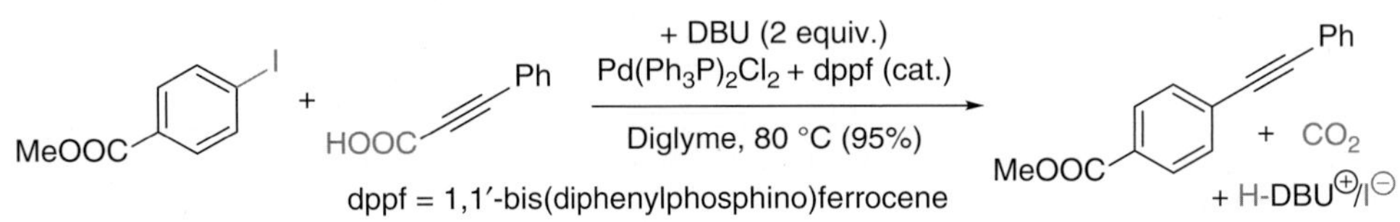

Answer

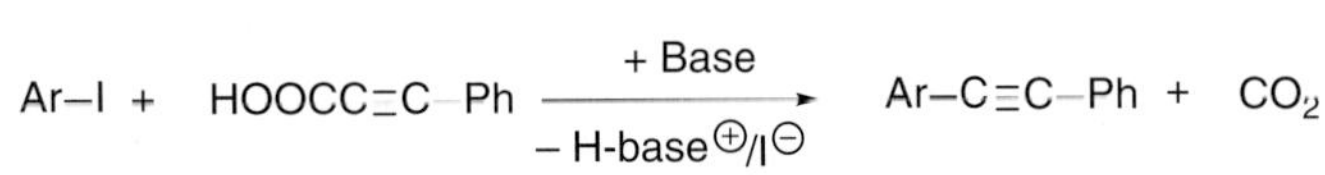

Possible mechanism I:

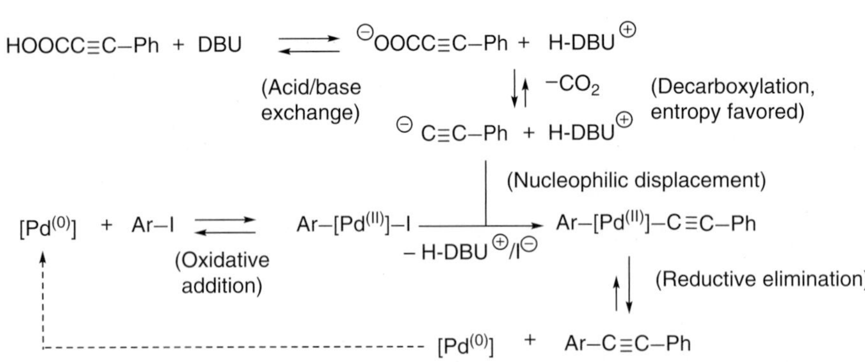

Possible mechanism II:

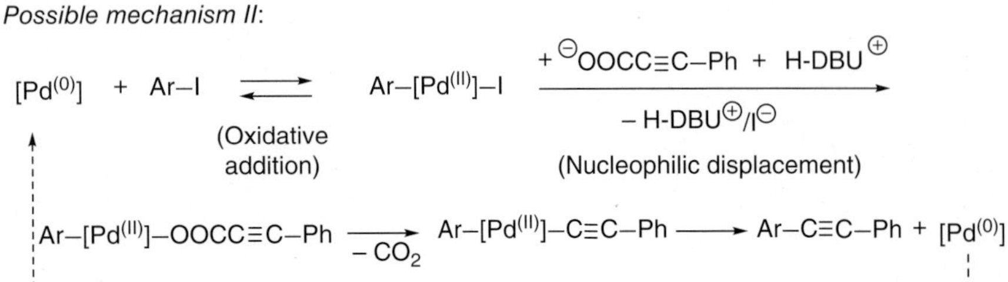

Problem 8.101 What is the main product **P** of the following reaction? [110]

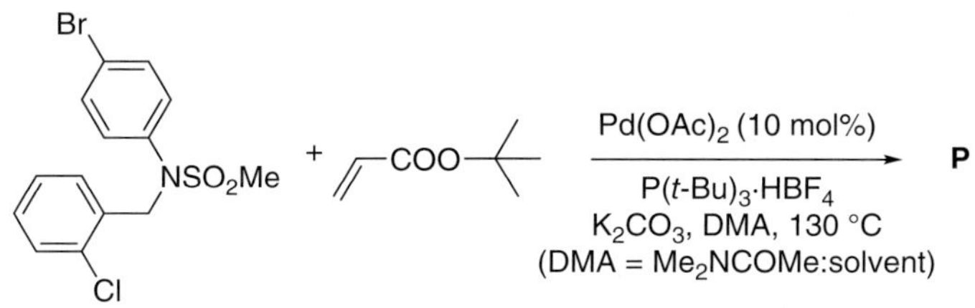

Answer

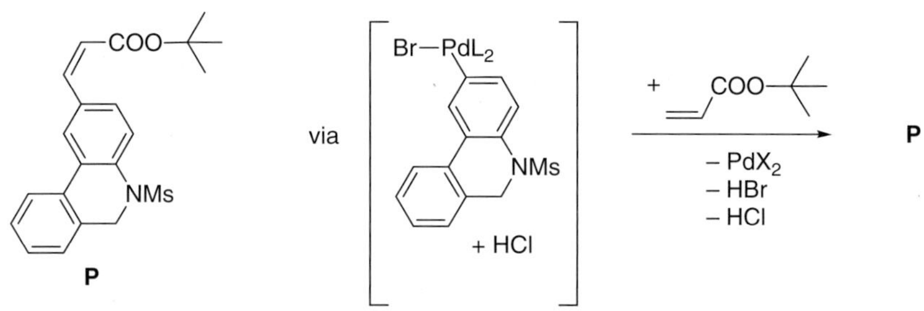

Problem 8.102 Devise a one-pot synthesis of benzofurans from 2-bromophenols and ketones [111],

Answer

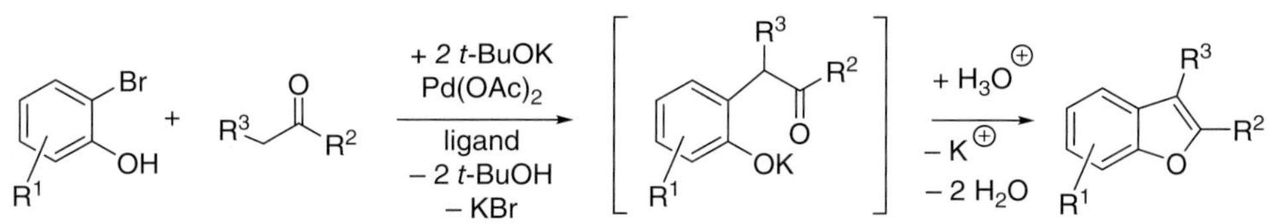

Problem 8.103 What is the main product **P** of the following reaction? [112]

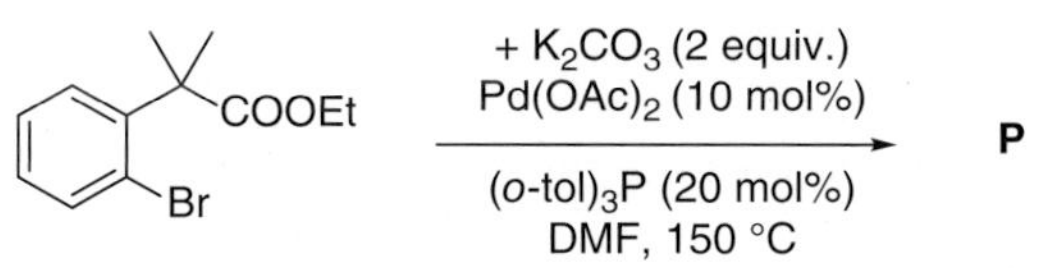

Answer

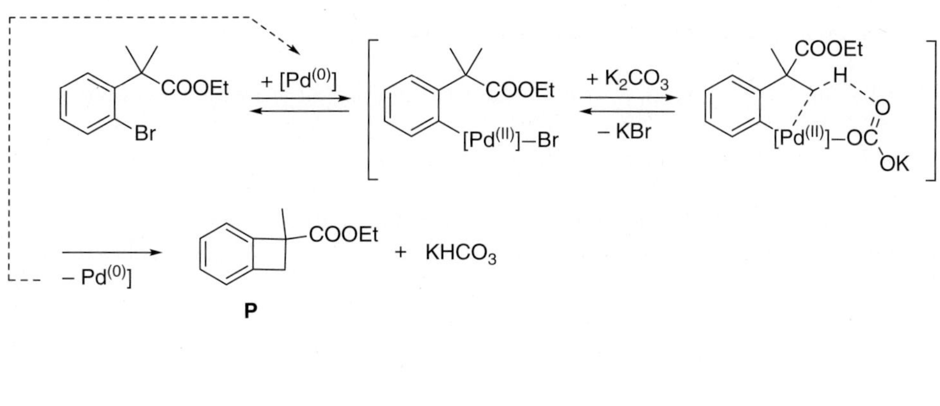

References

1 Yacob, S., Kilos, B.A., Barton, D.G., and Notestein, J.M. (2016). Vapor phase ethanol carbonylation over Rh supported on zeolite 13X. *Appl. Catal. A* 520: 122–131.

2 Serp, P., Chateau, L., Feurer, R. et al. (1998). Rhodium-catalyzed hydrocarbonylation of acetic acid into higher acids. *J. Mol. Catal. A Chem.* 136: 269–278.

3 Dong, Y. and Busacca, C.A. (1997). Indoles from *o*-haloanilines: Syntheses of tryptamines and tryptophols via regioselective hydroformylation of functionalized anilines. *J. Org. Chem.* 62: 6464–6465.

4 Schmidt, A.M. and Eilbracht, P. (2005). Synthesis of pharmacologically relevant Indoles with amine side chains via tandem hydroformylation/Fischer indole synthesis. *J. Org. Chem.* 70: 5528–5535.

5 Wright, M.E. and Cochran, B.B. (1996). Silylformylation of carbonyl compounds: a study of substrate, catalyst, and reaction conditions. *Organometallics* 15: 317–324.

6 Ojima, I., Tzamarioudaki, M., and Tsai, C.Y. (1994). Extremely chemoselective silylformylation and silylcarbocyclization of alkynals. *J. Am. Chem. Soc.* 116: 3643–3644.

7 Tsuji, J., Imamura, S., Morikawa, M., and Kiji, J. (1964). Organic syntheses by means of noble metal compounds. VI. Synthesis of muconic acid. *J. Am. Chem. Soc.* 86: 2095.

8 Tsuji, J. and Nogi, T. (1966). Organic syntheses by means of noble metal compounds. 22. Palladium-catalyzed carbonylation of diphenylacetylene. *J. Am. Chem. Soc.* 88: 1289–1292.

9 Gabriele, B., Salerno, G., Costa, M., and Chiusoli, G.P. (1995). Combined oxidative and reductive carbonylation of terminal alkynes with palladium iodide-thiourea catalysts. *J. Organomet. Chem.* 503: 21–28.

10 Gabriele, B., Salerno, G., Costa, M., and Chiusoli, G.P. (2003). Recent developments in the synthesis of heterocyclic derivatives by PdI_2-catalyzed oxidative carbonylation reactions. *J. Organomet. Chem.* 687: 219–228.

11 Ishikawa, T., Ishii, H., Shimizu, K. et al. (2004). Diastereoselective total synthesis of isocarbacyclin from L-ascorbic acid. *J. Org. Chem.* 69: 8133–8135.

12 Hirose, T., Miyakoshi, N., and Mukai, C. (2008). Total synthesis of (+)-achalensolide based on the Rh(I)-catalyzed allenic Pauson–Khand-type reaction. *J. Org. Chem.* 73: 1061–1066.

13 Yamamoto, A. (1995). Palladium-catalyzed double and single carbonylation of aryl halides and allylic compounds. *Bull. Chem. Soc. Jpn.* 68: 433–446.

14 Tamaru, Y., Ochiai, H., Yamada, Y., and Yoshida, Z. (1983). Unsymmetrical ketone synthesis via palladium catalyzed carbonylation of organic halides. *Tetrahedron Lett.* 24: 3869–3872.

15 Kramer, J.W., Joh, D.Y., and Coates, G.W. (2007). Carbonylation of epoxides to substituted 3-hydroxy-delta-lactones. *Org. Lett.* 9: 5581–5583.

16 Wang, M.D., Calet, S., and Alper, H. (1989). Regiospecific carbonylation and ring expansion of thietanes and oxetanes catalyzed by cobalt and or ruthenium carbonyls. *J. Org. Chem.* 54: 20–21.

17 Nakano, K., Katayama, M., Ishihara, S. et al. (2004). An alternative route to protected aldols: cobalt-catalyzed hydroformylation of epoxides and in situ protection of beta-hydroxyaldehydes by $HC(OMe)_3$. *Synlett* 1367–1370.

18 Fukumoto, Y., Chatani, N., and Murai, S. (1993). Rhodium-catalyzed ring-opening silylformylation of epoxides leading to beta-siloxy aldehydes. *J. Org. Chem.* 58: 4187–4188.

19 Cho, E.J. and Lee, D. (2007). Selectivity in the ruthenium-catalyzed Alder ene reactions of di- and triynes. *J. Am. Chem. Soc.* 129: 6692–6693.

20 Hansen, E.C. and Lee, D.S. (2005). Synthesis of beta,beta-disubstituted vinyl boronates via the ruthenium-catalyzed Alder ene reaction of borylated alkynes and alkenes. *J. Am. Chem. Soc.* 127: 3252–3253.

21 Kakiuchi, F., Sonoda, M., Tsujimoto, T. et al. (1999). The ruthenium-catalyzed addition of C–H bonds in aromatic nitriles to olefins. *Chem. Lett.* 1083–1084.

22 Kakiuchi, F., Yamamoto, Y., Chatani, N., and Murai, S. (1995). Catalytic addition of aromatic C–H bonds to acetylenes. *Chem. Lett.* 681–682.

23 Kakiuchi, F., Tanaka, Y., Sato, T. et al. (1995). Catalytic addition of olefinic C–H bonds to olefins. *Chem. Lett.* 679–680.

24 Zhou, B., Hou, W., Yang, Y.X., and Li, Y.C. (2013). Rhodium(III)-catalyzed amidation of aryl ketone *O*-methyl oximes with isocyanates by C–H activation: convergent synthesis of 3-methyleneisoindolin-1-ones. *Chem. Eur. J.* 19: 4701–4706.

25 Fürstner, A., Martin, R., and Majima, K. (2005). Cycloisomerization of enynes catalyzed by iron(0)-ate complexes. *J. Am. Chem. Soc.* 127: 12236–12237.

26 Jonas, K., Schieferstein, L., Krüger, C., and Tsay, Y.H. (1979). Tetrakis(ethylene)irondilithium and bis(eta-4-1,5-cyclooctadiene)irondilithium. *Angew. Chem. Int. Ed. Engl.* 18: 550–551.

27 Jonas, K. and Krüger, C. (1980). Alkali metal-transition metal pi-complexes. *Angew. Chem. Int. Ed. Engl.* 19: 520–537.

28 Lee, D.Y., Moon, C.W., and Jun, C.H. (2002). Synthesis of aliphatic ketones from allylic alcohols through consecutive isomerization and chelation-assisted hydroacylation by a rhodium catalyst. *J. Org. Chem.* 67: 3945–3948.

29 Neuhaus, J.D., Morrow, S.M., Brunavs, M., and Willis, M.C. (2016). Diversely substituted quinolines via rhodium-catalyzed alkyne hydroacylation. *Org. Lett.* 18: 1562–1565.

30 Neuhaus, J.D. and Willis, M.C. (2016). Homogeneous rhodium(I)-catalysis in de novo heterocycle syntheses. *Org. Biomol. Chem.* 14: 4986–5000.

31 Bäckvall, J.E. and Andell, O.S. (1986). Stereochemistry and mechanism of nickel-catalyzed hydrocyanation of olefins and conjugated dienes. *Organometallics* 5: 2350–2355.

32 Oda, S., Sam, B., and Krische, M.J. (2015). Hydroaminomethylation beyond carbonylation: allene-imine reductive coupling by ruthenium-catalyzed transfer hydrogenation. *Angew. Chem. Int. Ed.* 54: 8525–8528.

33 Chen, T.Y., Tsutsumi, R., Montgomery, T.P. et al. (2015). Ruthenium-catalyzed C–C coupling of amino alcohols with dienes via transfer hydrogenation: redox-triggered imine addition and related hydroaminoalkylations. *J. Am. Chem. Soc.* 137: 1798–1801.

34 Iverson, C.N. and Jones, W.D. (2001). Rhodium-catalyzed activation and functionalization of the C–C bond of biphenylene. *Organometallics* 20: 5745–5750.

35 Fang, J., Ren, J., and Wang, Z.W. (2008). $Sc(OTf)_3$-catalyzed smooth tandem [3+2] cycloaddition/ring opening of donor-acceptor cyclopropane 1,1-diesters with enol silyl ethers. *Tetrahedron Lett.* 49: 6659–6662.

36 Takasu, K., Nagao, S., and Ihara, M. (2006). Construction of highly-functionalized cyclopentanes from silyl enol ethers and activated cyclopropanes by [3+2] cycloaddition catalyzed by triflic imide. *Adv. Synth. Catal.* 348: 2376–2380.

37 Peace, B.W., Carman, F., and Wulfman, D.S. (1971). Soluble cooper(I)-catalyzed decomposition of dimethyl diazomalonate and di-tert-butyl diazomalonate in presence of some cycloalkenes. *Synthesis* 658–661.

38 Peace, B.W. and Wulfman, D.S. (1973). Preparation and reactions of diazomalonic esters. *Synthesis* 137–145.

39 Lautens, M., Klute, W., and Tam, W. (1996). Transition metal-mediated cycloaddition reactions. *Chem. Rev.* 96: 49–92.

40 Trost, B.M. and Chan, D.M.T. (1982). Intramolecular carbocyclic [3+2] cycloaddition via organopalladium intermediates. *J. Am. Chem. Soc.* 104: 3733–3735.

41 Trost, B.M., Grese, T.A., and Chan, D.M.T. (1991). Intramolecular palladium-catalyzed trimethylenemethane cycloadditions - initial studies. *J. Am. Chem. Soc.* 113: 7350–7362.

42 Meng, Q.X., Li, M., and Zhang, J.S. (2005). The computational study on the mechanism of rhodium(I)-catalyzed asymmetric carbonylative [4+1] cycloaddition with (R,R)-Me-DuPHOS-type ligand. A DFT study. *J. Mol. Struct. (THEOCHEM)* 726: 47–54.

43 Sarel, S., Felzenst, A., Victor, R., and Yovell, J. (1974). Photoreactions of a methylene-spirane and dispirane with $Fe(Co)_5$ – sigma, pi-complex formation from a double cyclopropane rearrangement. *J. Chem. Soc. Chem. Commun.* 1025–1026.

44 Liu, C.H., Zhuang, Z., Bose, S., and Yu, Z.X. (2016). $Fe_2(CO)_9$-mediated [5+1] cycloaddition of vinylcyclopropanes and CO for the synthesis of alpha, beta-cyclohexenones. *Tetrahedron* 72: 2752–2755.

45 Murakami, M., Itami, K., Ubukata, M. et al. (1998). Iridium-catalyzed [5+1] cycloaddition: Allenylcyclopropane as a five-carbon assembling unit. *J. Org. Chem.* 63: 4–5.

46 Kamitani, A., Chatani, N., Morimoto, T., and Murai, S. (2000). Carbonylative [5+1] cycloaddition of cyclopropyl imines catalyzed by ruthenium carbonyl complex. *J. Org. Chem.* 65: 9230–9233.

47 Hilt, G., Hess, W., and Harms, K. (2006). Asymmetric cobalt-catalyzed Diels–Alder reactions of a boron-functionalized 1,3-diene with alkynes and allylboration with aldehydes. *Org. Lett.* 8: 3287–3290.

48 Wender, P.A. and Smith, T.E. (1996). Transition metal-catalyzed intramolecular [4+2] cycloadditions: a novel method for the assembly of nitrogen heterocycles and its application to yohimban alkaloid synthesis. *J. Org. Chem.* 61: 824–825.

49 Kesenheimer, C. and Groth, U. (2006). Total synthesis of (–)-8-*O*-methyltetrangomycin (MM 47755). *Org. Lett.* 8: 2507–2510.

50 Huang, X. and Lin, Z.Y. (2003). Density functional theory studies of ruthenium-catalyzed bis-Diels–Alder cycloaddition of 1,5-cyclooctadiene with alkynes. *Organometallics* 22: 5478–5484.
51 Liu, L.X. and Sun, H.L. (2014). $[HCo(CO)_4]$-catalyzed three-component cycloaddition of epoxides, imines, and carbon monoxide: facile construction of 1,3-oxazinan-4-ones. *Angew. Chem. Int. Ed.* 53: 9865–9869.
52 Saito, S., Masuda, M., and Komagawa, S. (2004). Nickel-catalyzed intermolecular [3+2+2] cocyclization of ethyl cyclopropylideneacetate and alkynes. *J. Am. Chem. Soc.* 126: 10540–10541.
53 Saito, S. and Takeuchi, K. (2007). Nickel-catalyzed [4+3] cycloaddition of ethyl cyclopropylideneacetate and 1,3-dienes. *Tetrahedron Lett.* 48: 595–598.
54 Gulías, M., Duran, J., López, F. et al. (2007). Palladium-catalyzed [4+3] intramolecular cycloaddition of alkylidenecyclopropanes and dienes. *J. Am. Chem. Soc.* 129: 11026–11027.
55 Craft, D.T. and Gung, B.W. (2008). The first transannular [4+3] cycloaddition reaction: synthesis of the ABCD ring structure of cortistatins. *Tetrahedron Lett.* 49: 5931–5934.
56 Ni, Y. and Montgomery, J. (2006). Synthetic studies and mechanistic insight in nickel-catalyzed [4+2+1] cycloadditions. *J. Am. Chem. Soc.* 128: 2609–2614.
57 Trost, B.M., Hu, Y., and Horne, D.B. (2007). Total synthesis of (+)-frondosin A. Application of the Ru-catalyzed [5+2] cycloaddition. *J. Am. Chem. Soc.* 129: 11781–11790.
58 Ashfeld, B.L. and Martin, S.F. (2005). Enantioselective syntheses of tremulenediol A and tremulenolide A. *Org. Lett.* 7: 4535–4537.
59 Ashfeld, B.L., Miller, K.A., Smith, A.J. et al. (2007). Features and applications of $[Rh(CO)_2Cl]_2$-catalyzed alkylations of unsymmetrical allylic substrates. *J. Org. Chem.* 72: 9018–9031.
60 Hong, X., Stevens, M.C., Liu, P. et al. (2014). Reactivity and chemoselectivity of allenes in Rh(I)-catalyzed intermolecular (5+2) cycloadditions with vinylcyclopropanes: allene-mediated rhodacycle formation can poison Rh(I)-catalyzed cycloadditions. *J. Am. Chem. Soc.* 136: 17273–17283.
61 Zhou, M.B., Song, R.J., Wang, C.Y., and Li, J.H. (2013). Synthesis of azepine derivatives by silver-catalyzed [5+2] cycloaddition of gamma-amino ketones with alkynes. *Angew. Chem. Int. Ed.* 52: 10805–10808.
62 Kaloko, J.J., Gary, Y.H., and Ojima, T.I. (2009). One-step formation of fused tetracyclic skeletons from cyclohexene-diynes and carbon monoxide through Rh(I)-catalyzed [2+2+2+1] cycloaddition reaction. *Chem. Commun.* 4569–4571.
63 Aouf, C., El Abed, D., Giorgi, M., and Santelli, M. (2007). Unexpected Ti-catalyzed formal intramolecular [4+4] cycloaddition of 1,1 '-bi[(S*,S*)-6-(trimethylsilyl)cyclohepta-2,4-dien-1-yl]. *Tetrahedron Lett.* 48: 4969–4972.
64 Sahani, R.L. and Liu, R.S. (2016). Gold-catalyzed [4+3] and [4+4]-annulation reactions of t-butyl propiolate derivatives with epoxides and oxetanes for the construction of 1,4-dioxepane and 1,5-dioxocane cores. *Chem. Commun.* 52: 7482–7485.
65 Lautens, M., Tam, W., Lautens, J.C. et al. (1995). Cobalt-catalyzed [2-Pi+2-Pi+2-Pi] (homo Diels-Alder) and [2-Pi+2-Pi+4-Pi] cycloadditions of bicyclo[2.2.1]hepta-2,5-dienes. *J. Am. Chem. Soc.* 117: 6863–6879.
66 Varela, J.A., Castedo, L., and Saa, C. (2003). "Formal" ruthenium-catalyzed [4+2+2] cycloaddition of 1,6-diynes to 1,3-dienes: formation of cyclooctatrienes vs vinylcyclohexadienes. *Org. Lett.* 5: 2841–2844.
67 Lee, S.I., Park, S.Y., and Chung, Y.K. (2006). Rhodium-catalyzed [4+2+2] cycloaddition reaction of two enynes or diynes with one diene to give eight-membered ring compounds. *Adv. Synth. Catal.* 348: 2531–2539.
68 Achard, M., Mosrin, M., Tenaglia, A., and Buono, G. (2006). Cobalt(I)-catalyzed [6+2] cycloadditions of cyclooctatetra(tri)ene with alkynes. *J. Org. Chem.* 71: 2907–2910.
69 Ura, Y., Utsumi, T., Tsujita, H. et al. (2006). Formation of divalent ruthenacycles via oxidative cyclization of 1,3,5-cyclooctatriene with maleic anhydride or maleimides: an intermediate for the transition metal-mediated [6+2] cycloaddition of 1,3,5-trienes and alkenes. *Organometallics* 25: 2934–2942.
70 Oonishi, Y., Hosotani, A., and Sato, Y. (2011). Rh(I)-catalyzed formal [6+2] cycloaddition of 4-allenals with alkynes or alkenes in a tether. *J. Am. Chem. Soc.* 133: 10386–10389.
71 Oonishi, Y., Hosotani, A., and Sato, Y. (2012). Construction of monocyclic eight-membered rings: intermolecular rhodium(I)-catalyzed [6+2] cycloaddition of 4-allenals with alkynes. *Angew. Chem. Int. Ed.* 51: 11548–11551.
72 Rigby, J.H. and Fleming, M. (2002). Construction of the ingenane core using an Fe(III) or Ti(IV) Lewis acid-catalyzed intramolecular [6+4] cycloaddition. *Tetrahedron Lett.* 43: 8643–8646.

73 Bosch, M., Brookhart, M.S., Ilg, K., and Werner, H. (2000). The rhodium-catalyzed cyclotetramerization of butadiene: isolation and structural characterization of an intermediate. *Angew. Chem. Int. Ed.* 39: 2304–2307.

74 Tobisch, S. and Werner, H. (2004). [(RhL)-L-I]-catalyzed cyclotetramerization of 1,3-butadiene: a theoretical investigation of alternative mechanistic paths for the generation of the [Rh-III(octadienediyl)(PR3)](+) complex. *Dalton Trans.* 2963–2968.

75 Seoane, A., Casanova, N., Quinones, N. et al. (2014). Straightforward assembly of benzoxepines by means of a rhodium(III)-catalyzed C–H functionalization of *o*-vinylphenols. *J. Am. Chem. Soc.* 136: 834–837.

76 Sonobe, T., Oisaki, K., and Kanai, M. (2012). Catalytic aerobic production of imines en route to mild, green, and concise derivatizations of amines. *Chem. Sci.* 3: 3249–3255.

77 Nishimura, T., Araki, H., Maeda, Y., and Uemura, S. (2003). Palladium-catalyzed oxidative alkynylation of alkenes via C–C bond cleavage under oxygen atmosphere. *Org. Lett.* 5: 2997–2999.

78 Li, Z.P. and Li, C.J. (2005). Highly efficient copper-catalyzed nitro-mannich type reaction: cross-dehydrogenative-coupling between sp^3 C–H bond and sp^3 C–H bond. *J. Am. Chem. Soc.* 127: 3672–3673.

79 Wright, D.L., Usher, L.C., and Estrella-Jimenez, M. (2001). Unusual influence of substituents on ring-opening metathesis reactions. *Org. Lett.* 3: 4275–4277.

80 Zaminer, J., Stapper, C., and Blechert, S. (2002). A new and efficient synthesis of (−)-indolizidine 167B by tandem metathesis. *Tetrahedron Lett.* 43: 6739–6741.

81 Kalbarczyk, K.P. and Diver, S.T. (2009). Enyne metathesis/bronsted acid-promoted heterocyclization. *J. Org. Chem.* 74: 2193–2196.

82 Stragies, R., Schuster, M., and Blechert, S. (1999). A novel ruthenium-catalysed tandem diyne cycloisomerisation-cross metathesis process. *Chem. Commun.* 237–238.

83 Chachignon, H., Scalacci, N., Petricci, E., and Castagnolo, D. (2015). Synthesis of 1,2,3-substituted pyrroles from propargylamines via a one-pot tandem enyne cross metathesis-cyclization reaction. *J. Org. Chem.* 80: 5287–5295.

84 Zhang, W. and Moore, J.S. (2004). Arylene ethynylene macrocycles prepared by precipitation-driven alkyne metathesis. *J. Am. Chem. Soc.* 126: 12796.

85 Geyer, A.M., Wiedner, E.S., Gary, J.B. et al. (2008). Synthetic, mechanistic, and computational investigations of nitrile-alkyne cross-metathesis. *J. Am. Chem. Soc.* 130: 8984–8999.

86 Bos, P.H., Minnaard, A.J., and Feringa, B.L. (2008). Catalytic asymmetric conjugate addition of Grignard reagents to alpha,beta-unsaturated sulfones. *Org. Lett.* 10: 4219–4222.

87 Shirakawa, E., Ikeda, D., Masui, S. et al. (2012). Iron–copper cooperative catalysis in the reactions of alkyl Grignard reagents: exchange reaction with alkenes and carbometalation of alkynes. *J. Am. Chem. Soc.* 134: 272–279.

88 Ito, S., Itoh, T., and Nakamura, M. (2011). Diastereoselective carbometalation of oxa- and azabicyclic alkenes under iron catalysis. *Angew. Chem. Int. Ed.* 50: 454–457.

89 Zhu, G.G. and Negishi, E.I. (2008). 1,4-pentenyne as a five-carbon synthon for efficient and selective syntheses of natural products containing 2,4-dimethyl-1-penten-1,5-ylidene and related moieties by means of Zr-catalyzed carboalumination of alkynes and alkenes. *Chem. Eur. J.* 14: 311–318.

90 Kondakov, D.Y. and Negishi, E. (1995). Zirconium-catalyzed enantioselective methylalumination of monosubstituted alkenes. *J. Am. Chem. Soc.* 117: 10771–10772.

91 Kondakov, D.Y. and Negishi, E. (1996). Zirconium-catalyzed enantioselective alkylalumination of monosubstituted alkenes proceeding via noncyclic mechanism. *J. Am. Chem. Soc.* 118: 1577–1578.

92 Obora, Y., Tsuji, Y., and Kawamura, T. (1995). 1,4-Carbosilylation of 1,3-dienes via palladium-catalyzed 3-component coupling reaction. *J. Am. Chem. Soc.* 117: 9814–9821.

93 Dongol, K.G., Koh, H., Sau, M., and Chai, C.L.L. (2007). Iron-catalysed sp^3-sp^3 cross-coupling reactions of unactivated alkyl halides with alkyl Grignard reagents. *Adv. Synth. Catal.* 349: 1015–1018.

94 Guisan-Ceinos, M., Soler-Yanes, R., Collado-Sanz, D. et al. (2013). Ni-catalyzed cascade cyclization-Kumada alkyl–alkyl cross-coupling. *Chem. Eur. J.* 19: 8405–8410.

95 Jackson, R.F.W., Wishart, N., Wood, A. et al. (1992). Preparation of enantiomerically pure protected 4-oxo-alpha-amino acids and 3-aryl-alpha-amino acids from serine. *J. Org. Chem.* 57: 3397–3404.

96 Jackson, R.F.W., Wishart, N., and Wythes, M.J.A. (1993). New alpha-amino-acid gamma-anion equivalent – direct synthesis of enantiomerically pure protected amino-acids. *Synlett* 219–220.

97 Dunn, M.J., Jackson, R.F.W., Pietruszka, J. et al. (1993). Preparation of a serine-derived organozinc reagent in tetrahydrofuran – synthesis of novel, enantiomerically pure allenic, acetylenic and heteroaryl amino-acids. *Synlett* 499–500.
98 Dunn, M.J., Jackson, R.F.W., Pietruszka, J., and Turner, D. (1995). Synthesis of enantiomerically pure unsaturated alpha-amino-acids using serine-derived zinz/copper reagents. *J. Org. Chem.* 60: 2210–2215.
99 Cardenas, D.J., Mateo, C., and Echavarren, A.M. (1994). Synthesis of oxapalladacycles and azapalladacycles from organostannanes. *Angew. Chem. Int. Ed.* 33: 2445–2447.
100 Molander, G.A. and Wisniewski, S.R. (2012). Stereospecific cross-coupling of secondary organotrifluoroborates: potassium 1-(benzyloxy)alkyltrifluoroborates. *J. Am. Chem. Soc.* 134: 16856–16868.
101 Saeki, T., Son, E.C., and Tamao, K. (2004). Boron trifluoride induced palladium-catalyzed cross-coupling reaction of 1-aryltriazenes with areneboronic acids. *Org. Lett.* 6: 617–619.
102 Molander, G.A. and Sandrock, D.L. (2008). Orthogonal reactivity in boryl-substituted organotrifluoroborates. *J. Am. Chem. Soc.* 130: 15792–15793.
103 van Haaren, R.J., Oevering, H., Coussens, B.B. et al. (1999). On the influence of the bite angle of bidentate phosphane ligands on the regioselectivity in allylic alkylation. *Eur. J. Inorg. Chem.* 1237–1241.
104 Ozawa, F., Kubo, A., and Hayashi, T. (1992). Palladium-catalyzed asymmetric arylation of 2,3-dihydrofuran – 1,8-bis(dimethylamino)naphthalene as an efficient base. *Tetrahedron Lett.* 33: 1485–1488.
105 Sugihara, T., Satoh, T., Miura, M., and Nomura, M. (2003). Rhodium-catalyzed Mizoroki–Heck-type arylation of alkenes with aroyl chlorides under phosphane- and base-free conditions. *Angew. Chem. Int. Ed.* 42: 4672–4674.
106 Wu, G.Z., Lamaty, F., and Negishi, E. (1989). Metal-promoted cyclization. 26. Palladium-catalyzed cyclization of benzyl halides and related electrophiles containing alkenes and alkynes as a novel route to carbocycles. *J. Org. Chem.* 54: 2507–2508.
107 Liu, C., Tang, S., Liu, D. et al. (2012). Nickel-catalyzed Heck-type alkenylation of secondary and tertiary *a*-carbonyl alkyl bromides. *Angew. Chem. Int. Ed.* 51: 3638–3641.
108 Nishikata, T., Noda, Y., Fujimoto, R., and Sakashita, T. (2013). An efficient generation of a functionalized tertiary-alkyl radical for copper-catalyzed tertiary-alkylative Mizoroki–Heck type reaction. *J. Am. Chem. Soc.* 135: 16372–16375.
109 Pyo, A., Kim, Y.H., Park, K. et al. (2012). Mechanistic study of palladium-catalyzed decarboxylative coupling of phenylpropiolic acid and aryl iodide. *Appl. Organomet. Chem.* 26: 650–654.
110 Leclerc, J.P., André, M., and Fagnou, K. (2006). Heck, direct arylation, and hydrogenation: two or three sequential reactions from a single catalyst. *J. Org. Chem.* 71: 1711–1714.
111 Potukuchi, H.K., Spork, A.P., and Donohoe, T.J. (2015). Palladium-catalyzed alpha-arylation of carbonyls in the de novo synthesis of aromatic heterocycles. *Org. Biomol. Chem.* 13: 4367–4373.
112 Baudoin, O., Herrbach, A., and Gueritte, F. (2003). The palladium-catalyzed C–H activation of benzylic gem-dialkyl groups. *Angew. Chem. Int. Ed.* 42: 5736–5740.